Iterative Methods and Preconditioning for Large and Sparse Linear Systems with Applications

MONOGRAPHS AND RESEARCH NOTES IN MATHEMATICS

Published Titles

Actions and Invariants of Algebraic Groups, Second Edition, Walter Ferrer Santos and Alvaro Rittatore

Analytical Methods for Kolmogorov Equations, Second Edition, Luca Lorenzi

Application of Fuzzy Logic to Social Choice Theory, John N. Mordeson, Davender S. Malik and Terry D. Clark

Blow-up Patterns for Higher-Order: Nonlinear Parabolic, Hyperbolic Dispersion and Schrödinger Equations, Victor A. Galaktionov, Enzo L. Mitidieri, and Stanislav Pohozaev

Bounds for Determinants of Linear Operators and Their Applications, Michael Gil′

Complex Analysis: Conformal Inequalities and the Bieberbach Conjecture, Prem K. Kythe

Computation with Linear Algebraic Groups, Willem Adriaan de Graaf

Computational Aspects of Polynomial Identities: Volume l, Kemer's Theorems, 2nd Edition Alexei Kanel-Belov, Yakov Karasik, and Louis Halle Rowen

A Concise Introduction to Geometric Numerical Integration, Fernando Casas and Sergio Blanes

Cremona Groups and Icosahedron, Ivan Cheltsov and Constantin Shramov

Delay Differential Evolutions Subjected to Nonlocal Initial Conditions Monica-Dana Burlică, Mihai Necula, Daniela Roșu, and Ioan I. Vrabie

Diagram Genus, Generators, and Applications, Alexander Stoimenow

Difference Equations: Theory, Applications and Advanced Topics, Third Edition Ronald E. Mickens

Dictionary of Inequalities, Second Edition, Peter Bullen

Elements of Quasigroup Theory and Applications, Victor Shcherbacov

Finite Element Methods for Eigenvalue Problems, Jiguang Sun and Aihui Zhou

Integration and Cubature Methods: A Geomathematically Oriented Course, Willi Freeden and Martin Gutting

Introduction to Abelian Model Structures and Gorenstein Homological Dimensions Marco A. Pérez

Iterative Methods and Preconditioning for Large and Sparse Linear Systems with Applications, Anatoly Galperin *Iterative Methods without Inversion*, Daniele Bertaccini and Fabio Durastante

Iterative Optimization in Inverse Problems, Charles L. Byrne

Published Titles Continued

Line Integral Methods for Conservative Problems, Luigi Brugnano and Felice Iavernaro

Lineability: The Search for Linearity in Mathematics, Richard M. Aron, Luis Bernal González, Daniel M. Pellegrino, and Juan B. Seoane Sepúlveda

Mathematical Modelling of Waves in Multi-Scale Structured Media, A. B. Movchan, N. V. Movchan, I. S. Jones, and D. J. Colquitt

Modeling and Inverse Problems in the Presence of Uncertainty, H. T. Banks, Shuhua Hu, and W. Clayton Thompson

Monomial Algebras, Second Edition, Rafael H. Villarreal

Noncommutative Deformation Theory, Eivind Eriksen, Olav Arnfinn Laudal, and Arvid Siqveland

Nonlinear Functional Analysis in Banach Spaces and Banach Algebras: Fixed Point Theory Under Weak Topology for Nonlinear Operators and Block Operator Matrices with Applications, Aref Jeribi and Bilel Krichen

Nonlinear Reaction-Diffusion-Convection Equations: Lie and Conditional Symmetry, Exact Solutions, and Their Applications, Roman Cherniha, Mykola Serov, and Oleksii Pluikhin

Optimization and Differentiation, Simon Serovajsky

Partial Differential Equations with Variable Exponents: Variational Methods and Qualitative Analysis, Vicenţiu D. Rădulescu and Dušan D. Repovš

A Practical Guide to Geometric Regulation for Distributed Parameter Systems Eugenio Aulisa and David Gilliam

Reconstruction from Integral Data, Victor Palamodov

Signal Processing: A Mathematical Approach, *Second Edition*, Charles L. Byrne

Sinusoids: Theory and Technological Applications, Prem K. Kythe

Special Integrals of Gradshteyn and Ryzhik: the Proofs – Volume I, Victor H. Moll

Special Integrals of Gradshteyn and Ryzhik: the Proofs – Volume II, Victor H. Moll

Spectral and Scattering Theory for Second-Order Partial Differential Operators, Kiyoshi Mochizuki

Stochastic Cauchy Problems in Infinite Dimensions: Generalized and Regularized Solutions, Irina V. Melnikova

Submanifolds and Holonomy, Second Edition, Jürgen Berndt, Sergio Console, and Carlos Enrique Olmos

Symmetry and Quantum Mechanics, Scott Corry

The Truth Value Algebra of Type-2 Fuzzy Sets: Order Convolutions of Functions on the Unit Interval, John Harding, Carol Walker, and Elbert Walker

Variational-Hemivariational Inequalities with Applications, Mircea Sofonea and Stanislaw Migórski

Willmore Energy and Willmore Conjecture, Magdalena D. Toda

Forthcoming Titles

Groups, Designs, and Linear Algebra, Donald L. Kreher

Handbook of the Tutte Polynomial, Joanna Anthony Ellis-Monaghan and Iain Moffat

Microlocal Analysis on R^n and on NonCompact Manifolds, Sandro Coriasco

Practical Guide to Geometric Regulation for Distributed Parameter Systems, Eugenio Aulisa and David S. Gilliam

MONOGRAPHS AND RESEARCH NOTES IN MATHEMATICS

Iterative Methods and Preconditioning for Large and Sparse Linear Systems with Applications

Daniele Bertaccini
Fabio Durastante

CRC Press
Taylor & Francis Group
Boca Raton London New York

CRC Press is an imprint of the
Taylor & Francis Group, an **informa** business
A CHAPMAN & HALL BOOK

CRC Press
Taylor & Francis Group
6000 Broken Sound Parkway NW, Suite 300
Boca Raton, FL 33487-2742

Printed on acid-free paper
Version Date: 20180125

International Standard Book Number-13: 978-1-4987-6416-2 (Hardback)

Visit the Taylor & Francis Web site at
http://www.taylorandfrancis.com

and the CRC Press Web site at
http://www.crcpress.com

This book was supported in part by the 2018 GNCS-INDAM project "Tecniche innovative per problemi di algebra lineare", and by grant PRIN 20083KLJEZ, MIUR, Roma, Italia.

Contents

Preface

This book on iterative methods for linear equations can be used as a tutorial and a reference for those who need to solve sparse and/or structured large linear systems of algebraic equations. It can also be used as a textbook for courses in iterative methods or as source material for a course in numerical analysis at the graduate level. It is assumed that the reader is familiar with calculus, elementary numerical analysis, linear algebra, and the main ideas of direct methods for the numerical solution of dense linear systems as described in, e.g., [270]. This text is completed with a bibliography with more than 550 titles where the interested reader can find more details and proofs of results considered here.

The focus is on some real life applications and then on a certain number of methods suitable for them.

This book can be used for a Master's and/or Ph.D. level graduate course in science and engineering and can be useful to researchers, from any field of engineering, health, physical, chemical and biological sciences, etc.

Chapter 1 contains some basic notions of error analysis and sparse matrices. In Chapter 2 there are basic facts on some fundamental iterative algorithms for linear systems, focusing in particular on Krylov subspace methods. Chapter 3 recalls the definition and some general purpose preconditioning strategies. Finally, in Chapter 4, a few examples of preconditioners for Toeplitz and Toeplitz–like, complex symmetric and saddle point systems are considered, together with some introductory notes on multigrid preconditioners.

A short review of linear algebra is also included; see Appendix A.

A number of computational examples and exercises are provided. The former are intended to validate the theoretical results and to give a sense of how the algorithms perform and not to give a complete picture of performance or to be a suite of test problems. Some test matrices and existing *test sets* are mentioned in Appendix B1. Finally, a few numerical linear algebra software packages and repositories are cited in Appendix B2, together with a short description of the codes shipped with the book and how to get them.

D. Bertaccini and F. Durastante

List of Figures

List of Tables

List of Algorithms

Symbol Description

Symbol	Description
A, B, C	Latin upper case letters are used to denote matrices,
$\mathbf{u}, \mathbf{v}, \mathbf{w}$	Bold Latin lower case letters are used to denote vectors,
$\mathbf{a}_{i,:}, \mathbf{a}_{:,j}$	Denote respectively the ith row, jth column of matrix A,
α, β, h, k	Greek and Latin lower case letters are used to denote the scalar quantity,
$a_{i,j}$	Denote the element of position (i, j) of the matrix A,
$A^{(k)}, \mathbf{u}^{(k)}$	Elements of a sequence of matrices or vectors are denoted by a $(\cdot)$ superscript,
α_k, h_k	Elements of a sequence of scalar are denoted by a Latin lower subscript,
$\mathbb{R}, \mathbb{C}, \mathbb{Q}$	Numerical fields, respectively real, complex and rational,
$\mathbb{N}, \mathbb{Z}$	Sets of the natural and integer numbers, respectively,
$\mathcal{C}^k(\Omega)$	Space of the functions of class k over the set Ω,
$\mathcal{C}_0^k(\Omega)$	Space of the functions of class k over the set Ω with compact support,
$\mathbb{L}^p,\ \ell^p$	Space of the Lebesgue integrable functions/sequences of order p,
$\mathcal{B}(U, V)$	space of bounded functions from U to V, if $U = V$ then $\mathcal{B}(U, V) = \mathcal{B}(U)$,
$\mathbb{W}^{p,q}$	Sobolev space of order q with p weak derivatives,
$\\|\cdot\\|_p$	p-norm, either for functions, vectors or matrices $(p = 1, 2, \ldots, \infty)$,
$\\|\cdot\\|_F$	Fröbenius norm, either vectors or matrices,
$T_n(f)$	Toeplitz matrix of dimension $n \times n$ generated by the function f,
$C_n(f)$	Circulant matrix of dimension $n \times n$ generated by the function f,
$\mathbb{K}_n$	Space of polynomials of degree less or equal to k over the field $\mathbb{K}$,
$\langle \cdot, \cdot \rangle_A$	Denotes the scalar product induced by the square matrix A, if A is omitted is the standard scalar product,
$A \otimes B$	Denotes the Kronecker product between two matrices A and B of opportune size,
$\cdot \circ \cdot$	Denotes the Hadamard product (entrywise product) between two matrices or vectors of opportune size,
$\cdot^T, \cdot^H$	Denotes respectively the transpose and conjugate transpose operations, either for vectors and matrices,
$\det(\cdot)$	Determinant of a square matrix,
$\mathrm{tr}(\cdot)$	Trace of a square matrix,
$\mathrm{rank}(\cdot)$	Rank of a matrix,
$\kappa_p(\cdot)$	Condition number in the p norm, if p is omitted $p = 2$,
$\mathrm{span}\{\cdot\}$	Linear space generated by some vectors,
$\Lambda(\cdot), \sigma(\cdot)$	Spectrum of a matrix,
$\lambda_j(\cdot)$	jth eigenvalue of a matrix,
$\sigma_j(\cdot)$	jth singular value of matrix.

Chapter 1

Introduction and Motivations

The innermost computational kernel of many large–scale scientific applications and industrial numerical simulations, in particular where systems of partial differential equations are involved in the models, is often a sequence of large and sparse linear systems which typically consumes a significant portion of the overall computational time required by the simulation. Many of the matrix problems are in the form of systems of linear equations, although other matrix problems, such as eigenvalue calculations, can occur too. Recent advances in technology have led to a dramatic growth in the size of the matrices to be handled, and iterative techniques are often used in such circumstances, especially when decompositional approaches require prohibitive storage requirements. Computational experience accumulated in the past couple of decades indicates that a good preconditioner holds the key for an effective iterative solver. More recently, the use of parallel computing capabilities of the modern hardware architectures has increased the speed and the range of problems that can be solved; see also § 1.3.

In this book we concentrate on solving linear systems as

$$A\mathbf{x} = \mathbf{b},$$

where for us A will be (if not otherwise stated) a real nonsingular square matrix, $\mathbf{b}$ the known vector and $\mathbf{x}$ the vector of the unknowns.

In order to find an approximation for $\mathbf{x}$ we could use *direct methods*. However, often numerical approximations of many models, in particular of those based on partial differential equations, produce sparse and/or structured matrices A. In general, direct methods are not the best choice because

- they can destroy the underlying sparsity/structure during the resolution process.
- We could be interested in approximations with lower accuracy than the

one provided by, e.g., Gaussian Elimination with pivoting. This could be due to the low precision of the data or to a truncation error introduced by the approximation process generating the linear system(s).

- They cannot use the information given by the initial guess for the solution $\mathbf{x}$. The underlying information is often available (and precious) in many classes of problems.

Both these issues can be fulfilled by *iterative methods.*

In this book we will not consider direct methods. An overview can be found in, e.g., [176, 203, 271, 390].

1.1 Notes on error analysis

Everywhere in this book we refer to absolute and relative error between a vector x and its approximation $\tilde{x}$.

Definition 1.1 (Absolute and relative error). If $\tilde{\mathbf{x}} \in \mathbb{K}^n$ is an *approximation* to $\mathbf{x} \in \mathbb{C}^n$, the *absolute error* in $\tilde{\mathbf{x}}$ relative to the norm $\|\cdot\|$ is defined as $\|\tilde{\mathbf{x}} - \mathbf{x}\|$, while the *relative error* in $\tilde{\mathbf{x}}$ relative to the norm $\|\cdot\|$ is defined as

$$\frac{\|\tilde{\mathbf{x}} - \mathbf{x}\|}{\|\mathbf{x}\|}.$$

All numerical methods proposed here work with the limited precision data that can be handled by our computing machines: *floating point arithmetic* that is an example of *finite precision arithmetic.* At each step, including the simple storage of a floating point number, we can introduce first *rounding errors* that are inherent to the limited precision of the instrument used for our computations. In order to understand this source of errors, assume that, as usual, non-integer numbers are stored in our computers in binary (i.e., in basis 2) with at most t digits. Note that our PCs use a binary *IEEE 754 arithmetics*, see [315], with t up to 53 in Matlab and in the *double precision* arithmetic used by most compilers, including various version of compilers for *Fortran* and *C* languages. In more detail, the *machine numbers* in *IEEE arithmetic* are all and only those real that are 0 or can be written as

$$x = \pm 2^p \sum_{i=1}^{t} d_i\, 2^{-i}, \; d_i = 0 \text{ or } 1, \; d_1 = 1, \; -\min \leq p \leq \max, \tag{1.1}$$

where in the *double precision* IEEE 754 arithmetic min $= -1021$, max $= 1024$. Of course, not all non-integer numbers can be represented in the format (1.1). Those that cannot must be *rounded* (and therefore approximated and an error introduced) because exact representation can require more than t or even a not finite number of digits in binary even if in basis 10 they have a finite representation. As an example, 0.1 in basis 10, written as 0.1_{10} for short, in binary is periodic!

$$0.1_{10} = 0.0001100110011\ldots_2 .$$

However, an upper bound for the error made storing a non-machine number can be easily computed. Let $fl(x)$ be the machine number used by the computer to store x.

Theorem 1.1. *For all $x \in \mathbb{R}$,*

$$\left| \frac{fl(x) - x}{x} \right| \leq \frac{1}{2} 2^{-t+1}.$$

In order to understand the effects of rounding errors on the accuracy of computations, it is very important to introduce the notion of *Machine precision.*

Definition 1.2 (Machine precision). The number $eps = 2^{-t+1}$ is called here *machine precision*, and *eps* is the smallest positive machine number such that $fl(1 + eps) > 1$.

For example, *double precision* IEEE 754 arithmetic gives $eps = 2^{-53} \simeq 2.22 \cdot 10^{-16}$.

Usually the error due to the above representation is called *inherent* or *rounding* error. Observe that in general there are at least three other main sources of errors in numerical computations: *data uncertainty*, *truncation* or *analytic* and *algorithmic error*. Data uncertainty and its effects are not investigated here but just recall that often in problems in engineering and economics data is usually accurate to only a few digits and therefore it makes no sense to reduce the other errors much below.

To simplify, the error analysis is based on the assumption that, writing the total error as the sum of the three (four) above mentioned components and neglecting the nonlinear terms, is a reasonable approximation. By calling E_{total} the global error and assuming that the data are exact, we write:

$$E_{\text{total}} = E_{\text{inherent}} + E_{\text{truncation}} + E_{\text{algorithmic}} + \delta,$$

where δ includes the nonlinear terms and is considered negligible with respect to the other components.

In the following we suppose $A \in \mathbb{C}^{n \times n}$, $\mathbf{x}, \mathbf{b} \in \mathbb{C}^n$ and that the linear system

$$A\mathbf{x} = \mathbf{b}, \tag{1.2}$$

is consistent. Observe that in a direct method, if no rounding (i.e. representation) errors are present, the solution of the system (1.2) would be computed exactly. On the other hand, for an *iterative method*, even assuming the absence of rounding errors, we need to stop after some iterations, producing a truncation error.

Whatever method we choose, we can not ignore the presence of representation and rounding errors in our computations. It is possible to study these issues by introducing an abstract and precise formulation of the condition of transformations between certain spaces, see e.g. [445]. Here we resume briefly error analysis for problem (1.2). For a more complete treatment of this fundamental issue, see Bevilacqua et al. [81], Higham [304], and in particular the milestone by Wilkinson [542].

1.1.1 Inherent error

The *inherent error*, that *does not depend* on the algorithm selected for the solution of (1.2), measures the sensitivity of the solution to perturbations on the data of the problem, in our case the matrix A and the vector $\mathbf{b}$. Heuristically, the data $A, \mathbf{b}$ are *well–conditioned* if "small" perturbations on them **can** cause "small" changes in the results of the problem (1.2). Otherwise, the data are *ill–conditioned.* In particular, *conditioning* is considered as a characteristic of the **problem**, independently of the algorithm used. A more rigorous meaning of the term *conditioning* can be found in the theorem below.

An analysis of the inherent error can be performed by studying the effect on the solution of perturbations on the data. Following this idea for the case of the linear system in (1.2) gives an important result (see, e.g., [84]).

Theorem 1.2. *Let $\delta A \in \mathbb{C}^{n\times n}$ and $\delta\mathbf{b} \in \mathbb{C}^n$ be the perturbations on the data of problem* (1.2) *with* $\mathbf{b} \neq \mathbf{0}$. *Let $\|\cdot\|$ be any induced matrix norm (see appendix A.1). If A is not singular and $\|A^{-1}\|\|\delta A\| < 1$, then also $A + \delta A$ is not singular. If we denote by $\mathbf{x} + \delta\mathbf{x}$ the solution produced on the system by the perturbed data, we have:*

$$(A + \delta A)(\mathbf{x} + \delta\mathbf{x}) = \mathbf{b} + \delta\mathbf{b},$$

where

$$\frac{\|\delta\mathbf{x}\|}{\|\mathbf{x}\|} \leq \frac{\|A\|\|A^{-1}\|}{1 - \|A^{-1}\|\|\delta A\|}\left(\frac{\|\delta A\|}{\|A\|} + \frac{\|\delta \boldsymbol{b}\|}{\|\mathbf{b}\|}\right). \tag{1.3}$$

The scalar quantity $\|A\|\|A^{-1}\|$ is important enough to deserve a definition.

Definition 1.3 (Matrix Condition Number). Given a consistent matrix norm $\|\cdot\|$, the condition number of A is defined as $\kappa(A) = \|A^{-1}\|\|A\|$. Particularly, we denote with $\kappa_p(A) = \|A^{-1}\|_p\|A\|_p$ the matrix condition number with respect to the matrix induced p–norm.

Definition 1.3 permits us to rewrite the bound (1.3) as

$$\varepsilon_{\mathbf{x}} = \frac{\|\delta\mathbf{x}\|}{\|\mathbf{x}\|} \leq \frac{\kappa(A)}{1 - \kappa(A)\dfrac{\|\delta A\|}{\|A\|}}\left(\frac{\|\delta A\|}{\|A\|} + \frac{\|\delta\mathbf{b}\|}{\|\mathbf{b}\|}\right) = \frac{\kappa(A)(\varepsilon_A + \varepsilon_{\mathbf{b}})}{1 - \kappa(A)\varepsilon_A},$$

where $\varepsilon_A = \|\delta A\|/\|A\|$ and $\varepsilon_{\mathbf{b}} = \|\delta\mathbf{b}\|/\|\mathbf{b}\|$ are the perturbations on the (entries of the) matrix A and on the (entries of the) vector $\mathbf{b}$, respectively. Note that if we just stored A and $\mathbf{b}$ from exact data, then ε_A, $\varepsilon_{\mathbf{b}}$ can be considered of the order of $c \times eps$, $c \geq 1$ a function of the norm used and here we reasonably assume $c = 1$ to simplify the treatise.

Let us consider few properties of $\kappa(X)$, X a square nonsingular matrix.

1. $\kappa(A) \geq 1$,
2. $\kappa(AB) \leq \kappa(A)\kappa(B)$,
3. $\kappa(\alpha A) = \kappa(A)$ for any $\mathbb{K} \ni \alpha \neq 0$,
4. $\kappa_2(A) = 1$ if and only if A is a nonzero scalar multiple of an orthogonal matrix,
5. $\kappa_2(A) = \kappa_2(A^T)$,
6. $\kappa_2(A^T A) = \kappa_2(A)^2$,
7. $\kappa_2(A) = \sigma_1/\sigma_n$, where σ_1 and σ_n are the largest and smallest singular values of A.

Now, since by property 1 $\kappa(A) \geq 1$, if $\kappa(A)$ takes "small" values then "small" perturbations on the data are expected to give back "small" perturbations on the solution and the problem is said to be *well–conditioned*. On the other hand, if $\kappa(A)$ is "large", "small" perturbations on the data can give back "large" perturbations on the solution and the problem is *ill–conditioned*. In particular, for simplicity we could consider A well–conditioned if $eps \cdot \kappa(A) \ll 1$ and ill–conditioned otherwise because of the effects on the relative errors on $\mathbf{x}$, see Theorem 1.2. In general, we should use more care, especially if A is not just a spare matrix but is a term of a sequence of matrices, e.g., $\{A_j\}$, with size increasing with j, say, such as those generated by the discretization of a differential problem; see, e.g., [351], and consider also the behavior of $\kappa(\cdot)$ with respect to j.

Moreover, observe that in a base–b finite precision arithmetic, we can interpret $\log_b(\kappa(A))$ as an estimate of how many base–b digits can be lost in solving our linear system with that matrix independently on the algorithm used, i.e., an estimate of the worst–case loss of precision. Recall that if $\tilde{\mathbf{x}}$ is an approximation of $\mathbf{x}$ with a relative error less than β^{1-t}, we say that $\tilde{\mathbf{x}}$ has t significant digits in the basis β.

Note the differences of *well/ill–conditioned* (problem) with *well/ill-posed* (problem) by Hadamard.

Definition 1.4. A problem is said to be *well-posed* in the Hadamard's sense if (i) it has a solution, (ii) the solution is unique; (iii) the solution's behavior changes continuously with the initial conditions. Otherwise, it is *ill–posed* .

Example 1.1 (Hilbert matrices, [487, 512]). A famous class of ill–conditioned matrices is the Hilbert matrix $H_n(p) \in \mathbb{R}^{n\times n}$ with $p \neq -1, -2, \dots, -(2n-1)$. A Hilbert matrix, for a given value of p, is defined as

$$(H_n(p))_{i,j} = h_{i,j}^{(p)} = (p+i+j-1)^{-1} \text{ for all } i,j = 1,2,\dots,n.$$

These matrices naturally arise in various situations, e.g., if we want to approximate a continuous function $f(x)$ in the interval $[0,1]$ by a polynomial of degree $n-1$ in x such that the error

$$E = \int_0^1 (p_{n-1}(x) - f(x))^2 \, dx, \quad p_{n-1}(x) = \sum_{j=1}^{n} \alpha_j x^{j-1}$$

is minimized. By differentiating E within respect to the coefficients α_j and imposing the result to be 0 gives its minimum; we obtain the following system of n equation in the α_js for $i = 1,2,\dots,n$:

$$\sum_{j=1}^{n} \left(\int_0^1 x^{i+j-2} \, dx \right) \alpha_j = \left(\equiv \sum_{j=1}^{n} h_{i,j}^{(0)} = \right) \int_0^1 f(x) x^{i-1} \, dx \equiv b_i,$$

i.e., $H_n(0)\boldsymbol{\alpha} = \mathbf{b}$. For this particular class of matrices we can express, for nonnegative integer values of p, the inverse explicitly as

$$(H_n^{-1}(p))_{i,j} = \frac{(-1)^{i+j}}{p+i+j-1} \left[\frac{\prod_{k=0}^{n-1} (p+i+k)(p+j+k)}{(i-1)!(n-i)!(j-1)!(n-j)!} \right].$$

In particular, for $n = 4$:

$$H_4(1) = \begin{bmatrix} 1/2 & 1/3 & 1/4 & 1/5 \\ 1/3 & 1/4 & 1/5 & 1/6 \\ 1/4 & 1/5 & 1/6 & 1/7 \\ 1/5 & 1/6 & 1/7 & 1/8 \end{bmatrix},$$

$$H_4^{-1}(1) = \begin{bmatrix} 200 & -1200 & 2100 & -1120 \\ -1200 & 8100 & -15120 & 8400 \\ 2100 & -15120 & 29400 & -16800 \\ -1120 & 8400 & -16800 & 9800 \end{bmatrix}.$$

Therefore we can estimate the row (similarly for the column) sum as

$$\begin{aligned} \sum_{i=1}^{n} |(H_n(p))_{i,j}| &= \psi(j+n+p) - \psi(j+p), \\ \sum_{i=1}^{n} |(H_n^{-1}(p))_{i,j}| &= \frac{\Gamma(n+p+1)\Gamma(j+n+p)}{\Gamma(j)\Gamma(n)\Gamma(-j+n+1)} \cdot \dots \\ &\quad \dots \cdot {}_3\tilde{F}_2(1-n, j+p, n+p+1; p+1, j+p+1; -1), \end{aligned}$$

where $\psi(x)$ is the digamma function and ${}_3\tilde{F}_2(x, y, z)$ is the regularized generalized hypergeometric function, see, e.g., [2]. By using these expressions we can compute the ∞–norm of both A and A^{-1} and thus compute the condition number for the Hilbert matrix while the $\kappa_2(A)$ can be obtained by looking at the eigenvalues of $H_2(p)$ since it is symmetric, see the values in Table 1.1. Observe what happens when we attempt to solve the interpolation problem

Table 1.1: Example 1.1. Conditions number of Hilbert matrix $H_n(0)$ for various n

n	$\|H_n\|_2$	$\|H_n^{-1}\|_2$	$\kappa_2(H_n)$	$\|H_n\|_\infty$	$\|H_n^{-1}\|_\infty$	$\kappa_\infty(H_n)$
6	1.62	9.24×10^6	1.50×10^7	2.45	1.19×10^7	2.91×10^7
7	1.66	2.86×10^8	4.75×10^8	2.59	3.80×10^8	9.85×10^8
8	1.70	9.00×10^9	1.53×10^{10}	2.72	1.25×10^{10}	3.39×10^{10}
9	1.73	2.86×10^{11}	4.93×10^{11}	2.83	3.89×10^{11}	1.10×10^{12}
10	1.75	9.15×10^{12}	1.60×10^{13}	2.93	1.21×10^{13}	3.54×10^{13}
11	1.77	2.95×10^{14}	5.23×10^{14}	3.02	4.09×10^{14}	1.23×10^{15}
12	1.80	9.54×10^{15}	1.71×10^{16}	3.10	1.33×10^{16}	4.12×10^{16}
13	1.81	3.10×10^{17}	5.63×10^{17}	3.18	4.16×10^{17}	1.32×10^{18}
14	1.83	1.01×10^{19}	1.85×10^{19}	3.25	1.40×10^{19}	4.54×10^{19}
15	1.85	3.31×10^{20}	6.12×10^{20}	3.32	4.64×10^{20}	1.54×10^{21}

for $f(x) = \sin(2\pi x)$ and $n = 15$. To show the effect of the ill-conditioning we perturb only the right–hand side $\mathbf{b}$ with a random vector $\delta\mathbf{b}$ such that $\varepsilon_{\mathbf{b}} \approxeq 10^{-7}$. The results are reported in Figure 1.1. As expected, observe that

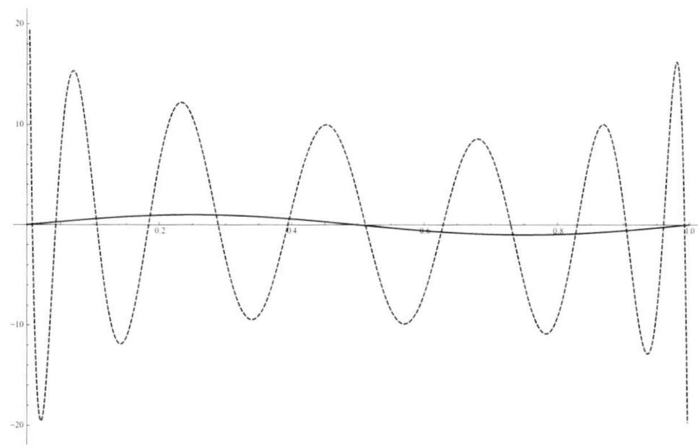

Figure 1.1: Example 1.1. Interpolation problem with Hilbert matrix, comparison between the solution retrieved from the perturbed problem (dashed line) and the objective function (solid line).

$\kappa_2(A) \approx 10^{20}$ and the small perturbations are strongly amplified. We stress again that this effect is independent from the solution algorithm, being only a characteristic of the problem. ✓

1.1.2 Algorithmic error

Let us focus on *algorithmic error* and thus on the algorithm used for computing the actual value $\mathbf{x} = A^{-1}\mathbf{b}$, considering the data exact and studying the rounding errors' propagation. Instead of focusing on the relative error on the quantity $\tilde{\mathbf{x}}$, called *forward* or *a–priori error*, we try to determine which data $\delta\mathbf{b}$ can give $\tilde{\mathbf{x}} = A^{-1}(\mathbf{b} + \delta\mathbf{b})$ instead of $\mathbf{x}$. Clearly, there can be many perturbations $\delta\mathbf{b}$ for which $\tilde{\mathbf{x}}$ can be obtained. We decide to take the perturbation with the smallest norm $\|\delta\mathbf{b}\|$, among the ones for which the equality is obtained. Then, we can consider the ratio $\|\delta\mathbf{b}\|/\|\mathbf{b}\|$ or the value of $\|\delta\mathbf{b}\|$, and call it *backward error*. See Figure 1.2.

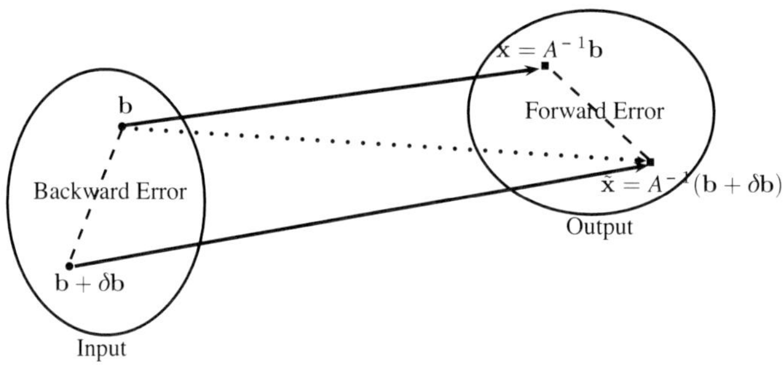

Figure 1.2: Backward and forward error. Continuous line arrows denote the exact computation while dotted line arrows denote the actual computation.

The *backward error analysis* reduces the analysis of *algorithmic error* to finding a bound, as small as possible, to the backward error relative to a computed solution. This goes along with the general observation we made at the end of the previous section about being interested in approximation with lower precision. If the bound from the backward analysis is no larger than the inherent uncertainties, e.g., uncertainties due to other measurements or rounding errors from finite precision arithmetic, then the solution we have obtained can not be distinguished from the solution we are looking for. As a secondary feature, and not less noticeable, *backward error analysis* gives a solid interpretation for the rounding errors as a perturbation in the data, that is a well understood process from Applied Mathematics.

Therefore, it is reasonable to say that an algorithm for computing $\mathbf{x} = A^{-1}\mathbf{b}$ is *backward stable* if, for any $\mathbf{b}$, it produces a computed $\tilde{\mathbf{x}}$ with a backward error that is comparable with the *machine precision*; Definition 1.2.

The relationship between *forward* and *backward error* for a given computational task is governed by the conditioning through the condition number as introduced in Definition 1.3. When all these pieces fit together we can observe that, qualitatively,

$$\text{forward error} \lesssim \text{condition number} \times \text{backward error},$$

with the approximate equality being actually possible.

For many computational tasks, that give reliable results anyway, backward stability can not be proved, therefore a relaxed relation is used. Heuristically we want to accept as *numerically stable* any method that provides almost the right answer for almost the right data. Mixing together backward and forward stability, one usually requires two bounds for the quantities $\delta\mathbf{x}$ and $\delta\mathbf{b}$ in

$$\tilde{\mathbf{x}} + \delta\mathbf{x} = A^{-1}(\mathbf{b} + \delta\mathbf{b}).$$

This weaker formulation is usually known as *mixed forward–backward stability* and whenever the quantities $\delta\mathbf{x}$ and $\delta\mathbf{b}$ can be proven to be "small", often the algorithm for computing f is said to be *numerically stable.*

Remark 1.1. Even if a numerically stable algorithm is applied to an ill–conditioned problem, like $A\mathbf{x} = \mathbf{b}$ for an ill–conditioned A, it can produce inaccurate results, i.e., even if we bound the backward error, a large condition number can make the forward error very large, that is the *inherent error* becomes dominant. For ill–conditioned problems, even the small bounded errors of stable algorithms can lead to large errors in the approximated solution, consider again Example 1.1.

There exist many well known numerically stable algorithms in Numerical Linear Algebra. Single floating point operations are both forward and backward stable, naïve scalar product algorithms are only backward stable, Gaussian elimination with complete pivoting, triangular backward substitution, QR factorization (both with Housolder and Givens rotations) and the Golub–Kahan algorithm for SVD decomposition are backward stable, see [271] for details. Quite surprisingly, Gaussian elimination with partial pivoting is *not* backward stable, even if the linear equations for which instability is exhibited are extraordinarily rare.

Exercise 1.1. Consider the upper triangular matrices $B_n \in \mathbb{R}^{n\times n}$ of the form,

$$B_n = \begin{bmatrix} 1 & -1 & \dots & -1 \\ 0 & \ddots & & \vdots \\ \vdots & \ddots & \ddots & -1 \\ 0 & \dots & 0 & 1 \end{bmatrix}.$$

Compute $\kappa_\infty(B_n)$, i.e., its condition number in ∞–norm. Are they ill–conditioned? What happens to the matrix B_n if we set $(B_n)_{n,1} = -2^{2-n}$? Discuss if the determinant of a given matrix A gives any information on its conditioning, does a "small" or a "big" determinant imply anything?

Exercise 1.2. Prove Theorem 1.1.

Exercise 1.3. Prove Theorem 1.2.

1.2 Sparse matrices

Throughout this book we distinguish between *dense* and *sparse* matrices, since in numerical linear algebra the computations made with them needs to be faced in different ways and presents indeed different costs. A first *operative notion* of sparse matrix, independent of the kind of problem, is the one given by Wilkinson in [544]:

> "The matrix may be sparse, either with the non–zero elements concentrated on a narrow band centered on the diagonal or alternatively they may be distributed in a less systematic manner. We shall refer to a matrix as dense if the percentage of zero elements or its distribution is such as to make it uneconomic to take advantage of their presence."

The above is a heuristic starting point that gives the idea: a *sparse* matrix is not *dense*. If we think about a sequence of matrices $A \in \mathbb{R}^{n\times n}$, where n is, e.g., a function of a discretization parameter from a PDE model, we can give a rigorous definition.

Definition 1.5 (Sparse matrix). A matrix A_n of a given sequence of matrices in $\mathbb{R}^{n\times n}$ is sparse if the number of nonzero entries of A, $\text{nnz}(A)$, is $O(n)$.

A useful tool in sparse matrix computation is *graph theory.*

Definition 1.6. (struct(A)) Given a sparse matrix $A \subset \mathbb{R}^{n\times n}$, consider the graph $G(A)$, called the structure of A, or $G(A) = \text{struct}(A)$, defined by the vertex set V and edge set E:

$$\begin{aligned} V =&\{i : 1 = 1, \dots, n\}, \\ E =&\{(i,j) : i \neq j \text{ and } A_{i,j} \neq 0\}. \end{aligned}$$

A graphical representation of both $G(A)$ and the sparsity pattern of a matrix A, i.e., a 2D plot in which the axes represents the rows and columns of A and for each nonzero entry of A a point is plotted, is given in Figure 1.3. Observe that in this discussion the *size* of the nonzeros entries of the matrix A does not play a role. This can be a limitation, as we will see in the following, see, e.g., § 3.5.1, because in many cases we need to handle matrices that are indeed dense but in which many of the entries are of negligible magnitude. Sometimes these matrices are called *numerically sparse* or *localized* because, by dropping the entries which fall below a threshold magnitude, we can approximate them by sparse matrices; see Benzi in [48] for an interesting overview of localization in matrix computations. Another useful graphic tool is the *city plot*, based on adding a 3rd dimension to the sparsity pattern, either with a color scale or a true zeta axis, on the basis of the magnitude of the elements, see Figure 1.4. Observe also that both *sparse* and *localized* matrices sometimes exhibit a distribution of their entries in some particular patterns.

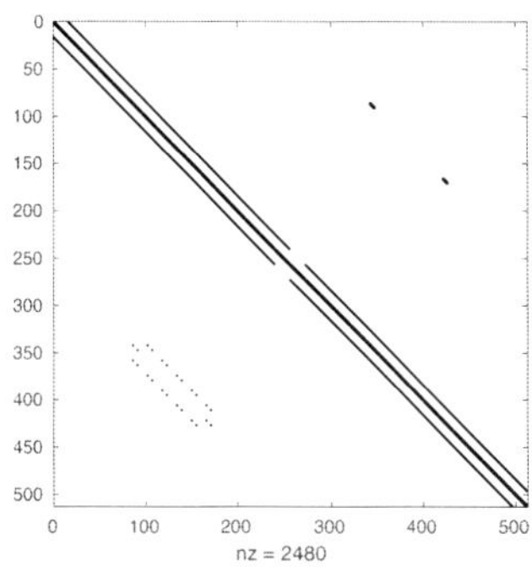

(a) Sparsity Pattern

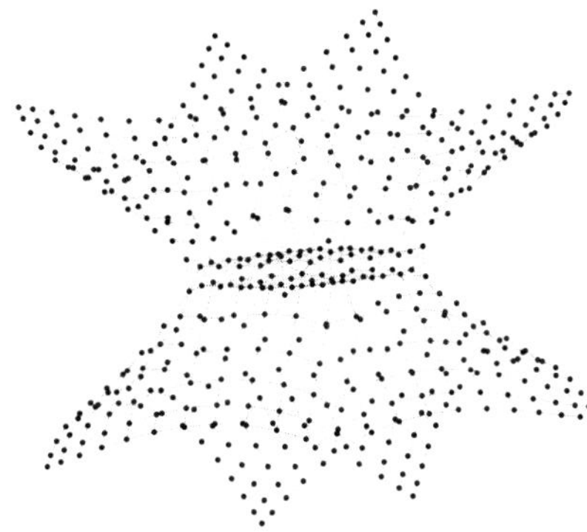

(b) Matrix Graph

Figure 1.3: Representing sparse matrix: pattern and matrix graph.

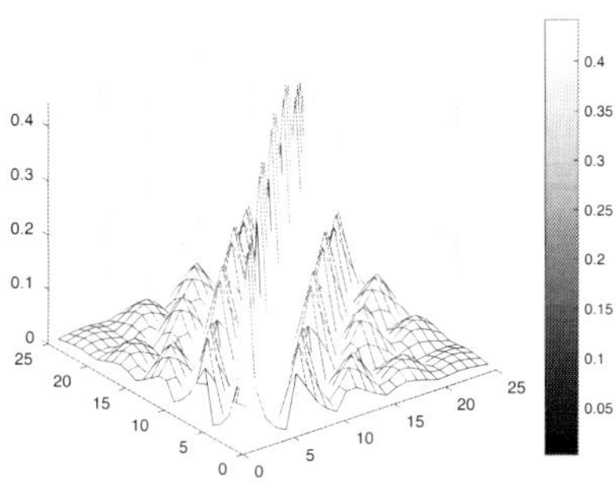

(a) City Plot

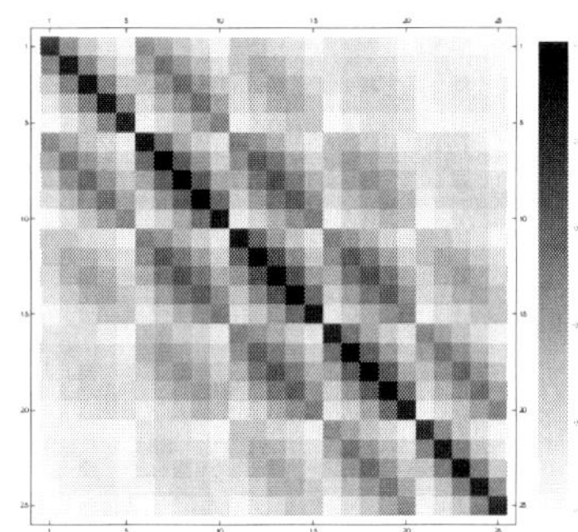

(b) City Plot Pattern

Figure 1.4: Representing sparse matrix: city plot.

We will come back to these topics by considering preconditioners for some *structured matrices* in Chapter 4.

Let us discuss how the presence of many zero entries in the underlying matrices can be used to save computational resources.

1.2.1 Ad-hoc storage and computing schemes for sparse matrices

One of the fundamental basic operations of iterative methods, and usually the most computationally expensive, is the matrix–vector product. Consider the product $A\mathbf{x}$, where A is a generic (and then potentially dense) matrix in $\mathbb{R}^{n\times n}$ and $\mathbf{x}$ a vector in $\mathbb{R}^n$. What is the cost of this operation? We need to perform n scalar products with each row of A and the vector $\mathbf{x}$, $< \mathbf{a}_{i,:}, \mathbf{x} >$

for $i = 1, \ldots, n$, that have both n elements. Hence, n^2 multiplications and $n(n-1)$ sums. By assuming that a sum and a multiplication take the same computer time to be performed, for short we conventionally call each of them a *flop*, and then we need $2n^2 - n$ flops. On the other hand, if the coefficient matrix A is *sparse*, this computation can be made much more efficiently if the zero entries of A are not stored. To this end, exploiting *sparse storage schemes* and specializing the matrix–vector operations would be the cheapest way to go. In the rest of this section, we use a different paradigm for storing and accessing entries of sparse matrices. In particular, nonzero entries are stored contiguously in an array and other arrays are used for the corresponding indexes. Observe that with this kind of approach, the matrix–vector product algorithm needs to know where the entries are located in the full matrix. Some specialized sparse storage schemes, optimized for the various linear algebra operations, have been investigated, see, e.g., [36, 203, 227, 262, 364, 450, 476, 504]. In the following we use few *storage schemes* that have a well established applicability.

Let us start from the simplest modification of the usual storage scheme, i.e., the *Coordinate Format* (COO). In this format the matrix A is stored into three arrays, e.g., `val`, `row` and `col`. The first is able to accommodate floating point entries of A and the others just the row and column index where the corresponding entry is stored, i.e., `val`(k) has value $a_{i,j}$ and `row`$(k) = i$ and `col`$(k) = j$, respectively. The matrix-vector product with this format is given by

$$y_i = (A\mathbf{x})_i = \sum_{j=1}^{n} a_{i,j} x_j \qquad \text{for } i = 1, \ldots, n, \tag{1.4}$$

as in Algorithm 1.1. The other useful operation we need to investigate is the

Algorithm 1.1: Matrix–vector product with A in COO format

Input: $A \in \mathbb{R}^{n\times n}$ in COO format: `val`, `col`, `row`, $\mathbf{x}$
Output: $\mathbf{y} = A\mathbf{x}$.

$y_i \leftarrow 0$ for $i = 1, \ldots, n$;
for $k = 1, \ldots, \mathrm{nnz}(A)$ **do**
  $y_{\texttt{row}(k)} \leftarrow y_{\texttt{row}(k)} + \texttt{val}(k) x_{\texttt{col}(k)}$; `/* Scatter and Gather */`

product $\mathbf{y} = A^T\mathbf{x}$,

$$y_i = (A^T\mathbf{x})_i = \sum_{j=1}^{n} a_{j,i} x_j, \qquad i = 1, 2, \ldots, n, \tag{1.5}$$

for a matrix A, stored in COO format, the algorithm is obtained simply by inverting the roles of the arrays `col` and `row` in Algorithm 1.1.

To reduce the number of stored indices, even if they need a relatively small space because they need just integer arrays, the *Compressed Row Storage* (CRS) and the *Compressed Column Storage* (CCS) are the most popular alternatives. In the CRS format the consecutive nonzero entries of the matrix rows are put in contiguous memory locations. Let us assume that A is sparse and *non*symmetric. We need to create 3 arrays: one able to store floating point (or complex) numbers for the nonzero entries of A, `val`, and `col_ind` and `row_ptr`, that contain the (integer) indexes of the entries. Thus, the nonzeros of A are stored in a row–wise fashion in `val`, while the indexes are defined according to the following rule

$$\begin{cases} \texttt{col\_ind}(k) \leftarrow j, & \text{if } \texttt{val}(k) = a_{i,j}, \\ \texttt{row\_ptr}(i) \leq k < \texttt{row\_ptr}(i+1), & \text{if } \texttt{val}(k) = a_{i,j}, \\ \texttt{row\_ptr}(n+1) \leftarrow \text{nnz}(A) + 1. \end{cases}$$

In this way, we need to store $2\,\text{nnz}(A) + n + 1$ integers instead of n^2. The Compressed Column Storage (CCS), also called the Harwell–Boeing sparse matrix format (see [204]), is defined in an analogue way, except that the columns of A are stored in `val` instead of the rows, i.e., the CCS format for A is the CRS format for A^T. In the case of a symmetric A matrix it is sufficient to store only the upper or the lower triangular part of A.

Example 1.2 (Compress Row/Column Storage). Let us consider the matrix $A \in \mathbb{R}^{4\times 4}$ given by

$$A = \begin{bmatrix} 1 & 2 & 0 & 0 \\ 1 & 0 & 0 & 3 \\ 4 & 0 & 5 & 0 \\ 0 & 1 & 0 & 0 \end{bmatrix}.$$

Its representation in Compress Row Storage is given by

$$\begin{cases} \texttt{val} = [1, 2, 1, 3, 4, 5, 1], \\ \texttt{col\_ind} = [1, 2, 1, 4, 1, 3, 2], \\ \texttt{row\_ptr} = [1, 3, 5, 7, 8], \end{cases},$$

while, in Compress Column Storage

$$\begin{cases} \texttt{val} = [1, 1, 4, 2, 1, 5, 3], \\ \texttt{row\_ind} = [1, 2, 3, 1, 4, 3, 2], \\ \texttt{col\_ptr} = [1, 4, 6, 7, 8], \end{cases},$$

and it easy to verify that the CCS can be obtained as the CRS of A^T. ✓

The matrix–vector product in CRS format is then obtained by means of the usual formula (1.4). Moreover, since we have stored the entries of A by accessing them row-wise, we can implement the matrix–vector multiplication as in Algorithm 1.2. Observe that these methods achieve exactly what we have proposed at the beginning, i.e., we only multiply the nonzero matrix entries. Moreover, the operation count is only $2\,\text{nnz}(A)$, a saving within respect to the $O(n^2)$ flops of the general *dense* case. This is because performing products

Algorithm 1.2: Matrix–vector product, A stored in CRS format

Input: $A \in \mathbb{R}^{n\times n}$ in CRS format: `val`, `col_ind`, `row_ptr`, $\mathbf{x}$
Output: $\mathbf{y} = A\mathbf{x}$.
1 **for** $i = 1, \ldots, n$ **do**
2 $\quad y_i \leftarrow 0$;
3 $\quad$ **for** $j = row\_ptr(i), \ldots, row\_ptr(i+1) - 1$ **do**
4 $\quad\quad y_i \leftarrow y_i + \mathtt{val}(j) x_{\mathtt{col\_ind}(j)}$; /* Gather */

with zero entries has a nonzero computational cost, that can be non negligible whenever done, e.g., for a hundred thousand times or more.

Unfortunately, we cannot directly use the standard formula (1.5) for computing the product with the transpose matrix. Indeed, this would imply accessing the column of the matrix, an operation that is quite inefficient for matrices stored in the CRS format. Thus, it is better to switch the indices and resort to Algorithm 1.3. The Algorithm 1.3 has the same overall structure of Algorithm 1.2 using the same indirect addressing. What is different between the two procedures is the memory access pattern: Algorithm 1.2 at line 1 loads two vectors of data, a row of A and the input vector, and writes out a single scalar. Algorithm 1.3 reads only one element of $\mathbf{x}$ and one row of the matrix A. Both algorithms read and write the vector $\mathbf{y}$. Thus, for A^T, the access to memory limits the overall performance. Moreover, the COO format admits a very fast conversion to and from CSR/CSC formats, thus COO format can be used as a fast way for constructing sparse matrices, while matrix–vector product can then be executed after the CSR/CSC conversion. The following modification of the CRS and CCS storage scheme can be put in place if A is composed by some square dense blocks of nonzeros in some regular pattern. Let us suppose that n_b is the dimension of each block and $\text{nnz}_b(A)$ is the number of nonzero block in $A \in \mathbb{R}^{n\times n}$. In this case the total storage that is needed for the compression is $\text{nnz}_b(A)n_b^2$ and the block dimension n_d of A is the obtained simply by $n_d = n/n_b$. As for the standard CRS scheme, we need 3 arrays for the *Block Compressed Row Storage* (BCRS); these are a rectangular array which access the nonzero block in a

Algorithm 1.3: Matrix–vector product with A^T in CRS format

Input: $A \in \mathbb{R}^{n\times n}$ in CRS format: `val`, `col_ind`, `row_ptr`, $\mathbf{x}$
Output: $\mathbf{y} = A\mathbf{x}$.
$y_i \leftarrow 0$ for $i = 1, \ldots, n$;
for $j = 1, \ldots, n$ **do**
$\quad$ **for** $i = row\_ptr(j), \ldots, row\_ptr(j+1) - 1$ **do**
$\quad\quad y_{\mathtt{col\_ind}(i)} \leftarrow y_{\mathtt{col\_ind}(i)} + \mathtt{val}(i) x_j$;

block–row–wise fashion $\texttt{val}(1,\ldots,\text{nnz}_b(A),1,\ldots,n_b,1,\ldots,n_b)$, an integer array $\texttt{col\_ind}(1,\ldots,\text{nnz}_b(A))$ which stores the actual column indices of the original matrix A of the first elements of the corresponding nonzero blocks, and a pointer array $\texttt{row\_blk}(1,\ldots,n_d+1)$ whose entries point to the beginning of each block row in the corresponding `val` and `col_ind` entry. This kind of storage becomes convenient whenever the matrices have a large n_b. Observe that a *Block Compressed Column Storage* (BCCS) can be considered by storing the sparse nonempty blocks with the CCS format.

If the underlying sparse matrix A is banded, i.e., there exist $p,q \geq 0$ such that $a_{i,j} \neq 0$ if and only if $i-p \leq j \leq i+q$, we can allocate the whole matrix by using the *Compressed Diagonal Storage* (CDS), a rectangular array $n \times (p+q+1)$ `val` that permits to store the matrix A by means of the mapping $\texttt{val}(i,j) \leftarrow a_{i,i+j}$. In the CDS storage scheme, some zeros corresponding to non–existing matrix elements are appended to arrays. In the standard CDS case, the matrix–vector product can be performed either by row or columns, see Algorithm 1.4, while the algorithm for the transpose product is just a minor variation of the latter. The above algorithm accesses the memory by

Algorithm 1.4: Matrix–vector product, A stored in CDS format

Input: $A \in \mathbb{R}^{n\times n}$: `val`, $p,q \in \mathbb{N}$, $\mathbf{x}$
Output: $\mathbf{y} = A\mathbf{x}$.

$y_i \leftarrow 0$ for $i = 1,\ldots,n$;
for $d = -p,\ldots,q$ **do**
  for $l = \max(1,1-d),\ldots,\min(n,n-d)$ **do**
    $y_l \leftarrow y_l + \texttt{val}(l,d)x_{l+d}$;

using three vectors per inner iteration. Thus, it can be heavier than the version with the CSR storage. Nevertheless, no indirect addressing is used, and then the algorithm can be easy vectorized. A modification of this format, especially tuned for working with parallel and vector processors is the *Jagged Diagonal Storage* (JDS) format, see [387, 429].

Exercise 1.4. Write an algorithm for computing the product $\mathbf{y} = A^T\mathbf{x}$ for $A \in \mathbb{R}^{n\times n}$ stored in CDS format.

Exercise 1.5. Transform the routine in Algorithms 1.1–1.4 for the matrix–vector product kernels, i.e., $\mathbf{y} = A\mathbf{x}$ and $y = A^T\mathbf{x}$, into kernels for the Basic Linear Algebra Subroutine (BLAS): *SGEMV*, i.e,

$$\mathbf{y} = \alpha A\mathbf{x} + \beta\mathbf{z}, \quad \text{and} \quad \mathbf{y} = \alpha A^T\mathbf{x} + \beta\mathbf{z},$$

for α, β two scalars and $\mathbf{z} \in \mathbb{R}^n$.

Remark 1.2. Sparse matrices are not the only matrices for which fast matrix–vector product routines can be devised. By means of *displacement formulas* and *fast transformations* it is possible to reduce the computational cost for matrix–vector operations to less than an $O(n^2)$. Two popular examples are Toeplitz and Circulant matrices, see § 4.1.

Example 1.3 (Matrix vector product). In Figure 1.5 we have the timings for the matrix vector products $\mathbf{y} = A\mathbf{x}$, where

$$A = \begin{bmatrix} T & D_1 & & \\ D_2 & T & \ddots & \\ & \ddots & \ddots & D_1 \\ & & D_2 & T \end{bmatrix}_{n^2 \times n^2}, T = \begin{bmatrix} -4 & 1 & & \\ 1 & -4 & \ddots & \\ & \ddots & \ddots & 1 \\ & & 1 & -4 \end{bmatrix}_{n \times n},$$

$$D_1 = \text{diag}(-1, \ldots, -1), D_2 = \text{diag}(3, \ldots, 3),$$

and $\mathbf{x} = (1, \ldots, 1)^T$, with A stored in the considered matrix storage formats. We observe that CSR and CSC achieve the best timings. Moreover, the BSR,

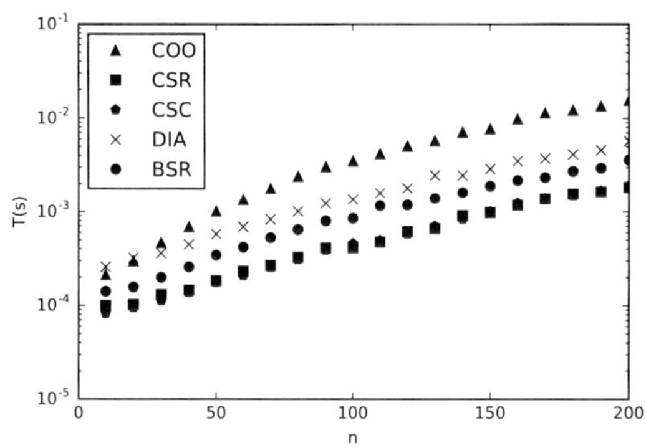

Figure 1.5: Example 1.3. Timings for the matrix–vector product with various sparse storage schemes.

even if A has a very clear block structure, does not beat the CSR/CSC formats since the blocks are itself too sparse to give an effective compression. ✓

1.3 On parallel computing and hardware acceleration

Traditionally we are used to write computer software in a *serial* way, i.e., we construct algorithms as serial streams of instructions in a way that only one instruction can be executed at a time. Nevertheless, many of the core operations that are involved with numerical linear algebra, e.g., matrix–vector products, matrix–matrix products, vector adds etc., can be thought of, and of course implemented, also in a *parallel* form. To accomplish this, one needs to break the underlying computational task into several independent sub–tasks that can be executed simultaneously. The architectures capable of parallel computing are of various type. We go from a single computer with multiple processors to several computers on the same network to the use of *specialized hardware*. Moreover, any combination of these frameworks is also viable, at least in principle. Why are we interested in moving in this direction? Until the first decade of the twenty-first century, the driving force in satisfying the prediction of Moore's Law[1] [392] has been the *Frequency Scaling* approach: reducing the time for cycle ratio. Nevertheless, this procedure is taxing in two directions: the effective possibility of reducing the size of components and the need of managing power consumption, that depends directly on the processor frequency. Parallel computing and in our specific case, *data parallel*[2] computing, provides an answer to this problem.

From the point of view of the underlying machinery, here we briefly focus on the approach of *hardware acceleration*, i.e., on the use of a *specialized* piece of hardware that works as an *accelerator*, and *co–processors*. Being more specific, hardware of this kind are the **G**raphic **P**rocessing **U**nits (GPUs) and pure processor/co–processor like the Xeon–Phi architectures. Each hardware comes with their own features in term of interfaces, programming languages and way of approaching the computations. Therefore, to make a quick and simple comparison, we look here only at two parameters: the theoretical bandwidth peak and the number of floating point operations per cycle (FLOPS/cycle).

Bandwidth differently from the CPUs, the hardware accelerators are optimized for increasing the **throughput** of the computation, i.e., the number of processes we complete per unit of time. The essential idea is generating a massive number of tasks that can be executed in parallel by each of the computing elements of the accelerator. A wider band gives the ability to perform stream on more data. We measure this quantity in GB/s.

Figure 1.6 reports this data, giving a perspective of the direction in which technology is developing.

Let us now look at a most tangible example of what we have discussed so far.

Example 1.4. Consider the core operations for iterative methods, the matrix–vector product $A\mathbf{v}$ with A a sparse matrix. We perform the computation for A of growing size on a CPU, an Intel(R) Xeon(R) CPU E5-2670 v2 @ 2.50GHz, and on a GPU, a NVIDIA GeForce GTX 860M, by using the MATLAB interfaces. As we see from Figure 1.7, the speed–up, measured in term of seconds needed to perform the tasks, obtained with the accelerator is substantial. ✓

---

[1] «The number of transistors in a dense integrated circuit doubles approximately every two years»

[2] Data parallelism is focused on the distribution of the data, e.g., matrices and vectors, across different computing nodes, which operate on them in parallel.

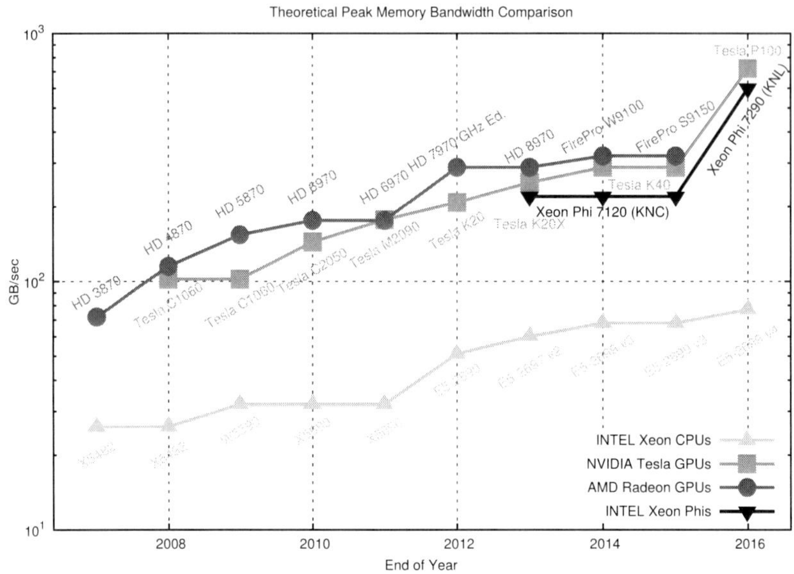

(a) Bandwidth

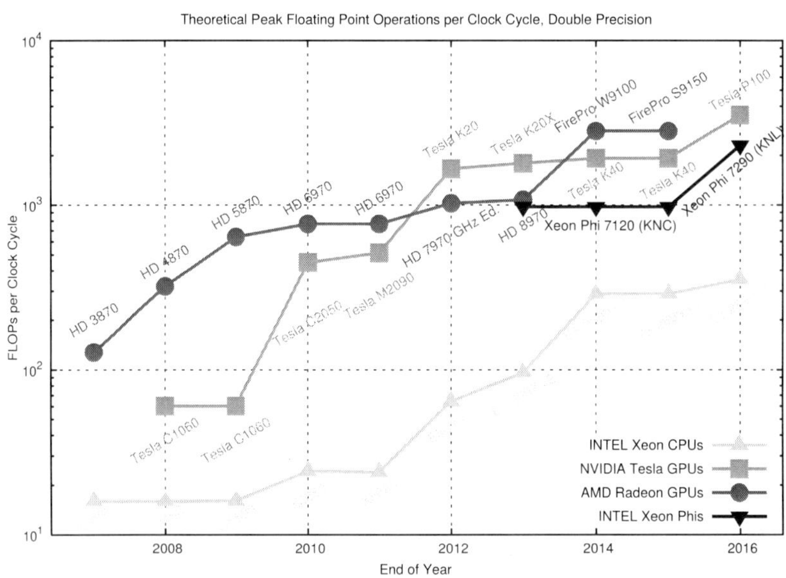

(b) FLOPS per Cycle

Figure 1.6: Benchmark over time of some commercial hardware accelerators. Data and elaboration are from *www.karlrupp.net*.

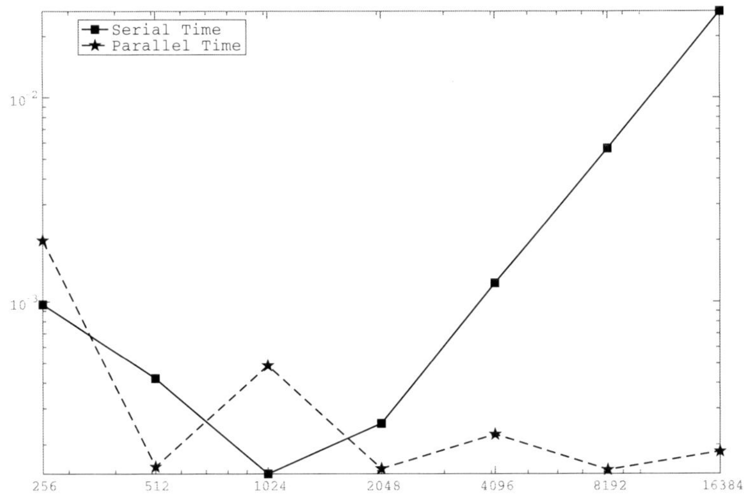

Figure 1.7: Acceleration of the sparse matrix vector product with the GPU in the MATLAB environment. As we see, while the size of the matrix (on the x axis) increases, the GPU accelerated version achieves better timing (on the y axis).

We can not cover the implementation details of the *basic linear algebra subroutines* nor the way to optimize the code and the algorithms for these type of architectures as it would take us out of the target of this book, see, e.g., [39, 120, 175, 366, 545]. Nevertheless, we will try to make it known whenever these strategies can be applied or are available for the material in later chapters.

We conclude this section with some references. For information regarding the news in this area and in large scale computing a good starting point is the website: *www.top500.org*. For the use of NVIDIA hardware for computations refer to the book by Sanders and Kandrot [463] and the website `developer.nvidia.com`, while for the use of the AMD hardware and, more in general, the Open Computing Language (openCL) refer to the book by Tay [505] and the websites `developer.amd.com` and `www.khronos.org/opencl`. For the INTEL architecture a reference is the book by Jeffers and Reinders [318]. For libraries that implement the BLAS operation in parallel, refer to [172, 200, 230, 516, 517]. Further information is given, when needed, throughout the book.

Chapter 2

Some iterative algorithms for linear systems

2.1 Iterative algorithms and simple one-dimensional techniques

Let us consider the linear system

$$A\mathbf{x} = \mathbf{b}, \tag{2.1}$$

where A is a real $n \times n$ matrix, $\mathbf{b}$ the known vector, and $\mathbf{x}$ the vector of the unknowns.

An iterative method for searching an approximate solution of the algebraic linear system (2.1) is a strategy that generates a sequence of candidate approximations $\mathbf{x}^{(k)}$, $k = 0, 1, \ldots$, for the solution, starting from a given initial guess $\mathbf{x}^{(0)}$. A sequence of candidate approximations $\{\mathbf{x}^{(k)}\}$ in $\mathbb{R}^n$ (or in $\mathbb{C}^n$) is said to be *convergent* to $\mathbf{x}$ if there exists a norm $||\cdot||$ such that

$$\lim_{k\to\infty} ||\mathbf{x} - \mathbf{x}^{(k)}|| = 0;$$

see also Definition A.12. The underlying iterative method generating $\{\mathbf{x}^{(k)}\}$ is said to be *convergent* if the sequence $\{\mathbf{x}^{(k)}\}$ is convergent for all $\mathbf{x}^{(0)}$. Moreover and very important, iterative methods considered here involve the matrix A only in the context of matrix-vector multiplications. Differently to the direct, in general, iterative methods do not terminate after a finite number of steps. They require some *stopping criteria* in order to become *algorithms.* We terminate the underlying methods when an estimate for the $||\cdot||$ norm of the (unknown!) relative error

$$\frac{||\mathbf{e}^{(k)}||}{||\mathbf{x}||}$$

is less than a user-prescribed quantity ε, i.e.,

$$\frac{||\mathbf{e}^{(k)}||}{||\mathbf{x}||} = \frac{||\mathbf{x}^{(k)} - \mathbf{x}||}{||\mathbf{x}||} < \varepsilon$$

and the number of steps done k is less than a maximum allowed N in order to avoid loops. Very often the estimate of the error is based on the *residual* $\mathbf{r}^{(k)}$:

$$\mathbf{r}^{(k)} = \mathbf{b} - A\mathbf{x}^{(k)},$$

a quantity that is easy to compute but...it is not always so close to the error. Indeed, by using the following straightforward identities (exercise: prove them!)

$$A\mathbf{e}^{(k)} = -\mathbf{r}^{(k)}, \quad ||b|| \leq ||A||\,||x||,$$

we obtain the following upper bound for the norm of the relative error

$$\frac{||\mathbf{e}^{(k)}||}{||\mathbf{x}||} \leq k(A)\,\frac{||\mathbf{r}^{(k)}||}{||\mathbf{b}||},$$

where $k(A) = \|A\|\,\|A^{-1}\|$ is the *condition number* of the matrix A. Therefore, if at step k we experience that

$$\frac{\|\mathbf{r}^{(k)}\|}{\|\mathbf{b}\|} < \varepsilon$$

and stop the iterative method, then our computed approximation $\mathbf{x}^{(k)}$ *can* be quite far from x if $k(A)$ is large, i.e., the relative error can be $k(A)$ times greater than ε, the desired accuracy.

In the next sections we quickly recall basic (based on a splitting of the matrix A) iterative and one-dimensional projection methods in order to discuss more in detail the more frequently used and powerful *Krylov subspace methods.*

For more details on those methods see, e.g., Wilkinson [543], Parlett [418], Golub and Van Loan [271], Saad [455]. Historical notes can be found in Saad and Van der Vorst [460] and in Bertaccini et al. [69].

2.1.1 Splitting methods

Let us consider the solution of a linear system

$$A\mathbf{x} = \mathbf{b},\quad A \in \mathbb{C}^{n\times n},\quad \mathbf{x}, \mathbf{b} \in \mathbb{C}^n,\quad \det(A) \neq 0,\quad a_{i,i} \neq 0,\ i = 1, \ldots, n,$$

i.e., we assume that matrix A is non–singular, and moreover that its diagonal entries $a_{i,i}$ are all non–zero complex numbers.

Let us call *splitting iterative methods*, or *splitting methods* for short, the iterative algorithms for linear systems $A\mathbf{x} = \mathbf{b}$ that are a linear fixed-point iteration of the form $\mathbf{x}^{(k+1)} = G\,\mathbf{x}^{(k)} + \mathbf{c}$, where G is based on a decomposition (or *splitting*) of the matrix A of the form $A = M - N$, where M is a nonsingular matrix. By substitution we find

$$A\mathbf{x} = \mathbf{b}\ \rightarrow M\mathbf{x} - N\mathbf{x} = \mathbf{b}\ \rightarrow \mathbf{x} = (M^{-1}N)\,\mathbf{x} + M^{-1}\,\mathbf{b},$$

and, by setting an initial vector $\mathbf{x}^{(0)}$, we can build a sequence of candidate approximations for the solution $\mathbf{x}$ by

$$\mathbf{x}^{(k+1)} = (M^{-1}N)\,\mathbf{x}^{(k)} + M^{-1}\mathbf{b} = G\,\mathbf{x}^{(k)} + \mathbf{c},\quad k = 0, 1, 2, \ldots$$

Let us focus on the convergence of these methods. In order to proceed, compute the norm of the error between the generic kth iterate of the fixed-point iterative method and the true (obviously not known) solution $\mathbf{x}^* = A^{-1}\mathbf{b}$:

$$\mathbf{e}^{(k)} = \mathbf{x}^{(k)} - \mathbf{x}^*.$$

By using the definition of G we obtain

$$\mathbf{e}^{(k)} = G\mathbf{e}^{(k-1)} = \ldots = G^k\mathbf{e}^{(0)},\ k \geq 0. \tag{2.2}$$

By observing that $\rho(G) < 1$ if and only if

$$\lim_{k\to\infty} G^k = 0,$$

(this can be easily proved by decomposing G in its Jordan canonical form; see [271] for details) we have

$$\lim_{k\to\infty} \mathbf{e}^{(k)} = 0,$$

thus proving the following fundamental convergence result for stationary iterative methods.

Theorem 2.1. *The stationary iterative method with iteration matrix G is convergent if and only if $\rho(G) < 1$.*

The following result can help very much if we are able to find that $\|G\| < 1$ for an induced matrix norm $\|\cdot\|$ that is easy to compute for the linear system we are considering.

Corollary 2.1. *If there exists an induced matrix norm $\|\cdot\|$ such that $\|G\| < 1$, then the stationary iterative method converges.*

Choosing a matrix norm over $\mathbb{C}^{n\times n}$ induced by a vector norm over $\mathbb{C}^n$ gives

$$\|\mathbf{e}^{(k)}\| \leq \|G^k\| \cdot \|\mathbf{e}^{(0)}\|,\ k \geq 0, \Rightarrow \frac{\|\mathbf{e}^{(k)}\|}{\|\mathbf{e}^{(0)}\|} \leq \|G^k\|,\ k \geq 0.$$

We can use the value of $\|G^k\|$ as a basis for comparing the ratio of convergence of the different fixed-point iterative methods, i.e., the *average* reduction factor for k iterations is given by

$$\rho = \left(\frac{\|\mathbf{e}^{(k)}\|}{\|\mathbf{e}^{(0)}\|}\right)^{1/k}$$

and this can be bounded by the *rate of convergence* of the iteration matrix G, i.e., by the **spectral radius**:

$$\rho(G) = \max_{\lambda\in\sigma(G)} |\lambda|,\ \sigma(G) = \{\lambda \in \mathbb{C} : \det(\lambda I - G) = 0\}.$$

For splitting methods we can always rewrite the matrix $G = M^{-1}N$ for the underlying splitting methods as:

$$G = M^{-1}N = M^{-1}(M - A) = I - M^{-1}A. \tag{2.3}$$

We can give the following definition, characterizing the convergence of the fixed-point splitting methods.

Definition 2.1 (Convergent Splitting). Given a non-singular matrix $A \in \mathbb{C}^{n\times n}$ we call a *convergent splitting* of A a regular splitting of the matrix A of the form $A = M - N$ where

$$\rho(M^{-1}N) = \rho(G) = \rho(I - M^{-1}A) < 1.$$

Remark 2.1. The *rate of convergence* is the simplest mean for comparing the behavior of two iterative methods of this form. However, in certain cases it can give misleading informations. Only for Hermitian (or normal) matrices G_1, G_2 it is true that $\rho(G_1) < \rho(G_2) < 1$ implies that $\|G_1^k\| < \|G_2^k\| < 1$ $\forall\, k > 0$. Therefore, for general matrices the condition on the spectral radii does not imply the one on the norms, i.e., the latter is going to be true only asymptotically.

Two very popular *splitting methods* are the *Jacobi* [316] and the *Gauss–Seidel* [257] algorithms. Although being two classic methods these are interesting as they can be applied as *building blocks* for more elaborate methods or as *preconditioners* for other iterative methods. Let us consider the following splitting of the matrix A.

$$A = \begin{pmatrix} d_{1,1} & -f_{1,2} & \text{---} & -f_{1,n} \\ -e_{2,1} & \ddots & \text{-F} & \vert \\ \vert & \text{-E} & \text{D} & -f_{n-1,n} \\ -e_{n,1} & \text{---} & -e_{n,n-1} & d_{n,n} \end{pmatrix} = D - E - F, \tag{2.4}$$

i.e., D is the diagonal part A, E is minus the strict lower triangular part of A and F is minus the upper strict triangular part of A:

$$(D - E - F)\mathbf{x} = \mathbf{b} \Rightarrow D\mathbf{x} = (E + F)\mathbf{x} + \mathbf{b} \Rightarrow \mathbf{x} = D^{-1}(E + F)\mathbf{x} + D^{-1}\mathbf{b},$$

then,

$$\mathbf{x}^{(k+1)} = D^{-1}(E + F)\mathbf{x}^{(k)} + D^{-1}\mathbf{b}, \;\; k \geq 0.$$

We can rewrite the previous equation in terms of the components of the vector $\mathbf{x}$ thus obtaining the expression

$$x_i^{(k+1)} = -\sum_{\substack{j=1 \\ j\neq i}}^{n} \left(\frac{a_{i,j}}{a_{i,i}}\right) x_j^{(k)} + \frac{b_i}{a_{i,i}}, \;\; 1 \leq i \leq n, \; k \geq 0. \tag{2.5}$$

We observe that, in general, we need to save all the components of the vector $\mathbf{x}^{(k)}$ while computing the components of the vector $\mathbf{x}^{(k+1)}$, i.e., we are solving component by component all equations of the linear system. In this way we have the standard *Jacobi method*. On the other hand, we could be tempted to use the latest estimates of the element of $\mathbf{x}^{(k+1)}$ to update the full vector, i.e., to use during the ith step all the $x_j^{(k+1)}$ for $j = 1, 2, \ldots, i-1$. In this way we get the following variant of (2.5).

$$x_i^{(k+1)} = -\sum_{j=1}^{i-1}\left(\frac{a_{i,j}}{a_{i,i}}\right)x_j^{(k+1)} - \sum_{j=i+1}^{n}\left(\frac{a_{i,j}}{a_{i,i}}\right)x_j^{(k)} + \frac{b_i}{a_{i,i}},\ 1 \leq i \leq n,\ k \geq 0. \tag{2.6}$$

From the computational point of view, this method has the advantage of not needing the space to store simultaneously both the approximation of $x_i^{(k)}$ and $x_i^{(k+1)}$. We have built in this way the *Gauss–Seidel* iterative method. Observe also that the latter is not *parallelizable* or *vectorizable*, since we need to compute one component at a time. Nevertheless, the vector form can be recovered, at least partially, for problems that admit a particular decoupling of the unknowns. If it is possible to decouple the system as

$$\begin{pmatrix} D_1 & C \\ C^T & D_2 \end{pmatrix}\begin{pmatrix} \mathbf{x}_1 \\ \mathbf{x}_2 \end{pmatrix} = \begin{pmatrix} \mathbf{b}_1 \\ \mathbf{b}_2 \end{pmatrix}, \text{ with } D_1, D_2 \text{ diagonal matrices}$$

where F represents the coupling between the unknown, the following block form of the method can be used

$$\begin{aligned} D_1\mathbf{x}_1^{(k+1)} =& \mathbf{b}_1 - C\mathbf{x}_2^{(k)}, \\ D_2\mathbf{x}_2^{(k+1)} =& \mathbf{b}_2 - C\mathbf{x}_1^{(k)}. \end{aligned}$$

For other details on these approaches, mainly related to PDEs discretizations, see, e.g., [555] and § 4.4.

Definition 2.2 (Regular Splitting). Given a nonsingular matrix $A \in \mathbb{C}^{n\times n}$ we call a *regular splitting* of A a splitting of the matrix $A = M - N$ such that

1. $\det(M) \neq 0$;
2. $M^{-1} \geq 0$ and $N \geq 0$.

We write $A \geq 0$ iff $\forall i, j\ a_{i,j} \geq 0$, so if $M - N \geq 0$ we write $M \geq N$.

Various splitting methods, using the notation established in (2.4), can be obtained for different values of M (and then of course of N). Some are listed in Table 2.1.

Let us look at an example showing why these methods, per se and without further refinements, can deliver poor performance when facing even a simple and well known problem.

Example 2.1. Consider the finite difference Laplacian on a square-shaped domain. This gives a symmetric and positive definite matrix A. The right-hand side is given by the constant function 1. The standard Jacobi (JA) method is used with the stopping criteria $\varepsilon = 10^{-6}$ and a maximum number of iteration `maxit = 1000` for an equispaced mesh with $n+1$ points. In Table 2.1 and Figure 2.1 we report the performance A † means that the method fails to converge in `maxit` iterations. ✓

Table 2.1: Splitting Methods

	M	Splitting of:
Jacobi (JA)	D	A
Damped Jacobi (DJA)	ωD	ωA
Gauss–Seidel (GS)	$D - E$	A
Backward Gauss–Seidel (BGS)	$D - F$	A
Successive Over Relaxation (SOR)	$D - \omega E$	ωA
Symmetric Gauss–Seidel (SGS)	$(D - E)D^{-1}(D - F)$	A
Successive Symmetric Over Relaxation (SSOR)	$(D - \omega E)D^{-1}(D - \omega F)$	ωA

Table 2.2: Example 2.1. Convergence of the Jacobi (JA) method for the 2D Laplacian discretized with centered differences

n	Iteration	Error	$T(s)$	$\rho(G)$
64	221	7.66e-06	8.76e-02	0.940
169	539	1.28e-05	4.33e-03	0.975
324	995	1.78e-05	8.59e-03	0.986
529	†	3.60e-03	1.20e-02	0.991
784	†	6.56e-02	1.85e-02	0.994

Improvement on these strategies can be devised substantially in two ways:

- by using methods reported in Table 2.1 as a building block for the *multigrid methods*, a way we will discuss in Chapter 4;
- exploiting acceleration strategies, both from the mathematical point of view and the implementation architecture, as in [551], in which the main application of the algorithm is in the solution of the elliptic partial differential equations.

Other popular methods of this class are the so–called relaxation method, devised initially to face with elliptic problems where *Jacobi* and *Gauss–Seidel* methods show poor convergence. Following the decomposition in Table 2.1 we can start from the *Successive Over Relaxation* method (SOR) from

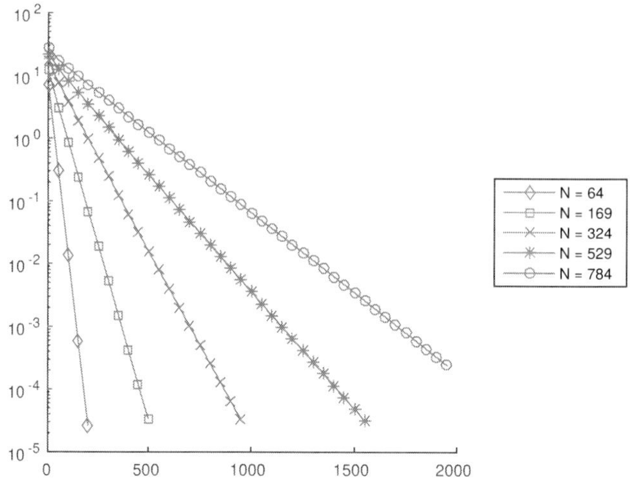

Figure 2.1: Example 2.1. Plot in logarithmic scale of the iteration versus the error for the solution of the linear system generated by the 2D discretization of the Laplacian operator.

Young's [554] Ph.D. thesis:

$$x_i^{(k+1)} = \frac{\omega}{a_{i,i}} \left(b_i - \sum_{j=1}^{i-1} a_{i,j} x_j^{(k+1)} - \sum_{j=i+1}^{n} a_{i,j} x_j^{(k)} \right) + (1-\omega) x_i^{(k)}, \qquad (2.7)$$

with $\omega \in \mathbb{R}$ and $\omega > 0$. A sharp bound for the convergence of these methods can be stated for a class of matrices that we will discuss in § 3.4.2, i.e., the so–called H–matrices (see Definition 3.8), see [395].

Proposition 2.1 ((SOR) Convergence). *Let A be a H–matrix. The SOR method converges if*

$$0 < \omega < \frac{2}{1 + \rho(|G|)},$$

where G is the matrix in (2.3) *for the SOR method (see Table 2.1) and $|\cdot|$ takes the absolute values of each element of the matrix*[1].

More in general,

Theorem 2.2 (Ostrowsky–Reich, [414, 443]). *Let A be symmetric with D, as in decomposition* (2.4)*, positive definite. Then (SOR) converges for* $0 < \omega < 2$ *if and only if A is positive definite.*

[1]This is necessary because the proof of the results needs the use of Perron–Frobenius theorem, see [252–254].

As we have observed for Gauss–Seidel methods, the ordering of the unknowns can be a crucial issue. Mixing together *Gauss–Seidel* (GS) and *Backward Gauss–Seidel* (BGS), in which the unknowns are ordered from n to 1, we can obtain the Symmetric Gauss–Seidel method (SGS). Then, we can apply the same relaxation technique of the (SOR) method obtaining the *Successive Symmetric Over Relaxation* method (SSOR). In this way, during the computation, the unknowns are ordered from 1 to n and then from n to 1 thus obtaining a symmetric iteration matrix. Similarly to what we have stated for the SOR method, the following results can be established.

Proposition 2.2 (Alefeld and Varga [6]). *If A is a H–matrix the SSOR method converges if*

$$0 < \omega < \frac{2}{1 + \rho(|G|)},$$

where G is the matrix in equation (2.3) *related to the SSOR method (see Table 2.1) and $|\cdot|$ takes the absolute values of each element of the matrix.*

Theorem 2.3. *Let A be a symmetric positive definite matrix with D, as in decomposition* (2.4), *positive definite. Then the SSOR method converges for* $0 < \omega < 2$.

The last feature we want to investigate is the *smoothing property*. We need to start again from the equation for the residuals (2.2). For the reader who is familiar with the numerical treatments of partial differential equations, these topics will sound very similar to *von Neumann's stability analysis*, that in some case is called also *local Fourier analysis*, [351]. The fundamental idea relies in considering the residual vector $\mathbf{r}$ in the span of the grid function Fourier series. In this way, smooth residuals are equivalent to Fourier coefficients that show a decay. We call one of the underlying methods a good *relaxation scheme* for a certain Fourier frequency if the maximal ratio by which this frequency is reduced per step of the method is less than 1. Observe that in this context the convergence ratio of the underlying method is irrelevant. To go deeper in this direction we need to focus on each particular problem. We will focus again on this direction in § 4.2.

Exercise 2.1. Determine the condition on the matrix A for the convergence of the methods of Table 2.1.

Exercise 2.2. Given A symmetric with D, as in the splitting of (2.4), positive definite. Prove that A is positive definite if and only if the Gauss–Seidel method converges.

Exercise 2.3. Find two nonHermitian matrices G_1 and G_2 such that $\rho(G_1) < \rho(G_2) < 1$ and for which there exists $k \in \mathbb{N}$ such that $\|G_1^k\| > \|G_2^k\|$.

2.1.2 Generalities on projection techniques

The idea of the *projection techniques* is based on the extraction of an approximate solution for

$$A\mathbf{x} = \mathbf{b},$$

$A \in \mathbb{R}^{n\times n}$, from $\mathbb{R}^n$. If $\mathcal{K}$ is the search subspace or the subspace of *candidate approximants* and is of dimension m, then in general m constraints should be imposed in order to have a hope to extract a unique solution. Usually these constraints are imposed by m independent orthogonality conditions. This requires defining another subspace of $\mathbb{R}^n$, $\mathcal{L}$, the *subspace of constraints.* This is shared by other mathematical frameworks and it is called the *Petrov–Galerkin* conditions.

As in [455], we call here a projection method *orthogonal* if $\mathcal{K} = \mathcal{L}$; oblique otherwise.

A projection technique onto the subspace $\mathcal{K}$ orthogonal to $\mathcal{L}$ for the underlying linear systems is a process which determines a candidate approximate solution $\tilde{\mathbf{x}}$ by imposing

$$\tilde{\mathbf{x}} \in \mathcal{K}, \quad \mathbf{b} - A\mathbf{x} \perp \mathcal{L},$$

where $\perp$ means orthogonal to. In order to include the information on an initial guess $\mathbf{x}^{(0)}$, we should extract $\tilde{\mathbf{x}}$ in the affine subspace $\mathbf{x}^{(0)} + \mathcal{K}$ and then the problem reformulates as finding $\tilde{\mathbf{x}}$ such that

$$\tilde{\mathbf{x}} \in \mathbf{x}^{(0)} + \mathcal{K}, \quad \mathbf{b} - A\mathbf{x} \perp \mathcal{L}.$$

Most standard techniques use a sequence of projections. A new projection uses a new pair of subspaces $\mathcal{K}$ and $\mathcal{L}$ at each iteration and an initial guess $\mathbf{x}^{(0)}$ as the most recent approximation obtained from the previous approximation step. Note that projections form an unifying framework for many iterative solvers, including the stationary methods of the previous section. As an example, if we take $\mathcal{K} = \mathcal{L} = \text{span}\{\mathbf{e}_i\}$, where $\mathbf{e}_i$ is the i-th unitary vector of $\mathbb{R}^n$, we derive the Gauss–Seidel method. The projections are cycled for $i = 1, \ldots, n$ until convergence.

When $\mathcal{K} = \mathcal{L}$ the Petrov–Galerkin conditions are called the Galerkin conditions.

Now, let us consider a very illustrative matrix representation of the above projection techniques.

Let $V = [\mathbf{v_1}, \ldots, \mathbf{v_m}]$ an $n \times m$ matrix whose columns are the vectors $\mathbf{v}_i$, $i = 1, \ldots, m$, which form a basis of $\mathcal{K}$ and $W = [\mathbf{w}_1, \ldots, \mathbf{w}_m]$ an $n \times m$ matrix whose columns are the vectors $\mathbf{w}_1, \ldots, \mathbf{w}_m$ which form a basis of $\mathcal{L}$. If the candidate approximation for the solution of the underlying linear system is

$$\tilde{\mathbf{x}} = \mathbf{x}^{(0)} + V\,\mathbf{y}, \tag{2.8}$$

then the orthogonality conditions for the residual vector r with respect to vectors $\mathbf{v}_i$

$$\mathbf{r} \perp \mathcal{L} \quad \Rightarrow \quad W^T(\mathbf{b} - A\tilde{\mathbf{x}}) = 0$$

lead to the following system of equations:

$$W^T A V \mathbf{y} = W^T \mathbf{r}^{(0)}, \tag{2.9}$$

where $\mathbf{r}^{(0)} = \mathbf{b} - A\,\mathbf{x}^{(0)}$ is the initial residual. If the matrix $W^T A V$ is nonsingular, then we can use the matrix-vector notation for writing the approximate solution $\tilde{\mathbf{x}}$

$$\tilde{\mathbf{x}} = \mathbf{x}^{(0)} + V \left(W^T A V\right)^{-1} W^T \mathbf{r}^{(0)}.$$

Usually, the matrix $M = W^T A V$ is not formed explicitly but algorithms can compute the product $\mathbf{w} = M\mathbf{v}$ for any vector $\mathbf{v}$ and this is enough for all iterative methods considered here.

We note that if m is small compared to n (supposed always large here) and $M = W^T A V$ is nonsingular, we could compute $\tilde{\mathbf{x}}$ by solving the $m \times m$ linear system $M\mathbf{y} = W^T \mathbf{r}^{(0)}$ by, e.g., a direct method. However, M, the "projected version" of A can be singular even if A is not. An example of this issue is given by the nonsingular A defined as

$$A = \begin{pmatrix} 0 & I \\ I & I \end{pmatrix}. \tag{2.10}$$

By taking $V = W = [\mathbf{e}_1, \ldots, \mathbf{e}_m]$, where the $\mathbf{e}_i$s are the canonical basis vectors of $\mathbb{R}^m$, we obtain a null $m \times m$ matrix M because $W^T A V$ is made of all zeros.

There are at least two very important cases where M is nonsingular.

Proposition 2.3. *If A, $\mathcal{K}$ and $\mathcal{L}$ satisfy one of the two conditions*

- *(a) A is symmetric and definite and $\mathcal{K} = \mathcal{L}$, or*
- *(b) A is nonsingular and $A\mathcal{K} = \mathcal{L}$,*

then $M = W^T A V$ is nonsingular for any bases of $\mathcal{K}$ e $\mathcal{L}$.

Proof. Let the column vectors of V be a basis for $\mathcal{K}$ and those of W for $\mathcal{L}$, respectively. (a) If the column vectors of V are a basis for $\mathcal{K}$ and W of $\mathcal{L}$, respectively, there exists an invertible matrix G of size $m \times m$ such that

$$M = W^T A V = (VG)^T A V = G^T V^T A V,$$

which is nonsingular too because $V^T A V$ is positive definite because A is also positive definite.

(b) Since $\mathcal{L} = A\mathcal{K}$, then $W = AVG$ with G nonsingular $m \times m$ and

$$M = W^T A V = G^T (AV)^T AV = G^T V^T A^T A V.$$

Finally, by observing that A is nonsingular, the $n \times m$ matrix AV has full (i.e., m supposed $m < n$) rank, we find that also $(AV)^T AV$ and therefore M are nonsingular. □

2.1.3 Four one-dimensional projection algorithms

2.1.3.1 Steepest descent

The first one-dimensional projection algorithm we consider is *steepest descent*, known also as *the gradient method* (for symmetric and definite linear systems) because it is a special version of the gradient method for unconstrained minimization of a function f; see, e.g., [401]. In particular, both methods are based on the search direction

$$\mathbf{d}^{(k)} = -\nabla f(\mathbf{x}^{(k)}),$$

the negative of the gradient of $f(\mathbf{x}^{(k)})$, where for steepest descent, at each iteration k, we minimize the A–norm of the error

$$f(\mathbf{x}^{(k)}) = \|\mathbf{x}^{(k)} - \mathbf{x}^*\|_A = (A(\mathbf{x}^{(k)} - \mathbf{x}^*), (\mathbf{x}^{(k)} - \mathbf{x}^*))).$$

The gradient of the f above is

$$\begin{aligned}\nabla f(\mathbf{x}^{(k)}) =& \nabla(\mathbf{x}^{(k)^T} A\mathbf{x}^{(k)} - 2\mathbf{x}^{(k)^T} A\mathbf{x}^* + \mathbf{x}^{(k)^T} A\mathbf{x}^*)\\ =& 2(A\mathbf{x}^{(k)} - \mathbf{b}) = -2\mathbf{r}^{(k)},\end{aligned}$$

where $\mathbf{r}^{(k)} = \mathbf{b} - A\,\mathbf{x}^{(k)}$, $k = 0, 1, \ldots$, is the residual vector at iteration k. This gives the Algorithm 2.1 which, at each step k, minimizes the residual $\mathbf{r}^{(k)}$ in the A–norm for all vectors of the form $\mathbf{x} + \alpha\,\mathbf{d}$.

Algorithm 2.1: Steepest–descent

Input: $A \in \mathbb{R}^{n\times n}$ SPD, $\mathbf{b} \in \mathbb{R}^n$, Maximum number of iteration $N_{\max}$, Initial Guess $\mathbf{x}^{(0)}$

Output: $\tilde{\mathbf{x}}$ candidate approximation.

for $k = 0, \ldots, N_{max}$ **do**
- $\mathbf{r}^{(k)} \leftarrow \mathbf{b} - A\,\mathbf{x}^{(k)}$;
- $\alpha_k \leftarrow (\mathbf{r}^{(k)^T}\mathbf{r}^{(k)})/(\mathbf{r}^{(k)^T} A\,\mathbf{r}^{(k)})$;
- $\mathbf{x}^{(k+1)} \leftarrow \mathbf{x}^{(k)} + \alpha_k\,\mathbf{r}^{(k)}$;
- **if** *<Stopping criteria satisfied>* **then**
  - **return** $\tilde{\mathbf{x}} = \mathbf{x}^{(k+1)}$

In order to see the algorithm of Steepest Descent as a projection one, at each iteration k, we note that the subspaces $\mathcal{K}$ and $\mathcal{L}$ are spanned by the search direction given by the vector $\mathbf{d}^{(k)} = -\nabla f(\mathbf{x}^{(k)}) = \mathbf{r}^{(k)}$, i.e.,

$$\mathcal{K} = \text{span}\{\mathbf{r}^{(k)}\}, \quad \mathcal{L} = \text{span}\{\mathbf{r}^{(k)}\}.$$

We get the candidate approximations $\mathbf{x}^{(k)}$ as

$$\mathbf{x}^{(k+1)} = \mathbf{x}^{(k)} + \alpha_k\,\mathbf{d}^{(k)} = \mathbf{x}^{(k)} + \alpha_k\,\mathbf{r}^{(k)}, \quad k = 0, 1, \ldots$$

where α_k is chosen in order to minimize $f(\mathbf{x}^{(k)})$. By observing that from $\mathbf{x}^{(k+1)} \leftarrow \mathbf{x}^{(k)}+\alpha_k \mathbf{r}^{(k)}$ we obtain $\mathbf{r}^{(k+1)} = \mathbf{r}^{(k)}-\alpha_k A\mathbf{r}^{(k)}$ and the orthogonality between $\mathbf{r}^{(k+1)}$ and $\mathbf{r}^{(k)}$, we find

$$\alpha_k = \frac{< \mathbf{r}^{(k)}, \mathbf{r}^{(k)} >}{< \mathbf{r}^{(k)}, A\mathbf{r}^{(k)} >}.$$

We note that the descent of $f(\mathbf{x}^{(k)}) = \|\mathbf{x}^{(k)} - \mathbf{x}^*\|_A$ is the "steepest" only locally and not globally and therefore the underlying algorithm is not so fast as expected; see, e.g., [455]. In particular, if $\kappa(A)$ is large, we could need many iterations in order to reduce the relative error to the required tolerance. This can be also argued by the following result.

Theorem 2.4. *Let A be Hermitian and positive definite. Then, for the A-norm of the error vector $\mathbf{x}^* - \mathbf{x}^{(k)}$ in the algorithm 2.1 we have:*

$$\begin{aligned} \|\mathbf{x}^{(k+1)} - \mathbf{x}^*\|_A \leq & \frac{\kappa_2(A)-1}{\kappa_2(A)+1}\|\mathbf{x}^{(k)} - \mathbf{x}^*\|_A \\ \leq & \left(\frac{\kappa_2(A)-1}{\kappa_2(A)+1}\right)^{k+1} \|\mathbf{x}^{(0)} - \mathbf{x}^*\|_A, \end{aligned} \tag{2.11}$$

and then we have convergence for any $\mathbf{x}^{(0)}$.

Proof. We set $\mathbf{e}^{(k)} = \mathbf{x}^* - \mathbf{x}^{(k)}$ and the A–norm of $\mathbf{e}^{(k+1)}$ is

$$\|\mathbf{e}^{(k+1)}\|_A^2 =< \mathbf{r}^{(k+1)}, \mathbf{e}^{(k+1)} - \alpha_k \mathbf{r}^{(k)} > .$$

By construction $\mathbf{r}^{(k+1)}$ must be orthogonal to the search direction $\mathbf{r}^{(k)}$ and therefore the second term in the right hand side should be zero. Then, by observing that

$$\|\mathbf{e}^{(k)}\|_A^2 =< \mathbf{r}^{(k)}, A^{-1}\mathbf{r}^{(k)} >, \tag{2.12}$$

and

$$\|\mathbf{e}^{(k+1)}\|_A^2 =< \mathbf{r}^{(k)} - \alpha_k A\mathbf{r}^{(k)}, \mathbf{e}^{(k)} >=< \mathbf{r}^{(k)}, A^{-1}\mathbf{r}^{(k)} > -\alpha_k < \mathbf{r}^{(k)}, \mathbf{r}^{(k)} >$$

we have

$$\|\mathbf{e}^{(k+1)}\|_A^2 = \|\mathbf{e}^{(k)}\|_A^2 \left(1 - \frac{< \mathbf{r}^{(k)}, \mathbf{r}^{(k)} >}{< \mathbf{r}^{(k)}, A\mathbf{r}^{(k)} >} \frac{< \mathbf{r}^{(k)}, \mathbf{r}^{(k)} >}{< \mathbf{r}^{(k)}, A^{-1}\mathbf{r}^{(k)} >}\right).$$

Then, from Kantorovich inequality (see, e.g., [455, Lemma 5.1] for a proof)

$$\frac{< B\mathbf{x}, \mathbf{x} >}{< \mathbf{x}, \mathbf{x} >}\frac{< B^{-1}\mathbf{x}, \mathbf{x} >}{< \mathbf{x}, \mathbf{x} >} \leq \frac{(\lambda_{\min} + \lambda_{\max})^2}{4(\lambda_{\max}\lambda_{\min})},$$

where $\lambda_{\min}$ e $\lambda_{\max}$ are the smallest and largest eigenvalues of the Hermitian and positive definite A, we get the result. □

Steepest Descent works only for Hermitian positive definite linear systems. In the sequel we recall a couple of simple generalizations that can manage more general problems: *Minres* for nonsymmetric but Hermitian and *Residual Norm Steepest Descent* for any square nonsingular matrices A, respectively.

2.1.3.2 Minimal Residual or MR iteration

The *Minimal Residual* or *MR iteration* method can be seen in a certain sense as a modification of *Steepest Descent* able to manage general positive definite matrices A; see [280] or [455]. It should not be confused with the *Minimum Residual* or MinRes by Paige and Saunders [416]. We recall that a matrix A is positive definite if its Hermitian part $\mathrm{Herm}(A)$,

$$\mathrm{Herm}(A) = \frac{A + A^H}{2},$$

is positive definite.

As for *Steepest Descent*, at each step, the *Minimal Residual* updates the iterations with

$$\mathbf{x}^{(k+1)} = \mathbf{x}^{(k)} + \alpha_k \mathbf{r}^{(k)}, \tag{2.13}$$

where now α_k is computed in order to minimize the function

$$\tilde{f}(\mathbf{x}^{(k)}) = \frac{1}{2}\|A\mathbf{x}^{(k)} - b\|_2^2 = \frac{1}{2}\mathbf{x}^{(k)^H} A^H A\mathbf{x}^{(k)} - \mathbf{b}^H A\mathbf{x}^{(k)} + \frac{1}{2}\mathbf{b}^H\mathbf{b}.$$

This is equivalent to minimize

$$\frac{1}{2}\|\mathbf{r}\|_2^2 = \tilde{f}(\mathbf{x}^{(k)}),$$

i.e., the 2-norm of the residual and not the A–norm as for the Steepest Descent. By observing that from (2.13) we obtain $\mathbf{r}^{(k+1)} = \mathbf{r}^{(k)} - \alpha_k A\mathbf{r}^{(k)}$ and the orthogonality between $\mathbf{r}^{(k+1)}$ and $A^H\mathbf{r}^{(k)}$, we get

$$\alpha_k = \frac{< \mathbf{r}^{(k)}, A\mathbf{r}^{(k)} >}{< A\mathbf{r}^{(k)}, A\mathbf{r}^{(k)} >}$$

because

$$\nabla\tilde{f}(\mathbf{x}^{(k)}) = A^H A\mathbf{x}^{(k)} - A^H\mathbf{b} = -A^H\mathbf{r}^{(k)}.$$

We note that now

$$\mathcal{K} = \mathrm{span}\{\mathbf{r}^{(k)}\}, \quad \mathcal{L} = \mathrm{span}\{A\,\mathbf{r}^{(k)}\},$$

i.e., the search and constraint spaces are not generated by the same vectors as it was for Steepest Descent. The sequence of the candidate approximations $\{\mathbf{x}^{(k)}\}$ generated by algorithm 2.2 converges provided that $\mathrm{Herm}(A)$ is positive definite.

Theorem 2.5. *Let A be a positive definite matrix, i.e.,* $\mathrm{Herm}(A)$ *is positive definite. The algorithm 2.2 converges for any $\mathbf{x}^{(0)}$ and*

$$\|\mathbf{r}^{(k+1)}\|_2 \leq \left(1 - \frac{\lambda_{\min}(\mathrm{Herm}(A))^2}{\|A\|_2^2}\right)^{1/2} \|\mathbf{r}^{(k)}\|_2.$$

Proof. Similar to the proof of Theorem 2.4. □

Similarly to Steepest Descent, if $\kappa_2(A)$ is large, we can expect a slow convergence of iterations.

Algorithm 2.2: Minimal Residual

Input: $A \in \mathbb{R}^{n\times n}$ SPD, $\mathbf{b} \in \mathbb{R}^n$, Maximum number of iteration $N_{\max}$, Initial Guess $\mathbf{x}^{(0)}$

Output: $\tilde{\mathbf{x}}$ candidate approximation.

for $k = 0, \dots, N_{max}$ **do**

  $\mathbf{r}^{(k)} \leftarrow \mathbf{b} - A\,\mathbf{x}^{(k)}$;

  $\mathbf{v} \leftarrow A\,\mathbf{r}^{(k)}$;

  $\alpha_k \leftarrow (\mathbf{v}^H\,\mathbf{r}^{(k)})/(\mathbf{v}^H\,\mathbf{v})$;

  $\mathbf{x}^{(k+1)} \leftarrow \mathbf{x}^{(k)} + \alpha_k\,\mathbf{r}^{(k)}$;

  if <*Stopping criteria satisfied*> **then**

    return $\tilde{\mathbf{x}} = \mathbf{x}^{(k+1)}$

2.1.3.3 Residual Norm Steepest Descent

The third projection method of this section can be viewed as a further generalization of *Steepest Descent* able to manage a generic nonsingular linear system and will be recalled here shortly.

The projection subspaces $\mathcal{K}$ and $\mathcal{L}$ are still unidimensional and now are

$$\mathcal{K} = \mathrm{span}\{A^H\,\mathbf{r}^{(k)}\}, \quad \mathcal{L} = \mathrm{span}\{A\,\mathbf{r}^{(k)}\}.$$

By using arguments similar to *Steepest Descent* (Algorithm 2.1), we derive Algorithm 2.3. As for *Minimal Residual*, the underlying algorithm minimizes the norm 2 of the residual at each step k, i.e., it minimizes the function

$$\tilde{f}(\mathbf{x}^{(k)}) = \|\mathbf{b} - A\,\mathbf{x}^{(k)}\|_2^2.$$

Note that Algorithm 2.3 is mathematically equivalent to *Steepest Descent* applied to the normal equations

$$A^H\,A\,\mathbf{x} = A^H\,\mathbf{b}. \tag{2.14}$$

2.1.3.4 Richardson and non-stationary Richardson

Also Richardson's method(s) can be derived by matrix splitting (see § 2.1.1) but we prefer to discuss it briefly here because it can be easily generalized to become one of the possible formulations of the *Conjugate Gradient*, a Krylov subspace method, and therefore not stationary at all. Let us begin with the "stationary" Richardson. If $A = P - N$ we can write

$$\mathbf{x}^{(k+1)} = P^{-1}N\mathbf{x}^{(k)} + P^{-1}\mathbf{b} = P^{-1}(P-A)\mathbf{x}^{(k)} + P^{-1}\mathbf{b} = \mathbf{x}^{(k)} + P^{-1}\mathbf{r}^{(k)},$$

where $\mathbf{r}^{(k)}$ is the residual defined as usual, $\mathbf{r}^{(k)} = \mathbf{b} - A\mathbf{x}^{(k)}$. We get the underlying iterative method (stationary) if we introduce a parameter $\alpha > 0$:

$$\mathbf{x}^{(k+1)} = \mathbf{x}^{(k)} + \alpha\,P^{-1}\mathbf{r}^{(k)}, \tag{2.15}$$

Algorithm 2.3: Residual Norm Steepest Descent

Input: $A \in \mathbb{R}^{n\times n}$,$\mathbf{b} \in \mathbb{R}^n$, Maximum number of iteration $N_{\max}$, Initial Guess $\mathbf{x}^{(0)}$

Output: $\tilde{\mathbf{x}}$ candidate approximation.

$\mathbf{r}^{(0)} \leftarrow \mathbf{b} - A\mathbf{x}^{(0)}$;
for $k = 0, \dots, N_{max}$ **do**
    $\mathbf{v}^{(k)} \leftarrow A^H\, \mathbf{r}^{(k)}$;
    $\mathbf{w}^{(k)} \leftarrow A\, \mathbf{v}^{(k)}$;
    $\alpha_k \leftarrow \|\mathbf{v}^{(k)}\|_2^2 / \|\mathbf{w}^{(k)}\|_2^2$;
    $\mathbf{x}^{(k+1)} \leftarrow \mathbf{x}^{(k)} + \alpha_k\, \mathbf{w}^{(k)}$;
    $\mathbf{r}^{(k)} \leftarrow \mathbf{b} - \alpha\, \mathbf{w}^{(k)}$;
    if *<Stopping criteria satisfied>* **then**
        return $\tilde{\mathbf{x}} = \mathbf{x}^{(k+1)}$

where $\mathbf{x}^{(0)}$ as usual is the initial approximation. The matrix P is called *preconditioner*. The Richardson stationary method is based on the recurrence relation

$$\begin{aligned}\mathbf{x}^{(k+1)} =& \mathbf{x}^{(k)} + \alpha\, P^{-1}\mathbf{r}^{(k)} = \mathbf{x}^{(k)} + \alpha\, P^{-1}(\mathbf{b} - A\mathbf{x}^{(k)})\\ =& (I - \alpha\, P^{-1}A)\mathbf{x}^{(k)} + \alpha\, P^{-1}\mathbf{b},\end{aligned}$$

where the iteration matrix is $M = I - \alpha\, P^{-1}A$. Therefore, given an initial guess $\mathbf{x}^{(0)}$, we can apply Theorem 2.1 to the spectral radius of $I - \alpha\, P^{-1}A$. The spectral radius of M is given by

$$\rho(M) = \max_{1\leq i\leq n} |1 - \alpha\lambda_i|,$$

where $\lambda_1, \dots, \lambda_n$ are the eigenvalues of the preconditioned matrix $P^{-1}A$. We get that the optimal value of the parameter α is given by

$$\alpha_* = \frac{2}{\min_i\{\lambda_i\} + \max_j\{\lambda_j\}}. \tag{2.16}$$

For this value the spectral radius of the iteration matrix is

$$\rho(\alpha_*) = \frac{\kappa_2(P^{-1}A) - 1}{\kappa_2(P^{-1}A) + 1}.$$

From the convergence results of the previous section, we observe that when $\kappa_2(P^{-1}A)$ is not far from 1, i.e., when the matrix is well conditioned (recall that $\kappa_2(\cdot) \geq 1$), we expect a fast convergence of the iterations. Therefore, a proper choice of the preconditioned matrix P can be of paramount importance. Nonetheless, the auxiliary linear system with matrix P should be computationally easy to solve, surely much less than those with matrix A, otherwise it has no sense to apply this method!

We now consider *non-stationary Richardson*. Let us suppose A symmetric and positive definite and we modify (2.15) by choosing a different parameter for each iteration k, i.e. α_k instead of α in (2.15). More precisely, let us determine α_k in

$$\mathbf{x}^{(k+1)} = \mathbf{x}^{(k)} + \alpha_k P^{-1}\mathbf{r}^{(k)} \tag{2.17}$$

by minimizing the distance $\|\mathbf{x}^{(k)} - \mathbf{x}\|_A$ at each iteration k.

Recall that $\|\cdot\|_A$ is the so–called *energy norm* that is defined for a vector $\mathbf{x}$ as $\|\mathbf{x}\|_A = \mathbf{x}^H A\mathbf{x}$, A Hermitian and definite (positive or negative). Otherwise it could be not a norm.

By rewriting $\|\mathbf{x} - \mathbf{x}_*\|$ using the expression of $\mathbf{x}_*$, $\mathbf{x}_* = \mathbf{x}^{(k)} + \alpha_k P^{-1} r^{(k)}$, we obtain a second degree polynomial in α_k:

$$\|\mathbf{x} - \mathbf{x}_*\| = \alpha_k^2 (\tilde{\mathbf{r}}^{(k)})^T A\tilde{\mathbf{r}}^{(k)} - 2\alpha_k (\tilde{\mathbf{r}}^{(k)})^T \mathbf{r}^{(k)} + (\mathbf{x} - \mathbf{x}^{(k)})^T A(\mathbf{x} - \mathbf{x}^{(k)}) \tag{2.18}$$

where for brevity we set $\tilde{\mathbf{r}}^{(k)} = P^{-1}\mathbf{r}^{(k)}$, the preconditioned residual. By differentiating with respect to α_k, we compute the minimum of (2.18) which is reached for

$$\alpha_k = \frac{(\tilde{\mathbf{r}}^{(k)})^T\, \mathbf{r}^{(k)}}{(\tilde{\mathbf{r}}^{(k)})^T\, A\, \tilde{\mathbf{r}}^{(k)}}. \tag{2.19}$$

We note that the recurrence relation (2.17) with α_k defined as in (2.19) defines a new and nice method with respect to the previous stationary version: it does not require information on the extremal eigenvalues of $P^{-1}A$. The latter information is often very difficult to get or computationally unfeasible.

Finally, we observe that (2.17) and (2.19) are the main building blocks defining a particular type of one-dimensional projection algorithm: a *Krylov subspace method*, that will be studied in the next section: the *Conjugate Gradient.*

Exercise 2.4. Write an algorithm for the (stationary) Richardson's method (hint: see, e.g., the algorithms 2.1 and 2.2). What about non-stationary Richardson?

2.2 Krylov subspace algorithms, orthogonal projections: CG and GMRES

2.2.1 Krylov subspace iterative algorithms

As it is easy to argue, it is much more effective projecting on subspaces of increasing dimension instead of on one-dimensional ones. In this section we consider orthogonal, and in the next, oblique projection techniques for algebraic linear systems.

We introduced projection methods as an attempt to generate approximations $\{\mathbf{x}^{(m)}\}$ for the system of n linear algebraic equations in n unknowns:

$$M\mathbf{x} = \mathbf{b}$$

all belonging to the affine subspace $\mathbf{x}^{(0)} + \mathcal{K}_m$, where $\mathcal{K}_m$ is a subspace of $\mathbb{R}^n$ of dimension m, $1 \leq m \leq n$, and $\mathbf{x}^{(0)}$ is an initial guess. The candidate approximation $\mathbf{x}^{(m)}$ for $\mathbf{x}$ is determined such that the residual $\mathbf{r}^{(m)}$ satisfies the Petrov–Galerkin conditions

$$\mathbf{r}^{(m)} \equiv \mathbf{b} - M\,\mathbf{x}^{(m)} \perp \mathcal{L}_m, \quad M \in \mathbb{R}^{n\times n}, \quad \mathbf{b} \in \mathbb{R}^n \tag{2.20}$$

where $\mathcal{L}_m$ is another subspace of $\mathbb{R}^n$ of dimension m. With this process, the solution $\mathbf{x} = M^{-1}\mathbf{b}$ is approximated by $\mathbf{x}^{(0)} + \mathsf{Q}_{m-1}(M)\,\mathbf{b}$, where $\mathsf{Q}_{m-1}(z)$ is a suitable polynomial of degree at most $m-1$.

A *Krylov subspace* $\mathcal{T}$ for the matrix M related to a non null vector v is defined as

$$\mathcal{T}_m(M, \mathbf{v}) = \mathrm{span}\{\mathbf{v}, M\mathbf{v}, M^2\mathbf{v}, \ldots, M^{m-1}\mathbf{v}\}. \tag{2.21}$$

We note two important issues that characterize the Krylov subspace methods.

Remark 2.2.

- The Krylov subspace are nested, i.e. $\mathcal{T}_m(M, \mathbf{v}) \subseteq \mathcal{T}_{m+1}(M, \mathbf{v})$. A *Krylov subspace method* is an algorithm that at step $m \geq 1$ uses Krylov subspaces for $\mathcal{K}_m$ and for $\mathcal{L}_m$.

- The property $\mathcal{T}_m(M, \mathbf{v}) \subseteq \mathcal{T}_{m+1}(M, \mathbf{v})$, $m = 1, 2, \ldots$ (and $\mathcal{T}_m(M, \mathbf{v}) = \mathbb{R}^n$ for $m \geq n$ because $\mathcal{T}_m$ is a subspace) of the Krylov subspaces implies that any method for which a Petrov-Galerkin condition holds will, in exact arithmetic, terminate in at most n steps. In practice one wants the methods to produce the desired approximation to the solution of the underlying linear system in many fewer than n iterations.

By choosing $\mathcal{K}_m$ and $\mathcal{L}_m$ as different Krylov subspaces, we can have different projection methods. Here we mention only some of them; more can be found in [455], [280] and [328], the main sources of inspiration of this book for iterative methods and their analysis. In particular, we consider Krylov subspace iterative algorithms derived from the *Lanczos* and *Arnoldi* famous algorithms. The first was introduced in 1950 by Lanczos [343] for estimating eigenvalues of sparse symmetric matrices. It generates a sequence of symmetric tridiagonal matrices whose eigenvalues, under suitable hypotheses, converge to those of M. The second is due to Arnoldi [11], introduced in 1951 to extend the above Lanczos strategy to nonsymmetric matrices. It is based on the Hessenberg reduction of the matrix M. For other details see, e.g., Wilkinson [543], Parlett [418], Golub and Van Loan [271], Saad [455] and [280].

2.2.2 The Conjugate Gradient (CG)

The *Conjugate Gradient* algorithm, or CG for short, is one of the most used well known and effective iterative techniques for solving linear systems with Hermitian and (positive or negative) definite matrices. CG is an orthogonal projection technique on the Krylov subspace $\mathcal{K}_m(M, \mathbf{r}^{(0)})$ (2.21), where $\mathbf{r}^{(0)} = \mathbf{b} - M\mathbf{x}^{(0)}$ is the initial residual. In theory, the underlying algorithm can be derived in at least two different ways:

1. by the minimization of a quadratic form, see, e.g., the seminal paper by Hestenes and Stiefel [303] and Golub and van Loan [271];

2. by tridiagonal reduction of the original matrix generated by the Lanczos algorithm; see Saad [455]).

In this section we consider both approaches. The first is also the first appeared, introduced by [303] in 1952 and probably the most popular in the literature. The second is very useful in order to give a unifying view of all iterative methods of projection type considered in this book.

Let $A \in \mathbb{R}^{n\times n}$ be a Hermitian and definite positive definite matrix. Here in order to simplify notation, we consider A with real entries and therefore A will be symmetric and definite, or *SPD* for short. If A is negative definite, all reasonings can be applied to $-A$ by considering the equivalent linear system $-A\,\mathbf{x} = -\mathbf{b}$.

2.2.3 CG by minimizing a quadratic functional

Let A be $n \times n$ and SPD, x a nonzero vector with n real entries. We define the A–norm as

$$\|\mathbf{x}\|_A = \sqrt{\mathbf{x}^T A\,\mathbf{x}}.$$

The CG algorithm 2.4 is build up in order to let $\mathbf{x}^{(m)}$ minimizing the quadratic functional

$$\phi(\mathbf{x}) = \frac{1}{2}\mathbf{x}^T A\,\mathbf{x} - \mathbf{x}^T\,\mathbf{b}, \quad \mathbf{x} \in \mathbb{R}^n \tag{2.22}$$

for $\mathbf{x} \in \mathbf{x}^0 + \mathcal{K}_m$, $\mathcal{K}_m = \mathcal{K}_m(A, \mathbf{r}^{(0)})$ while $\mathbf{r}^{(0)} = \mathbf{b} - A\,\mathbf{x}^{(0)}$ and $\mathbf{x}^{(0)}$ are the initial residual and guess, as usual. See [303, Theorem 6.1].

If $\mathbf{x} = \tilde{\mathbf{x}}$ is the value that minimizes $\phi(\mathbf{x})$ in (2.22), then its gradient is zero: $\nabla\phi(\tilde{\mathbf{x}}) = A\tilde{\mathbf{x}} - \mathbf{b} = 0$. This means that $\tilde{\mathbf{x}}$ is the solution of the underlying linear system.

We do not terminate iterations in the CG Algorithm 2.4 until an exact solution is found because this can happen only in exact arithmetics (i.e., without rounding errors, a very special and exceptionally rare situation). Instead, we provide a maximum number of iterations $N_{\max}$ that the algorithm can do and we can exit this loop once one or some criteria are satisfied. Of course we would like to stop whenever the error in a certain norm useful for our selected application, like the 2 or the max-norm, is less than a prescribed quantity

Algorithm 2.4: Conjugate Gradient Method

Input: $A \in \mathbb{R}^{n\times n}$ SPD, Maximum number of iterations N_{max}, Initial Guess $\mathbf{x}^{(0)}$

Output: $\tilde{\mathbf{x}}$, candidate approximation.

$\mathbf{r}^{(0)} \leftarrow \|\mathbf{b} - A\mathbf{x}^{(0)}\|_2$, $\mathbf{r} = \mathbf{r}^{(0)}$, $\mathbf{p} \leftarrow \mathbf{r}$;
$\rho_0 \leftarrow \|\mathbf{r}^{(0)}\|^2$;
for $k = 1, \ldots, N_{max}$ **do**
    if $k = 1$ **then**
        $\mathbf{p} \leftarrow \mathbf{r}$;
    else
        $\beta \leftarrow \rho_1/\rho_0$;
        $\mathbf{p} \leftarrow \mathbf{r} + \beta\,\mathbf{p}$;
    $\mathbf{w} \leftarrow A\,\mathbf{p}$;
    $\alpha \leftarrow \rho_1/\mathbf{p}^T\mathbf{w}$;
    $\mathbf{x} \leftarrow \mathbf{x} + \alpha\mathbf{p}$;
    $\mathbf{r} \leftarrow \mathbf{r} - \alpha\mathbf{w}$;
    $\rho_1 \leftarrow \|\mathbf{r}\|_2^2$;
    if *<Stopping criteria satisfied>* **then**
        Return: $\tilde{\mathbf{x}} = \mathbf{x}$;

$\varepsilon > 0$ and some order of magnitude less than the machine precision. But, as usual, knowing the error means in general knowing the exact solution and this is not feasible. Therefore, we usually check some easy-to-compute quantity that can be related to the solution's error like checking that the relative residual is less than ε.

$$\frac{\|\mathbf{r}^{(m)}\|_2}{\|\mathbf{r}^{(0)}\|_2} = \frac{\|\mathbf{b} - A\,\mathbf{x}^{(m)}\|_2}{\|\mathbf{b} - A\,\mathbf{x}^{(0)}\|_2} < \varepsilon.$$

Let us see how $\|\mathbf{r}^{(m)}\|_2/\|\mathbf{r}^{(0)}\|_2$ is far from the error.

Theorem 2.6. *Let A be SPD and $k_2(A) = \lambda_n/\lambda_1$ be the 2–norm condition number of A. We have:*

$$\frac{\|\mathbf{r}^{(m)}\|_2}{\|\mathbf{r}^{(0)}\|_2} \leq \sqrt{k_2(A)}\frac{\|\mathbf{x}^* - \mathbf{x}^{(m)}\|_A}{\|\mathbf{x}^* - \mathbf{x}^{(0)}\|_A}. \tag{2.23}$$

Proof. First, note that for any vector z we have

$$\|\mathbf{z}\|_A^2 = \mathbf{z}^T A\mathbf{z} = \mathbf{z}^T(A^{1/2}\,A^{1/2})\mathbf{z} = (A^{1/2}\mathbf{z})^T\,(A^{1/2}\mathbf{z}) = \|A^{1/2}\,\mathbf{z}\|_2^2. \tag{2.24}$$

Then it follows that

$$\frac{\|\mathbf{r}^{(m)}\|_2}{\|\mathbf{r}^{(0)}\|_2} = \frac{\|\mathbf{b} - A\mathbf{x}^{(m)}\|_2}{\|\mathbf{b} - A\mathbf{x}^{(0)}\|_2}\frac{\|A(\mathbf{x}^* - \mathbf{x}^{(m)})\|_2}{\|A(\mathbf{x}^* - \mathbf{x}^{(0)})\|_2}.$$

Moreover, since the spectral decomposition of A gives $\|A\|_2 = \lambda_1$ and $\|A^{-1}\|_2 = \lambda_n$, by (2.24) we obtain (2.23). □

Now, let us consider what CG does by analyzing its convergence. We start by recalling that $\mathbf{x}^{(m)}$ minimizes ϕ over the affine space $\mathbf{x}^{(0)} + \mathcal{K}_m$.

2.2.4 Convergence analysis of CG

Theorem 2.7. *Let $\mathbf{x}^*$ be such that $A\,\mathbf{x}^* = \mathbf{b}$, A be symmetric and positive definite. Then, if $\mathbb{P}_m$ is the set of polynomials of degree at most m, the m–th iteration of CG produces an approximation $\mathbf{x}^{(m)}$ such that*

$$\begin{aligned}\|\mathbf{x}^* - \mathbf{x}^{(m)}\|_A &= \min_{\substack{p\in\mathbb{P}_m\\ p(0)=1}} \|p(A)\,(\mathbf{x}^* - \mathbf{x}^{(0)})\|_A \\ &\leq \|\mathbf{x}^* - \mathbf{x}^{(0)}\|_A \left| \min_{\substack{p\in\mathbb{P}_m\\ p(0)=1}} \max_{\lambda\in\lambda(A)} p(z) \right|.\end{aligned} \tag{2.25}$$

Proof. $\mathbf{x}^{(m)}$ minimizes ϕ over $\mathbf{x}^{(0)} + \mathcal{K}_m$ because

$$\|\mathbf{x} - \mathbf{x}^*\|_A = 2\phi(\mathbf{x}) + (\mathbf{x}^*)^T A\,\mathbf{x}^*,$$

where $(\mathbf{x}^*)^T A\,\mathbf{x}^*$ is a constant term so there is nothing to minimize there and therefore

$$\|\mathbf{x}^* - \mathbf{x}^{(m)}\|_A \leq \|\mathbf{x}^* - \mathbf{w}\|_A, \quad \mathbf{w} \in \mathbf{x}^{(0)} + \mathcal{K}_m. \tag{2.26}$$

By expanding the above $\mathbf{w}$ we find

$$\mathbf{w} = \sum_{j=0}^{m-1} \gamma_j A^j \mathbf{r}^{(0)} + \mathbf{x}^{(0)},$$

$\gamma_j \in \mathbb{R}$, and therefore

$$\mathbf{x}^* - \mathbf{w} = \mathbf{x}^* - \mathbf{x}^{(0)} - \sum_{j=0}^{m-1} \gamma_j A^j (\mathbf{x}^* - \mathbf{x}^{(0)}) = p(A)(\mathbf{x}^* - \mathbf{x}^{(0)}),$$

where $p \in \mathbb{P}_m$ and $p(0) = 1$ by construction. Finally, by observing that

$$\min_{\substack{p\in\mathbb{P}_m\\ p(0)=1}} \|p(A)\,(\mathbf{x}^* - \mathbf{x}^{(0)})\|_A \leq \min_{\substack{p\in\mathbb{P}_m\\ p(0)=1}} \|p(A)\|\,\|\mathbf{x}^* - \mathbf{x}^{(0)}\|_A$$

and, by expanding the initial error by the eigenvectors of A, that

$$\|p(A)\| = \max_{\lambda\in\lambda(A)} |p(\lambda)|,$$

$\lambda(A)$ the set of the (real) eigenvalues of A, we have the result. □

Corollary 2.2. *If A is symmetric and positive definite and the eigenvalues of A are such that $0 < \lambda_1 \leq \ldots \leq \lambda_n$, we have*

$$\frac{\|\mathbf{x}^* - \mathbf{x}^{(m)}\|_A}{\|\mathbf{x}^* - \mathbf{x}^{(0)}\|_A} \leq 2 \left(\frac{\sqrt{k_2(A)} - 1}{\sqrt{k_2(A)} + 1} \right)^m, \tag{2.27}$$

where $k_2(A) = \lambda_n/\lambda_1$ is the 2–norm condition number of A.

Proof. We start from Theorem 2.25. By the expansion of $(\mathbf{x}^* - \mathbf{x}^{(0)})$ by the eigenvector basis of A and using Chebyshev polynomials of degree m of the first kind for p (a well known result from approximation theory), we derive the exact expression for $\min\limits_{\substack{p \in \mathbb{P}_m \\ p(0)=1}} \|p(A)\|$ in (2.27). The right hand side of (2.27) is readily obtained by recalling that $k_2(A) = \lambda_n/\lambda_1$. □

A detailed proof for Corollary 2.2 can be found in [455, Sections 6.10 and 6.11].

Corollary 2.3. *Under the same hypotheses of Corollary 2.2, the Conjugate Gradient converges n iterations in exact arithmetic.*

Proof. It is enough to consider the polynomial

$$p(z) = \prod_{j=1}^{n} (z - \lambda_j)/\lambda_j \tag{2.28}$$

in (2.25) and we find $p(z) = 0$, i.e., CG converges in n iterations. □

In general we can observe the following issues:

- in most applications where using an iterative solver has sense, the number of unknown n is large and one cannot afford to perform n iterations. Using CG in this way could cost more than a good direct solver. Therefore, it is better to use CG as an iterative method.
- Conjugate gradient can converge in few iterations if many of the eigenvalues of A are clustered far away from zero. The same effect can be forced by transforming it in an equivalent linear system $\tilde{A}\mathbf{x} = \tilde{\mathbf{b}}$ by *preconditioning* techniques; see Chapter 3.

Convergence analysis of CG when the spectrum is clustered

Theorem 2.8. *Let A be Hermitian and positive definite. Let m an integer, $1 < m < n$ and $c > 0$ a constant such that for the eigenvalues of A it holds*

$$0 < \lambda_1 \leq \lambda_2 \leq \lambda_3 \leq \ldots \leq \lambda_{n-m+1} \leq \ldots \leq \lambda_n.$$

Fixed $\varepsilon > 0$ and supposing exact arithmetic, an upper bound for the minimum number of iterations k reducing the relative error in energy norm for the approximation $\mathbf{x}^{(k)}$ produced by CG of ε is given by

$$\min\left\{\left\lceil \frac{1}{2}\sqrt{c/\lambda_1}\log\left(\frac{2}{\varepsilon}\right)+m+1\right\rceil, n\right\} \tag{2.29}$$

Proof. Let us partition the eigenvalues of A in two sets,

$$\sigma_1 = \{\lambda_1,\ldots,\lambda_{n-m}\},\quad \sigma_2 = \bigcup_{i=n-m+1}^{n} \lambda_i$$

and consider the polynomial

$$p_k(\lambda) = \prod_{i=n-m+1}^{n}\left(1-\frac{\lambda}{\lambda_i}\right)\frac{T_{k-m}\left(c+\lambda_1-2\lambda\right)/(c-\lambda_1)}{T_{k-m}\left(c+\lambda_1\right)/(c-\lambda_1)}$$

that belongs to $\mathbb{P}_k$ and $p_k(0)=1$, where $T_k(x)$ is the Chebyshev polynomial that, for all x real can be expressed as

$$T_k(x) = \left((x+\sqrt{x^2-1})^k + (x-\sqrt{x^2-1})^k\right)/2, \tag{2.30}$$

see [386] or [455] for more details on real and complex arguments Chebyshev polynomials.

By observing that

$$\max_{\lambda\in\sigma_1}|p_k(\lambda)| \le \max_{\lambda\in\sigma_1}\frac{T_{k-m}\left(c+\lambda_1-2\lambda\right)/(c-\lambda_1)}{T_{k-m}\left(c+\lambda_1\right)/(c-\lambda_1)} =$$

$$1/T_{k-m}\left(\frac{c+\lambda_1}{c-\lambda_1}\right)$$

and then the thesis follows by observing that the contribution of the eigenvalues greater than c (those outside the *cluster*) can only delay the convergence by a number of iterations at most equal to their number by the same arguments used in the proof of Corollary 2.3. □

2.2.5 CG by Lanczos Algorithm

As promised, let us now derive the Conjugate Gradient algorithm as a Krylov subspace method, i.e., by imposing that the approximate solution $\mathbf{x}^{(m)}$ belongs to the affine subspace $\mathbf{x}^{(0)}+\mathcal{K}_m$ where $\mathcal{K}_m$ is chosen to be a Krylov subspace, while the residual $\mathbf{r}^{(m)} = \mathbf{b}-A\,\mathbf{x}^{(m)}$ is orthogonal to another Krylov subspace, the *subspace of constraints*; see [455].

In order to build the method, we pass through the Lanczos Algorithm, that, under our hypotheses, transforms the underlying linear system generating a sequence of linear systems of increasing size whose matrices are tridiagonal, i.e., the only nonzero entries are on the main, sub and super diagonals.

A sequence $\{\mathbf{x}^{(j)}\}$, converging to the solution of the linear system $A\,\mathbf{x} = \mathbf{b}$, can be generated by a sequence of m orthonormal vectors $\{\mathbf{v}_1, \ldots, \mathbf{v}_m\}$, $m \leq n$, such that

$$\mathbf{x}^{(j)} - \mathbf{x}^{(0)} \in \text{span}\{\mathbf{v}_1, \ldots, \mathbf{v}_j\},\ j \leq n,$$

$\mathbf{x}^{(0)}$ initial guess of $\mathbf{x}$ such that the residual $\mathbf{r}^{(j)} = \mathbf{b} - A\,\mathbf{x}^{(j)}$ is minimum in some norm to be specified later. The Lanczos Algorithm, introduced in 1950 for computing eigenvalues of symmetric matrices, see [343], neglecting rounding errors, generates a sequence of orthonormal vectors $\{\mathbf{v}_1, \ldots, \mathbf{v}_j\}$ by the three term relation

$$\beta_{j+1}\mathbf{v}_{j+1} = A\mathbf{v}_j - \alpha_j\mathbf{v}_j - \beta_j\mathbf{v}_{j-1}. \tag{2.31}$$

After j steps we find

$$\text{span}\{\mathbf{v}_1, \ldots, \mathbf{v}_j\} = \text{span}\{\mathbf{v}_1, A\mathbf{v}_1, \ldots, A^{j-1}\mathbf{v}_1\} = \mathcal{K}_j(A, \mathbf{v}_1).$$

If V_j is the $n \times j$ matrix whose columns are $\mathbf{v}_1, \ldots, \mathbf{v}_j$, we have:

$$\begin{gathered} \mathbf{x}^{(j)} - \mathbf{x}^{(0)} = V_j\mathbf{y}^{(j)},\ \mathbf{y}^{(j)} \in \mathbb{R}^j, \\ V_j^T V_j = I_j,\ AV_j = V_jT_j + \tilde{\mathbf{r}}_j\mathbf{e}_j^T,\ \mathbf{e}_j^T = (0\ \cdots\ 0\ 1)_j. \end{gathered} \tag{2.32}$$

where

$$\tilde{\mathbf{r}}_j = (A - \alpha_j I)\mathbf{v}_j - \beta_j\mathbf{v}_{j-1}.$$

The matrix T_j is symmetric tridiagonal

$$T_j = \begin{bmatrix} \alpha_1 & \beta_2 & & & \\ \beta_2 & \alpha_2 & \beta_3 & & \\ & \ddots & \ddots & \ddots & \\ & & \ddots & \ddots & \beta_j \\ & & & \beta_j & \alpha_j \end{bmatrix}.$$

If, at the mth iteration,, $m \leq n$, we have $\|\tilde{\mathbf{r}}_m\| = 0$, we find that

$$AV_m = V_mT_m \ \Rightarrow\ T_m = V_m^T AV_m,$$

i.e., A can be considered reduced to the tridiagonal T_m. By the previous steps and from (2.32) we obtain

$$\begin{gathered} A\mathbf{x} = \mathbf{b} \Rightarrow A(\mathbf{x} - \mathbf{x}^{(0)}) = \mathbf{b} - A\mathbf{x}^{(0)} = \mathbf{r}^{(0)}, \\ \Rightarrow V_m^T AV_m\mathbf{y}_m = V_m^T\mathbf{r}^{(0)} \Rightarrow T_m\mathbf{y}_m = V_m^T\mathbf{r}^{(0)} \end{gathered}$$

and therefore

$$\mathbf{y}_m = T_m^{-1}V_m^T\mathbf{r}^{(0)} \Rightarrow \mathbf{x} = \mathbf{x}^{(0)} + V_mT_m^{-1}(V_m^T\mathbf{r}^{(0)}). \tag{2.33}$$

From (2.33), we build the sequence $\{\mathbf{x}^{(j)}\}$ approximating $\mathbf{x}$:

$$\mathbf{x}^{(j)} = \mathbf{x}^{(0)} + V_j T_j^{-1}(V_j^T \mathbf{r}^{(0)}). \tag{2.34}$$

We do not need to store vectors $\mathbf{v}_j$, $j < i-1$. Indeed, at step m, a factorization $L_m U_m$ for T_m can be generated dynamically:

$$L_m = \begin{bmatrix} 1 & & & \\ \lambda_2 & 1 & & \\ & \ddots & \ddots & \\ & & \lambda_m & 1 \end{bmatrix}, \quad U_m = \begin{bmatrix} \eta_1 & \beta_2 & & \\ & \ddots & \ddots & \\ & & \ddots & \beta_m \\ & & & \eta_m \end{bmatrix},$$

From (2.34) and $T_m^{-1} = U_m^{-1} L_m^{-1}$

$$\mathbf{x}^{(m)} = \mathbf{x}^{(0)} + V_m U_m^{-1} L_m^{-1}(V_m^T \mathbf{r}^{(0)}),$$

by posing

$$P_m = V_m U_m^{-1}, \quad \mathbf{z}_m = L_m^{-1}(V_m^T \mathbf{r}^{(0)}),$$

we express

$$\mathbf{x}^{(m)} = \mathbf{x}^{(0)} + P_m \mathbf{z}_m.$$

$\mathbf{p}_m$, the last column of P_m, can be computed from the previous one:

$$\mathbf{p}_m = \frac{\mathbf{v}_m - \beta_m \mathbf{p}_{m-1}}{\eta_m}, \quad \lambda_m = \frac{\beta_m}{\eta_{m-1}}, \ \eta_m = \alpha_m - \lambda_m \beta_m$$

by observing that $\mathbf{z}_m = (\mathbf{z}_{m-1}\ \zeta_m)^T$, where ζ_m is a scalar and

$$\mathbf{x}^{(m)} = \mathbf{x}^{(m-1)} + \zeta_m \mathbf{p}_m$$

that is the candidate approximation generated by updating the one of the previous step. Note that the residual vector $\mathbf{r}^{(m)} = \mathbf{b} - A\mathbf{x}^{(m)}$ is in the direction of $\mathbf{v}_{m+1}$, because (see [455])

$$\mathbf{r}^{(m)} = \mathbf{b} - M\mathbf{x}^{(m)} = -\beta_{m+1}(\mathbf{e}_m^T T_m^{-1} V_m^T \mathbf{r}^{(0)})\mathbf{v}_{m+1}. \tag{2.35}$$

Moreover, we have the following result.

Proposition 2.4. *The vectors $\mathbf{p}_1, \ldots, \mathbf{p}_m$, where $P_m = [\mathbf{p}_1, \ldots, \mathbf{p}_m]$, are "A–orthogonal" or* conjugate, *i.e.,* $< A\,\mathbf{p}_i, \mathbf{p}_j >= 0$, $i \neq j$.

Proof. Follows by

$$P_m^T M P_m = (U_m^{-1})^T V_m^T M V_m U_m^{-1} = (U_m^{-1})^T T_m U_m^{-1} = (U_m^{-1})^T L_m$$

that is lower triangular but also symmetric because $P_m^T A P_m$ is symmetric, therefore is a diagonal matrix. □

As a consequence of (2.35), we provide a version of CG from Lanczos Algorithm 2.5.

Let us express solution and residual vectors at step jth as

$$\mathbf{x}^{(j+1)} = \mathbf{x}^{(j)} + \alpha_j \mathbf{p}_j, \Leftarrow \mathbf{r}^{(j+1)} = \mathbf{r}^{(j)} - \alpha_j A \mathbf{p}_j$$

The mutual orthogonality of $\mathbf{r}^{(j)}$s give us the α_js:

$$0 =< \mathbf{r}^{(j+1)}, \mathbf{r}^{(j)} >=< \mathbf{r}^{(j)} - \alpha_j M \mathbf{p}_j, \mathbf{r}^{(j)} >$$

$$\alpha_j = \frac{< \mathbf{r}^{(j)}, \mathbf{r}^{(j)} >}{< M\mathbf{p}_j, \mathbf{r}^{(j)} >}. \tag{2.36}$$

The next search direction, $\mathbf{p}_{j+1}$, is a linear combination of $\mathbf{r}^{(j+1)}$ and $\mathbf{p}_j$, i.e., $\mathbf{p}_{j+1} = \mathbf{r}^{(j+1)} + \beta_j \mathbf{p}_j$. Therefore, $< A\mathbf{p}_j, \mathbf{r}^{(j)} >=< A\mathbf{p}_j, \mathbf{p}_j - \beta_{j-1}\mathbf{p}_{j-1} >=< A\mathbf{p}_j, \mathbf{p}_j >$ that can be substituted in (2.36). Moreover, from previous relations

$$\begin{aligned} \beta_j &= -\frac{< \mathbf{r}^{(j+1)}, M\mathbf{p}_j >}{< \mathbf{p}_j, M\mathbf{p}_j >} = \frac{< \mathbf{r}^{(j+1)}, (\mathbf{r}^{(j+1)} - \mathbf{r}^{(j)}) >}{\alpha_j < \mathbf{p}_j, M\mathbf{p}_j >} \\ &= \frac{< \mathbf{r}^{(j+1)}, \mathbf{r}^{(j+1)} >}{< \mathbf{r}^{(j)}, \mathbf{r}^{(j)} >}. \end{aligned} \tag{2.37}$$

that gives *Conjugate Gradient* algorithm 2.4.

Algorithm 2.5: Lanczos Algorithm

Input: $A \in \mathbb{R}^{n \times n}$, $\mathbf{v}_1 \in \mathbb{R}^n$ such that $\|\mathbf{v}_1\|_2 = 1$

Output: $V_m = [\mathbf{v}_1, \ldots, \mathbf{v}_m]$, $T_m = \mathrm{trid}(\beta, \alpha, \beta) \in \mathbb{R}^{m \times m}$

for $j = 1, \ldots, m$ **do**
- $\mathbf{w}_j = A\mathbf{v}_j - \beta_j \mathbf{v}_{j-1}$;
- $\alpha_j =< \mathbf{w}_j, \mathbf{v}_j >$;
- $\mathbf{w}_j = \mathbf{w}_j - \alpha_j \mathbf{v}_j$;
- $\beta_{j+1} = \|\mathbf{w}_j\|_2$;
- **if** $\beta_{j+1} = 0$ **then**
  - **return**;
- $\mathbf{v}_{j+1} = \mathbf{w}_j / \|\mathbf{w}_j\|_2$;

Computational complexity

The CG algorithm, as the other Krylov subspace methods, has the nice property that the matrix A itself need not be formed or stored, only a routine for matrix-vector products is required in order to use the algorithm. This is the reason why Krylov subspace methods are often called *matrix-free*.

We need store only four vectors $\mathbf{x}$, $\mathbf{w}$, $\mathbf{p}$, and $\mathbf{r}$. Each iteration requires a single matrix–vector product to compute $\mathbf{w} = A\mathbf{p}$, two scalar products (one for $\mathbf{p}^T\mathbf{w}$ and one to compute $\|\mathbf{r}\|^2$), and three operations of the form $\alpha\,\mathbf{x}+\mathbf{y}$, where $\mathbf{x}$ and $\mathbf{y}$ are vectors and α is a scalar. It is remarkable that the iteration can progress without storing a basis for the entire Krylov subspace thanks to the existence of a three-term relation for the Hessenberg matrix, that becomes symmetric and tridiagonal, in the Arnoldi process. In particular, the symmetric Lanczos algorithm can be viewed as a simplification of Arnoldi's algorithm when the matrix is symmetric.

It is clear that the CG-type algorithms, i.e., algorithms defined through short-term recurrences, are more desirable than those algorithms which require storing entire sequences of vectors as in the GMRES Algorithm. The former algorithms require less memory and operations per step. An optimal Krylov subspace projection based on a short-term recurrence is a technique which minimizes a certain norm of the error, or residual, on the Krylov subspace independently from the starting vector. Such methods can be defined from the Arnoldi process. However, it was shown by Faber and Manteuffel that this cannot happen if A is nonHermitian; see, e.g., [280, Chapter 6].

2.2.6 The Arnoldi Algorithm

The Arnoldi algorithm is an orthogonal projection method on $\mathcal{K}_m$ for general nonHermitian matrices. It was introduced in 1951 as a way to reduce a matrix in Hessenberg form and this is the main use of it here. Moreover, Arnoldi suggested in [11] that the eigenvalues of the Hessenberg matrix generated from an $n \times n$ matrix A, after a number of steps less than n, can give approximations for extremal eigenvalues of A. Later, the underlying algorithm was discovered to be an effective (cheap under appropriate assumptions) and powerful tool for approximating eigenvalues of large and sparse matrices. A shift-and-invert strategy is needed if one is searching other eigenvalues; see, e.g., [456] and references therein.

Algorithm 2.6: Arnoldi

Input: $A \in \mathbb{R}^{n\times n}$, $\mathbf{v}_1 \in \mathbb{R}^n$ such that $\|\mathbf{v}_1\|_2 = 1$
Output: $V_m = [\mathbf{v}_1, \dots, \mathbf{v}_m]$, $H_m \in \mathbb{R}^{(m+1)\times m}$

for $j = 0, \dots, m$ **do**
- $h_{i,j} \leftarrow < A\,\mathbf{v}_j, \mathbf{v}_i >$, $i = 1, \dots, j$;
- $\mathbf{w}_j \leftarrow A\,\mathbf{v}_j - \sum_{i=1}^{j} h_{i,j}\,\mathbf{v}_i$;
- $h_{j+1,j} \leftarrow \|\mathbf{w}_j\|_2$;
- **if** $h_{j+1,j} = 0$ **then**
  - **exit**;
- $\mathbf{v}_j \leftarrow \mathbf{w}_j / h_{j+1,j}$;

At each step in Algorithm 2.6 A is multiplied by $\mathbf{v}_j$ and the resulting vector

$\mathbf{w}_j$ is orthonormalized against all previous vectors $\mathbf{v}_j$ by a Gram-Schmidt-like process. The process stops if $\mathbf{w}_j$ computed in line 4 is the zero vector.

Note that we need to store the (usually full) matrix $n \times m$ V_m, whose columns are given by $\mathbf{v}_1, \ldots, \mathbf{v}_m$ at each step m, and the $(m+1) \times m$ upper Hessenberg matrix $H_m = (h_{i,j})$ with $m^2 + 1$ nonzero entries.

Theorem 2.9. *If algorithm 2.6 does not terminate before step m, vectors $\mathbf{v}_1$, $\mathbf{v}_2$, ..., $\mathbf{v}_m$ form an orthonormal basis for the Krylov subspace*

$$\mathcal{K}_m(A, \mathbf{v}_1) = \text{span}\{\mathbf{v}_1,\ A\,\mathbf{v}_1,\ \ldots,\ A^{m-1}\,\mathbf{v}_1\}.$$

Proof. The vectors $\mathbf{v}_j$, $j = 1, \ldots, m$ are orthonormal. They span $\mathcal{K}_m(A, \mathbf{v}_1)$ because every vector $\mathbf{v}_j$ can be written in the form $q_{j-1}(A)\mathbf{v}_1$, where q_{j-1} is a polynomial of degree $j-1$. By induction: trivially true for for $j = 1$ because $\mathbf{v}_1 = q_0(A)\mathbf{v}_1$, $q_0 = 1$. Let us suppose that the thesis holds true for all integer less or equal to j. We finish by observing that for $\mathbf{v}_{j+1}$ we have

$$\begin{aligned} h_{j+1,j}\,\mathbf{v}_{j+1} =& A\,\mathbf{v}_j - \sum_{i=1}^{j} h_{i,j}\,\mathbf{v}_i \\ =& A\,q_{j-1}(A)\,\mathbf{v}_1 - \sum_{i=1}^{j} h_{i,j}\,q_{i-1}(A)\mathbf{v}_1 = q_j(A)\,\mathbf{v}_1. \end{aligned}$$ □

Let us state a result that will be very important also for some method in the next sections such as GMRES.

Theorem 2.10. *Let V_m an $n \times m$ with columns $\mathbf{v}_1, \ldots, \mathbf{v}_m$, $\overline{H}_m = (h_{i,j})$ the $(m+1) \times m$ upper Hessenberg matrix whose entries are computed in the Arnoldi algorithm 2.6 and H_m the $m \times m$ submatrix extracted from $\overline{H}_m$ by deleting its last line.*

$$A\,V_m = V_m\,\overline{H}_m\, +\, \mathbf{w}_m\,\mathbf{e}_m^T = V_{m+1}\,H_m, \tag{2.38}$$

where $\mathbf{e}_m$ is $[0, \ldots, 0, 1]^T \in \mathbb{R}^m$ and

$$V_m^H\,A\,Vm = H_m. \tag{2.39}$$

Proof. The (2.38) come from the Arnoldi algorithm and by observing that $V_m = [\mathbf{v}_1, \mathbf{v}_2, \ldots, \mathbf{v}_m]$ in

$$A\,\mathbf{v}_j = \sum_{i=1}^{j+1} h_{i,j}\,\mathbf{v}_i, \quad j = 1, \ldots, m.$$ □

Theorem 2.11. *The Arnoldi algorithm 2.6 stops at step $j < n$ if and only if the minimal polynomial of $\mathbf{v}_1$ has degree j. In this case the subspace $\mathcal{K}_j$ is invariant for A.*

Proof. If the degree of the minimal polynomial of A is j, then $\mathbf{w}_j$ should be necessarily the zero vector because of the Gram-Schmidt orthogonalization process. Suppose that the Arnoldi algorithm 2.6 stops at step $j < n$ because $\|\mathbf{w}_j\| = 0$. Then, the degree μ of the minimal polynomial of $\mathbf{v}_1$ is such that $\mu \leq j$ but it is not possible that $\mu < j$ otherwise the algorithm could stop before step jth. □

Exercise 2.5. Explicit the proof of (2.38).

Exercise 2.6. Write an algorithm based on the Arnoldi algorithm 2.6 to solve a linear system.

2.2.7 Generalized minimal residual: GMRES

Generalized Minimum Residual, or GMRES is a projection method approximating the solution of linear system $A\mathbf{x} = \mathbf{b}$ on the affine subspace $\mathbf{x}^{(0)} + \mathcal{K}_m(A, \mathbf{v}_1)$, $\mathbf{x}^{(0)}$ starting guess for the solution. The Petrov–Galerkin conditions (2.20) for GMRES can be written as $\mathcal{L} = A\mathcal{K}_m(A, \mathbf{r}^{(0)})$, $\mathbf{r}^{(0)} = \mathbf{b} - A\,\mathbf{x}^{(0)}$. GMRES, at each iteration, determines $\mathbf{x}^{(j+1)}$ such that $\|\mathbf{r}^{(j+1)}\|_2$ is minimum. Let us write the GMRES algorithm starting from its properties:

$$A\,V_m = V_m H_m + \mathbf{w}_m \mathbf{e}_m^T = V_{m+1} \overline{H}_m, \tag{2.40}$$

$V_m = [\mathbf{v}_1, \ldots, \mathbf{v}_m]$ is the orthonormal basis of $\mathcal{K}_m(M, \mathbf{v}_1)$ and H_m is the $m \times m$ Hessenberg submatrix extracted from $\overline{H}_m$ by deleting the $(m+1)$th line. At step m, the candidate solution $\mathbf{x}^{(m)}$ will be the vector minimizing the residual in the 2–norm:

$$\|\mathbf{r}^{(m)}\|_2 = \|\mathbf{b} - A\,\mathbf{x}^{(m)}\|_2.$$

From the previous relations,

$$\begin{aligned}
\mathbf{b} - A\,\mathbf{x}^{(m)} &= \mathbf{b} - A\,(\mathbf{x}^{(0)} + V_m\mathbf{y}) \\
&= \mathbf{r}^{(0)} - A\,V_m\mathbf{y} \\
&= V_{m+1}\left(V_{m+1}^T\mathbf{r}^{(0)} - \overline{H}_m\mathbf{y}\right) \\
&= V_{m+1}(\beta\mathbf{e}_1 - \overline{H}_m\mathbf{y})
\end{aligned}$$

where $\beta = \|\mathbf{r}^{(0)}\|_2$, $\mathbf{e}_1 = [1, 0, \ldots, 0]_{m+1}^T$. To minimize the expression

$$\|\beta\mathbf{e}_1 - \overline{H}_m\mathbf{y}\|_2, \quad \mathbf{y} \in \mathbb{R}^m \tag{2.41}$$

we need to solve a linear least squares problem $(m+1) \times m$. Its approximate solution at step m is given by

$$\mathbf{x}^{(m)} = \mathbf{x}^{(0)} + V_m\mathbf{y}^{(m)},$$

where $\mathbf{y} = \mathbf{y}^{(m)} \in \mathbb{R}^m$ minimizes (2.41). We use modified Gram–Schmidt orthogonalization in GMRES algorithm 2.7.

To solve the least squares problem (2.41), we need to transform the upper Hessenberg matrix into an upper triangular one by Givens plane rotations Q_j (see, e.g., [271]):

$$Q = Q_{m-1} \cdot \ldots \cdot Q_1 \in \mathbb{R}^{(m+1)\times(m+1)}$$

$$\begin{aligned} \min_{\mathbf{y}} \|\beta \mathbf{e}_1 - \overline{H}_m \mathbf{y}\|_2 &= \min_{\mathbf{y}} \|Q\beta \mathbf{e}_1 - Q\overline{H}_m \mathbf{y}\|_2 \\ &= \min_{\mathbf{y}} \|\mathbf{g}^{(m)} - \overline{R}_m \mathbf{y}\|_2 \end{aligned}$$

where $\mathbf{g}^{(m)} \in \mathbb{R}^{m+1}$, $\overline{R}_m \in \mathbb{R}^{(m+1)\times m}$, $\mathbf{y} \in \mathbb{R}^m$. It can be easily checked that the last component of $\mathbf{g}^{(m)}$ is the norm of the residual $\mathbf{r}^{(m)}$; see Saad and Schultz in [457]. By observing that

$$\begin{aligned} \|\beta \mathbf{e}_1 - \overline{H}_m \mathbf{y}\|_2^2 &= \|Q\beta \mathbf{e}_1 - Q\overline{H}_m \mathbf{y}\|_2^2 \\ &= \|\mathbf{g}^{(m)} - \overline{R}_m \mathbf{y}\|_2^2 \\ &= |\mathbf{e}_{m+1}^T \mathbf{g}^{(m)}|^2 + \|\hat{\mathbf{g}}^{(m)} - R_m \mathbf{y}\|_2^2, \end{aligned}$$

we find

$$\mathbf{y}^{(m)} = R_m^{-1} \hat{\mathbf{g}}^{(m)},$$

where $R_m \in \mathbb{R}^{m\times m}$ is the upper triangular matrix extracted by deleting the last line from $\overline{R}_m$ and $\hat{\mathbf{g}}^{(m)}$ is the vector whose entries are the first m of $\mathbf{g}^{(m)}$. The plane rotations Q_j, $j = 1, \ldots, m-1$, can be applied, for each iteration of GMRES, at the mth new of $\overline{H}_m$. This gives also the residual norm without computing explicitly $\mathbf{r}^{(m)} = \mathbf{b} - A\,\mathbf{x}^{(m)}$.

GMRES stops at step jth if and only if $h_{j+1,j} = 0$. In this case (and in exact arithmetic), the computed solution $\mathbf{x}^{(j)}$ can be considered exact because $\mathbf{b} - A\,\mathbf{x}^{(j)} = 0$ and $\mathbf{x} = \mathbf{x}^{(j)} \in \mathcal{K}_j(A, \mathbf{v}_1)$. It is the only possible forced stop for GMRES, and is a *Lucky Breakdown* differently to what we will see for bi–Lanczos, BiCG, BiCGstab, etc. Note that, for the same reasons of CG, GMRES terminates in at most n iterations in exact arithmetic and that GMRES can stagnate, i.e., no significant residual error reduction until step nth; see [280] for more details and considerations on convergence issues.

Computational complexity

There are various possibilities to modify GMRES to overcome the memory issues mentioned in the previous section.
Among the most common GMRES variants we recall (see, e.g., Saad [455, section 6.5]):

- Quasi–GMRES: instead of the orthogonalization procedure Arnoldi, we keep in memory at most a fixed number k of vectors $\mathbf{v}_{m-k+1}, \ldots, \mathbf{v}_m$. In this way, H_m becomes a band Hessenberg matrix.

Algorithm 2.7: GMRES

Input: $A \in \mathbb{R}^{n \times n}$,$\mathbf{b} \in \mathbb{R}^n$, Maximum number of iteration m, Initial Guess $\mathbf{x}^{(0)}$

Output: $\tilde{\mathbf{x}}$ candidate approximation.

$\mathbf{r}^{(0)} \leftarrow \mathbf{b} - A\,\mathbf{x}^{(0)}$;
$\beta \leftarrow \|\mathbf{r}^{(0)}\|_2$;
$\mathbf{v}_1 \leftarrow \mathbf{r}^{(0)}/\beta$;
for $j = 1, \ldots, m$ **do**
  $\mathbf{w}_j \leftarrow A\,\mathbf{v}_j$;
  for $i = 1, \ldots, j$ **do**
    $h_{i,j} \leftarrow < \mathbf{w}_j, \mathbf{v}_i >$;
    $\mathbf{w}_j \leftarrow \mathbf{w}_j - h_{i,j}\,\mathbf{v}_i$;
  $h_{j+1,j} \leftarrow \|\mathbf{w}_j\|_2$;
  if $h_{j+1,j} = 0$ *or <Stopping criteria satisfied>* **then**
    $m = j$;
    break;
  $\mathbf{v}_{j+1} = \mathbf{w}_j / \|\mathbf{w}_j\|_2$;
Compute $\mathbf{y}^{(m)}$ such that
$\|\mathbf{r}^{(m)}\|_2 = \|\mathbf{b} - A\,\mathbf{x}^{(m)}\|_2 = \|\beta\mathbf{e}_1 - \underline{H}_m\mathbf{y}\|_2 = \min_{\mathbf{y}\in\mathbb{R}^m}$;
Build candidate approximation $\tilde{\mathbf{x}}$;

- Restarted GMRES (see algorithm 2.8): after a max number of iterations k, usually from 10 up to 50, if GMRES has not reached convergence, we set $\mathbf{r}^{(0)} = \mathbf{r}^{(m)}$, $\mathbf{x}^{(0)} = \mathbf{x}^{(m)}$, and return to the beginning of the cycle.

The limit of these strategies is that we must pay something for the loss of information due to partial orthogonalization, in the case of the Quasi-GMRES, or to restarting in the case of Restarted-GMRES. The price we pay are the following issues: the possibility of stagnation of the algorithms if the matrix of the underlying linear system is not positive definite, less robustness, and the absence of the nice and complete convergence theory of GMRES.

2.2.8 Convergence analysis of GMRES and of GMRES(m)

Proposition 2.5. *If A is positive definite, i.e., $\mathbf{x}^T A\,\mathbf{x} > 0$, $\|\mathbf{x}\| > 0$, then GMRES(m) (and then GMRES that can be considered GMRES(m) with $m = \infty$) converges for all $m \geq 1$.*

Proof. The Krylov subspace $\mathcal{K}_m(A, \mathbf{v}_1)$ includes the initial residual vector for all values of $m \geq 1$. Recall that the algorithm minimizes the norm-2 of the

Algorithm 2.8: Restarted GMRES or GMRES(m)

Input: $A \in \mathbb{R}^{n\times n}$, $\mathbf{b} \in \mathbb{R}^n$, Maximum number of iteration m, Initial Guess $\mathbf{x}^{(0)}$

Output: $\tilde{\mathbf{x}}$ candidate approximation.

1 $\mathbf{r}^{(0)} \leftarrow \mathbf{b} - A\,\mathbf{x}^{(0)}$;
2 $\beta \leftarrow \|\mathbf{r}^{(0)}\|_2$;
3 $\mathbf{v}_1 \leftarrow \mathbf{r}^{(0)}/\beta$;
4 **for** $j = 1, \ldots, m$ **do**
5     $\mathbf{w}_j \leftarrow A\,\mathbf{v}_j$;
6   **for** $i = 1, \ldots, j$ **do**
7         $h_{i,j} \leftarrow <\mathbf{w}_j, \mathbf{v}_i>$;
8         $\mathbf{w}_j \leftarrow \mathbf{w}_j - h_{i,j}\,\mathbf{v}_i$;
9     $h_{j+1,j} \leftarrow \|\mathbf{w}_j\|_2$;
10   **if** $h_{j+1,j} = 0$ *or <Stopping criteria satisfied>* **then**
11         $m = j$;
12   **exit**;
13     $\mathbf{v}_{j+1} = \mathbf{w}_j/\|\mathbf{w}_j\|_2$;
14 Compute $\mathbf{y}^{(m)}$ such that $\|\mathbf{r}^{(m)}\|_2 = \|\mathbf{b} - A\,\mathbf{x}^{(m)}\|_2 = \|\beta\mathbf{e}_1 - \underline{H}_m\mathbf{y}\|_2 = \min_{\mathbf{y}\in\mathbb{R}^m}$;
15 **if** *<Stopping criteria* not *satisfied>* **then**
16     $\mathbf{r}^{(0)} \leftarrow \mathbf{r}^{(m)}$;
17     $\mathbf{x}^{(0)} \leftarrow \mathbf{x}^{(m)}$;
18   **Goto 2**;
19 Build candidate approximation $\tilde{\mathbf{x}}$;

residual in the subspace $\mathcal{K}_m(A, \mathbf{v}_1)$. Then, at each iteration we will have

$$\|\mathbf{r}^{(j+1)}\|_2 \leq \left(1 - \frac{\lambda_{min}(A)^2}{\|A\|_2^2}\right)^{1/2} \|\mathbf{r}^{(j)}\|_2$$

where $\lambda_{min}(A)$ is the smallest eigenvalue of A, if it is symmetric. Otherwise, $\lambda_{min}(A)$ is the smallest eigenvalue of $(A + A^T)/2$, i.e., of its symmetric part. □

If $\mathbf{r}^{(j)} = \mathbf{b} - A\,\mathbf{x}^{(j)}$, then at step jth of GMRES, we have

$$\begin{aligned}\|\mathbf{r}^{(j)}\|_2 &= \|(I - A\,\mathsf{q}_{j-1}(A))\mathbf{r}^{(0)}\|_2\\ &= \min_{\mathsf{q}(z)\in\mathbb{P}_{j-1}} \|(I - A\mathsf{q}(A))\mathbf{r}^{(0)}\|_2,\end{aligned}$$

where $\mathsf{q}_j(z)$ is a polynomial of degree at most $j-1$ in z. Then,

Theorem 2.12. *If A can be diagonalized, i.e., if we can find $X \in \mathbb{R}^{n\times n}$ non*

singular and such that

$$A = X\,\Lambda\,X^{-1},\ \ \Lambda = diag(\lambda_1, \ldots, \lambda_n),\ \ K_2(X) = \|X\|_2\,\|X^{-1}\|_2,$$

$K_2(X) = \|X\|_2\,\|X^{-1}\|_2$ *condition number of* X*, then at step* m*, we have*

$$\|r\|_2 \le K_2(X)\|\mathbf{r}^{(0)}\|_2 \min_{\mathsf{p}(z)\in\mathbb{P}_m,\mathsf{p}(0)=1}\ \max_{i=1,\ldots,n} |\mathsf{p}(\lambda_i)|, \tag{2.42}$$

where $\mathsf{p}(z)$ *is the polynomial of degree less or equal to* m *such that* $\mathsf{p}(0) = 1$ *and the expression in the right hand side in* (2.42) *is minimum.*

Proof. Let $\mathsf{p}(z)$ be a polynomial in z of degree less or equal to m such that $\mathsf{p}(0) = 1$ and take $\mathbf{x} \in \mathbf{x}^{(0)} + \mathcal{K}_m(A, \mathbf{v}_1)$. Then, by the latter, we can write

$$\mathbf{b} - A\,\mathbf{x} = \mathsf{p}(A)\mathbf{r}^{(0)}.$$

By recalling that A can be diagonalized by X we write

$$\|\mathbf{b} - A\,\mathbf{x}\|_2 = \|\mathsf{p}(X\Lambda X^{-1})\mathbf{r}^{(0)}\|_2 = \|X\mathsf{p}(\Lambda)X^{-1}\mathbf{r}^{(0)}\|_2,$$

where Λ is a diagonal matrix and

$$\|\mathsf{p}(\Lambda)\|_2 = \max_{i=1,\ldots,n} |\mathsf{p}(\lambda_i)|.$$

At the mth step, $\mathbf{x}^{(m)}$ is determined such that the Euclidean norm of the residual is minimal; therefore

$$\|\mathbf{b} - A\,\mathbf{x}^{(m)}\| \le \|\mathbf{b} - A\,\mathbf{x}\|_2 \le \|X\|_2\|X^{-1}\|_2\|\mathbf{r}^{(0)}\|_2 \cdot \max_{i=1,\ldots,n} |\mathsf{p}(\lambda_i)|,$$

and, by using the polynomial of degree m $\mathsf{p}(\mathsf{z})$ that minimizes the right hand side of the previous bound, we proved the theorem. □

From the previous theorem, when A is a preconditioned matrix and the eigenvalues form $\hat{n}$ clusters in the complex plane far away from the origin, with $\hat{n} \ll n$, we expect that GMRES converges in $k + \hat{n}$ iterations, $k \ll n$, if $K_2(X)$ is not large. But we can state something more precise in the next discussion.

Convergence of GMRES when the spectrum is clustered

Let us consider a linear system with a matrix A having a spectrum $\sigma(A)$ clustered around a certain point in the complex plane far from the origin[2]. It is natural to partition $\sigma(A)$ as follows

$$\sigma(A) = \sigma_c(A) \cup \sigma_0(A) \cup \sigma_1(A),$$

[2]If the spectrum of A is not clustered and we consider the use of a *preconditioner* with a given matrix P, the following analysis will focus on, e.g., $K = P^{-1}A$ or $K = AP^{-1}$; see Chapters 3 and 4.

where $\sigma_c(A)$ denotes the clustered set of eigenvalues of A and $\sigma_0(A) \cup \sigma_1(A)$ denotes the set of the outliers. Here we assume that the clustered set $\sigma_c(A)$ of eigenvalues is contained in a convex set Ω.

Now, let us consider in more detail the sets

$$\sigma_0(A) = \{\hat{\lambda}_1, \hat{\lambda}_2, \ldots, \hat{\lambda}_{j_0}\} \quad \text{and} \quad \sigma_1(A) = \{\tilde{\lambda}_1, \tilde{\lambda}_2, \ldots, \tilde{\lambda}_{j_1}\}$$

denoting two sets of j_0 and j_1 outliers, respectively. The sets σ_0 and σ_1 are defined such that, if $\hat{\lambda}_j \in \sigma_0(A)$, we have

$$1 < \left|1 - \frac{z}{\hat{\lambda}_j}\right| \leq c_j, \quad \forall z \in \Omega,$$

while, for $\tilde{\lambda}_j \in \sigma_1(A)$,

$$0 < \left|1 - \frac{z}{\tilde{\lambda}_j}\right| < 1, \quad \forall z \in \Omega,$$

respectively.

Under the above assumptions, we can state the following bound.

Theorem 2.13. *The number of full GMRES iterations j needed to attain a tolerance ε on the relative residual in the 2-norm $\|\mathbf{r}^{(j)}\|_2/\|\mathbf{r}^{(0)}\|_2$ for the linear system $A\mathbf{x} = \mathbf{b}$, where A is diagonalizable, is bounded above by*

$$\min\left\{ j_0 + j_1 + \left\lceil \frac{\log(\varepsilon) - \log(\kappa_2(X))}{\log(\rho)} - \sum_{\ell=1}^{j_0} \frac{\log(c_\ell)}{\log(\rho)} \right\rceil, n \right\}, \tag{2.43}$$

where

$$\rho^k = \frac{\left(a/d + \sqrt{(a/d)^2 - 1}\right)^k + \left(a/d + \sqrt{(a/d)^2 - 1}\right)^{-k}}{\left(c/d + \sqrt{(c/d)^2 - 1}\right)^k + \left(c/d + \sqrt{(c/d)^2 - 1}\right)^{-k}}, \tag{2.44}$$

and the set $\Omega \in \mathbb{C}^+$ is the ellipse with center c, focal distance d and major semi axis a.

Proof. From (2.42), $\|\mathbf{r}^{(j)}\|_2/\|\mathbf{r}^{(0)}\|_2 \leq \varepsilon (\ll 1)$ is satisfied (in the exact arithmetic) if

$$\kappa_2(X) \cdot \min_{\mathbf{p}_j(0)=1} \max_{\lambda \in \sigma(A)} |\mathbf{p}_j(\lambda)| \leq \varepsilon, \tag{2.45}$$

where $\sigma(A)$ contains the set of eigenvalues of A and $\mathbf{p}_j(z)$ is a j-degree polynomial. We have:

$$\min_{\mathbf{p}_j(0)=1} \max_{z \in \sigma(A)} |\mathbf{p}_j(z)| \leq \max_{z \in \sigma(A)} |\hat{p}(z) \cdot q(z) \cdot \tilde{p}(z)|, \tag{2.46}$$

where

$$\hat{p}(z) = \left(1 - \frac{z}{\hat{\lambda}_1}\right) \cdots \left(1 - \frac{z}{\hat{\lambda}_{j_0}}\right), \quad \tilde{p}(z) = \left(1 - \frac{z}{\tilde{\lambda}_1}\right) \cdots \left(1 - \frac{z}{\tilde{\lambda}_{j_1}}\right)$$

are the polynomials whose roots are the outlying eigenvalues in $\sigma_0 \cup \sigma_1$ and $q(z)$ is a polynomial of degree at most $j - j_0 - j_1 \geq 0$ such that $q(0) = 1$. Using the notations above, we have

$$|\hat{p}(z)| \leq \prod_{\ell=1}^{j_0} \left|1 - \frac{z}{\hat{\lambda}_\ell}\right| \leq \prod_{\ell=1}^{j_0} c_\ell \quad \text{and} \quad |\tilde{p}(z)| \leq \prod_{\ell=1}^{j_1} \left|1 - \frac{z}{\tilde{\lambda}_\ell}\right| \leq \prod_{\ell=1}^{j_1} 1 \leq 1, \; \forall z \in \Omega.$$

Therefore,

$$\max_{z \in \sigma(A)} |\hat{p}(z) \cdot q(z) \cdot \tilde{p}(z)| \leq \left(\prod_{\ell=1}^{j_0} c_l\right) \max_{z \in \Omega} |q(z)|. \tag{2.47}$$

The polynomial $q(z)$ can be chosen to be the shifted and scaled complex Chebyshev polynomial $q(z) = C_k((c-z)/d)/C_k(c/d)$ which is small on the set containing $\sigma_c(A)$. Indeed, by using results on Chebyshev polynomials (see [455, Sections 6.11.2,6.11.4]), we can derive the following bound

$$\begin{aligned} \max_{z \in \Omega} |q(z)| &= \frac{C_k(a/d)}{|C_k(c/d)|} \\ &= \frac{\left(a/d + \sqrt{(a/d)^2 - 1}\right)^k + \left(a/d + \sqrt{(a/d)^2 - 1}\right)^{-k}}{\left(c/d + \sqrt{(c/d)^2 - 1}\right)^k + \left(c/d + \sqrt{(c/d)^2 - 1}\right)^{-k}} \end{aligned} \tag{2.48}$$

An upper bound on j now easily follows from (2.46), (2.47), (2.48) and by observing that, in exact arithmetics, GMRES converges in at most n iterations:

$$j - j_0 - j_1 = \left\lceil \frac{\log(\varepsilon)}{\log(\rho)} - \frac{\log(\kappa_2(X))}{\log(\rho)} - \frac{\sum_{\ell=1}^{j_0} \log(c_\ell)}{\log(\rho)} \right\rceil$$

from which (2.43) follows. □

We stress that

$$\rho \simeq \tilde{\rho} = \frac{a + \sqrt{a^2 - d^2}}{c + \sqrt{c^2 - d^2}}.$$

In particular, when the major axis is parallel to the imaginary axis, is centered in $(c, 0)$, $c > 0$, and has length $2a$, while the minor axis $2b$, respectively, we have $a \geq b$ and

$$\tilde{\rho} = \frac{a + \sqrt{a^2 - |a^2 - b^2|}}{c + \sqrt{c^2 + |a^2 - b^2|}} = \frac{a + b}{c + \sqrt{c^2 + a^2 - b^2}},$$

and we use this expression to approximate ρ in the bound (2.43) in practice for our model problems.

According to Theorem 2.13, the outliers do not affect the asymptotic convergence rate of the GMRES method, but rather they introduce a latency effect of $j_0 + j_1$ iterations plus the term $\sum_{l=1}^{j_0} \log(c_l)/\log(\rho)$; see (2.43).

We stress that the condition number of X, the matrix that diagonalizes A, cannot be neglected in the above bound, otherwise the eigenvalues alone can give highly misleading information on the convergence process, see [282]. On the other hand, if X has a huge condition number (e.g., growing exponentially with the size of the matrix), the underlying bound is useless.

Remark 2.3. By [280], let us consider some GMRES issues for A highly nonnormal and/or A non definite. In these two cases, the results based on the Chebyshev polynomials cannot be used any more and we have to reconsider the bound (2.42) on the residual and its sharpness.

- **Non definite matrices.** If the eigenvalues are clustered around the origin, finding a minimizing polynomial that has value 1 on the origin and is less than 1 everywhere on some closed curve around the origin containing the eigenvalues, is impossible by the maximum principle. Then, the convergence analysis becomes less useful and the bound is not sharp at all. Similarly, it is impossible to get a polynomial having value 1 on the origin and with small absolute value in all the scattered eigenvalues, unless we let the polynomial degree grows. But in this way we find a slow convergence[3].

- **Non normal matrices.** the bound depends on the condition number of the eigenvector matrix X; if it is large, then the convergence analysis can produce again a non sharp bound.

We can say more: any convergence curve is possible; see the next theorem proposed in Greenbaum et al. [281], Greenbaum and Strakoš [282].

Theorem 2.14 (Greenbaum et al. [281])**.** *Given a non-increasing positive sequence* $\{f_k\}_{k=0,\ldots,n-1}$ *with* $f_{n-1} > 0$ *and a set of non–zero complex numbers* $\{\lambda_i\}_{i=1,2,\ldots,n} \subset \mathbb{C}$, *there exists a matrix* A *with eigenvalues* $\lambda_1, \lambda_2, \ldots, \lambda_n$ *and a right-hand side* $\boldsymbol{b}$ *with* $\|\boldsymbol{b}\| = f_0$ *such that the residual vectors* $\mathbf{r}^{(k)}$ *at each step of the GMRES algorithm applied to solve* $A\boldsymbol{x} = \boldsymbol{b}$ *with* $\boldsymbol{x}^{(0)} = \boldsymbol{0}$, *satisfy* $\|\boldsymbol{r}^{(k)}\| = f_k$, $\forall\, k = 1, 2, \ldots, n-1$.

Remark 2.4. Observe that the assumption $f_{n-1} > 0$ in the above theorem means that GMRES converges to the solution exactly at iteration n, and both the dimension of the Krylov space and the degree of the minimizing polynomial is n. The result can be modified to obtain every iteration/residual graph.

[3] A high polynomial degree implies that the underlying Krylov subspace has a of large dimension, possibly approaching the degree of freedom of the problem or even more and therefore more iterations are required for GMRES algorithm.

2.3 Krylov subspace algorithms, oblique projections: BiCG, QMR, CGS, BiCGstab, BiCGstab(l)

We observed that GMRES for non-Hermitian linear systems requires an amount of storage increasing with the iterations. For this reason we considered GMRES(m). Note that for Hermitian problems we can prove that GMRES produces a tridiagonal H_m (and not just an Hessenberg) matrix and it reduces to a three-term relation, so there is no need of any restart or GMRES(m) and the operation count of a GMRES iteration is independent of the iteration number.

Therefore, here we briefly recall some other popular and effective Krylov method with a computational cost per iteration that does not increase, even though they can require more matrix-vector products to reduce the relative norm of the residual of a prescribed quantity. Moreover, all of them have the possibly of failure even for a nonsingular matrix, although that can be alleviated by the use of (costly) look-ahead strategies.

The methods we are going to discuss do not have nice convergence theories like those of CG and GMRES. However, often the convergence behavior is reasonable for wide classes of important problems like several from, e.g., fluid dynamics, chemistry and others; therefore they are widely used, especially in industrial and engineering applications.

2.3.1 The bi–Lanczos algorithm and BiCG

In addition to his famous algorithm, generalized in the Algorithm called *bi–Lanczos* [344], Lanczos suggested a Krylov projection algorithm for linear systems $Ax = b$ for nonsymmetric real or non-Hermitian matrices; see Algorithm 2.9. It produces two sequences of vectors that are bi-orthogonal, i.e.,

$$\{\mathbf{v}_j\},\ \{\mathbf{w}_j\},\ \mathbf{v}_j, \mathbf{w}_j \in \mathbb{R}^n,$$

such that $< \mathbf{v}_i, \mathbf{w}_j >= 0$, $i \neq j$, $< \mathbf{v}_i, \mathbf{w}_i > \neq 0$. $\mathbf{v}_j$, $\mathbf{w}_j$ are determined to satisfy the three-term recurrence:

$$\begin{aligned}\delta_{j+1}\mathbf{v}_{j+1} &= A\mathbf{v}_j - \alpha_j\mathbf{v}_j - \beta_j\mathbf{v}_{j-1},\\ \beta_{j+1}\mathbf{w}_{j+1} &= A^T\mathbf{w}_j - \alpha_j\mathbf{w}_j - \delta_j\mathbf{w}_{j-1},\end{aligned} \tag{2.49}$$

β_j, δ_j should be such that $< \mathbf{v}_j, \mathbf{w}_j >= 1$ for all j; see, e.g., [455] and Algorithm 2.9. At step j under appropriate conditions, $\mathbf{v}_1, \ldots, \mathbf{v}_m$ and $\mathbf{w}_1, \ldots, \mathbf{w}_m$ form two orthonormal basis, one for each of the Krylov subspaces

$$\mathcal{K}_m(A, \mathbf{v}_1) = \text{span}\{\mathbf{v}_1, A\,\mathbf{v}_1, \ldots, A^{m-1}\mathbf{v}_1\},$$

and

$$\mathcal{K}_m(A^T, \mathbf{w}_1) = \text{span}\{\mathbf{w}_1, A^T\mathbf{w}_1, \ldots, (A^T)^{m-1}\mathbf{w}_1\}.$$

Algorithm 2.9: Lanczos Biorthogonalization or bi–Lanczos

Input: $\mathbf{v}_1$, $\mathbf{w}_1 \in \mathbb{R}^n$ such that $< \mathbf{v}_1, \mathbf{w}_1 >= 1$, $A \in \mathbb{R}^{n\times n}$, $m \in \mathbb{N}$ such that $m \leq n$

Output: $V = [\mathbf{v}_1, \ldots, \mathbf{v}_j], W = [\mathbf{w}_1, \ldots, \mathbf{w}_j] \in \mathbb{R}^{n\times j}$, $j \leq m$, $T_j \in \mathbb{R}^{j\times j}$

$\beta_1 \leftarrow 0$, $\delta 1 \leftarrow 0$;
$\mathbf{w}_0 \leftarrow (0, \ldots, 0)^T$, $\mathbf{v}_0 \leftarrow (0, \ldots, 0)^T$;
for $j = 1, \ldots, m$ **do**
  $\alpha_j \leftarrow < A\mathbf{v}_j, \mathbf{w}_j >$;
  $\hat{\mathbf{v}}_{j+1} \leftarrow A\mathbf{v}_j - \alpha_j \mathbf{w}_j - \beta_j \mathbf{v}_{j-1}$;
  $\hat{\mathbf{w}}_{j+1} \leftarrow A^T\mathbf{w}_j - \alpha_j \mathbf{w}_j - \delta_j \mathbf{w}_{j-1}$;
  $\delta_{j+1} \leftarrow |< \hat{\mathbf{v}}_{j+1}, \hat{\mathbf{w}}_{j+1} >|^{1/2}.$;
  if $\delta_{j+1} = 0$ **then**
    return
  $\beta_{j+1} \leftarrow < \hat{\mathbf{v}}_{j+1}, \hat{\mathbf{w}}_{j+1}) > /\delta_{j+1}$;
  $\mathbf{v}_{j+1} \leftarrow \hat{\mathbf{v}}_{j+1}/\delta_{j+1}$;
  $\mathbf{w}_{j+1} \leftarrow \hat{\mathbf{w}}_{j+1}/\beta_{j+1}$;

Let us summarize some properties that will be used in the sequel; see [344] and [543], [455] for more details.

Theorem 2.15. *If $\delta_{m+1} > 0$, then at step m for bi–Lanczos, we have:*

i. $< \mathbf{v}_j, \mathbf{w}_i >= 0$, $i \neq j$, *i.e.,* $\mathbf{v}_1, \ldots, \mathbf{v}_j$, $\mathbf{w}_1, \ldots, \mathbf{w}_j$ *are a bi-orthogonal set of vectors;*

ii. $\mathbf{v}_1, \ldots, \mathbf{v}_m$ *is a basis for* $\mathcal{K}_m(A, \mathbf{v}_1)$ *and* $\mathbf{w}_1, \ldots, \mathbf{w}_m$ *for* $\mathcal{K}_m(A^T, \mathbf{w}_1)$*;*

iii. If $V_m = [\mathbf{v}_1, \ldots, \mathbf{v}_m]$, $W_m = [\mathbf{w}_1, \ldots, \mathbf{w}_m]$, *we have:*

$$W_m^T A V_m = T_m,$$

$$T_m = \begin{pmatrix} \alpha_1 & \beta_2 & & & \\ \delta_2 & \alpha_2 & \beta_3 & & \\ & \ddots & \ddots & \ddots & \\ & & \ddots & \ddots & \beta_m \\ & & & \delta_m & \alpha_m \end{pmatrix}.$$

Proof. *i*) By hypothesis, $< \mathbf{v}_1, \mathbf{w}_1 >= 1$. Let us suppose that $\mathbf{v}_1, \ldots, \mathbf{v}_j$ and $\mathbf{w}_1, \ldots, \mathbf{w}_j$ are bi–orthogonal and let us prove that also $\mathbf{v}_1, \ldots, \mathbf{v}_{j+1}$ and $\mathbf{w}_1, \ldots, \mathbf{w}_{j+1}$ are.

$$< \mathbf{v}_{j+1}, \mathbf{w}_j >= \frac{1}{\delta_{j+1}} (< A\mathbf{v}_j, \mathbf{w}_j > -\alpha_j < \mathbf{v}_j, \mathbf{w}_j > -\beta_j < \mathbf{v}_{j-1}, \mathbf{w}_j >)$$

Algorithm 2.10: Bi–Conjugate Gradients, or BiCG

Input: $A \in \mathbb{R}^{n\times n}$, $\mathbf{b} \in \mathbb{R}^n$, Maximum number of iterations $N_{\max}$
Output: $\tilde{\mathbf{x}}$ candidate approximation.

$\mathbf{r}^{(0)} \leftarrow \mathbf{b} - A\mathbf{x}^{(0)}$;
chose $\mathbf{r}^{(0)*}$ such that $< \mathbf{r}^{(0)}, \mathbf{r}^{(0)*} > \neq 0$;
$\mathbf{p}^{(0)} \leftarrow \mathbf{r}^{(0)}$, $\mathbf{p}^{(0)*} \leftarrow \mathbf{r}^{(0)*}$;
for $j = 1, \ldots, N_{max}$ **do**
  $\alpha_j \leftarrow < \mathbf{r}^{(j)}, \mathbf{r}^{(j)*} > / < A\mathbf{p}^{(j)}, \mathbf{p}^{(j)*} >$;
  $\mathbf{x}^{(j+1)} \leftarrow \mathbf{x}^{(j)} + \alpha_j \mathbf{p}^{(j)}$;
  $\mathbf{r}^{(j+1)} \leftarrow \mathbf{r}^{(j)} - \alpha_j A\mathbf{p}^{(j)}$;
  if <*Stopping criteria satisfied*> **then**
    return $\tilde{\mathbf{x}} = \mathbf{x}^{(j+1)}$
  $\mathbf{r}^{(j+1)*} \leftarrow \mathbf{r}^{(j)*} - \alpha_j A^T \mathbf{p}^{(j)*}$;
  $\beta_j \leftarrow < \mathbf{r}^{(j+1)}, \mathbf{r}^{(j+1)*} > / < \mathbf{r}^{(j)}, \mathbf{r}^{(j)*} >$;
  $\mathbf{p}^{(j+1)} \leftarrow \mathbf{r}^{(j+1)} + \beta_j \mathbf{p}^{(j)}$;
  $\mathbf{p}^{(j+1)*} \leftarrow \mathbf{r}^{(j+1)*} + \beta_j \mathbf{p}^{(j)*}$;

By inductive hypothesis, $< \mathbf{v}_{j-1}, \mathbf{w}_j >= 0$, and the remaining terms in the right hand side of the previous expression are zero by the definition of α_j. For $i < j$, by considering the inductive hypothesis and the second 3–term recurrence relation

$$\begin{aligned}
\delta_{j+1}(\mathbf{v}_{j+1}, \mathbf{w}_i) &= (< A\mathbf{v}_j, \mathbf{w}_i > -\alpha_j < \mathbf{v}_j, \mathbf{w}_i > -\beta_j < \mathbf{v}_{j-1}, \mathbf{w}_i >) \\
&= \left(< \mathbf{v}_j, A^T \mathbf{w}_i > -\beta_j < \mathbf{v}_{j-1}, \mathbf{w}_i >\right) \\
&= (< \mathbf{v}_j, \beta_{i+1}\mathbf{w}_{i+1} + \alpha_i \mathbf{w}_i + \delta_i \mathbf{w}_{i-1} > -\beta_j < \mathbf{v}_{j-1}, \mathbf{w}_i >).
\end{aligned}$$

The last expression is zero by induction if $i < j - 1$. On the other hand, if $i = j - 1$ we compute

$$\begin{aligned}
\delta_{j+1} &= < \mathbf{v}_{j+1}, \mathbf{w}_{j-1} >= \\
&= (< \mathbf{v}_j, \beta_j \mathbf{w}_j + \alpha_{j-1}\mathbf{w}_{j-1} + \delta_{j-1}\mathbf{w}_{j-2} > -\beta_j < \mathbf{v}_{j-1}, \mathbf{w}_{j-1} >) \\
&= \beta_j (< \mathbf{v}_j, \mathbf{w}_j > - < \mathbf{v}_{j-1}, \mathbf{w}_{j-1} >) = 0
\end{aligned}$$

Similarly we can see that $< \mathbf{v}_i, \mathbf{w}_{j+1} >= 0$ for $i \leq j$, while $< \mathbf{v}_{j+1}, \mathbf{w}_{j+1} >= 1$.

The *ii*) and *iii*) can be derived directly from the algorithm because they are a compact way to write the 3-term recurrence relations (2.49). □

Note that, if at step m we find $\delta_{m+1} = 0$, then we have the desired basis for the subspace $\mathcal{K}_m(A, \mathbf{v}_1)$. On the other hand, if $\beta_{m+1} = 0$ with $\delta_{m+1} \neq 0$, then the basis of $\mathcal{K}_m(A^T, \mathbf{w}_1)$ is determined, *bi–Lanczos* stops, but we do not have enough info for $\mathcal{K}_m(A, \mathbf{v}_1)$. This event is called *serious breakdown* (Wilkinson [543]) and is a pathology shared by all algorithms derived from *bi–Lanczos*; see [107, 108] for algorithmic way to avoid breakdowns by jumping over the singular denominators by the block bordering method for orthogonal polynomials.

From the previous proposition we discover that operator T_m is the oblique projection of A on the Krylov subspace $\mathcal{K}_m(A, \mathbf{v}_1)$ orthogonal to $\mathcal{K}_m(A^T, \mathbf{w}_1)$. Similarly, T_m^T represents the projection of A^T on $\mathcal{K}_m(A^T, \mathbf{w}_1)$ orthogonal to $\mathcal{K}_m(A, \mathbf{v}_1)$. We stress that *bi–Lanczos* explicitly needs the transpose matrix A^T.

The $\mathbf{x}^{(m)}$ generated by *bi–Lanczos* is searched in the affine subspace $\mathbf{x}^{(0)} + \mathcal{K}_m(A, \mathbf{v}_1)$ and its residual is parallel to $\mathbf{v}_{m+1}$ and orthogonal to $\text{span}\{\mathbf{w}_1, \ldots, \mathbf{w}_m\}$. From Theorem 2.15, we have:

$$
\begin{aligned}
\mathbf{x}^{(j)} - \mathbf{x}^{(0)} &= V_j \mathbf{y}^{(j)}, \ \mathbf{y}^{(j)} \in \mathbb{R}^j \\
\Rightarrow AV_m \mathbf{y}^{(m)} = \mathbf{b} - A\mathbf{x}^{(0)} &= \mathbf{r}^{(0)} \Rightarrow W_m^T A V_m \mathbf{y}^{(m)} = W_m^T \mathbf{r}^{(0)} \\
\Rightarrow T_m \mathbf{y}^{(m)} &= W_m^T \mathbf{r}^{(0)} \Rightarrow \mathbf{y}^{(m)} = T_m^{-1}\left(W_m^T \mathbf{r}^{(0)}\right) \\
\Rightarrow \mathbf{x}^{(m)} &= \mathbf{x}^{(0)} + V_m T_m^{-1}\left(W_m^T \mathbf{r}^{(0)}\right).
\end{aligned}
$$

Let $L_m U_m$ be the *LU* factorization for T_m. By defining $P_m = V_m U_m^{-1}$ and $P_m^* = W_m L_m^{-1}$, $\mathbf{p}_1, \ldots, \mathbf{p}_m$, $\mathbf{p}_1^*, \ldots, \mathbf{p}_m^*$ column vectors of P_m and of P_m^*, respectively, we can express $\mathbf{x}^{(m)}$ as follows:

$$
\begin{aligned}
\mathbf{x}^{(m)} &= \mathbf{x}^{(0)} + V_m T_m^{-1} V_m^T \mathbf{r}^{(0)} \\
&= \mathbf{x}^{(0)} + V_m U_m^{-1} L_m^{-1} V_m^T \mathbf{r}^{(0)} \\
&= \mathbf{x}^{(0)} + P_m L_m^{-1} V_m^T \mathbf{r}^{(0)} \\
&= \mathbf{x}^{(m-1)} + \alpha_{m-1} \mathbf{p}_{m-1}
\end{aligned}
$$

The sequences $\mathbf{p}_\mathbf{j}^*$ and $\mathbf{p}_i$ are A-orthogonal, i.e., $(\mathbf{p}_i^*, A\,\mathbf{p}_j) = 0$ FOR $i \neq j$. Therefore, we derive the *Bi-Conjugate Gradient* Algorithm 2.10, or BiCG for short, by the bi–Lanczos simply generalizing the steps we did for the Lanczos' Algorithm to get CG two times: one for each directions $\mathbf{p}_j$, $\mathbf{p}_j^*$. Note also that the residual vectors $\mathbf{r}^{(j)}$, $\mathbf{r}_*^{(j)}$ are parallel to $\mathbf{v}_{j+1}$, w_{j+1}, respectively, and $\mathbf{r}^{(1)}, \ldots, \mathbf{r}^{(m)}$, $\mathbf{r}_*^{(1)}, \ldots, \mathbf{r}_*^{(m)}$ are such that $< \mathbf{r}^{(i)}, \mathbf{r}_*^{(j)} >= \delta_{i,j}$, $i \neq j$, $i, j \leq m$, where $\delta_{j,j} = 1$, $\delta_{i,j} = 0$, $i \neq j$.

BiCG can become unstable for two reasons:

1. possible breakdown of its *bi–Lanczos* core;

2. breakdown of the *LU* factorization (without pivoting) for T_m.

The first issue can occur for a breakdown of *bi–Lanczos* because $\mathbf{r}_*^{(m+1)} = 0$ but $\mathbf{r}^{(m+1)} \neq 0$ or $\|\mathbf{w}_{m+1}\| = 0$ with $\|\mathbf{v}_{m+1}\| \neq 0$. In practice, $\text{span}\{\mathbf{w}_1, \ldots, \mathbf{w}_m\}$ is invariant for A^T, but $\text{span}\{\mathbf{v}_1, \ldots, \mathbf{v}_m\}$ it is not for A.

There exist *look-ahead* techniques to overcome this issue; see [105, 106,

248], but they are computationally expensive and are more useful when *bi–Lanczos* is used for computing (some) eigenvalues of a given matrix. The *look-ahead* techniques destroy the tridiagonal structure of T_m. Therefore, often it is simpler to choose a different $\mathbf{r}_*^{(0)}$ and start again.

Computational complexity

The computationally most expensive operations of *bi–Lanczos*–based algorithms are the two matrix-vector product per each iteration and some vector sums and scalar products. For BiCG we need approximately

$$m(2\,\mathrm{nnz}(M) + 8n)$$

flop for m iterations, where $\mathrm{nnz}(M)$ are the nonzero entries of A. Note that, as for all methods in this section and differently from GMRES, the vectors that must be kept in memory do not depend on the iteration number m.

2.3.2 Quasi Minimal Residual or QMR

The *Quasi Minimal Residual* or QMR algorithm is conceptually similar to GMRES but the basis for the underlying Krylov subspace $\mathcal{K}_m(A, \mathbf{r}^{(0)})$ is built by the bi-Lanczos' instead of the Arnoldi's algorithm. We recall that for the bi-Lanczos Algorithm we have:

$$A\,V_m = V_{m+1}\tilde{T}_m,$$

where $\tilde{T}_m$ is a $(m+1) \times m$ tridiagonal matrix (compare with the upper Hessenberg matrix $\overline{H}_m$ of the Arnoldi's algorithm) and the column vectors in V_{m+1} are not orthonormal as those generated by Arnoldi; see Theorem 2.9.

At the step m, for QMR's Algorithm, the candidate approximation $\mathbf{x}^{(m)}$ is still given by

$$\mathbf{x}^{(m)} = \mathbf{x}^{(0)} + V_m\,\mathbf{y}. \tag{2.50}$$

Recall that for GMRES, y in (2.50) is determined by minimizing the 2–norm of the residual $\mathbf{r}^{(m)}$

$$\begin{aligned}\mathbf{r}^{(m)} =& \mathbf{b} - A\,\mathbf{x}^{(m)} = \mathbf{b} - A(\mathbf{x}^{(0)} + V_m\,\mathbf{y}) = \mathbf{r}^{(0)} - AV_m\mathbf{y}\\ =& V_{m+1}(\beta\mathbf{e}_1 - \tilde{H}_m\mathbf{y}),\end{aligned}$$

where $\|\mathbf{r}^{(0)}\|_2 = \beta$, $\beta\mathbf{v}_1 = \mathbf{r}^{(0)}$, and therefore of $\beta\mathbf{e}_1 - \tilde{T}_m\mathbf{y}$ because the columns of V_{m+1} are orthonormal:

$$\|\mathbf{r}^{(m)}\|_2 = \|\mathbf{b} - A\,\mathbf{x}^{(m)}\|_2 = \|V_{m+1}(\beta\mathbf{e}_1 - \tilde{T}_m\mathbf{y})\|_2 = \|\beta\mathbf{e}_1 - \tilde{T}_m\mathbf{y}\|_2.$$

On the other hand, by using Lanczos instead of Arnoldi to build $\mathcal{K}_m(A, \mathbf{r}^{(0)})$ as in QMR's Algorithm 2.11, we still have

$$\begin{aligned}\mathbf{r}^{(m)} =& \mathbf{b} - A\,\mathbf{x}^{(m)} = \mathbf{b} - A(\mathbf{x}^{(0)} + V_m\,\mathbf{y}) = \mathbf{r}^{(0)} - AV_m\mathbf{y}\\ =& V_{m+1}(\beta\mathbf{e}_1 - \tilde{T}_m\mathbf{y}),\end{aligned} \tag{2.51}$$

where $\|\mathbf{r}^{(0)}\|_2 = \beta$, $\beta\mathbf{v}_1 = \mathbf{r}^{(0)}$, but now the columns of V_{m+1} in (2.51) are not orthonormal anymore. But we can still reasonably minimize $F(y)$

$$F(y) = \|\beta e_1 - \tilde{T}_m y\|_2,$$

by hoping that this will reduce also

$$\|\mathbf{r}^{(m)}\|_2 = \|V_{m+1}(\beta e_1 - \tilde{T}_m y)\|_2$$

even though this does not mean that we are minimizing the 2-norm of the residual $\mathbf{r}^{(m)}$ anymore. Note that the QMR iterates are always defined provided the bi-Lanczos Algorithm does not break down.

It is important to observe that the norm of the QMR's residual can be directly related to that of GMRES. This is a result due to Freund and Nachtigal [249].

Theorem 2.16. *If the Lanczos Algorithm does not break down before step m, we have:*

$$\|\mathbf{r}^{(m)}_{QMR}\|_2 \leq k_2(V_{m+1})\|\mathbf{r}^{(m)}_{GMRES}\|_2.$$

Unfortunately, the condition number of V_{m+1}, the matrix whose columns are the basis vectors of $\mathcal{K}_m(A, \mathbf{r}^{(0)})$ produced by the (bi-)Lanczos' Algorithm, cannot be bounded a priori and can be ill-conditioned even if the the Lanczos vectors $\mathbf{v}_1, \ldots, \mathbf{v}_{m+1}$ are well defined.

Note that the QMR Algorithm requires matrix-vector products either with A and A^T because of the bi-Lanczos' Algorithm. This can be uncomfortable because, e.g., in some applications (i) the matrix A^T can be not explicitly available because it is known only as a result of a sequence of operations and/or (ii) in parallel computing this can slow down the computations. There are transpose-free alternatives for QMR. The most popular is TFQMR, not described here; see Freund [247].

2.3.3 BiCGstab

BiCGstab and BiCGstab(l), introduced by Van der Vorst [527, 528], can be considered as transpose-free variants of BiCG. However, BiCGstab is so popular that it is better to take an entire section to describe it in some detail.

In order to use it as an iterative solver for the linear system $A\mathbf{x} = \mathbf{b}$, the key idea of BiCGstab is expressing the residual and direction vectors at step m directly as polynomials in the matrix A:

$$\mathbf{r}^{(m)} = p_m(A)\mathbf{r}^{(0)}, \quad \mathbf{p}_m = q_m(A)\mathbf{r}^{(0)} \tag{2.52}$$

where $p_m(z)$, $q_m(z)$ are polynomials of degree m in z that assume value 1 for $z = 0$. Similarly,

$$\mathbf{r}_*^{(m)} = p_m(A^T)\mathbf{r}_*^{(0)}, \quad \mathbf{p}_m^* = q_m(A^T)\mathbf{r}_*^{(0)}. \tag{2.53}$$

Algorithm 2.11: Quasi–Minimal Residual (QMR)

Input: $A \in \mathbb{R}^{n\times n}$, $\mathbf{b} \in \mathbb{R}^n$, Maximum number of iterations $N_{\max}$, Initial Guess $\mathbf{x}^{(0)}$, Tolerance $\varepsilon > 0$.

Output: $\tilde{\mathbf{x}}$ candidate approximation.

`/* Version without lookahead */`

$\mathbf{r}^{(0)} \leftarrow \mathbf{b} - A\mathbf{x}^{(0)}$;

$\gamma_1 \leftarrow \|\mathbf{r}^{(0)}\|_2$, $\mathbf{w}_1 \leftarrow \mathbf{v}_1 \leftarrow \mathbf{r}^{(0)}/\gamma_1$;

for $j = 1, \ldots, N_{max}$ **do**

- $\alpha_j \leftarrow < A\mathbf{v}_j, \mathbf{w}_j >$; `/* Biorthogonalization procedure */`
- $\hat{\mathbf{v}}_{j+1} \leftarrow A\mathbf{v}_j - \alpha_j \mathbf{v}_j - \beta_j \mathbf{v}_{j-1}$;
- $\hat{\mathbf{w}}_{j+1} \leftarrow A^T\mathbf{w}_j - \alpha_j \mathbf{w}_j - \delta_j \mathbf{w}_{j-1}$;
- $\delta_{j+1} \leftarrow \sqrt{| < \hat{\mathbf{v}}_{j+1}, \hat{\mathbf{w}}_{j+1} > |}$;
- **if** $\delta_{j+1} = 0$ **then**
  - **Stop**; `/* Serious breakdown */`
- $\beta_{j+1} \leftarrow < \hat{\mathbf{v}}_{j+1}, \hat{\mathbf{w}}_{j+1} > / \delta_{j+1}$;
- $\mathbf{v}_{j+1} \leftarrow \hat{\mathbf{v}}_{j+1}/\beta_{j+1}$;
- $\mathbf{w}_{j+1} \leftarrow \hat{\mathbf{v}}_{j+1}/\delta_{j+1}$;
- $T_{j,j} \leftarrow \alpha_j$, $T_{j,j+1} \leftarrow \beta_j$, $T_{j+1,j} \leftarrow \delta_j$; `/* QR update */`
- **if** $j > 2$ **then**
  - $T_{j-2,j} \leftarrow s_{j-2} T_{j-1,j}$;
  - $T_{j-1,j} \leftarrow c_{j-2} T_{j-1,j}$;
- **if** $j > 1$ **then**
  - $T_{j-1,j} \leftarrow c_{j-1} T_{j-1,j} + s_{j-1} T_{j,j}$;
  - $T_{j,j} \leftarrow -s_{j-1} T_{j-1,j} - c_{j-1} T_{j,j}$;
- $c_j \leftarrow |T_{j,j}| / \sqrt{T_{j,j}^2 + T_{j+1,j}^2}$; `/* Use robust Givens rotation! */`
- $s_j \leftarrow c_j T_{j+1,j}/T_{j,j}$;
- $\gamma_{j+1} \leftarrow -s_j \gamma_j$;
- $\gamma_j \leftarrow c_j \gamma_j$;
- $\alpha_j \leftarrow c_j \alpha_j + s_j \delta_{j+1}$;
- $\mathbf{p}^{(j)} \leftarrow \left(\mathbf{v}_j - T_{j-2,j}\mathbf{p}^{(j-2)} - T_{j-1}\mathbf{p}^{(j-1)}\right)/T_{j,j}$;
- $\mathbf{x}^{(j)} \leftarrow \mathbf{x}^{(j-1)} + \gamma_j \mathbf{p}^{(j)}$;
- **if** $|\gamma_{j+1}| < \varepsilon$ **then**
  - **return** $\tilde{\mathbf{x}} = \mathbf{x}^{(j)}$;

From BiCG Algorithm 2.10, by using (2.52), (2.53) we obtain

$$\begin{aligned}\alpha_m &= \frac{< p_m(A)\mathbf{r}^{(0)}, p_m(A^T)\mathbf{r}_*^{(0)} >}{< Aq_m(A)\mathbf{r}^{(0)}, q_m(A^T)\mathbf{r}_*^{(0)} >} \\ &= \frac{< p_m^2(A)\mathbf{r}^{(0)}, \mathbf{r}_*^{(0)} >}{< Aq_m^2(A)\mathbf{r}^{(0)}, \mathbf{r}_*^{(0)} >}. \end{aligned} \tag{2.54}$$

The parameter β_m is derived similarly. Sonneveld [488] observed that, by using (2.54), we can directly build, at each step of BiCG, the vectors

$$\mathbf{r}^{(m)} = p_m^2(A)\mathbf{r}^{(0)}, \quad \mathbf{p}_m = Aq_m^2(A)\mathbf{r}^{(0)}. \tag{2.55}$$

In this way, we can skip the construction of $\mathbf{r}_*^{(m)}$, $\mathbf{p}_m^*$, where $p_m(z)$, $q_m(z)$ are the same polynomials used for (2.52) and (2.53). The recurrence relations for the polynomials (2.55) are

$$\begin{aligned} p_{m+1}(z) &= \ p_m(z) - \alpha_m z q_m(z), \\ q_{m+1}(z) &= \ p_m(z) - \beta_m q_m(z). \end{aligned} \tag{2.56}$$

If we write the vectors $\mathbf{r}^{(m)}$ and $\mathbf{p}_m$, $\mathbf{q}_m$ as

$$\mathbf{r}^{(m)} = p_m^2(A)\mathbf{r}^{(0)}, \quad \mathbf{p}_m = q_m^2(A)\mathbf{r}^{(0)}, \quad \mathbf{q}_m = p_{m+1}(A)q_m(A)\mathbf{r}^{(0)},$$

we can adapt the recurrence relations (2.56) in order to get $\mathbf{r}^{(m)}$ and $\mathbf{x}^{(m)}$ as in BiCG's Algorithm and thus we derive the so called *Conjugate Gradient Squared*, or CGS; see Algorithm 2.12 ([104, 455, 488] for more details). Here we will not spend much time on CGS because it is less robust than BiCGstab.

Van der Vorst in [527] observed that vector $\mathbf{r}^{(m)}$ can be expressed not only by squaring $p_m(z)$, but also with the product of $p_m(z)$ and $s_m(z)$, where the latter is a polynomial of degree m, called *stabilizing polynomial*, written as m linear factors in z, $g_{m,1}(z) = (1 - \omega_m z)$. Therefore,

$$\mathbf{r}^{(m)} = s_m(A)p_m(A)\mathbf{r}^{(0)},$$

and $s_m(z)$ can be defined recursively as

$$s_{m+1}(z) = (1 - \omega_m z)s_m(z) = g_{m,1}(z)s_m(z), \tag{2.57}$$

where ω_m is a scalar, determined at each iteration to minimize $\|r_{m+1}\|_2$.

Therefore,

$$\begin{aligned} s_{m+1}(z)p_{m+1}(z) &= (1 - \omega_m z)s_m(z)p_{m+1}(z) \\ &= (1 - \omega_m z)(s_m(z)p_m(z) - \alpha_m z s_m(z)q_m(z)), \\ s_m(z)q_m(z) &= s_m(z)(p_m(z) + \beta_{m-1}q_{m-1}(z)) \\ &= s_m(z)p_m(z) + \beta_{m-1}(1 - \omega_{m-1}z)s_{m-1}(z)q_{m-1}(z), \end{aligned}$$

Algorithm 2.12: Conjugate Gradient Squared Method (CGS)

Input: $A \in \mathbb{R}^{n\times n}$, $\mathbf{b} \in \mathbb{R}^n$, Maximum number of iterations $N_{\max}$, Initial Guess $\mathbf{x}^{(0)}$.

Output: $\tilde{\mathbf{x}}$ candidate approximation.

$\mathbf{r}^{(0)} \leftarrow \mathbf{b} - A\mathbf{x}^{(0)}$;

Choose an arbitrary $\mathbf{r}_*^{(0)}$ vector;

$\mathbf{p}_0 \leftarrow \mathbf{u}_0 \leftarrow \mathbf{r}^{(0)}$;

for $j = 1, \ldots, N_{max}$ **do**

- $\alpha_j \leftarrow < \mathbf{r}^{(j)}, \mathbf{r}_*^{(0)} > / < A\mathbf{p}^{(j)}, \mathbf{r}_*^{(0)} >$;
- $\mathbf{q}^{(j)} \leftarrow \mathbf{u}^{(j)} - \alpha_j A\mathbf{p}^{(j)}$;
- $\mathbf{x}^{(j+1)} \leftarrow \mathbf{x}^{(j)} + \alpha_j(\mathbf{u}^{(j)} + \mathbf{q}^{(j)})$;
- $\mathbf{r}^{(j+1)} \leftarrow \mathbf{r}^{(j)} - \alpha_j(\mathbf{u}^{(j)} + \mathbf{q}^{(j)})$;
- **if** <*Stopping criteria satisfied*> **then**
  - **return** $\tilde{\mathbf{x}} = \mathbf{x}^{(j+1)}$
- $\beta_j \leftarrow < \mathbf{r}^{(j+1)}, \mathbf{r}_*^{(0)} > / < \mathbf{r}^{(j)}, \mathbf{r}_*^{(0)} >$;
- $\mathbf{u}^{(j+1)} \leftarrow \mathbf{r}^{(j+1)} + \beta_j \mathbf{q}^{(j)}$;
- $\mathbf{p}_{j+1} \leftarrow \mathbf{u}^{(j+1)} + \beta_j(\mathbf{q}^{(j)} + \beta_j \mathbf{p}^{(j)})$;

$p_m(z)$ is the polynomial of the residual of BiCG Algorithm. Then, from (2.56), we find $p_{m+1}(z) = p_m(z) - \alpha_m z q_m(z)$. By writing the direction vector $\mathbf{p}_m$ as

$$\mathbf{p}_m = q_m(A)s_m(A)\mathbf{r}^{(0)},$$

we obtain the recurrence relations for $\mathbf{r}^{(m+1)}$ and $\mathbf{p}_{m+1}$:

$$\begin{aligned} \mathbf{r}^{(m+1)} &= (I - \omega_j A)(\mathbf{r}^{(m)} - \alpha_m A\mathbf{p}_m), \\ \mathbf{p}_{m+1} &= \mathbf{r}^{(m+1)} + \beta_m (I - \omega_m A)\mathbf{p}_m. \end{aligned}$$

The computation of the scalar parameter β_m in BiCG requires $\mathbf{r}_*^{(m)}$:

$$\beta_m = \frac{< \mathbf{r}^{(m+1)}, \mathbf{r}_*^{(m+1)} >}{< \mathbf{r}^{(m)}, \mathbf{r}_*^{(m)} >},$$

while in BiCGstab the previous expression is obtained by using

$$\rho_m =< \mathbf{r}^{(m)}, \mathbf{r}_*^{(0)} >=< p_m(A)\mathbf{r}^{(0)}, s_m(A^T)\mathbf{r}_*^{(0)} >=< s_m(A)p_m(A)\mathbf{r}^{(0)}, \mathbf{r}_*^{(0)} >,$$

$$\frac{\rho_{m+1}}{\rho_m} = \frac{\omega_m < \mathbf{r}^{(m+1)}, \mathbf{r}_*^{(m+1)} >}{\alpha_m < \mathbf{r}^{(m)}, \mathbf{r}_*^{(m)} >} \Rightarrow \beta_m = \frac{\alpha_m}{\omega_m}\frac{\rho_{m+1}}{\rho_m}. \tag{2.58}$$

The (2.58) is obtained by considering that $\mathbf{r}^{(m)} = p_m(A)\mathbf{r}^{(0)}$ is orthogonal to all vectors of the form $(A^T)^i\mathbf{r}_*^{(0)}$ for $i < m$. From the previous relations we

Algorithm 2.13: BiCGstab method

Input: $A \in \mathbb{R}^{n\times n}$, $\mathbf{b} \in \mathbb{R}^n$, Maximum number of iterations $N_{\max}$, Initial Guess $\mathbf{x}^{(0)}$.

Output: $\tilde{\mathbf{x}}$ candidate approximation.

$\mathbf{r}^{(0)} \leftarrow \mathbf{b} - A\mathbf{x}^{(0)}$;
choose $\hat{\mathbf{r}}^{(0)}$ such that $< \mathbf{r}^{(0)}, \hat{\mathbf{r}}^{(0)} > \neq 0$;
$\mathbf{p}_0 \leftarrow \mathbf{r}^{(0)}$;
for $j = 1, \ldots, N_{max}$ **do**
    $\alpha_j \leftarrow < \mathbf{r}^{(j)}, \hat{\mathbf{r}}^{(0)} > / < A\mathbf{p}_j, \hat{\mathbf{r}}^{(0)} >$;
    $\mathbf{s}_j \leftarrow \mathbf{r}^{(j)} - \alpha_j A\mathbf{p}_j$;
    if <*Stopping criteria satisfied*> **then**
        $\mathbf{x}^{(j+1)} \leftarrow \mathbf{x}^{(j)} + \alpha_j \mathbf{p}_j$;
        return $\tilde{\mathbf{x}} = \mathbf{x}^{(j+1)}$
    $\omega_j \leftarrow < A\mathbf{s}_j, \mathbf{s}_j > / < A\mathbf{s}_j, A\mathbf{s}_j >$;
    $\mathbf{x}^{(j+1)} \leftarrow \mathbf{x}^{(j)} + \alpha_j \mathbf{p}_j + \omega_j \mathbf{s}_j$;
    $\mathbf{r}^{(j+1)} \leftarrow \mathbf{s}_j - \omega_j A\mathbf{s}_j$;
    if <*Stopping criteria satisfied*> **then**
        return $\tilde{\mathbf{x}} = \mathbf{x}^{(j+1)}$
    $\beta_j \leftarrow \frac{\alpha_j < \mathbf{r}^{(j+1)}, \hat{\mathbf{r}}^{(0)} >}{\omega_j < \mathbf{r}^{(j)}, \hat{\mathbf{r}}^{(0)} >}$;
    $\mathbf{p}_{j+1} \leftarrow \mathbf{r}^{(j+1)} + \beta_j (\mathbf{p}_j - \omega_j A\mathbf{p}_j)$;

compute:

$$\begin{aligned} < \mathbf{r}^{(m)}, \mathbf{r}_*^{(m)} > &= < p_m(A)\mathbf{r}^{(0)}, p_m(A^T)\mathbf{r}_*^{(0)} >= \\ &= < p_m(A)\mathbf{r}^{(0)}, p_m(A^T)\mathbf{r}_*^{(0)} >= \\ &= < p_m(A)\mathbf{r}^{(0)}, \frac{\delta_m^0}{\tau_j^0} s_m(A^T)\mathbf{r}_*^{(0)} >= \\ &= \frac{\delta_m^0}{\tau_m^0}\rho_m, \end{aligned}$$

where

$$p_m(z) = \sum_{i=0}^{m} \tau_{m-i}^{(m)} z^i, \quad s_m(z) = \sum_{i=0}^{m} \delta_{m-i}^{(m)} z^i.$$

From (2.57) we find

$$\delta_{m+1}^0 = -\omega_j \delta_m^0, \quad \tau_{m+1}^0 = \alpha_j \tau_0^m.$$

Finally,

$$\beta_m = \frac{\alpha_m}{\omega_m}\frac{\rho_{m+1}}{\rho_m} = \frac{\alpha_m}{\omega_m}\frac{(\mathbf{r}^{(m+1)}, \mathbf{r}_*^{(0)})}{(\mathbf{r}^{(m)}, \mathbf{r}_*^{(0)})}.$$

From BiCG Algorithm we have

$$
\begin{aligned}
\alpha_m &= \frac{< \mathbf{r}^{(m)}, \mathbf{r}_*^{(m)} >}{< A\mathbf{p}_m, \mathbf{p}_m^* >} \\
&= \frac{< p_m(A)\mathbf{r}^{(0)}, p_m(A^T)\mathbf{r}_*^{(0)} >}{< Aq_m(A)\mathbf{r}^{(0)}, q_m(A^T)\mathbf{r}_*^{(0)} >} \\
&= \frac{< p_m(A)\mathbf{r}^{(0)}, s_m(A^T)\mathbf{r}_*^{(0)} >}{< Aq_m(A)\mathbf{r}^{(0)}, s_m(A^T)\mathbf{r}_*^{(0)} >} \\
&= \frac{< s_m(A)p_m(A)\mathbf{r}^{(0)}, \mathbf{r}_*^{(0)} >}{< As_m(A)q_m(A)\mathbf{r}^{(0)}, \mathbf{r}_*^{(0)} >} \\
&= \frac{< \mathbf{r}^{(m)}, \mathbf{r}_*^{(0)} >}{< A\,\mathbf{p}_m, \mathbf{r}_*^{(0)} >} \\
&= \frac{\rho_m}{< A\mathbf{p}_m, \mathbf{r}_*^{(0)} >}.
\end{aligned} \tag{2.59}
$$

By setting

$$q_m(z) = \sum_{i=0}^{m} \gamma_{m-i}^{(m)} z^i,$$

we derive the second line in (2.59) by observing that $\tau_m^0 \equiv \gamma_m^0$ and considering the orthogonality of $p_m(A)\mathbf{r}^{(0)}$ and $(A^T)^i\mathbf{r}_*^{(0)}$, $i < m$. The stabilization scalar ω_m is chosen to minimize the residual norm of

$$\mathbf{r}^{(m+1)} = (I - \omega_m A)s_m(A)p_{m+1}(A)\mathbf{r}^{(0)}.$$

By writing

$$\mathbf{s}_m = \mathbf{r}^{(m)} - \alpha_m A\mathbf{p}_m,$$

we have

$$\mathbf{r}^{(m+1)} = (I - \omega_m A)\mathbf{s}_m. \tag{2.60}$$

Therefore, minimizing the norm of (2.60), we find

$$\omega_m = \frac{< A\mathbf{s}_m, \mathbf{s}_m >}{< A\mathbf{s}_m, A\mathbf{s}_m >}.$$

Moreover, the update of the candidate approximation $\mathbf{x}^{(m+1)}$ can be computed from $\mathbf{r}^{(m+1)}$ as

$$
\begin{aligned}
\mathbf{r}^{(m+1)} &= \mathbf{s}_m - \omega_m A\mathbf{s}_m = \mathbf{r}^{(m)} - \alpha_m A\mathbf{p}_m - \omega_m A\mathbf{s}_m \\
&= \mathbf{r}^{(m)} - \alpha_m A\mathbf{p}_m - \omega_m A\mathbf{s}_m.
\end{aligned}
$$

By substituting $\mathbf{r}^{(m)} = \mathbf{b} - A\mathbf{x}^{(m)}$ in the previous expression we finally get

$$
\begin{aligned}
-A\left(\mathbf{x}^{(m+1)} - \mathbf{x}^{(m)}\right) = \mathbf{r}^{(m+1)} - \mathbf{r}^{(m)} &= -\alpha_m A\mathbf{p}_m - \omega_m A\mathbf{s}_m, \\
\Rightarrow \quad \mathbf{x}^{(m+1)} &= \mathbf{x}^{(m)} + \alpha_m \mathbf{p}_m + \omega_m \mathbf{s}_m.
\end{aligned}
$$

Note that, if ω_m is relatively small or zero, we can have instability. In this case it is better to stop the algorithm because the update of the vectors is not reliable. Otherwise, if we continue, we would get a stagnating convergence; see [527]. A similar behavior can be observed if se $| < \mathbf{r}^{(m)}, \mathbf{r}_*^{(0)} > |$ is very small or zero. In this case it is much better to start again by providing a new $\mathbf{r}_*^{(0)}$ such that $< \mathbf{r}^{(0)}, \mathbf{r}_*^{(0)} > \neq 0$.

2.3.3.1 BiCGstab(l)

We have seen that BiCGstab can break down also for nonsingular matrices. Fortunately, there are some variants that can help.

- BiCGstab(l): choose a polynomial of degree l greater than one, $g_{m,l}(z)$, in (2.57);
- BiCGstab(l)^{OR}: choose a polynomial $g_{m,l}(z)$ of degree $l \geq 1$ in order to determine implicitly a more stable basis for $\mathcal{K}_m(A^T, \mathbf{r}_*^{(0)})$.

Here we discuss briefly only the first attempt. The second often gives less impressive improvements to the convergence; see [484, 486]. Moreover, in our experience, a more effective convergence improvement in difficult problems is much better faced with *preconditioning*; see next chapters.

Let us recall the update of the residual vectors in BiCGstab: at step m, we have $\mathbf{r}^{(m+1)}$ from $\mathbf{r}^{(m)}$ by the linear term $(I - \omega_m A)$:

$$\mathbf{r}^{(m+1)} = (I - \omega_m A) s_m(A) p_{m+1}(A) \mathbf{r}^{(0)}. \tag{2.61}$$

The matrices we are interested in here are not Hermitian, thus their eigenvalues can be complex. The polynomials that minimize the 2 norm of (2.61) can have complex roots; see Van der Vorst [528]. Unfortunately, the polynomial in (2.61) can only have real roots because it is a product of linear (real) factors. We can overcome this issue by increasing the degree of the polynomial on the right–hand side of (2.61). By providing the polynomial in A for $\mathbf{r}^{(m+1)}$ (2.61) as a product of quadratic factors, we obtain BiCGstab(2), Algorithm 2.14. At each iteration, it provides two BiCG steps instead of one. As in [486, 528], in order to simplify the notation, the indexes of the vectors are not displayed in BiCGstab(2)'s Algorithm 2.14.

Note that one of the two BiCG' steps is not used to build the approximations at the current step, but only to build the second order polynomial $g_{m,2}(z)$ in the update for $\mathbf{r}^{(m+1)}$. Therefore, in BiCGstab(2), $\mathbf{r}^{(m+1)}$ is updated from $\mathbf{r}^{(m)}$ as follows:

$$\mathbf{r}^{(m+1)} = s_{m,2}(M) p_{m,2}(M) \mathbf{r}^{(0)},$$

where $s_{m,2}(z)$, $p_{m,2}(z)$ are polynomials in z given in the form of the product of quadratic factors.

Similarly to BiCGstab, BiCGstab(2) become unstable when at least one of the scalars ω_1, ω_2 or $|(\mathbf{r}^{(m)}, \mathbf{r}_*^{(0)})|$ is numerically zero.

Algorithm 2.14: BiCGstab(2)

Input: $A \in \mathbb{R}^{n \times n}$, $\mathbf{b} \in \mathbb{R}^n$, Maximum number of iterations $N_{\max}$,
Initial Guess $\mathbf{x}^{(0)}$.

Output: $\tilde{\mathbf{x}}$ candidate approximation.

$\mathbf{r}^{(0)} = \mathbf{b} - A, \mathbf{x}^{(0)}$
choose $\mathbf{r}_*^{(0)}$ such that $(\mathbf{r}^{(0)}, \mathbf{r}_*^{(0)}) \neq 0$
$\mathbf{u}_0 = 0$, $\rho_0 = 1$, $\alpha = 0$, $\omega_2 = 1$
for $j = 1 : N_{max}$ **do**
  $\rho_1 = (\mathbf{r}, \mathbf{r}_*^{(0)})$;
  $\beta = \dfrac{\alpha \rho_1}{-\omega_2 \rho_0}$;
  $\rho_0 = \rho_1$;
  $\mathbf{u} = \mathbf{r} - \beta \mathbf{u}$;
  $\mathbf{v} = A\mathbf{u}$;
  $\gamma = (\mathbf{v}, \mathbf{r}_*^{(0)})$;
  $\alpha = \rho_0 / \gamma$;
  $\mathbf{r} = \mathbf{r} - \alpha \mathbf{v}$;
  $\mathbf{x} = \mathbf{x} + \alpha \mathbf{u}$;
  if <*Stopping criteria satisfied*> **then**
    $\mathbf{x} = \mathbf{x} + \alpha \mathbf{u}$;
    return $\tilde{\mathbf{x}} = \mathbf{x}$
  $\mathbf{s} = A\mathbf{r}$;
  $\rho_1 = (\mathbf{s}, \mathbf{r}_*^{(0)})$;
  $\beta = \alpha \rho_1 / \rho_0$;
  $\rho_0 = \rho_1$;
  $\mathbf{v} = \mathbf{s} - \beta \mathbf{v}$;
  $\mathbf{w} = A\mathbf{v}$;
  $\gamma = (\mathbf{w}, \mathbf{r}_*^{(0)})$;
  $\alpha = \rho_0 / \gamma$;
  $\mathbf{u} = \mathbf{r} - \beta \mathbf{u}$;
  $\mathbf{r} = \mathbf{r} - \alpha \mathbf{v}$;
  $\mathbf{s} = \mathbf{s} - \alpha \mathbf{w}$;
  $\mathbf{t} = A\mathbf{s}$;
  $\omega_1 = (\mathbf{r}, \mathbf{s})$, $\mu = (\mathbf{s}, \mathbf{s})$, $\nu = (\mathbf{s}, \mathbf{t})$;
  $\omega_2 = (\mathbf{r}, \mathbf{t})$, $\tau = (\mathbf{t}, \mathbf{t})$;
  $\tau = \tau - \nu^2/\mu$, $\omega_2 = (\omega_2 - \nu\omega_1/\mu)/\tau$;
  $\omega_1 = (\omega_1 - \nu\omega_2)/\mu$;
  $\mathbf{x} = \mathbf{x} + \omega_1 \mathbf{r} + \omega_2 \mathbf{s} + \alpha \mathbf{u}$;
  $\mathbf{r} = \mathbf{r} - \omega_1 \mathbf{s} - \omega_2 \mathbf{t}$;
  if <*Stopping criteria satisfied*> **then**
    return $\tilde{\mathbf{x}} = \mathbf{x}$
  $\mathbf{u} = \mathbf{u} - \omega_1 \mathbf{v} - \omega_2 \mathbf{w}$;

The use of polynomials made of factors of degree 2 can be easily generalized giving BiCGstab(l), where, at each iteration, one performs l steps of BiCG; see [485, 486] for more details. However, we experienced that $l = 2$ in BiCGstab(l) usually is enough.

Computational complexity

For one iteration of BiCGstab we need 2 matrix-vector products, 4 scalar products and 6 updates of vectors. On the other hand, one iteration of BiCGstab(2) requires slightly less than approximately twice the above. On the other hand, often (but not always, see the numerical example 2.3) BiCGstab and BiCGstab(2) require more or less the same number of matrix-vector products to converge to a prescribed tolerance.

2.3.4 The convergence of BiCG, CGS, BiCGstab and BiCGstab(2)

We cannot give a convergence analysis for BiCG, CGS, BiCGstab and BiCGstab(l). Also the useful relations between the residuals we gave for QMR cannot be provided for the former methods because, differently to CG, for the algorithms derived from BiCG, nothing was minimized to get the candidate approximations and therefore no optimality relations hold; see also [280]. Some hints can be given. In particular, if:

- the matrix A of the underlying linear system is not very ill–conditioned;
- A can be diagonalized with a matrix of eigenvectors not too ill-conditioned and the eigenvalues are clustered far away from the origin,

then BiCGstab often requires approximately the same order of matrix-vector products of BiCG and BiCGstab(2) to converge to a prescribed tolerance. Otherwise, the convergence behavior of the iterative methods mentioned can be very different. It is interesting to note that BiCGstab(l), for $l > 1$, is often more robust than BiCGstab.

2.4 Which Krylov subspace method should I use?

After this brief review of the main classical and Krylov iterative methods for linear systems, with the aid of some simple tests, we can give some suggestions and remarks on the various methods and point the reader to the bibliography for other schemes not discussed here. Note that there is another Krylov method, *Flexible GMRES* or FGMRES, that is discussed in the next chapter because its use makes sense with (variable) preconditioning.

For generic real and symmetric or Hermitian definite (positive or negative) problems, we should use CG; see also the comparison with *Steepest Descent* in the next numerical test. For generic Hermitian problems (and therefore also, e.g., symmetric and indefinite) we should use MinRes, a method developed by Paige and Saunders in 1975 [416] not described here without preconditioning because it is substantially a special case of GMRES if A is Hermitian (or symmetric if it has real entries). In the chapters devoted to preconditioning we report a preconditioned version of MinRes; see Algorithm 3.2. For a more recent discussion of MinRes, see [280]. Both methods have a nice convergence theory, satisfy a three-term recurrence relation, and therefore we just need the *preconditioning* techniques of the next chapter to end the discussion. We should stress again that *CG* minimizes in exact arithmetic the A–norm of the error, while MinRes/GMRES minimize the 2-norm of the residual, a remarkable difference because residual and error are not the same thing; see also the discussion at the beginning of this chapter.

For nonsymmetric–real and non-Hermitian linear systems the choice is more delicate and depends on the problem. If the matrix has no special properties to enable fast matrix-vector multiplication then GMRES (not restarted) will be fine if it converges in few iterations, as is the case of matrices with clustered spectrum. Otherwise, if many iterations are required to GMRES, then use GMRES(m) or BiCGstab. For certain problems, BiCGstab(l) with $l = 2$ can be even faster; see also § 2.3.3.1. Sometimes QMR and TFQMR (Transpose-Free Quasi Minimal Residual, see [247]) have interesting performances but BiCGstab and BiCGstab(l) are often not worse, see also the tests below.

BiCG, CGS and CGNE/CGNR (CGNE: CG for the normal equations of $A\mathbf{x} = \mathbf{b}$, i.e., for $AA^H\mathbf{y} = \mathbf{b}$, $\mathbf{x} = A^H\mathbf{y}$ while CGNR is CG applied to $A^HA\mathbf{x} = A^H\mathbf{b}$) are in general not much used because of their more erratic convergence behavior with respect to the other methods mentioned above. Of course, it should be noted that if we use CGNE or CGNR, then we can apply the convergence analysis developed for CG to the normal equations and this can make the difference in some contexts. For more information on CGNE, CGNR and other schemes for the normal equations, see [455, Chapter 8].

Remark 2.5. In the following tests and related convergence plots, we do not report iterations BiCG, CGS, BiCGstab, BiCGstab(l) because

- it is not fair to compare iterations when the underlying algorithm could include multiple steps of some basic procedure as is the case of CGS, BiCGstab, BiCGstab(l) including one (CGS and BiCGstab) or more (BiCGstab(l), $l > 1$) BiCG steps, where one step of BiCG needs two matrix-vector products (just one is required for a CG, GMRES or QMR step).

- At each iteration of any Krylov method, the main computational load is given by the matrix-vector products, in general.

Table 2.3: Summary of Krylov subspace methods

Method	A	Type	Recurrence
CG	symmetric definite	optimal	three–terms
MinRes/SYMMLQ	symmetric indefinite	optimal	three–terms
Symmetric QMR	symmetric indefinite	non–optimal	three–terms
GMRES	general	optimal	full
GMRES(m)	general	non–optimal	m–terms
QMR/BiCGstab/TFQMR	general	non–optimal	three–terms

Example 2.2 (CG and one–dimensional projection methods). Let us consider again the problem in Example 2.1, i.e., the second order centered finite difference uniform discretization for the Laplace equation on a square domain. See Table 2.4 and Figure 2.2 for a comparison of the number of matrix–vector products and the timing needed for achieving a tolerance on the residual of $\varepsilon = 10^{-6}$ ✓

Table 2.4: Example 2.2. Comparison in terms of matrix vector products of *CG* and *Steepest Descent* for the finite difference discretization of the Laplace equation

	CG			Steepest Descent		
N	IT	$\|r\|_2$	T(s)	IT	$\|r\|_2$	T(s)
64	10	1.72e-15	6.65e-04	227	9.54e-07	2.11e-03
169	21	5.48e-07	7.30e-04	637	9.61e-07	3.33e-03
324	28	8.89e-07	9.19e-04	†	1.35e-05	6.61e-03
529	37	7.88e-07	9.73e-04	†	4.01e-03	7.92e-03
784	45	6.28e-07	1.60e-03	†	7.26e-02	1.02e-02

Example 2.3 (Non–Hermitian matrices). Let us consider one of the most popular test problems: the steady state convection–diffusion equation, that is

$$\begin{cases} -K\nabla^2 u + \nu\nabla \cdot u = f, & \mathbf{x} \in \Omega, \\ u \equiv g, & \mathbf{x} \in \partial\Omega. \end{cases}$$

We consider the sequence of linear systems

$$A_h \mathbf{u}_h = \mathbf{f}_h,$$

generated by the second order centered finite differences discretization on a square domain $\Omega = [0,1]^2$ with an equispaced mesh of $n+1$ knots, step h in each direction and homogeneous Dirichlet boundary conditions, i.e., $g \equiv 0$; see, e.g., [351]. The latter produces a sequence of linear systems $\{A_n\mathbf{x}_n = \mathbf{b}_n\}_n$ parametrized by the mesh parameter h. Regarding the coefficients $K, \nu \in \mathbb{R}^+$,

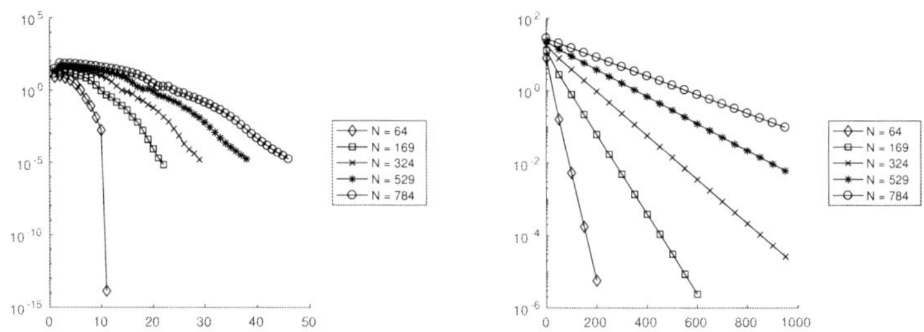

Figure 2.2: Example 2.2. Convergence history for the CG (left) and Steepest Descent method (right), second order centered finite difference discretization of the Laplace equation.

we consider the case in which $\nu > K$, that is, the case in which the sequence has a skew symmetric part $(A - A^T)/2$, given here by the discretization of $\nabla \cdot u$, dominant over the symmetric part $(A + A^T)/2$, given here by the discretization of $-\nabla^2 u$. In Figure 2.3 we compare the number of matrix–vector products versus the relative residuals, for BiCGstab, GMRES and GMRES(20). All the methods are set to achieve a tolerance on the relative residual of $\varepsilon = 10^{-6}$. In Figure 2.4 we compare the number of matrix–vector products versus

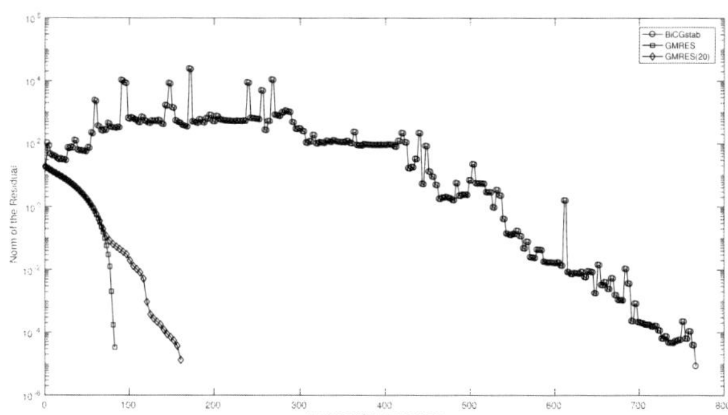

Figure 2.3: Example 2.3. Convergence history for BiCGstab, GMRES and GMRES(20) for a finite difference discretization of the steady state convection–diffusion equation with $\nu = 2$ and $K = 0.5$.

the relative residuals for BiCG, QMR, BiCGstab and BiCGstab(2). All the methods are set to achieve a tolerance on the residual of $\varepsilon = 10^{-6}$. Finally, we compare the number of matrix–vectors for all methods in Table 2.3 for increasing values of the grid parameter n. Note that the size of the underlying linear systems grows as n^2. ✓

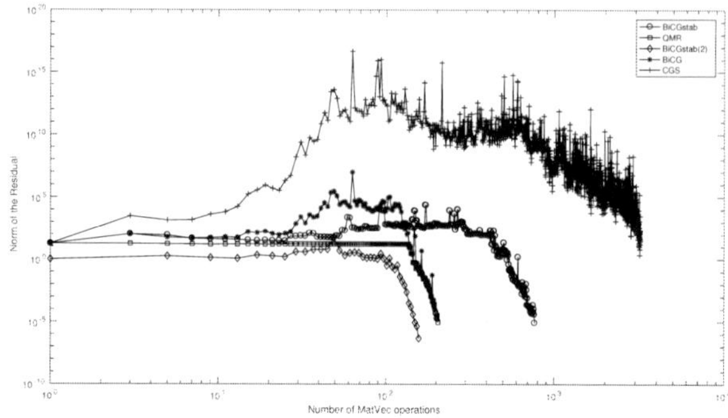

Figure 2.4: Example 2.3. Convergence history for BiCG, QMR, BiCGstab and BiCGstab(2) solving a finite difference discretization of the steady state convection–diffusion equation with $\nu = 2$ and $K = 0.5$.

Table 2.5: Example 2.3. Number of matrix–vector (MatVec) operations for the same discretization of the steady state convection–diffusion equation with $\nu = 2$ and $K = 0.5$

	BiCGstab		BiCGstab(2)		BiCG	
n	MatVec	T(s)	MatVec	T(s)	MatVec	T(s)
10	92	0.0467	44	0.0124	66	0.0042
50	996	0.1025	200	0.0637	300	0.0412
150	426	0.2730	620	0.9267	15294	12.4696
200	6450	6.5278	820	1.8090	80002	110.4035

	QMR		GMRES		GMRES(20)	
n	MatVec	T(s)	MatVec	T(s)	MatVec	T(s)
10	33	0.0042	28	0.0084	34	0.0081
50	158	0.0590	103	0.2402	171	0.1047
150	22501	51.5922	300	21.7015	403	2.1637
200	40001	158.8635	400	103.4093	507	4.6597

	CGS	
n	MatVec	T(s)
10	40	0.0179
50	5002	0.7749
150	45002	41.3386
200	80002	124.738

Chapter 3

General purpose preconditioning strategies

The iterative solvers discussed in the previous chapter can be effective for some problems. However, lack of robustness and sometimes erratic convergence behavior are recognized weakness of iterative solvers. These issues hamper the acceptance of iterative methods despite their intrinsic appeal for very large linear systems. Both the efficiency and robustness of iterative techniques can be very much improved by using *preconditioning*. Preconditioning is simply a means of transforming the original linear system into one equivalent, i.e., with the same solution, but which is faster and cheaper to solve with an appropriate iterative method. Quoting Saad in [455], "In general, the reliability of iterative techniques, when dealing with various applications, depends much more on the quality of the preconditioner than on the particular Krylov subspace accelerators used".

We cover some general purpose preconditioners in this chapter and a few others for systems with special structures in the next.

Here we first discuss the preconditioned versions of the Krylov subspace algorithms already seen, using a generic preconditioner, and FGMRES, the most popular example of an iterative method tailored for using another iterative method as a preconditioner.

3.1 Generalities on preconditioning

Let us consider the linear algebraic system

$$A\mathbf{x} = \mathbf{b}, \qquad \mathbf{x}, \mathbf{b} \in \mathbb{R}^n, \ A \in \mathbb{R}^{n\times n}. \tag{3.1}$$

We would like to speed up the convergence of the iterative methods we considered in the previous chapter, i.e., roughly speaking, getting the desired approximation in less time. To this end, we look for a mathematically (but not computationally!) equivalent[1] linear system with more favorable spectral properties for its matrix $\tilde{A}$:

$$\tilde{A}\mathbf{x} = \tilde{\mathbf{b}}, \qquad \mathbf{x}, \tilde{\mathbf{b}} \in \mathbb{R}^n, \ \tilde{A} \in \mathbb{R}^{n\times n}. \tag{3.2}$$

A similar argument can be made for systems with complex-valued entries.

[1] Mathematically equivalent: with the same solution in exact arithmetic.

Let us consider an invertible matrix M with properties detailed later. There are three basic equivalent systems we can consider. We could perform a *left preconditioning*:

$$M^{-1}A\mathbf{x} = M^{-1}\mathbf{b}, \quad \tilde{A} = M^{-1}A; \tag{3.3}$$

a *right preconditioning*:

$$\begin{cases} AM^{-1}\mathbf{u} = \mathbf{b}, & \tilde{A} = AM^{-1}, \\ \mathbf{x} = M^{-1}\mathbf{u}. \end{cases} \tag{3.4}$$

or a *split preconditioning* scheme, for which we seek the matrix M in a factored form $M = M_1M_2$, supposed well defined:

$$\begin{cases} M_1^{-1}AM_2^{-1}\mathbf{u} = M_1^{-1}\mathbf{b}, & \tilde{A} = M_1^{-1}AM_2^{-1}, \\ \mathbf{x} = M_2^{-1}\mathbf{u}. \end{cases} . \tag{3.5}$$

However what kind of transformation we are looking for? Potentially, M should be invertible, approximating A in some way and, if A is sparse, it would be nice to have a sparse M as well in order to reduce the computational cost of the *matrix-vector* products. Our selection would be fine if we achieve a faster and cheaper convergence to the desired accuracy for the iterative solver used for the equivalent linear system with respect to what happens applying it directly to (3.1). The first problem here is that these basic requirements are mutually contradictory, so let's start looking at them singularly in search of some reasonable compromise.

The first thing that might come to mind could be trying $M = A$. in this way the product $M^{-1}A$ would be the simple $A^{-1}A = I$ and then we could obtain the solution in zero iterations. This trivial approach has two flaws: (a) the time needed for computing the inverse of A, that is usually (much) more expensive than solving (3.1) with, say, the popular sparse Gaussian–elimination (see, e.g., [271]) and (b) the inverse of a sparse matrix is dense in general.

Theorem 3.1. *The inverse of a sparse matrix can be dense.*

Proof. The proof is long and a little bit technical. It uses the graph theory. See [260] for details. □

Clearly we have to account the time for computing our transformation and the time needed for the application to the iterative solver of our preconditioner M^{-1}. We stress that the application **is not intended as computing the matrix-matrix product**, too expensive to be taken into account, but is intended as computing the effect of the application of that product on a generic column vector. Summarizing the above, we can express the time required to calculate the solution T_{slv} of the linear system $A\mathbf{x} = \mathbf{b}$ with a preconditioned iterative method as:

$$T_{\text{slv}} = T_{\text{setup}} + N_{\text{it}} \times T_{\text{it}},$$

where T_{setup} is the time for computing our transformation, N_{it} is the number of iterations of the iterative solver needed to obtain the solution within the correct tolerance and T_{it} is the time needed for each of the iterations.

Now we can express the first issue of our demands; we have to find a balance between the request of having M a good approximation for A, i.e., roughly speaking $||M^{-1}A - I|| \approx 0$ or $||M - A|| \approx 0$ in some norm, and the T_{setup}, setup time for (often implicitly, i.e., we do not form M) building the preconditioner M and the T_{it}[2], time needed for the iterations, otherwise we are simply wasting our computational resources.

Definition 3.1 ([269, 523] – Cluster (proper)). A sequence of matrices $\{A_n\}_{n\geq 0}$, $A_n \in \mathbb{C}^{n\times n}$, has a **proper cluster** of eigenvalues in $p \in \mathbb{C}$ if, $\forall \varepsilon > 0$, the number of eigenvalues of A_n **not in** the ε–neighborhood $D(p,\varepsilon) = \{z \in \mathbb{C} \mid |z-p| < \varepsilon\}$ is bounded by a constant r that does not depend on n. Eigenvalues not in the *proper cluster* are called **outliers** or **outlying eigenvalues**. Furthermore, $\{A_n\}_n$ is **properly clustered** at a nonempty closed set $S \subseteq \mathbb{C}$ if for any $\varepsilon > 0$

$$q_\varepsilon(n,S) = \sharp\left\{\lambda_j(A_n)\ :\ \lambda_j \notin D(S,\varepsilon) = \bigcup_{p\in S} D(p,\varepsilon)\right\} = O(1),\ n \to +\infty,$$

in which $D(S,\varepsilon)$ is now the ε–neighborhood of the set S, and we have denoted by $q_\varepsilon(n,S)$ the cardinality of the set of the outliers.

Remark 3.1. Some requirements in Definition 3.1 can be relaxed. To ensure the *superlinear convergence* of a Krylov method for symmetric matrices (see § 2.2.4) the cluster should be proper. However, fast convergence can also be obtained if the number of eigenvalues *not* in the cluster is bounded by a function of the size that is an $o(n)$ for $n \to +\infty$. Generalizations for dealing with the non Hermitian cases are also available but the role of the eigenvalues in this case is not anymore so crucial to determine convergence properties; see § 2.2.8.

3.2 Krylov iterative methods for preconditioned iterations

Assume that a preconditioner M that approximates A in a way that will be explained from the next section is available. To apply M efficiently in order to perform left, right and split preconditioning as in (3.3), (3.4) or (3.5) and in a way to reduce as much as possible the rounding errors, we need to modify slightly the Krylov iterative algorithms discussed in Chapter 2.

[2]The T_{it} increase as a consequence of the loss of sparsity of the product $M^{-1}A$.

3.2.1 Preconditioned CG and MinRes method

Let us start from the description of the *preconditioned Conjugate Gradient method (PCG)*. Since we are dealing with a matrix A that is symmetric and definite, we should preserve this for the preconditioned system (3.2). Assume that also M is symmetric and positive definite. In this case, we can form the *Cholesky factorization* (see, e.g., [270]) for M with lower triangular factor L as $M = LL^T$ and apply the *split* preconditioning from (3.5). However, to preserve symmetry we can do this implicitly. Observe that $M^{-1}A$ is self–adjoint for both the $< \cdot, \cdot >_M$ and $< \cdot, \cdot >_A$ inner products:

$$\begin{aligned} < M^{-1}A\mathbf{x}, \mathbf{y} >_M =& < A\mathbf{x}, \mathbf{y} > = < \mathbf{x}, A\mathbf{y} > = < \mathbf{x}, MM^{-1}A\mathbf{y} > \\ =& < \mathbf{x}, M^{-1}A\mathbf{y} >_M, \\ < M^{-1}A\mathbf{x}, \mathbf{y} >_A =& < AM^{-1}A\mathbf{x}, \mathbf{y} > = < \mathbf{x}, AM^{-1}A\mathbf{y} > \\ =& < \mathbf{x}, M^{-1}A\mathbf{y} >_A . \end{aligned} \tag{3.6}$$

Therefore, we can modify the original CG from Algorithm 2.4, by substituting the usual Euclidean product with the M–inner product and observing that this has not to be computed explicitly, since

$$\begin{aligned} < \mathbf{z}^{(j)}, \mathbf{z}^{(j)} >_M =& < M^{-1}\mathbf{r}^{(j)}, M^{-1}\mathbf{r}^{(j)} >_M = < \mathbf{r}^{(j)}, \mathbf{z}^{(j)} >, \\ < M^{-1}A\mathbf{p}^{(j)}, \mathbf{p}^{(j)} >_M =& < A\mathbf{p}^{(j)}, \mathbf{p}^{(j)} > . \end{aligned}$$

Therefore the substitution of the inner product can be implemented as summarized in Algorithm 3.1. Similarly one can observe that for the right preconditioning (3.4) the matrix AM^{-1} is not Hermitian with either the Euclidean or the M–inner product. On the other hand, we can retain symmetry (Hermitianity if A and M have complex entries) with respect to the M^{-1}–inner

---

Algorithm 3.1: Preconditioned Conjugate Gradient Method

Input: $A \in \mathbb{R}^{n\times n}$ SPD, Maximum number of iterations N_{max}, Initial Guess $\mathbf{x}^{(0)}$, $M \in \mathbb{R}^{n\times n}$ SPD preconditioner

Output: $\tilde{\mathbf{x}}$, candidate approximation.

$\mathbf{r}^{(0)} \leftarrow \mathbf{b} - A\mathbf{x}^{(0)}$, $\mathbf{z}^{(0)} \leftarrow M^{-1}\mathbf{r}^{(0)}$, $\mathbf{p}^{(0)} \leftarrow \mathbf{z}^{(0)}$;

for $j = 0, \ldots, N_{max}$ **do**

- $\alpha_j \leftarrow {<\mathbf{r}^{(j)},\mathbf{z}^{(j)}>}/{A\mathbf{p}^{(j)},\mathbf{p}^{(j)}}$;
- $\mathbf{x}^{(j+1)} \leftarrow \mathbf{x}^{(j)} + \alpha_j \mathbf{p}^{(j)}$;
- $\mathbf{r}^{(j+1)} \leftarrow \mathbf{r}^{(j)} - \alpha_j A\mathbf{p}^{(j)}$;
- **if** *<Stopping criteria satisfied>* **then**
  - **Return:** $\tilde{\mathbf{x}} = \mathbf{x}^{(j+1)}$;
- $\mathbf{z}^{(j+1)} \leftarrow M^{-1}\mathbf{r}^{(j+1)}$;
- $\beta_j \leftarrow {<\mathbf{r}^{(j+1)},\mathbf{z}^{(j+1)}>}/{<\mathbf{r}^{(j)},\mathbf{z}^{(j)}>}$;
- $\mathbf{p}^{(j+1)} \leftarrow \mathbf{z}^{(j+1)} + \beta_j \mathbf{p}^{(j)}$;

---

product

product. Let us rewrite Algorithm 3.1 by introducing two sequences of auxiliary vectors. It is surprising that the latter reformulation is mathematically equivalent to Algorithm 3.1; see [455] for a proof.

Observe that in Algorithm 3.1 we perform at each iteration operations of the form

$$\mathbf{v} \mapsto M^{-1}A\mathbf{v}.$$

An efficient way to perform this operation is to first do the matrix-vector product $A\mathbf{v}$, store its intermediate result in $\mathbf{v}$ and then solve the linear system $M\mathbf{w} = \mathbf{v}$, storing again the result in $\mathbf{v}$.

Let us now focus on a preconditioned version of the MinRes Algorithm; see, e.g., [280] and [455]. This methods works on the usual Krylov subspace $\mathcal{K}_m(A, \mathbf{r}^{(0)})$ and A can be symmetric and indefinite. Thus, from the point of view of the convergence analysis, we need polynomials with zeros on both sides of the real axis and that take value 1 in 0 minimizing the residual $\mathbf{r}$ over the affine space $\mathbf{r}^{(0)} + A\mathcal{K}_m(A, \mathbf{r}^{(0)})$.

Since MinRes can reduce the Euclidean norm of the residual rather than the A–norm as CG does, we can not exploit the second part of (3.6). Thus, in the following Algorithm 3.2, MinRes is in a right preconditioned version (3.4) for a symmetric and positive definite preconditioner M.

Let $M = HH^T$ be a factorization for M, supposed well defined. By means of the Sylvester Law of Inertia (Theorem A.10), we know that the equivalent symmetric system

$$H^{-1}AH^{-T}\mathbf{y} = H^{-1}\mathbf{b},\ \mathbf{y} = H^T\mathbf{x},$$

has the same number of positive, zero and negative eigenvalues of A. Moreover, the corresponding residual for any approximation $\mathbf{x}^{(k)}$ is given by

$$H^{-1}(\mathbf{b} - M\mathbf{x}^{(k)}) = H^{-1}\mathbf{r}^{(k)} = H^T\mathbf{z}^{(k)},$$

where $\mathbf{r}^{(k)}$ is the original residual and $\mathbf{z}^{(k)} = M^{-1}\mathbf{r}^{(k)}$ is the preconditioned one. Therefore, we are minimizing the quantity $\|H^{-1}\mathbf{r}^{(k)}\|_2 = \|\mathbf{r}^{(k)}\|_{M^{-1}}$ over the preconditioned affine space $H^{-1}\left(\mathbf{r}^{(0)} + AM^{-1}\mathcal{K}_k(AM^{-1}, \mathbf{r}^{(0)})\right)$.

Remark 3.2. From the above discussion, the reduction of the residual in the preconditioned MinRes is in the $\|\cdot\|_{M^{-1}}$ norm, thus dependent on the preconditioner M. Therefore, one should be careful using a preconditioner that does not use the relative residual in its stopping criteria set or should at least compute $\|\mathbf{b} - A\mathbf{x}\|/\|\mathbf{r}^{(0)}\|$ before accepting $\mathbf{x}$ as an approximation for the solution of the underlying linear system.

3.2.2 Preconditioned GMRES/GMRES(m)

In case of GMRES or GMRES(m), or any other solver for non–Hermitian linear systems, the same three preconditioning options (left–, right–, split–) are available, even though we will see that the mathematical equivalence

Algorithm 3.2: Preconditioned MinRes

Input: $A \in \mathbb{R}^{n\times n}$, Maximum number of iterations N_{max}, Initial Guess $\mathbf{x}^{(0)}$, $M \in \mathbb{R}^{n\times n}$ preconditioner

Output: $\tilde{\mathbf{x}}$ candidate approximation.

$\mathbf{v}^{(0)} \leftarrow 0$, $\mathbf{w}^{(0)} \leftarrow 0$, $\mathbf{w}^{(1)} \leftarrow 0$;
$\mathbf{v}^{(1)} \leftarrow \mathbf{b} - A\mathbf{x}^{(0)}$;
$\mathbf{z}^{(1)} \leftarrow M^{-1}\mathbf{v}^{(1)}$;
$\gamma_1 \leftarrow < \mathbf{z}^{(1)}, \mathbf{v}^{(1)} >^{1/2}$;
$\eta \leftarrow \gamma_1$, $s_0 = s_1 = 0$, $c_0 = c_1 = 1$;
for $j = 1, \ldots, N_{max}$ **do**
  $\mathbf{z}^{(j)} \leftarrow \mathbf{z}^{(j)}/\gamma_j$;
  $\delta_j = < A\mathbf{z}^{(j)}, \mathbf{z}^{(j)} >$;
  $\mathbf{v}^{(j+1)} \leftarrow A\mathbf{z}^{(j)} - \delta_j/\gamma_j \mathbf{v}^{(j)} - \gamma_j/\gamma_{j-1}\mathbf{v}^{(j-1)}$;
  $\mathbf{z}^{(j+1)} \leftarrow M^{-1}\mathbf{v}^{(j+1)}$;
  $\gamma_{j+1} \leftarrow < \mathbf{z}^{(j+1)}, \mathbf{v}^{(j+1)} >^{1/2}$;
  $\alpha_0 \leftarrow c_j\delta_j - c_{j-1}s_j\gamma_j$; /* Givens rotation */
  $\alpha_1 \leftarrow (\alpha_0^2 + \gamma_{j+1}^2)^{1/2}$;
  $\alpha_2 \leftarrow s_j\delta_j + c_{j-1}c_j\gamma_j$;
  $\alpha_3 \leftarrow s_{j-1}\gamma_j$;
  $c_{j+1} \leftarrow \alpha_0/\alpha_1$;
  $s_{j+1} \leftarrow \gamma_{j+1}/\alpha_1$;
  $\mathbf{w}^{(j+1)} \leftarrow \mathbf{z}^{(j)} - \alpha_3\mathbf{w}^{(j-1)} - \alpha_2\mathbf{w}^{(j)}/\alpha_1$;
  $\mathbf{x}^{(j+1)} = \mathbf{x}^{(j-1)} + c_{j+1}\eta\mathbf{w}^{(j+1)}$;
  $\eta \leftarrow -s_{j+1}\eta$;
  if <*Stopping criteria satisfied*> **then**
    Return: $\tilde{\mathbf{x}} = \mathbf{x}^{(j+1)}$;

between them is lost. Moreover, if the preconditioner is ill–conditioned, these differences will become substantial.

Left–preconditioned GMRES

Let us build a GMRES algorithm for the solution of (3.3), i.e., GMRES using the Krylov subspace $\mathcal{K}_m(M^{-1}A, \mathbf{r}^{(0)})$ or the *left–preconditioned Krylov subspace*, i.e.,

$$\mathcal{K}_m(M^{-1}A, \mathbf{r}^{(0)}) = \text{span}\left\{\mathbf{r}^{(0)}, M^{-1}A\mathbf{r}^{(0)}, \ldots, \left(M^{-1}A\right)^{m-1}\mathbf{r}^{(0)}\right\}.$$

This is obtained simply by using the *Arnoldi* Algorithm 2.6 with the modified Gram–Schmidt process and is summarized in Algorithm 3.3. Observe that in this algorithm all the the residual vectors and their norms are computed with respect to the preconditioned residuals. Therefore, to enforce a stopping

---

Algorithm 3.3: GMRES with left preconditioning

Input: $A \in \mathbb{R}^{n\times n}$, Maximum number of iterations m, Initial Guess $\mathbf{x}^{(0)}$, $M \in \mathbb{R}^{n\times n}$ preconditioner

Output: $\tilde{\mathbf{x}}$ candidate approximation.

$\mathbf{r}^{(0)} \leftarrow M^{-1}(\mathbf{b} - A\mathbf{x}^{(0)})$; `/* Arnoldi process */`

$\beta \leftarrow \|\mathbf{r}^{(0)}\|_2$;

$\mathbf{v}^{(1)} = \mathbf{r}^{(0)}/\beta$;

for $j = 1, \ldots, m$ **do**

  $\mathbf{w} \leftarrow M^{-1}A\mathbf{z}^{(j)}$;

  for $i = 1, \ldots, j$ **do**

    $h_{i,j} \leftarrow < \mathbf{w}, \mathbf{v}^{(i)} >$;

    $\mathbf{w} \leftarrow \mathbf{w} - h_{i,j}\mathbf{v}^{(i)}$;

  $h_{j+1,j} \leftarrow \|\mathbf{w}\|_2$;

  $\mathbf{v}^{(j+1)} \leftarrow \mathbf{w}/h_{j+1,j}$;

$V_m \leftarrow [\mathbf{v}^{(1)}, \ldots, \mathbf{v}^{(m)}]$; `/* Build the Krylov subspace basis */`

$\mathbf{y}^{(m)} \leftarrow \arg\min_{\mathbf{y}} \|\beta\mathbf{e}_1 - \overline{H}_m\mathbf{y}\|_2$;

$\mathbf{x}^{(m)} \leftarrow \mathbf{x}^{(0)} + V_m\mathbf{y}^{(m)}$;

`// Convergence check and possibly a restart`

if <*Stopping criteria satisfied*> **then**

  Return: $\tilde{\mathbf{x}} = \mathbf{x}^{(m)}$;

else

  $\mathbf{x}^{(0)} \leftarrow \mathbf{x}^{(m)}$; `/* Restart */`

  goto 1;

---

criterion based on the (unpreconditioned) residual, in theory we need to multiply the preconditioned residuals by M and store them separately from the preconditioned ones. Alternatively, at each step, compute and use the true residual $\mathbf{r}^{(j)}_{\text{true}} = \mathbf{b} - A\mathbf{x}^{(j)}$, which has the same cost as the previous approach (one more matrix-vector product per iteration), and compare it in norm with $\varepsilon\|\mathbf{r}^{(0)}_{\text{true}}\|$.

3.2.2.1 Right–preconditioned GMRES

Let us build GMRES for the solution of (3.4), i.e., GMRES using the Krylov subspace $\mathcal{K}_m(AM^{-1}, \mathbf{r}^{(0)})$, the *right–preconditioned Krylov subspace*:

$$\mathcal{K}_m(AM^{-1}, \mathbf{r}^{(0)}) = \operatorname{span}\left\{\mathbf{r}^{(0)}, AM^{-1}\mathbf{r}^{(0)}, \ldots, \left(AM^{-1}\right)^{m-1}\mathbf{r}^{(0)}\right\}.$$

Similarly to GMRES Algorithm 2.7, we obtain Algorithm 3.4.

Line 14 of Algorithm 3.4 forms the approximate solution of the linear system as a linear combination of the preconditioned vectors $\mathbf{z}^{(i)}$, for $i = 1, \ldots, m$. Since these $\mathbf{v}^{(i)}$ vectors are obtained by using the same preconditioner M^{-1},

---

Algorithm 3.4: GMRES with right preconditioning

Input: $A \in \mathbb{R}^{n \times n}$, Maximum number of iterations m, Initial Guess $\mathbf{x}^{(0)}$, $M \in \mathbb{R}^{n \times n}$ preconditioner
Output: $\tilde{\mathbf{x}}$ candidate approximation.

$\mathbf{r}^{(0)} \leftarrow b - A\mathbf{x}^{(0)}$; `/* Arnoldi process */`
$\beta \leftarrow \|\mathbf{r}^{(0)}\|_2$;
$\mathbf{v}^{(1)} = \mathbf{r}^{(0)}/\beta$;
for $j = 1, \ldots, m$ **do**
  $\mathbf{z}^{(j)} \leftarrow M^{-1}\mathbf{v}^{(j)}$;
  $\mathbf{w} \leftarrow A\mathbf{z}^{(j)}$;
  for $i = 1, \ldots, j$ **do**
    $h_{i,j} \leftarrow < \mathbf{w}, \mathbf{v}^{(i)} >$;
    $\mathbf{w} \leftarrow \mathbf{w} - h_{i,j}\mathbf{v}^{(i)}$;
  $h_{j+1,j} \leftarrow \|\mathbf{w}\|_2$;
  $\mathbf{v}^{(j+1)} \leftarrow \mathbf{w}/h_{j+1,j}$;
$V_m \leftarrow [\mathbf{v}^{(1)}, \ldots, \mathbf{v}^{(m)}]$; `/* Build the Krylov subspace basis */`
$\mathbf{y}^{(m)} \leftarrow \arg\min_{\mathbf{y}} \|\beta\mathbf{e}_1 - \overline{H}_m\mathbf{y}\|_2$;
$\mathbf{x}^{(m)} \leftarrow \mathbf{x}^{(0)} + M^{-1}V_m\mathbf{y}^{(m)}$;
`// Convergence check and possibly a restart`
if *<Stopping criteria satisfied>* **then**
  Return: $\tilde{\mathbf{x}} = \mathbf{x}^{(m)}$;
else
  $\mathbf{x}^{(0)} \leftarrow \mathbf{x}^{(m)}$; `/* Restart */`
  goto 1;

---

i.e., $M^{-1}V_m\mathbf{y}^{(m)}$, we do not need to store them. On the other hand, if we use a non constant preconditioner, as is the case of using some iterations of another method instead of a fixed approximation M for A, we need to store all the $\mathbf{z}^{(i)}$s explicitly to provide the candidate approximation and we find Flexible GMRES/FGMRES, see § 3.2.3.

Split–preconditioned GMRES

If our preconditioner M is a positive definite matrix, or if M can be factored as $M = LU$, we can use split-preconditioning. In this case we need to work with the following system equivalent to $A\mathbf{x} = \mathbf{b}$:

$$L^{-1}AU^{-1}\mathbf{u} = L^{-1}\mathbf{b}, \ \ \mathbf{x} = U^{-1}\mathbf{u}.$$

We initialize the algorithm with the residual $\mathbf{r}^{(0)} = L^{-1}(\mathbf{b} - A\mathbf{x}^{(0)})$ and, to assemble the candidate approximate solution, multiply the linear combination $V_m\mathbf{y}^{(m)}$ by U^{-1}. Note that also in this case the residual vectors and their norms are computed with respect to the preconditioned residuals. Therefore, to enforce a stopping criterion based on the (unpreconditioned) residual, we need to multiply the preconditioned residuals by L (instead of M for the left preconditioned GMRES) and store them separately from the preconditioned ones. Alternatively and more computationally reliable, at each step one can compute and use the true residual $\mathbf{r}_{\text{true}}^{(j)} = \mathbf{b} - A\mathbf{x}^{(j)}$, which has the same cost as the previous approach (one more matrix-vector product per iteration), and compare it in norm with $\varepsilon\|\mathbf{r}_{\text{true}}^{(0)}\|$.

Comparison and analysis

So what kind of preconditioning (left/right/split) approach is appropriate? In principle, for a fixed preconditioner M, the spectra of the eigenvalues of the three associated linear systems (3.3), (3.4) and (3.5) is exactly the same (but the eigenvectors are different), see Exercise 3.1. One should expect them to behave at least similarly, even if a convergence analysis based only on the eigenvalues can be misleading for non Hermitian problems, see § 2.2.8.

Exercise 3.1. Prove that the spectrum of the eigenvalues of the left/right/split preconditioning is the same. What about the eigenvectors?

Proposition 3.1 (Saad [455]). *The approximate solution obtained by left or right preconditioned GMRES is of the form*

$$\begin{aligned}\mathbf{x}^{(m)} &= \mathbf{x}^{(0)} + s_{m-1}(M^{-1}A)M^{-1}\mathbf{r}^{(0)} \\ &= \mathbf{x}^{(0)} + M^{-1}s_{m-1}(M^{-1}A)\mathbf{r}^{(0)}\end{aligned}$$

where s_{m-1} is a polynomial of degree $m-1$ that minimizes the residual norm $\|\mathbf{b} - A\mathbf{x}^{(m)}\|_2$ in the right preconditioning case (Algorithm 3.4) and the preconditioned residual norm $\|M^{-1}(\mathbf{b} - A\mathbf{x}^{(m)})\|_2$ in the left preconditioning case (Algorithm 3.3).

Proof. Left preconditioned GMRES minimizes the residual norm, $\|M^{-1}(\mathbf{b} - A\mathbf{x})\|_2$, over the affine subspace $\mathbf{x}^{(0)} + \mathcal{K}_m(M^{-1}A, M^{-1}\mathbf{r}^{(0)})$ generating an approximate solution of the form

$$\mathbf{x}^{(m)} = \mathbf{x}^{(0)} + M^{-1}s_{m-1}(M^{-1}A)M^{-1}\mathbf{r}^{(0)},$$

with $s_{m-1}(x)$ the polynomial of degree $m-1$ which minimizes the norm

$$\|M^{-1}(\mathbf{r}^{(0)} - As(M^{-1}A)M^{-1}\mathbf{r}^{(0)})\|_2 = \|M^{-1}(\mathbf{r}^{(0)} - AM^{-1}s(M^{-1}A)\mathbf{r}^{(0)})\|_2$$

over all the set of polynomials s of degree less or equal than $m-1$.

On the other hand, for the right preconditioned GMRES we need to make a distinction between the auxiliary variable $\mathbf{u}$ and the original variable $\mathbf{x} = M^{-1}\mathbf{u}$. For the auxiliary variable Algorithm 3.4 minimizes the 2–Norm of the residual $\mathbf{r} = \mathbf{b} - AM^{-1}\mathbf{u}$ where $\mathbf{u}$ belongs to the affine space $\mathbf{u}^{(0)} + \mathcal{K}_m(AM^{-1}, \mathbf{r}^{(0)})$

Multiplying to the left by M^{-1} we find that the original variable $\mathbf{x}$ belongs to the space $M^{-1}\mathbf{u}^{(0)} + \mathcal{K}_m(M^{-1}A, M^{-1}\mathbf{r}^{(0)})$. Thus also the approximate solution in this case can be expressed as $\mathbf{x}^{(m)} = \mathbf{x}^{(0)} + s_{m-1}(AM^{-1})\mathbf{r}^{(0)}$. However now polynomial s_{m-1} minimizes the norm

$$\|\mathbf{r}^{(0)} - AM^{-1}s(AM^{-1})\mathbf{r}^{(0)})\|_2$$

among all polynomials of degree less or equal than $m-1$. □

Therefore, the two preconditioning approaches differ for only a multiplication by M^{-1} in the optimized quantity, while the residual are taken in the same affine space. In many practical situations the difference in the convergence behavior of the two approaches is often not great if M is not too ill conditioned.

3.2.3 Flexible GMRES (FGMRES): GMRES with variable preconditioning

Saad [451] introduced in 1993 the *Flexible GMRES*, or FGMRES for short, a generalization of GMRES that allows changing the preconditioner at each step. This is exploited to use few iterations of another iterative method as a preconditioner. The first paper describing a Krylov subspace method with a variable preconditioning strategy was the one by Axelsson and Vassilevski [23] that introduced the *Generalized Conjugate Gradient method* (GCG). FGMRES can be expressed as a particular case of the GCG, but with an implementation producing a more efficient computational scheme, see, e.g., [528]. In this framework we will see that iterative solvers such as SOR, SSOR, see, e.g., [179, 430], and *Multigrid* or *domain decomposition methods*, see, e.g., [274], can be used as a preconditioner with good parallel potentialities.

It is natural to consider preconditioners based on iterations of other iterative methods, possibly of another Krylov subspace method. The latter case provides an *inner–outer Krylov method*, that can be viewed as having a polynomial preconditioner with a polynomial that changes from one step to the next and is defined implicitly by the polynomial generated by the (inner) Krylov subspace method.

Changing the preconditioner at every step of the GMRES algorithm with right preconditioning can be incorporated in line 5 of Algorithm 3.5 as

$$\mathbf{z}^{(j)} \leftarrow M_j^{-1}\mathbf{v}^j, \quad j = 1, ..., m,$$

and store the results for updating the $\mathbf{x}^{(m)}$ vector in line 14. These are the main simple modifications producing the *Flexibile GMRES* (FGMRES) described in Algorithm 3.5. As remarked in [482, 483], the FGMRES is defined not only by the fact that we use a sequence of preconditioners $\{M_j^{-1}\}_j$, i.e., we change it from one iteration to the next, but also because the solution is obtained directly from the new preconditioned basis Z_m neglecting the basis V_m, that we need to store anyway to perform the orthogonalization steps inside

Algorithm 3.5: FGMRES: GMRES with variable preconditioning

Input: $A \in \mathbb{R}^{n\times n}$, Maximum number of iterations m, Initial Guess $\mathbf{x}^{(0)}$, $\{M_j \in \mathbb{R}^{n\times n}\}_j$ preconditioners
Output: $\tilde{\mathbf{x}}$ candidate approximation.

$\mathbf{r}^{(0)} \leftarrow b - A\mathbf{x}^{(0)}$; /* Arnoldi process */
$\beta \leftarrow \|\mathbf{r}^{(0)}\|_2$;
$\mathbf{v}^{(1)} = \mathbf{r}^{(0)}/\beta$;
for $j = 1, \dots, m$ **do**
    $\mathbf{z}^{(j)} \leftarrow M_j^{-1}\mathbf{v}^{(j)}$;
    $\mathbf{w} \leftarrow A\mathbf{z}^{(j)}$;
    for $i = 1, \dots, j$ **do**
        $h_{i,j} \leftarrow <\mathbf{w}, \mathbf{v}^{(i)}>$;
        $\mathbf{w} \leftarrow \mathbf{w} - h_{i,j}\mathbf{v}^{(i)}$;
    $h_{j+1,j} \leftarrow \|\mathbf{w}\|_2$;
    $\mathbf{v}^{(j+1)} \leftarrow \mathbf{w}/h_{j+1,j}$;
$Z_m \leftarrow [\mathbf{z}^{(1)}, \dots, \mathbf{z}^{(m)}]$; /* Build the Preconditioned subspace basis */
$\mathbf{y}^{(m)} \leftarrow \arg\min_{\mathbf{y}} \|\beta\mathbf{e}_1 - \overline{H}_m\mathbf{y}\|_2$;
$\mathbf{x}^{(m)} \leftarrow \mathbf{x}^{(0)} + Z_m\mathbf{y}^{(m)}$;
// Convergence check and possibly a restart
if *<Stopping criteria satisfied>* **then**
    Return: $\tilde{\mathbf{x}} = \mathbf{x}^{(m)}$;
else
    $\mathbf{x}^{(0)} \leftarrow \mathbf{x}^{(m)}$; /* Restart */
    goto 1;

the Arnoldi process. Thus, the difference between FGMRES and the usual GMRES algorithm is that the action of AM_j^{-1} on a vector $\mathbf{v}$ of the Krylov subspace is no longer in the space generated by the column of V_{m+1}. Therefore, the subspace $\text{span}(Z_m)$ is not necessarily a Krylov subspace. Nevertheless, the basis Z_m and V_{m+1} can be related by an expression similar to the one in equation (2.40):

$$AZ_m = V_{m+1}\overline{H}_m. \tag{3.7}$$

Moreover, if we denote, as usual, by H_m the $m \times m$ matrix obtained from $\overline{H}_m$ by deleting its last row and denoting $\tilde{\mathbf{v}}_{j+1}$ the $(j+1)$th vector, we obtain (see also the similarities with (2.40))

$$AZ_m = V_mH_m + \tilde{\mathbf{v}}_{j+1}\mathbf{e}_m^T. \tag{3.8}$$

We now have the tools to state and prove an optimality property of FGMRES as well.

Proposition 3.2 (Saad [451]). *The candidate approximate solution $\mathbf{x}^{(m)}$ obtained at step m of Algorithm 3.5 minimizes the residual norm $\|\mathbf{b} - A\mathbf{x}^{(m)}\|_2$ over the affine subspace $\mathbf{x}^{(0)} + \text{span}(Z_m)$.*

Proof. We start considering an arbitrary vector $\mathbf{z} = \mathbf{x}^{(0)} + Z_m\mathbf{y}$ in the affine subspace $\mathbf{x}^{(0)} + \text{span}(Z_m)$, then we have

$$\begin{aligned}\mathbf{b} - A\mathbf{z} =& \mathbf{b} - A(\mathbf{x}^{(0)} + Z_m\mathbf{y}) = \mathbf{r}^{(0)} - AZ_m\mathbf{y}\\ =& \beta\mathbf{v}^{(1)} - V_{m+1}\overline{H}_m\mathbf{y} = V_{m+1}\left(\beta\mathbf{e}_1 - \overline{H}_m\mathbf{y}\right).\end{aligned}$$

Since $\mathbf{y}^{(m)}$ is $\mathbf{y}^{(m)} = \arg\min_{\mathbf{y}} \|\beta\mathbf{e}_1 - \overline{H}_m\mathbf{y}\|_2$, it follows that the approximate solution $\mathbf{x}^{(m)} = \mathbf{x}^{(0)} + Z_m\mathbf{y}^{(m)}$ achieves the smallest residual norm in $\mathbf{x}^{(0)} + \text{span}(Z_m)$. □

Note that, differently from the GMRES and right–preconditioned GMRES, the nonsingularity of A no longer implies the non singularity of the H_j matrices.

Proposition 3.3 (Saad [451]). *Assume that $\beta = \|\mathbf{r}^{(0)}\|_2 \neq 0$ and that $j-1$ steps of FGMRES (Algorithm 3.5) have been successfully performed. In addition, let H_j be nonsingular. Then, $\mathbf{x}^{(m)}$ is exact if and only if $h_{j+1,j} = 0$.*

Details on FGMRES using as a variable preconditioner another Krylov subspace method can be found in [481].

A useful modification of FGMRES is based on the so–called *deflated restarting.*

Example 3.1. Consider a fluid dynamics problem using finite elements from the collection [177]: the *FIDAP/ex20.* This is a 2D attenuation problem of a surface disturbance, thus the fully coupled Navier–Stokes equations are discretized with $Q2$ element for the velocity and $P1$ for the pressure over the domain $[0, 7] \times [0, 5]$ on a regular structured but not uniform mesh. The density is set to 1.0 while the viscosity is set to 0.1. We compare GMRES, GMRES(10), GMRES(20), FGMRES+GMRES(10) and FGMRES+GMRES(20) algorithms for the solution of the linear system from this problem. All methods are set to achieve a tolerance on the Euclidean norm of the relative residual of $\varepsilon = 10^{-6}$. As we see from Figure 3.1, Restarted GMRES loses information ending with a stagnation, while the Flexible-GMRES achieves convergence in few iterations. ✓

Remark 3.3. There exist also other *flexible* versions of Krylov subspace algorithms that cover the use of variable preconditioning strategies. We refer to [404] for a flexible variant for CG, to [537] for flexible variants of the BiCG and BiCGstab algorithms, to [501] for a flexible variant of the QMR method. These kind of approaches has also been used for methods based on the induced dimension reduction (IDR) from [541] in [259, 489].

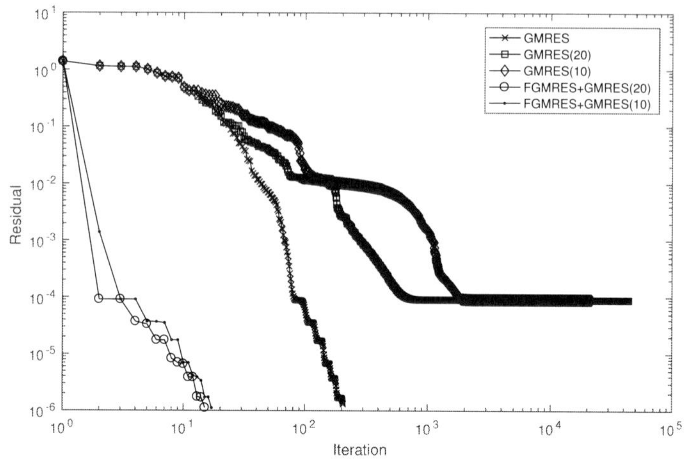

Figure 3.1: Example 3.1. Comparison of the convergence history for some variation of the GMRES algorithm.

3.2.4 Preconditioned oblique projection: CGS and BiCGstab

Consider now CGS and BiCGstab. To shorten the presentation, since many of the issues are similar to the one already discussed for GMRES, we focus on the case of split preconditioning $M = M_1 M_2$ only, by allowing the possibility of either M_1 or M_2 being the identity matrix. The left and right preconditioning readily follow.

Let us start from CGS, the *Conjugate Gradient Squared* (Algorithm 2.12). We need to perform the following change of variables:

$$\begin{array}{lll} \tilde{\mathbf{p}}^{(i)} \mapsto M_1^{-1}\mathbf{p}^{(i)}, & \tilde{\mathbf{q}}^{(i)} \mapsto M_1^{-1}\mathbf{q}^{(i)}, & \tilde{\mathbf{v}}^{(i)} \mapsto M_1^{-1}\mathbf{v}^{(i)}, \\ \tilde{\mathbf{r}}^{(i)} \mapsto M_1^{-1}\mathbf{r}^{(i)}, & \tilde{\mathbf{u}}^{(i)} \mapsto M_1^{-1}\mathbf{u}^{(i)}, & \tilde{\mathbf{x}}^{(i)} \mapsto M_2\mathbf{p}^{(i)}, \\ \tilde{\mathbf{r}}_*^{(0)} \mapsto M_1^T\mathbf{r}_*^{(0)}. & & \end{array}$$

Thus, instead of studying the effects of the left, right or split preconditioning, we look for the effects on the algorithm given by the choice for the vector $\mathbf{r}_*^{(0)}$.

Note that, differently from the case of left or split preconditioned GMRES, this algorithm works with the original variables $\mathbf{x}$ and residual $\mathbf{r}$. Therefore, the stopping criteria can be enforced on the residuals.

With the same approach we can write down the change of variables for preconditioning BiCGstab as

$$\begin{array}{lll} \tilde{\mathbf{p}}^{(i)} \mapsto M_1^{-1}\mathbf{p}^{(i)}, & \tilde{\mathbf{v}}^{(i)} \mapsto M_1^{-1}\mathbf{v}^{(i)}, & \tilde{\mathbf{r}}^{(i)} \mapsto M_1^{-1}\mathbf{r}^{(i)}, \\ \tilde{\mathbf{s}} \mapsto M_1^{-1}\mathbf{s}, & \tilde{\mathbf{t}} \mapsto M_1^{-1}\mathbf{t}, & \tilde{\mathbf{x}}^{(i)} \mapsto M_2\mathbf{x}^{(i)}, \\ \tilde{\mathbf{r}}_*^{(0)} \mapsto M_1^T\mathbf{r}_*^{(0)}. & & \end{array}$$

Algorithm 3.6: Preconditioned CGS Method (PCGS)

Input: $A \in \mathbb{R}^{n\times n}$, Maximum number of iterations $N_{\max}$, Initial Guess $\mathbf{x}^{(0)}$, $M \in \mathbb{R}^{n\times n}$ preconditioner

Output: $\tilde{\mathbf{x}}$ candidate approximation.

$\mathbf{r}^{(0)} \leftarrow \mathbf{b} - A\mathbf{x}^{(0)}$;
Chose $\mathbf{r}_*^{(0)}$ such that $< \mathbf{r}^{(0)}, \mathbf{r}_*^{(0)} > \neq 0$;
$\rho_0 \leftarrow 1$, $\mathbf{p}_0 \leftarrow 0$, $\mathbf{q}^{(0)} \leftarrow 0$;
for $i = 1, 2, 3, \ldots, N_{max}$ **do**
  $\rho_i \leftarrow < \mathbf{r}_*^{(0)}, \mathbf{r}^{(i-1)} >$;
  $\beta \leftarrow \rho_i / \rho_{i-1}$;
  $\mathbf{u} \leftarrow \mathbf{r}^{(i-1)} + \beta\mathbf{q}^{(i-1)}$;
  $\mathbf{p}_i \leftarrow \mathbf{u} + \beta(\mathbf{q}^{(i-1)} + \beta\mathbf{p}_{i-1})$;
  $\mathbf{y} \leftarrow M^{-1}\mathbf{p}^{(i)}$;
  $\mathbf{v} \leftarrow A\mathbf{y}$;
  $\alpha \leftarrow \rho_i / <\mathbf{r}_*^{(0)}, \mathbf{v}>$;
  $\mathbf{q}^{(i)} \leftarrow \mathbf{u} + \alpha\mathbf{v}$;
  $\mathbf{z} \leftarrow M^{-1}(\mathbf{u} + \mathbf{q}^{(i)})$;
  $\mathbf{x}^{(i)} \leftarrow \mathbf{x}^{(i-1)} + \alpha\mathbf{z}$;
  $\mathbf{r}^{(i)} \leftarrow \mathbf{r}^{(i-1)} - \alpha A\mathbf{z}$;
  if *<Stopping criteria satisfied>* **then**
    Return: $\tilde{\mathbf{x}} = \mathbf{x}^{(j)}$;

and modify the standard Algorithm 2.13 to obtain its preconditioned version given in Algorithm 3.7. However, for BiCGstab there is a difference between left, right and split preconditioning, a difference that can not be characterized through the choice of the starting residual as in *PCGS* (Algorithm 3.6). This depends on the expression for the ω_i at Line 15. Observe that if we use the right preconditioning scheme, i.e., $M_1 = 1$, we are minimizing the residual for the original system rather than for the preconditioned one.

A similar approach can be used for the preconditioned version of QMR (Algorithm 2.11), BiCGstab(2) (Algorithm 2.14) and, in general, for all the BiCGstab(l) versions discussed in the previous chapter.

3.3 On Jacobi, SOR, and SSOR preconditioners

Let us consider some splitting (fixed-point) iterative methods used as preconditioners. In particular, here we focus briefly on iterative methods introduced in § 2.1.1.

Algorithm 3.7: Preconditioned BiCGstab method

Input: $A \in \mathbb{R}^{n\times n}$, Maximum number of iterations $N_{\max}$, Initial Guess $\mathbf{x}^{(0)}$, $M = M_1 M_2 \in \mathbb{R}^{n\times n}$ preconditioner

Output: $\tilde{\mathbf{x}}$ candidate approximation.

$\mathbf{r}^{(0)} \leftarrow \mathbf{b} - A\mathbf{x}^{(0)}$;

choose $\mathbf{r}_*^{(0)}$ such that $< \mathbf{r}^{(0)}, \hat{\mathbf{r}}^{(0)} > \neq 0$;

$\rho_0 \leftarrow 1,\ \alpha \leftarrow 1,\ \omega_0 = \leftarrow 1$;

$\mathbf{v}^{(0)} \leftarrow 0,\ \mathbf{p}_0 \leftarrow 0$;

for $j = 1, \ldots, N_{max}$ **do**

- $\rho_i \leftarrow < \mathbf{r}_*^{(0)}, \mathbf{r}^{(i-1)} >$;
- $\beta \leftarrow (\rho_i/\rho_{i-1})(\alpha/\omega_{i-1})$;
- $\mathbf{p}_i \leftarrow \mathbf{r}^{(i-1)} + \beta(\mathbf{p}_{i-1} - \omega_{i-1}\mathbf{v}^{(i-1)})$;
- $\mathbf{y} \leftarrow M^{-1}\mathbf{p}_i$;
- $\mathbf{v}^{(i)} \leftarrow A\mathbf{y}$;
- $\alpha \leftarrow \rho_i/{<\mathbf{r}_*^{(0)}, \mathbf{v}^{(i)}>}$;
- $\mathbf{s} \leftarrow \mathbf{r}^{(i-1)} - \alpha\mathbf{v}^{(i)}$;
- $\mathbf{z} \leftarrow M^{-1}\mathbf{s}$;
- $\mathbf{t} \leftarrow A\mathbf{z}$;
- $\omega_i \leftarrow {<M_1^{-1}\mathbf{t}, M_1^{-1}\mathbf{s}>}/{<M_1^{-1}\mathbf{t}, M_1^{-1}\mathbf{t}>}$;
- $\mathbf{x}^{(i)} \leftarrow \mathbf{x}^{(i-1)} + \alpha\mathbf{y} + \omega_i\mathbf{z}$;
- $\mathbf{r}^{(i)} = \mathbf{s} - \omega_i\mathbf{t}$;
- **if** <*Stopping criteria satisfied*> **then**
  - **Return:** $\tilde{\mathbf{x}} = \mathbf{x}^{(j)}$;

Consider a matrix $A \in \mathbb{R}^{n\times n}$ either diagonally dominant and/or positive definite. In these cases, its diagonal entries can contain precious information. Therefore, the simplest preconditioning approach is taking M as the diagonal matrix whose entries are the diagonal entries of A:

$$M = \text{diag}(a_{1,1}, a_{2,2}, \ldots, a_{n,n}).$$

Doing this means taking the *Jacobi method* from § 2.1.1 as a preconditioner. In general, we can use within the same approach any *splitting method* of Table 2.1 by choosing as the preconditioner M the matrix of the regular splitting of the matrix A, i.e., $A = M - N$. As a popular example, we can use the SSOR stationary iterative method with the PCG method for solving linear systems from some simple elliptic problems; see [555]. To improve the performances of this preconditioner, both their *block formulations* and their combination with the *decomposition technique* were considered more recently in the literature; see, e.g., [452, 458, 459].

Example 3.2. We consider the solution of the problem Sherman1 from the Davis and Hu [177] collection. This is represented by a symmetric and negative definite matrix of size $n = 1000$. We transform the problem in $-A\mathbf{x} = -\mathbf{b}$ with the PCG algorithm and the Jacobi, SOR and SSOR preconditioners (with optimal parameter ω). Results are collected in Figure 3.2. Observe that

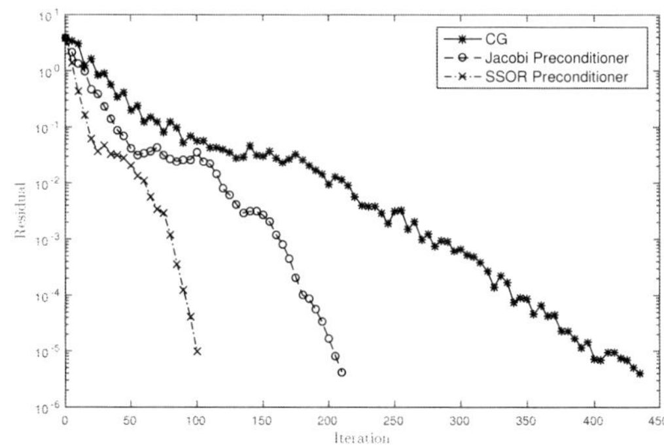

Figure 3.2: PCG with SSOR and Jacobi preconditioner.

PCG with the optimal SSOR preconditioner returns better performance than with the Jacobi, as we expected from the discussion in § 2.1.1. ✓

3.4 Incomplete factorization preconditioners

One of the most popular techniques for solving linear systems represented by a direct method, the *Gaussian–elimination* algorithm. The central idea is obtaining a factorization of the matrix A of the system as the product of two matrices L and U, lower and upper triangular, respectively, i.e., $A = LU$. The solution of the system $A\mathbf{x} = \mathbf{b}$ is obtained first by computing forward substitution for the lower triangular system and then using the solution of the latter for the upper triangular. The upper triangular linear system is solved by backward substitution and we get $\mathbf{x}$:

$$A\mathbf{x} = \mathbf{b}, \Rightarrow LU\mathbf{x} = \mathbf{b}, \Rightarrow \text{Solve } L\mathbf{y} = \mathbf{b} \rightarrow \mathbf{y}, \Rightarrow \text{Solve } U\mathbf{x} = \mathbf{y} \rightarrow \mathbf{x}. \quad (3.9)$$

In this section we modify this strategy in order to get an approximate factorization that, for a sparse matrix A (in a sense specified later), gives sparse triangular factors and then use these as a preconditioner for iterative solvers. As a matter of fact, computing a factorization of the sparse matrix A produces, in general, dense (or almost dense) triangular factors. See the pattern of the LU factorization in Figure 3.3. This is clearly a situation we want to avoid due to the remarks made earlier about effectiveness of preconditioners: augmenting the number of non-zeros entries makes computing (and storing) matrix-vector products more expensive.

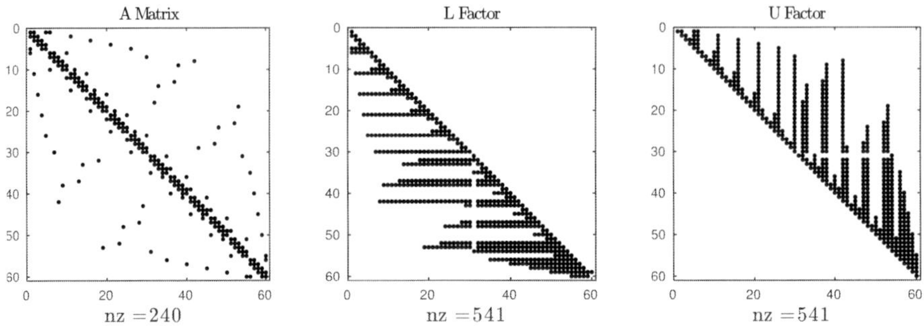

Figure 3.3: $A = LU$ factorization of a sparse matrix. The LU factor are sensibly denser than the A.

Definition 3.2. Preconditioners obtained by building an approximation of the matrix A of the linear system are sometimes called **implicit preconditioners**.

The name is due to the fact that for applying such preconditioners an auxiliary linear system must be solved.

General comments on explicit preconditioners and comparisons with implicit preconditioners are at the beginning of § 3.5.

3.4.1 Incomplete Cholesky factorization or IC

When the matrix A in the linear system

$$A\mathbf{x} = \mathbf{b},\ A \in \mathbb{R}^{n\times n},\ \mathbf{x}, \mathbf{b} \in \mathbb{R}^n$$

is symmetric and positive definite, i.e.

$$A = A^T,\ \mathbf{y}^T A\mathbf{y} > 0,\ \forall \mathbf{0} \neq \mathbf{y} \in \mathbb{R}^n,$$

we can simplify the LU factorization produced by the Gaussian–elimination (3.9) in the *Cholesky Factorization*:

$$A = LL^T,\ L \in \mathbb{R}^{n\times n}, \tag{3.10}$$

i.e., $U = L^T$ in the LU factorization (3.9), with L a lower triangular matrix. This factorization can be obtained in a column–by–column way by computing for $j = 1, \ldots, n$:

$$\begin{aligned} L_{i,j} &= \frac{A_{i,j} - \sum_{k=1}^{i-1} L_{k,i} L_{k,j}}{L_{i,i}},\ i = 1, \ldots, j-1 \\ L_{j,j} &= \left[A_{j,j} - \sum_{k=1}^{i-1} L_{k,j}^2 \right]^{1/2}. \end{aligned} \tag{3.11}$$

To avoid the computation of the square roots in the previous formula we can reformulate the latter as a *root–free Cholesky factorization*:

$$A = LDL^T,$$

where D is a diagonal matrix and L is a lower triangular matrix with all ones on the main diagonal, sometimes called *unit lower triangular*; see, e.g., [271, Algorithm 4.1.2]. As we have observed, the computation of the Cholesky factorization, both in the standard and root–free formulations, starting from a matrix A that is sparse, can give rise to a denser or also a full matrix L. Therefore, following the idea presented in [329, 384], under appropriate assumptions on A, an *Incomplete Cholesky factorization* or *IC*, i.e.,

$$A = \tilde{L}\tilde{L}^T + C \text{ or } A = \tilde{L}\tilde{D}\tilde{L}^T + C,$$

can be produced, with $\tilde{L}$ a lower triangular matrix, $\tilde{D}$ a diagonal matrix and C a matrix containing elements that are dropped along the computation of the factors $\tilde{L}$ to maintain them sparse and such that $\|C\|_F$ is small in the sense we discussed at the beginning of § 3.1.

The entries that will be discarded in computing the incomplete factorization process can be selected through several strategies. We postpone the details in § 3.4.2 for the case of *Incomplete LU factorization* for generic (and thus also nonsymmetric) systems. Here we consider only the incomplete Cholesky factorization with zero fill–in, or $IC(0)$, Algorithm, i.e., the algorithm that produces incomplete incomplete Cholesky factorization such that its lower triangular factor $\tilde{L}$ has nonzero entries only in the same lower triangular position of the matrix A. Note that all algorithms that do not introduce nonzero entries in their factors where in A they are zero are also called zero *fill–in* algorithms. We start considering the pattern P defined as

$$P = \{(i,j) \in \{1,\ldots,n\}^2 \,:\, A = (a_{i,j}),\, a_{i,j} \neq 0\},$$

and set to zero all the entries that are not in the sparsity pattern P of the matrix A. In this way we produce $\tilde{L}$ that has the same sparsity pattern of the matrix A. Obviously, we need to check whenever this procedure is feasible, i.e., given a generic sparse symmetric positive definite matrix A, is its $IC(0)$ factorization always well defined? The answer is "no"in general, as the following example shows.

Example 3.3. Consider the symmetric and positive definite matrix

$$A = \begin{bmatrix} 3 & -2 & 0 & 2 \\ -2 & 3 & -2 & 0 \\ 0 & -2 & 3 & -2 \\ 2 & 0 & -2 & 3 \end{bmatrix},$$

for which we can compute explicitly the Cholesky factorization:

$$A = LL^T = \begin{bmatrix} \sqrt{3} & 0 & 0 & 0 \\ -\frac{2}{\sqrt{3}} & \sqrt{\frac{5}{3}} & 0 & 0 \\ 0 & -2\sqrt{\frac{3}{5}} & \sqrt{\frac{3}{5}} & 0 \\ \frac{2}{\sqrt{3}} & \frac{4}{\sqrt{15}} & -\frac{2}{\sqrt{15}} & \frac{1}{\sqrt{3}} \end{bmatrix} \begin{bmatrix} \sqrt{3} & -\frac{2}{\sqrt{3}} & 0 & \frac{2}{\sqrt{3}} \\ 0 & \sqrt{\frac{5}{3}} & -2\sqrt{\frac{3}{5}} & \frac{4}{\sqrt{15}} \\ 0 & 0 & \sqrt{\frac{3}{5}} & -\frac{2}{\sqrt{15}} \\ 0 & 0 & 0 & \frac{1}{\sqrt{3}} \end{bmatrix}.$$

While the incomplete Cholesky factorization obtained by imposing the same sparsity pattern of A, i.e., the $IC(0)$ algorithm, gives us

$$\tilde{L}^T = \begin{bmatrix} \sqrt{3} & 0 & 0 & 0 \\ -\frac{2}{3}\sqrt{3} & \frac{1}{3}\sqrt{15} & 0 & 0 \\ 0 & \frac{2}{5}\sqrt{15} & \frac{1}{5}\sqrt{15} & 0 \\ \frac{2}{3}\sqrt{3} & 0 & -\frac{2}{3}\sqrt{15} & \sqrt{-5} \end{bmatrix}.$$

Note that the last diagonal entry of $\tilde{L}$ is equal to $\sqrt{-5}$ for a positive definite matrix A! Therefore the algorithm fails, having encountered a non positive pivot for a positive definite matrix A. ✓

To prove that the $IC(0)$ algorithm is well posed, we need to introduce a particular class of matrices: the M–matrices, see [62][3].

Definition 3.3 (M–matrix def#1). A matrix A is called M–matrix if it can be written as:

$$A = sI - B, \quad s > 0, \quad B \geq 0$$

with $s > \rho(B)$.

Definition 3.4 (M–matrix def#2). A matrix $A \in \mathbb{R}^{n\times n}$ is called a non-singular M–matrix if:

1. $a_{i,i} > 0 \; \forall i = 1, 2, \ldots, n$;
2. $a_{i,j} \leq 0 \; \forall i \neq j,\; i, j = 1, 2, \ldots, n$;
3. $\det(A) \neq 0$;
4. the entries of the inverse of A are positive, i.e., $A^{-1} \geq 0$.

Sometimes the property $A^{-1} \geq 0$ is also called *inverse positive.*

Definition 3.5 (M–matrix def#3). A matrix A is an M–matrix if it has all positive diagonal entries, and there exists a diagonal matrix $D = diag(d_1, \ldots, d_n)$ such that AD is strictly diagonally dominant, i.e.:

$$a_{i,i}d_i > \sum_{j\neq i} |a_{i,j}|d_j, \quad i = 1, \ldots, n.$$

[3]In [62] are given fifty equivalent definitions of M–matrix. We use only those required to prove the existence of incomplete LU factorization. Other information can bve found in [432] and [546].

Definition 3.6 (M–matrix def#4). A matrix A is an M–matrix if all its principal submatrices are inverse positive.

Definition 3.7 (M–matrix def#5). A matrix A is an M–matrix if it has a regular splitting, see Definition 2.2, i.e., $A = M - N$ with $M^{-1} \geq 0$ and $N \geq 0$ such that $\rho(M^{-1}N) < 1$.

Proposition 3.4. *The definitions 3.3, 3.4, 3.5, 3.6 and 3.7 are equivalent.*

Proof. We give a proof of some of the above equivalences.

(3.3) ⇒ (3.4) if A is written as $A = sI - B$ with $s > \rho(B)$ then $a_{i,i} > 0$ $\forall i = 1, 2, \ldots, n$ because the diagonal elements of B are all less than B's spectral radius, i.e., $b_{i,i} \leq \rho(B)$ $\forall i = 1, 2, \ldots, n$, for the Perron–Frobenius theorem[4].

From Definition (3.3) we have $a_{i,j} = -b_{i,j}$ $\forall i \neq j$, $i, j = 1, 2, \ldots, n$ and this implies $a_{i,j} \leq 0$ $\forall i \neq j$, $i, j = 1, 2, \ldots, n$ because $B \geq 0$. The matrix A is non–singular because $A = sI - B = s(I - s^{-1}B)$ and $\rho(s^{-1}B) < 1$ because $s > \rho(B)$. In this way we can write

$$A^{-1} = \frac{1}{s}\left(I - \frac{B}{s}\right)^{-1} = \frac{1}{s}\sum_{k=0}^{\infty}\left(\frac{B}{s}\right)^{k}$$

which also shows that $A^{-1} \geq 0$.

(3.4) ⇒ (3.3) Now let $A = (a_{i,j})$ be an M–matrix in the sense of Definition 3.4 for $s > \max\{0, \max_{i=1,\ldots,n} a_{i,i}\}$, let $A = sI - B$, where $B = sI - A \geq 0$. Then, $A = sI - B$ is a regular splitting of the matrix A, and $s > \rho(B)$ because $A^{-1} \geq 0$.

(3.3) ⇒ (3.7) It is sufficient to take $M = sI$ and $N = B$.

(3.7) ⇒ (3.4) We define $G = M^{-1}N$ and assume $\rho(G) < 1$, then since $A = M(I - G)$, then A is non singular. On the other hand, we have

$$A^{-1} = (I - G)^{-1}M^{-1} = \sum_{k=0}^{+\infty} G^k M^{-1} \geq 0$$

because G and M^{-1} are non negative.

(3.4) ⇒ (3.5) $A^{-1} \geq 0$ and $a_{i,i} > 0$ $\forall i = 1, \ldots, n$. Now let $\mathbf{x} = A^{-1}\mathbf{e}$, where $\mathbf{e} = (1, 1, \ldots, 1)^T$, then we can consider the positive diagonal matrix D obtained putting $d_{i,i} = x_i$ $\forall i = 1, \ldots, n$. In this way,

$$AD\mathbf{e} = A\mathbf{x} = AA^{-1}\mathbf{e} = \mathbf{e} >> 0$$

[4]Information on the Perron–Frobenius Theorem can be found in [390], the original material comes from [425], [252–254].

and thus AD has all positive row sum, but we also have $a_{i,j} \leq 0$ for $i \neq j$ and $\forall i, j = 1, \ldots, n$ so we have shown that AD is strictly diagonally dominant. □

Remark 3.4. Definition 3.6 implies that all principal submatrices of an M–matrix are themselves M-matrices.

Then we state the following proposition on *regular splitting* from [532]:

Proposition 3.5. *Let M,N be a regular splitting (Definition 2.2) of $A \in \mathbb{R}^{n\times n}$. Then we have:*

$$\rho(M^{-1}N) < 1 \Leftrightarrow \begin{cases} \det(A) \neq 0, \\ A^{-1} \geq 0. \end{cases} .$$

Proof. $\Leftarrow$ This implication is the same we use to prove the implication between (3.7) and (3.4) in Proposition 3.4.

$\Rightarrow$ Now we define $G = M^{-1}N$ and in this way we have $A = M(I - G)$ and from this it follows that $(I - G)$ is also nonsingular. Moreover:

$$\begin{aligned} A^{-1}N &= \left(M\left(I - M^{-1}N\right)\right)^{-1} N = \\ &= \left(I - M^{-1}N\right)^{-1} M^{-1}N = \\ &=(I - G)^{-1}G. \end{aligned} \tag{3.12}$$

For the definition of regular splitting (Definition 2.2), G is nonnegative. The Perron-Frobenius Theorem[5] implies that there exists a non negative eigenvector $\mathbf{x}$ whose eigenvalue is $\rho(G)$, i.e. $G\mathbf{x} = \rho(G)\mathbf{x}$. Using (3.12) we have:

$$A^{-1}N\mathbf{x} = \frac{\rho(G)}{1-\rho(G)}\mathbf{x},$$

and therefore, since also $A^{-1}N$ is nonnegative, we have

$$\frac{\rho(G)}{1-\rho(G)} \geq 0 \Rightarrow 0 \leq \rho(G) \leq 1$$

but $I - G$ is nonsingular, hence $\rho(G) < 1$. □

Now we can state an important existence result for incomplete Cholesky decompositions.

Proposition 3.6 (Meijerink and Van der Vorst [384]). *Let A be a symmetric nonsingular M-matrix. Then, given a set of indexes P, the related incomplete Cholesky decomposition $\tilde{L}\tilde{D}\tilde{L}^T$, where*

$$A = \tilde{L}\tilde{D}\tilde{L}^T + C,$$

is a regular splitting.

---

[5] Information on the theorem is in [390], while the original material comes from [252–254, 425].

We will prove this result, together with its analogous for the incomplete LU factorization, in the next section.

In this way we have the existence of the incomplete Cholesky decomposition for a class more narrow than that of the SPD matrices. We discuss some way to circumvent this issue with the $IC(0)$ algorithm in § 3.4.4.

Let us now consider a robust variant of the incomplete Cholesky decomposition algorithm by [325, 498, 511]. This variant is more computationally expensive, but computes an incomplete factorization without pivot breakdowns. We start again from the relation (3.11). Denote the matrices generated at the kth step of the complete algorithm as $A_{(k)}$ and $L_{(k)}$, i.e., $A_{(k)} = L_{(k)}L_{(k)}^T$ with $A_{(k)} \in \mathbb{R}^{k\times k}$ and $k < n$ and therefore at the $(k+1)$th step of the algorithm, we need to determine the decomposition

$$A_{(k+1)} = \begin{bmatrix} L_{(k)} & 0 \\ \mathbf{y}_{(k)}^T & \gamma_{k+1} \end{bmatrix} \begin{bmatrix} L_{(k)}^T & \mathbf{y}_{(k)} \\ 0 & \gamma_{k+1} \end{bmatrix} = \begin{bmatrix} A_{(k)} & \mathbf{b}_{(k)} \\ \mathbf{b}_{(k)}^T & \alpha_{k+1} \end{bmatrix}$$

where

$$L_{(k)}\mathbf{y}_{(k)} = \mathbf{b}_{(k)}, \; \gamma_{k+1} = \sqrt{\alpha_{k+1} - \mathbf{y}_{(k)}^T \mathbf{y}_{(k)}}.$$

Therefore, a breakdown occurs when, after having dropped the elements from $\mathbf{y}_{(k)}$, the quantity $\mathbf{y}_{(k)}^T\mathbf{y}_{(k)} \geq \alpha_{k+1}$. To ensure that no breakdowns occur, we need to recast the relation above. Let $L_{(j)}$ be the elementary Gauss transformation relative to the jth column. The diagonal entries, i.e., the square of the γ_k, can be obtained by the recursion:

$$\begin{cases} A^{(0)} = A, \\ A^{(j)} = L_{(j)}^{-1} A^{(j-1)} L_{(j)}^{-T}, \quad j = 1,\ldots,n. \end{cases} \Rightarrow A^{(n)} = \operatorname{diag}(\gamma_j^2)_{j=1}^n. \tag{3.13}$$

We can now express the inverse of the matrix $L_{(j)}$ as,

$$L_{(j)}^{-1} = I - \hat{L}^{(j)}, \; \hat{L}^{(j)} = L_{(j)} - I, \tag{3.14}$$

and express (3.13) as

$$\begin{cases} A^{(0)} = A, \\ A^{(j)} = A^{(j-1)} - \dfrac{1}{A_{j,j}^{(j-1)}} A_{:,j}^{(j-1)} A_{j,:}^{(j-1)} + A_{j,j}^{(j-1)} \mathbf{e}_j \mathbf{e}_j^T, \quad j = 1,\ldots,n. \end{cases}$$

Now the dropping procedure amounts to discarding some entries from the vectors $A_{:,j}^{(j-1)}$, $A_{j,:}^{(j-1)}$. We can express the computation of the pivot, putting it all together, as

$$\begin{aligned} \gamma_{k+1} &= \sqrt{\alpha_{k+1} - 2\mathbf{y}_{(k)}^T\mathbf{y}_{(k)} + \mathbf{y}_{(k)}^T L_{(k)}^{-1} A^{(k)} L_{(k)}^{-T} \mathbf{y}_{(k)}} \\ &= \left(\begin{bmatrix} \mathbf{y}_{(k)}^T L_{(k)}^{-1}, -1 \end{bmatrix} \begin{bmatrix} A_{(k)} & \mathbf{b}_{(k)} \\ \mathbf{b}_{(k)}^T & \alpha_{k+1} \end{bmatrix} \begin{bmatrix} L_{(k)}^{-T}\mathbf{y}_{(k)} \\ -1 \end{bmatrix} \right)^{1/2}, \end{aligned}$$

obtaining that the pivot is greater than 0 by the property of A being a symmetric and positive definite matrix. Clearly this strategy is more expensive, both in construction time and intermediate storage with respect to the standard $IC(0)$ algorithm. Variants of this algorithm proposed in [357, 358] can reduce the memory requirements of the approach we discuss above.

All discussions in this section are made for matrices with real entries but can be easily extended.

Exercise 3.2. Prove that if A is an M–matrix then the Gauss–Seidel method (§ 2.1.1) converges.

Exercise 3.3. Use (3.13) and (3.14) to get a relation for the diagonal entries of the root–free Cholesky factorization matrix D.

3.4.2 Nonsymmetric systems and ILU

In the previous section we focused on symmetric matrices. For a generic nonsingular matrix A we consider *incomplete LU* factorizations, or *ILU* for short, that are derived by performing Gaussian–elimination and dropping some elements in predetermined nondiagonal positions and/or whose moduli are smaller than a prescribed quantity.

Let us state a sufficient condition for the existence of incomplete LU factorizations, a preconditioning procedure that was introduced by Meijerink and Van der Vorst [384]. This discussion is also enough to prove a sufficient condition for the existence of the $IC(0)$ factorization discussed in the previous section as a particular case.

To get this result we need the following lemmas by Fan [226]:

Lemma 3.1 (Fan [226]). *Gaussian–elimination preserves the M–matrix property.*

Proof. Thanks to Definition 3.6 and Remark 3.4, without loss of generality, it is enough to prove that, if $L^{(1)}$ is an elementary Gauss transformation on A, then $A^{(1)} = L^{(1)}A$ is still an M–matrix. We have

$$L^{(1)} = \begin{pmatrix} 1 & & & & \\ -\frac{a_{2,1}}{a_{1,1}} & 1 & & 0 & \\ -\frac{a_{3,2}}{a_{1,1}} & & \ddots & & \\ \vdots & 0 & & \ddots & \\ -\frac{a_{n,1}}{a_{1,1}} & & & & 1 \end{pmatrix} \geq 0.$$

Moreover, $A^{(1)}$ is invertible as a product of invertible matrices. The entries of $A^{(1)}$ are

$$\begin{cases} a^{(1)}_{i,j} = a_{i,j} - \frac{a_{i,1}}{a_{1,1}} a_{1,j} & i > 1 \\ a^{(1)}_{i,i} = a_{1,1} & i = 1. \end{cases}$$

By the sign properties of the M–matrix A, we have

$$a_{i,i}^{(1)} > 0, \text{ and } a_{i,j}^{(1)} \leq 0 \text{ for } i \neq j.$$

Let us now consider $\left(A^{(1)}\right)^{-1}$. To ensure that A is an M–matrix we need to show that $\left(A^{(1)}\right)^{-1} \geq 0$. Since the first column of $A^{(1)}$ has only one nonzero, i.e. $a_{1,1}^{(1)} = a_{1,1}$, then $\left(A^{(1)}\right)^{-1}$ has also only one nonzero in the first column, which is equal to $1/a_{1,1}$:

$$\left(A^{(1)}\right)^{-1} \mathbf{e}_1 = \frac{1}{a_{1,1}} \mathbf{e}_1 > 0.$$

On the other hand, we also have:

$$\left(A^{(1)}\right)^{-1} \mathbf{e}_j = A^{-1} \left(L^{(1)}\right)^{-1} \mathbf{e}_j = A^{-1}\mathbf{e}_j \geq 0, \;\; j > 1$$

and we conclude that $\left(A^{(1)}\right)^{-1} \geq 0$, and so $A^{(1)}$ is an M–matrix. □

Remark 3.5. Lemma 3.1 implies that in the LU decomposition of A, if A is an M–matrix, then so is U.

Lemma 3.2 (Fan [226]). *Dropping off-diagonal entries preserves the M–matrix property.*

Proof. Using Definition 3.6 gives that dropping $a_{i,j}$ elements for $i \neq j$ does not alter M–matrix property. □

Remark 3.6. The two previous lemmas 3.1 and 3.2 imply that an incomplete LU factorization for A, say $\tilde{L}\tilde{U}$, if A is an M–matrix, then also $\tilde{U}$ is an M–matrix.

Lemma 3.3 (Fan [226]). *If A is an M–matrix the inverse of the elementary Gaussian transformation $L^{(i)}$ is also an M–matrix.*

Proof. Without loss of generality we can restrict ourselves to the $L^{(1)}$ matrix; it differs from the identity matrix just for the entry

$$l_{i,1}^{(1)} = -\frac{a_{i,1}}{a_{1,1}} \geq 0,$$

by the property of M matrix of A, hence $L^{(1)} \geq 0$. The inverse of $L^{(1)}$ is the $L^{(1)}$ matrix with the off-diagonal element changed in sign, so $(L^{-1})_{i,1}^{(1)} \leq 0$, and thus the matrix is an M–matrix. □

Lemma 3.4 (Fan [226]). *If A is an M–matrix, then so is the factor L in the LU factorization: $A = LU$.*

Proof. Formalizing the Gauss elimination process as:

$$L^{(n)}L^{(n-1)}\cdot\ldots\cdot L^{(1)}A = U,$$

we find

$$L = \left(L^{(1)}\right)^{-1}\left(L^{(2)}\right)^{-1}\cdot\ldots\cdot\left(L^{n)}\right)^{-1} \Rightarrow l_{i,j} = \left(L^{(j)}\right)^{-1}_{i,j},$$

and in this way L satisfies the Definition of M–matrix 3.4. □

3.4.3 Fill–in strategies: threshold and levels of fill

Putting all the results together we can formalize the incomplete LU factorization existence as:

Theorem 3.2 (Incomplete LU). *Let A be an M–matrix and P a given non–zero pattern. Then, the Algorithm 3.8 produces an incomplete factorization $\tilde{L}\tilde{U}$ for A in which both $\tilde{L}$ and $\tilde{U}$ are non–singular M-matrices. Moreover, $A = \tilde{L}\tilde{U} - R$ is also a regular splitting of A.*

Algorithm 3.8: General Incomplete LU Factorization.

Input: Matrix $A = (a_{i,j})_{i,j=1,\ldots,n}$.
Output: Matrix A factored in incomplete LU form.

for $i = 2, 3, \ldots, n$ **do**
  for $k = 1, \ldots, i-1$ **do**
    if $(i,k) \in P$ **then**
      $a_{i,k} \leftarrow a_{i,k}/a_{k,k}$;
      for $j = k+1, \ldots, n$ **do**
        if $(i,j) \in P$ **then**
          $a_{i,j} \leftarrow a_{i,j} - a_{i,k} \cdot a_{k,j}$;

As a corollary of the previous result, we have the existence of the incomplete LU factorization also for a slightly more general set of matrices. Following [376], we start defining H–matrices.

Definition 3.8 (H–matrix). Let $A = (a_{i,j}) \in \mathbb{R}^{n\times n}$, B such that $B = (b_{i,j}) \in \mathbb{R}^{n\times n}$, $b_{i,i} = a_{i,i}\ \forall\, i = 1, \ldots, n$ and $b_{i,j} = -|a_{i,j}|$ for $i \neq j$. A is an H–**matrix** if B is an M–matrix.

The following corollary to Theorem 3.2 is then straightforward.

Corollary 3.1. *Let $A \in \mathbb{R}^{n\times n}$ be an H–matrix with positive diagonal entries* *and $\hat{A}$ be the associate M–matrix. The* ILU *factorization with any nonzero pattern P gives*

$$\begin{aligned} A &= LU - R,\ \Sigma = \text{diag}(U) = \text{diag}(\sigma_1, \sigma_2, \ldots, \sigma_n), \\ \hat{A} &= \hat{L}\hat{U} - \hat{R},\ \hat{\Sigma} = \text{diag}(\hat{U}) = \text{diag}(\hat{\sigma}_1, \hat{\sigma}_2, \ldots, \hat{\sigma}_n), \end{aligned}$$

with $0 < \hat{\sigma}_i \leq \sigma_i$, $i = 1, 2, \ldots, n$, *and:*

$$\begin{array}{l} \hat{l}_{j,i} \leq -|l_{j,i}| \leq 0, \\ \hat{u}_{i,j} \leq -|u_{i,j}| \leq 0, \end{array} \quad j = i+1, \ldots, n.$$

Now we can concern ourselves with the choice of the pattern for this kind of factorization. The first simple idea can be choosing P as the sparsity pattern of the matrix A. In this way, we have the $ILU(0)$ algorithm.

The accuracy of the ILU(0) incomplete factorization can not be sufficient to yield an adequate rate of convergence; see examples at the end of the section. More accurate Incomplete LU factorizations are often more efficient as well as more reliable. These more accurate factorizations differ from ILU(0) because they allow a certain fill–in, i.e., some entries in the incomplete factorization are also nonzero in places where A has only zeros. In order to increase the fill–in, two main strategies can be used:

- allow entries in certain prescribed places in the incomplete factors (we need the concept of *level of fill*);
- discard entries during the computation of the incomplete factorization (we need the concept of *threshold*).

Both strategies can be combined in order to increase fill–in by levels and threshold. Let us start by describing the *levels of fill* in an ILU factorization first.

Fill–in strategies: levels of fill

The general version of the ILU algorithm with respect to a pattern can be formulated by defining the concept of *level of fill*, that is an integer quantity given to each element that is processed by standard Gaussian–elimination and represents its magnitude or importance in order to discard it or not. At the beginning of the factorization routine we say that an entry (i, j) of A has a level of fill 0 if in the original matrix A we have $a_{i,j} \neq 0$, otherwise we give to that entry a level of fill of ∞. In compact form:

$$\text{Inizialization: lev}_{i,j} = \begin{cases} 0 & a_{i,j} \neq 0, \text{ or i = j} \\ \infty & \text{otherwise} \end{cases} \tag{3.15}$$

Thus, entries with a lower level of fill are more important. Having completed the initialization step, we need to propagate the levels by following the pattern of access of the elimination algorithm. If we look at line 7 of Algorithm 3.8, we observe that a generic element $a_{i,j}$ is updated by the formula:

$$a_{i,j} \leftarrow a_{i,j} - a_{i,k} \cdot a_{k,j},$$

if $\text{lev}_{i,j}$ is the current level of the element $a_{i,j}$, we need to update it considering the quantities relative to the update, i.e., $\text{lev}_{i,j}$, $\text{lev}_{i,k}$ and $\text{lev}_{k,j}$. It is natural to define the new level of fill as the minimum between the actual level of fill of the element (i, j) and the increase in level obtained by accessing the elements (i, k) and (k, j). We can propagate the level of the element (i, j) by taking

$$\text{lev}_{i,j} = \min\{\text{lev}_{i,j}, \text{lev}_{i,k} + \text{lev}_{k,j} + 1\}.$$

This is done by executing a level update as shown in Algorithm 3.9. Then we obtain the general ILU(p) algorithm with p level of fill by defining the pattern:

$$P_p = \{(i, j) \,:\, \text{lev}_{i,j} \leq p\}. \tag{3.16}$$

and using it in Algorithm 3.8. We stress that this is a purely topological strategy, i.e., we do not take into account the values of entries being treated, even if it is reasonable to assume that the higher the level of fill is, the smaller are the computed elements. Moreover, we have now described the complete algorithm as a procedure in two steps, first computing the pattern P_p, then appling the general Incomplete LU Factorization Algorithm 3.8 with the pattern P_p from equation (3.16). The entire procedure can be combined, resulting in a single–step procedure that propagates the level and computes the actual factorization.

Algorithm 3.9: Level Computation

Input: A starting matrix for pattern P initialized as in (3.15)
Output: A matrix with the complete level of computation.

for $i = 2, \ldots, n$ **do**
    for $k = 1, \ldots, i - 1$ **do**
        for $j = k + 1, \ldots, n$ **do**
            $\text{lev}_{i,j} = \min\{\text{lev}_{i,j}, \text{lev}_{i,k} + \text{lev}_{k,j} + 1\}$;

Example 3.4. Example of the application of the ILU(p) preconditioner combined with GMRES to solve the linear system $A\mathbf{x} = \mathbf{b}$ with the matrix A as the HB/sherman1, from the Harwell-Boeing Collection[6], and the $\mathbf{b}$ vector associated. The size of the problem is $\dim(A) = 1000 \times 1000$, $\text{nnz}(A) = 3750$, and the GMRES is set to reach a $\varepsilon_{\text{GMRES}} = 1.e - 9$. The sparsity pattern of the various preconditioners are in Figure 3.4, while in Table 3.1 are the details of the solution of the linear system.

✓

[6] Information about this matrix can be found in [177].

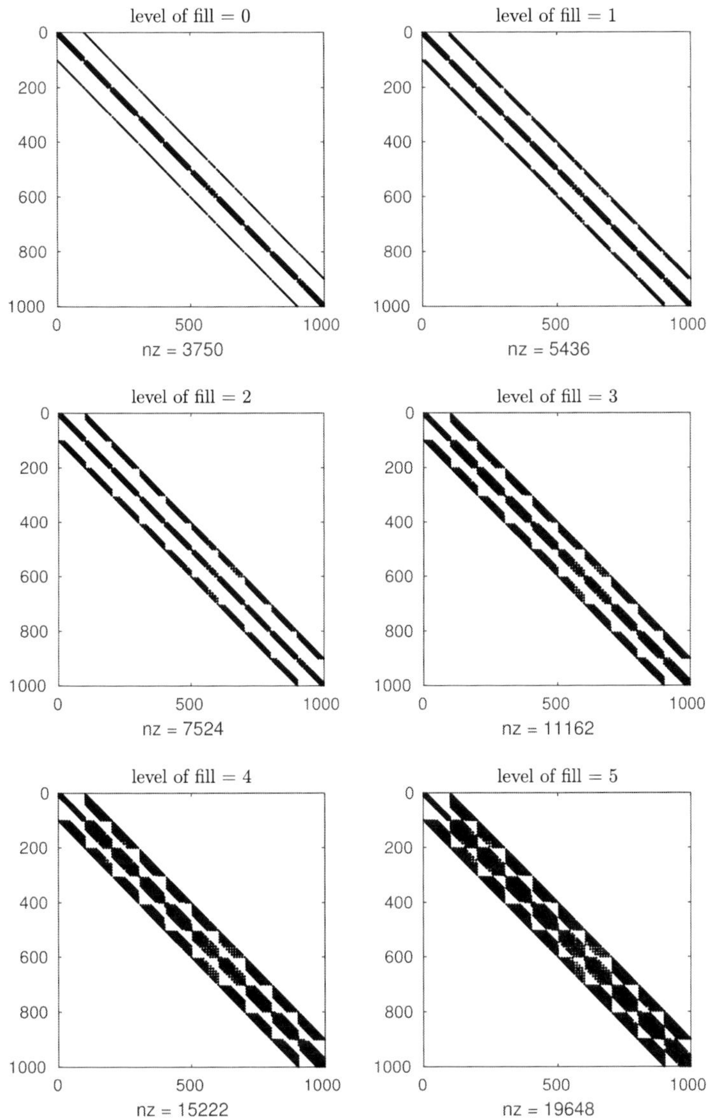

Figure 3.4: Example 3.4. Sparsity pattern of preconditioner ILU(p) for various values of p.

Table 3.1: Example 3.4. Performance of GMRES with ILU(p) and HB/sherman1

Preconditioner	Preconditioner Time (s)	Iteration	GMRES Time (s)
ILU(0)	7.08e-04	39	1.96e-02
ILU(1)	7.26e-04	21	9.83e-03
ILU(2)	1.10e-03	17	8.90e-03
ILU(3)	2.04e-03	12	6.47e-03
ILU(4)	2.63e-03	10	6.37e-03
ILU(5)	3.92e-03	8	5.11e-03
–	–	305	5.52e-01

More information on the levels of fill can be found in [455] and references therein.

Fill–in strategies: threshold

Let us consider the incomplete LU factorization using a dropping strategy based on a *threshold*. References to this can be found in [21, 453, 455]. The idea is replacing an element with zero if it satisfies a set of dropping criteria related to its absolute value.

We can apply a dropping rule row-wise by the same rule to all the elements of the row, as in Algorithm 3.10.

Algorithm 3.10: Algorithm ILUT, row–wise.

Input: A sparse matrix $A \in \mathbb{R}^{n\times n}$.
Output: L, U computed with threshold dropping.

for $i = 1, \dots, n$ **do**
    $\mathbf{w} \leftarrow \mathbf{a}_{i,:}$;
    for $k = 1, \dots, i-1$ *and* $w_k \neq 0$ **do**
        $w_k \leftarrow w_k / a_{k,k}$;
        Apply a *dropping rule (1)* to w_k;
        if $w_k \neq 0$ **then**
            $\mathbf{w} \leftarrow \mathbf{w} - w_k \cdot \mathbf{u}_{k,:}$;
    Apply a *dropping rule (2)* to row $\mathbf{w}$;
    for $j = 1, \dots, i-1$ **do**
        $l_{i,j} \leftarrow w_j$;
    for $j = i, \dots, n$ **do**
        $u_{i,j} \leftarrow w_j$;
    $\mathbf{w} \leftarrow 0$;

Depending on the choice of the dropping strategies in Algorithm 3.10, we can construct different incomplete factorization algorithms, and also the ILU(0) can be reinterpreted.

Definition 3.9 (ILUT(p,τ)). The ILUT(p,τ) algorithm is based on the following rules:

(1) define the relative tolerance $\tau_i = \tau \cdot \mathbf{a}_{i,:}$. If $\omega_k < \tau_i \Rightarrow \omega_k = 0$.

(2) Apply the dropping rule to the whole vector $\mathbf{w}$, then keep only the p largest modulus element in the L part of the row and the p largest modulus element in the U part of the row.

(3) The diagonal elements are *never* dropped.

The parameter p in the algorithm controls the fill-in of the matrix, i.e., the memory usage. The idea is similar to the level of fill-in of the ILU(p) algorithm, however it relies upon the value of the elements and not upon their position.

Now we want to prove the existence of this factorization, in a similar way to what we have done for the ILU(p). To achieve this, we have to do some preliminary work. Firstly, define a class of matrices, similar to what done with the M-matrices:

Definition 3.10 ($\hat{M}$-matrix). A matrix $H \in \mathbb{R}^{n\times n}$ is an $\hat{M}$-matrix if it satisfies the following conditions:

1. $h_{i,i} > 0$, $1 \leq i < n$ and $h_{n,n} \geq 0$;
2. $h_{i,j} \leq 0$, $i, j = 1, 2, \ldots, n$ and $i \neq j$;
3. $\sum_{j=i+1}^{n} h_{i,j} < 0$, $1 \leq i < n$.

Definition 3.11 (Row Sum). Given an $\hat{M}$-matrix $H \in \mathbb{R}^{n\times n}$ we define the **row sum** of the i-th row as:

$$\text{rs}(\mathbf{h}_{i,:}) = < \mathbf{h}_{i,;}, \mathbf{e} > = \sum_{j=1}^{n} h_{i,j}.$$

Definition 3.12 (Dominance). Given an $\hat{M}$-matrix $H \in \mathbb{R}^{n\times n}$, the *row* $\mathbf{h}_{i,:}$ is *diagonally dominant* if $\text{rs}(\mathbf{h}_{i,:}) \geq 0$. An *$\hat{M}$-matrix* $H \in \mathbb{R}^{n\times n}$ is *diagonally dominant* if all rows are diagonally dominant.

To prove the existence of the factorization, a slight different dropping rule should be given, selecting some particular elements that are not to be dropped in any case.

Definition 3.13 (Drop strategy II).

$$a_{i,j_i} = \max_{j=i+1,\ldots,n} |a_{i,j}|, \quad \forall i < n$$

The elements generated in position (i, j_i) during the ILUT procedure (Algorithm 3.10) are not subject to the dropping rule in Definition 3.9.

We also need to establish the following notation relative to the Algorithm 3.10.

Remark 3.7. The row vector $\mathbf{w}$ resulting from line 4 of the Algorithm 3.10 is denoted as $\mathbf{u}_{i,:}^{(k+1)}$ in the following[7]. In this way the algorithm at the generic step $k = 1, \ldots, i-1$ becomes:

$$l_{i,k} = u_{i,k}^{(k)}/u_{k,k};$$

if $|l_{i,k}|$ meets the dropping rule $l_{i,k} = 0$, else:

$$u_{i,j}^{(k+1)} = u_{i,j}^{(k)} - l_{i,k} \cdot u_{k,j} - r_{i,j}^{k}, \quad j = k+1, \ldots, n,$$

where, following the new notation, $\mathbf{u}_{i,:}^{1} = \mathbf{a}_{i,:}$ and where $r_{i,j}^{(k)}$ is based on the dropping strategy, i.e.,

$$\begin{array}{ll} r_{i,j}^{(k)} = 0 & \Rightarrow \text{No dropping,} \\ r_{i,j}^{(k)} = u_{i,j}^{(k)} - l_{i,k} \cdot u_{k,j} & \Rightarrow u_{i,j}^{(k+1)} \text{ dropped.} \end{array}$$

In this way the i-th row of U at the i-th step of the Gaussian–elimination is:

$$\mathbf{u}_{i,:} = \mathbf{u}_{i-1,:}^{(i)}$$

and this satisfies the relation:

$$\mathbf{a}_{i,:} = \sum_{k=1}^{i} l_{k,j} \cdot u_{i,:}^{(k)} + \mathbf{r}_{i,:}.$$

where $\mathbf{r}_{i,:} = (r_{i,j})_{j \leq k}$.

Theorem 3.3 ($ILUT(p, \tau)$ existence). *Given a matrix $A \in \mathbb{R}^{n \times n}$ such that A is a diagonally dominant $\hat{M}$-matrix (Definition 3.12), then the rows $\mathbf{u}_{i,:}^{(k)}$, $k = 0, 1, 2, \ldots, i$ defined by:*

$$\begin{array}{l} u_{i,j}^{(k+1)} = u_{i,j}^{(k)} - l_{i,k} \cdot u_{k,j} - r_{i,j}^{k}, \quad j = k+1, \ldots, n; \\ \mathbf{u}_{i,:}^{(0)} = \mathbf{0}, \\ \mathbf{u}_{i,:}^{(1)} = \mathbf{a}_{i,:}, \end{array}$$

satisfy the following relation for $k = 1, \ldots, l$:

$$u_{i,j}^{(k)} \leq 0 \ j \neq i, \tag{3.17}$$

$$\operatorname{rs}(\mathbf{u}_{i,:}^{(k)}) \geq \operatorname{rs}(\mathbf{u}_{i,:}^{(k-1)}) \geq 0, \tag{3.18}$$

$$u_{i,i}^{(k)} > 0 \text{ when } i < n \text{ and } \mathbf{u}_{n,n}^{k} \geq 0. \tag{3.19}$$

[7]Note that the elements $u_{i,j}^{(k+1)} = 0$ for $j \leq k$

Proof. By induction over k. The result is trivially true for $k = 0$. To prove relation (3.17), start from

$$u_{i,j}^{(k+1)} = u_{i,j}^{(k)} - \underbrace{l_{i,k}}_{\leq 0} \cdot \underbrace{u_{k,j}}_{\leq 0} - \underbrace{r_{i,j}^{k}}_{\substack{=0, \text{ or} \\ =u_{i,j}^{(k)} - l_{i,k} \cdot u_{k,j}}},$$

Then, $u_{i,j}^{(k+1)} \leq u_{i,j}^{(k)} \leq 0$ or, being replaced by 0, $u_{i,j}^{(k+1)} \leq 0$. Thus, (3.17) is proved.

Now, to prove (3.18), suppose that it holds true for k. By the previous arguments $\mathbf{r}_{i,j}^{(k)} = 0$ except when the j-th element in the row is dropped, in which case $u_{i,j}^{(k+1)} = 0$, and $\mathbf{r}_{i,j}^{(k)} = u_{i,j}^{(k)} - l_{i,k} \cdot u_{k,j} \leq 0$. Therefore, $\mathbf{r}_{i,j}^{(k)} \leq 0$ always. Moreover, when an element in position (i, j) is not dropped,

$$\mathbf{u}_{i,j}^{(k+1)} = u_{i,j}^{(k)} - l_{i,k} \cdot u_{k,j} \leq u_{i,j}^{(k)},$$

and in particular by the dropping rule (Definition 3.13), for $i < n$, always have for $j = j_i$: $u_{i,j_i}^{k+1} \leq u_{i,j_i}^{(k)}$. Now, consider the row sums:

$$\begin{aligned} \text{rs}(\mathbf{u}_{i,:}^{(k)}) &= \text{rs}(\mathbf{u}_{i,:}^{(k)}) - l_{i,k} \cdot \text{rs}(\mathbf{u}_{k,:}) - \text{rs}(\mathbf{r}_{i,:}^{(k)}) \\ &\geq \text{rs}(\mathbf{u}_{i,:}^{(k)}) - l_{i,k} \cdot \text{rs}(\mathbf{u}_{k,:}) \\ &\geq \text{rs}(\mathbf{u}_{i,:}^{(k)}), \end{aligned}$$

and this establishes (3.18).

To prove the last relation (3.19), by the latter (3.18) for $i < n$:

$$\begin{aligned} u_{i,i}^{(k+1)} &\geq \sum_{j=k+1}^{n} \left(-u_{i,j}^{(k+1)}\right) = \sum_{j=k+1}^{n} \left|u_{i,j}^{(k+1)}\right| \\ &\geq \left|u_{i,j_i}^{(k+1)}\right| \geq \left|u_{i,j_i}^{(k)}\right| \geq \dots \\ &\geq \left|u_{i,j_i^{(1)}}\right| = |a_{i,j_i}|. \end{aligned}$$

By the definition of the dropping rule (Definition 3.13) and the property of A being a $\hat{M}$-matrices the proof is over. □

In an analogue way to what we have done for the left-looking row-oriented, Algorithm 3.10 can be arranged in a right-looking row-oriented way as in [229], giving rise to an analogue proof of the existence of the incomplete factorization.

Example 3.5. As an example we will consider the solution with GMRES algorithm of the linear system $A\mathbf{x} = \mathbf{b}$, with A as the HB/Sherman4 matrix[8] and $\mathbf{b}$

[8]Information about this matrix can be found in [177].

the relative right–hand side. The size of the matrix is 1104 and $\text{nnz}(A) = 3786$. GMRES is set to achieve the convergence with a tolerance of $\varepsilon_{\text{GMRES}} = 1e-9$, and the maximum number of inner iteration set to 20. The pattern of the ILUT matrix for various tolerance are in Figure 3.5, while the data relative to the solution are in Table 3.2. ✓

Table 3.2: Example 3.5. GMRES(20) and ILUTP with various tolerances

Tolerance	Preconditioner time (s)	Iteration	GMRES time (s)
ILUT(1.0e-01)	1.15e-03	37	2.26e-02
ILUT(1.0e-02)	2.82e-03	15	1.07e-02
ILUT(1.0e-03)	8.67e-03	7	8.05e-03
ILUT(1.0e-04)	1.67e-02	4	7.86e-03
ILUT(1.0e-05)	2.53e-02	3	8.30e-03
		115	9.49e-02

Remark 3.8. All the different choices of the dropping rules shown for the ILU factorization can be extended naturally to the case of the IC factorization; in this way the IC(P), IC(ε) algorithms can be easily defined as special cases. They are the incomplete Cholesky factorization with an arbitrary pattern P and the incomplete Cholesky factorization with a drop tolerance ε, respectively.

3.4.4 Existence and stability issues

The ILUT(p,τ) can fail for many of the matrices that do not satisfy the condition of being a diagonally dominant $\hat{M}$-matrix, encountering problems like zero pivots or exponentially growing (decaying) of the entries of the factors causing overflow/underflow condition or numerical instability. To face these issues, partial pivoting can be taken into account, giving rise to the *ILUTP* variant, where P stays indeed for pivoting.

Recall that *pivoting* is the process that consists of findind a nonzero entry in the order k head submatrix or at least in the kth row or column, and placing it in the (k,k)-diagonal position by interchanging rows/columns during the incomplete factorization process. An equivalent decomposition is recovered by using appropriate permutation matrices. Because of the data structure used in ILUT, row pivoting is not practical. Instead, column pivoting can be viable. See [455, Section 10.4.4] for the application to incomplete factorizations, and, e.g., [271] for generalities on pivoting techniques for direct methods.

It can be useful to specialize ILU algorithms for handling symmetric matrices because if A is symmetric, an ILU preconditioner can be nonsymmetric in general. In this case, an approach to achieve the LDU incomplete factorization can be used. In a way that is the complete analogue of what is done

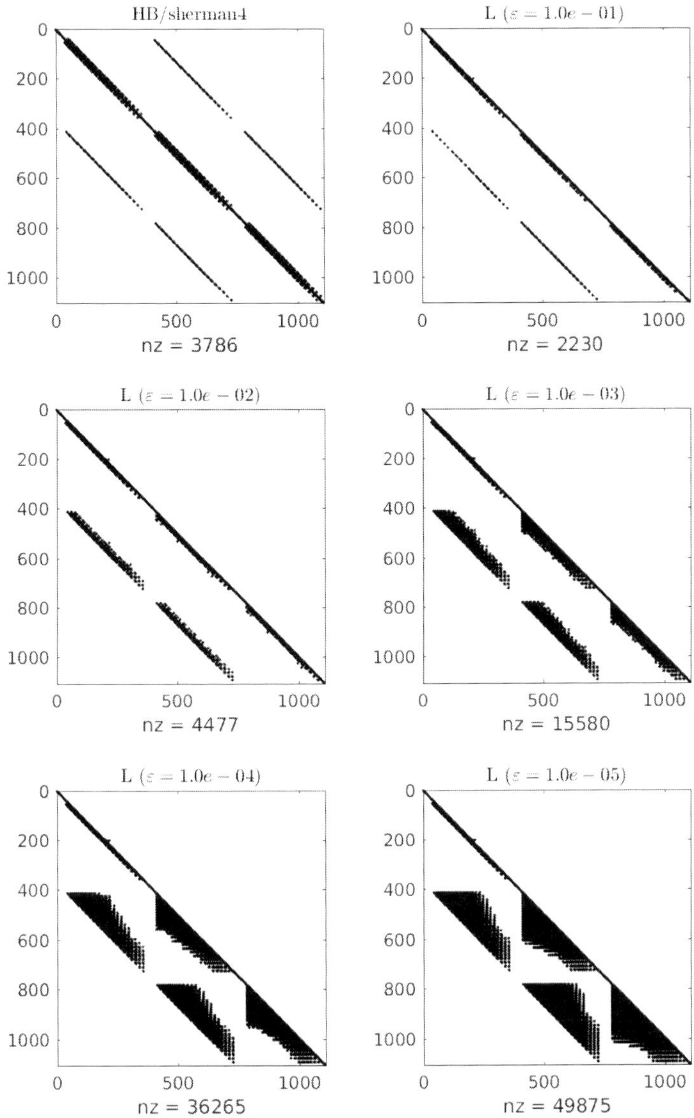

Figure 3.5: Example 3.5 (matrix HB/Sherman4). Patterns for ILUTP with various drop tolerances

with the IC factorization in § 3.4.1, let us start by defining $A^{(n)} = A$ and considering the sequence of matrices:

$$A^{(k+1)} = \begin{pmatrix} A^{(k)} & \mathbf{v}^{(k)} \\ \mathbf{w}^{(k)} & \alpha^{(k+1)} \end{pmatrix}, \quad A^{(1)} = D^{(1)} = a_{1,1}, \; L^{(1)} = U^{(1)} = 1.$$

When $A^{(k)}$ is non singular and has the *LDU* factorization $A^{(k)} = L^{(k)}D^{(k)}U^{(k)}$, then the *LDU* factorization of $A^{(k+1)}$ is given by:

$$\begin{aligned} A^{(k+1)} =& L^{(k+1)}D^{(k+1)}U^{(k+1)}, \\ L^{(k+1)} =& \begin{pmatrix} L^{(k)} & 0 \\ \mathbf{w}^{(k)}U_{(k)}^{-1}D_{(k)}^{-1} & 1 \end{pmatrix}, \\ D^{(k+1)} =& \begin{pmatrix} D^{(k)} & 0 \\ 0 & \alpha^{(k+1)} - \mathbf{y}^{(k)}D^{(k)}\mathbf{z}^{(k)} \end{pmatrix}, \\ U^{(k+1)} =& \begin{pmatrix} U^{(k)} & D_{(k)}^{-1}L_{(k)}^{-1}\mathbf{v}^{(k)} \\ 0 & 1 \end{pmatrix}. \end{aligned}$$

The implementation of this formula gives rise to the *ILUS* algorithm, where indeed *S* stand for Symmetric.

The other variant considered is the Crout variant of the ILU, i.e., the Crout strategy applied to the ILUT: the Algorithm 3.11 called ILUC.

Algorithm 3.11: ILUC - Crout version of ILU

Input: Matrix A, dropping tolerance.
Output: Matrix L and matrix U.

for $k = 1, \ldots, n$ **do**
  Initialize row $\mathbf{z}$: $\mathbf{z}_{1:k-1} = 0$, $\mathbf{z}_{k:n} = \mathbf{a}_{k,k:n}$;
  for $\{i \,|\, 1 \le i \le k-1 \textit{ and } l_{ki} \neq 0\}$ **do**
    $\mathbf{z}_{k:n} = \mathbf{z}_{k:n} - l_{k,i} \cdot \mathbf{u}_{i,k:n}$;
  Initialize column $\mathbf{w}$: $\mathbf{w}_{1:k} = 0$, $\mathbf{w}_{k+1:n} = \mathbf{a}_{k+1:n,k}$;
  for $\{i \,|\, 1 \le i \le k-1 \textit{ and } u_{i,k} \neq 0\}$ **do**
    $\mathbf{w}_{k+1:n} = \mathbf{w}_{k+1:n} - u_{i,k} \cdot \mathbf{l}_{k+1:n,i}$;
  Apply a dropping rule to row $\mathbf{z}$;
  Apply a dropping rule to column $\mathbf{w}$;
  $\mathbf{u}_{k,:} = \mathbf{z}$;
  $\mathbf{l}_{:,k} = \mathbf{w}/u_{k,k}$;
  $l_{k,k} = 1$;

The idea on the latter algorithm is based on computing the L factor starting from the bottom-left and the U from the upper-right. The last ILU variant considered here is the *MILU* approach, that is based on not completely discarding the dropped elements, but subtracting the discarded from the corresponding diagonal entry at the end of the iteration in the ILU(p) Algorithm 3.8. To obtain this, add the instruction: $u_{i,i} = u_{i,i} - < \mathbf{r}_{i,:}, \mathbf{e} >$, in which $\mathbf{e} = (1, 1, \ldots, 1)^T$ that in row form becomes $\mathbf{u}_{i,:} = \mathbf{u}_{i,:} - < \mathbf{r}_{i,:}, \mathbf{e} > \mathbf{e}_i^T$ modifying the algorithm as:

$$\mathbf{a}_{i,:} = \sum_{k=1}^{i} l_{i,k} \cdot \mathbf{u}_{k,:} + < \mathbf{r}_{i,:}, \mathbf{e} > \mathbf{e}_i^T - \mathbf{r}_{i,:}. \tag{3.20}$$

This variant of the algorithm sometimes recover the loss of performance of the classical ILU algorithm when applied to the discretization of *elliptic problems*. This behavior has already been observed in [217] and in [208], where a first proposal of solution was presented. The variants described with equation (3.20) have been introduced in [286], in which the strategy was also applied to a general matrix A. For a discussion on the differences between the two approaches when applied to matrices coming from finite differences in discretization of elliptic equations, see [153].

What we need to stress is that the incomplete factorization algorithm can fail for general symmetric and positive definite matrices, or SPD for short as usual, and, more frequently, for matrices that are non symmetric or indefinite.

In the SPD case, i.e., when thinking about the computation of the incomplete Cholesky factorization (§ 3.4.1) what can be done is acting on the matrix A. For example one can try to increase the diagonal dominance of the matrix A, either in a local or global way, by suitable permutations. On the other hand, one can compute the incomplete Cholesky factorization for a diagonal shift of the matrix $\tilde{A} = A + \alpha I$, I being the identity matrix and $\alpha \in \mathbb{R}$, see [376]. Obviously there exists a value of $\alpha^* > 0$ such that, for $\alpha \geq \alpha^*$, the factorization of $\tilde{A}$ surely exists and it is the one for which $\tilde{A}$ becomes diagonally dominant and therefore an H–matrix. Nevertheless, numerical experiments show that the optimal global α is usually smaller than the α^*. On the other hand, the other approach that can be viable is based on substituting the zero (or numerically zero) or negative pivots with an arbitrarily or heuristically determined constant, see [375, 465, 526]. Although this change ensures the generation of a well-defined factorization, it can result in a poor preconditioner, having increased the quantity $\|A - \tilde{L}\tilde{L}^T\|_F$ without control.

Example 3.6 (Clustering of the eigenvalues and optimality). Consider the same test problem from Example 2.3 based on the finite difference discretization of the convection–diffusion equation. In this case, various ILU preconditioners are considered. Table 3.3 reports the number of matrix–vector products needed for achieving a relative residual norm less than $\varepsilon = 1e - 6$ for the GMRES, GMRES(20) and BiCGstab methods, unpreconditioned and used with the ILU(0) preconditioner and the ILUT(τ), with a drop tolerance of $\tau = 1e - 2$. A modified ILU approach is used on the row (see [153] for a discussion about coupling modified ILU approach and finite difference discretizations) for preconditioner and using the Crout variant of the ILUT algorithm. When the size of A_h increases, it grows as N^2, the number of matrix vector products, and thus the computing time for the unpreconditioned methods grows accordingly. On the other hand, when using preconditioning strategies, two facts can be

Table 3.3: Example 3.6. Matrix–vector multiplications (IT) and computing time (T(s)) for the solution of the steady state convection–diffusion equation for several preconditioners and methods. A † is reported when the method do not converge in N^2 matrix vector products

preconditioner	*method*					
–	GMRES		GMRES(20)		BiCGSTAB	
N	IT	T(s)	IT	T(s)	IT	T(s)
10	45	0.3428	87	0.0337	186	0.0821
30	87	0.1996	180	0.2137	737	0.1407
110	238	7.1761	355	0.9737	†	†
210	431	191.6226	555	5.1138	†	†
ILU(0)	GMRES		GMRES(20)		BiCGSTAB	
N	IT	T(s)	IT	T(s)	IT	T(s)
10	16	0.1024	36	0.0634	21	0.0231
30	17	0.2260	37	0.0116	24	0.0069
110	19	0.9300	39	0.0663	32	0.0546
210	19	112.8163	39	1.3670	31	0.9089
ILUT(1e-2)	GMRES		GMRES(20)		BiCGSTAB	
N	IT	T(s)	IT	T(s)	IT	T(s)
10	4	0.0033	24	0.0032	4	0.0025
30	5	0.0201	25	0.0121	5	0.0038
110	6	0.8905	26	0.0176	7	0.0105
210	6	109.8935	26	1.3121	8	0.6051

observed: a drop in the overall number of iterations (that has to be expected since we are performing a preconditioning of the system) and the number of iterations becomes substantially independent from the size of the problem. This is what is usually called a condition of *optimality* of the preconditioner. This behavior goes along with the spectral analysis since, see Figure 3.6. The application of incomplete LU factorizations preconditioners produces a cluster in the spectrum of the preconditioned matrix, see § 2.2.8. Finally, let us look at the timings of the standard GMRES algorithm. Even if preconditioning reduces substantially the number of iterations, timings are far greater than for the preconditioned version of the other methods. As in § 2.2.7, this is due to the fact that full Krylov basis must be stored, and thus each new iteration becomes computationally heavier than the previous. On the other hand, a smart usage of memory for both the restarted GMRES or using BiCGstab, produce much more reasonable timings. ✓

Detailed information about these techniques and implementation issues can be found in [355, 437, 455].

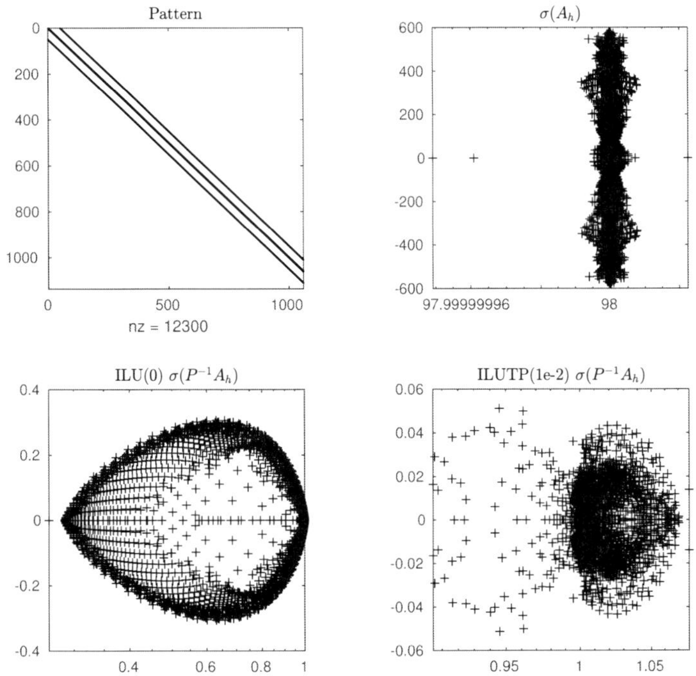

Figure 3.6: Example 3.6. Pattern of the A_h matrix, spectrum and clusters of the preconditioned versions.

Finally, note that the ILU factorization techniques were developed originally for M–matrices which arise from the discretization of elliptic PDEs. However, in general A is not an M– or H–matrix. For some important problems A can be definite (often nonsymmetric), such as in some linear convection-diffusion but quite rarely A can be, e.g., diagonally dominant and very poor performances can be observed for ILU preconditioners. So, be careful.

Remark 3.9. When A is indefinite but often just highly nonsymmetric, standard ILU may face several difficulties, mainly:

- breakdown due to a zero pivot;
- the quality of the preconditioner becomes very poor.

The first issue is known but the second is more delicate. Let us write A as $A = \tilde{L}\tilde{U} + \Delta$, where $\tilde{L}$ and $\tilde{U}$ are nonsingular and Δ is a residual matrix. By using $M = \tilde{L}\tilde{U}$ as a left preconditioner (but the same effect is produced for a right or split) we obtain

$$M^{-1}A = I + \tilde{U}^{-1}\tilde{L}^{-1}\Delta, = I + \tilde{\Delta}.$$

If A is, e.g., diagonally dominant, then it can be proved that $||\tilde{\Delta}||$ is small. In general, $||\tilde{\Delta}||$ can be very large when $||\tilde{U}^{-1}||$ and/or $||\tilde{L}^{-1}||$ are large even if $||\Delta||$ is small. Therefore, $I + \tilde{\Delta}$, the preconditioned matrix, cannot be considered at all a perturbation of the identity matrix and sometimes preconditioned methods can perform (much) worse than nonpreconditioned.

3.5 Approximate inverse preconditioners

The so–called *Approximate inverses* or *Approximate inverse preconditioners* are preconditioners approximating directly A^{-1} and not A, as usual. They have been intensively studied recently, in particular in the past twenty years; see the review [47] and, e.g., the more recent [73] and references therein. They do not require solving a linear system to be applied and often have very interesting parallel potentialities. On the other hand, they usually face a computational cost sensibly higher with respect to, e.g., ILU techniques considered in the previous paragraphs.

However, as observed in Remark 3.9, ILU can perform very poorly for important applications where A is not an M- or an H-matrix and/or that require a parallel implementation.

One possible remedy is to try to find a preconditioner that does not require solving a linear system. For example, the original system can be preconditioned by a matrix M which is a direct approximation to the inverse of A, thus requiring just matrix-vector multiplications for its application.

Main approximate inverse preconditioners

There exist several completely different algorithms for computing a sparse approximate inverse, with each approach having its own strengths and limitations. In the following, the discussion is limited to those methods that have been shown to be reasonably competitive in terms of performance, flexibility and robustness.

Usually two main basic types of approximate inverses exist, depending on whether the preconditioner M, approximating A^{-1}, is expressed as a single matrix or as product of two or more matrices. The latter type of preconditioners are known as *factored sparse approximate inverses*, and they are of the form $M = Z\,D^{-1}\,W^T$ or $M = Z\,W^T$, where Z and W are lower triangular matrices and D is diagonal, assuming that M admits a LU factorization. Within each class, there are several different techniques, depending on the algorithm used to compute the approximate inverse or approximate inverse factors. At present, there are three approaches that are more used:

- *Residual norm minimization,*

- *inversion and sparsification of an ILU*, and
- *incomplete bi-conjugation.*

However, before discussing them, some considerations on the entries of the inverse of a given matrix are essential.

3.5.1 On the decay of the entries of A^{-1}

At the beginning of the discussion on preconditioners, in Theorem 3.1 it was observed that the inverse A^{-1} of a sparse matrix A has often significantly more entries than A. Nevertheless, under suitable conditions, the magnitude of the elements of the inverse can play a fundamental role. Prior to analyzing this problem, consider an example that guides us.

Example 3.7. Consider the tridiagonal symmetric positive definite matrix

$$A = \begin{pmatrix} 2 & -1 & 0 & \dots & 0 \\ -1 & 2 & -1 & \ddots & \vdots \\ 0 & \ddots & \ddots & \ddots & 0 \\ \vdots & \ddots & -1 & 2 & -1 \\ 0 & \dots & 0 & -1 & 2 \end{pmatrix}_{n\times n} .$$

Instead of representing A with its pattern, observe its *cityplot*, see Figure 3.7. Even if the elements of the inverse are all different from zero, note the

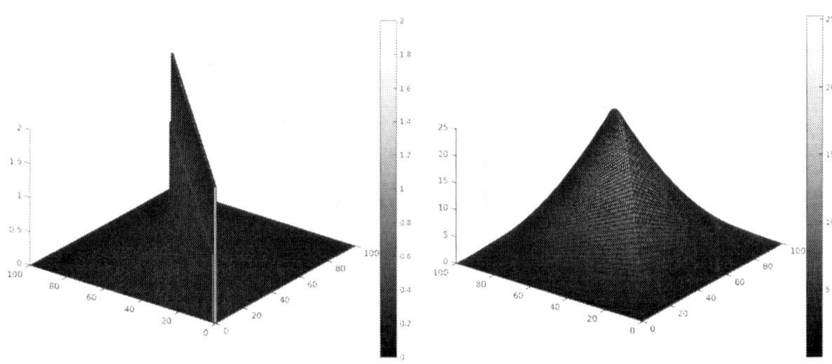

Figure 3.7: Example 3.7. Cityplots of the A matrix (on the left) and of A^{-1} on the right.

presence of a *decay*, i.e., a decreasing of their absolute value away from the main diagonal of the matrix. ✓

Let us start investigating this decay in the inverse matrices in order to use it, whenever it exists, for preconditioning strategies.

The first result that can help is due to Demko, Moss and Smith.

Theorem 3.4 (Demko et al. [185]). *Let A and A^{-1} be in $\mathcal{B}(\ell^2(s))$. Then if A is positive definite and m-banded,*

$$(|A^{-1}|)_{i,j=1}^{n} = |a_{i,j}^{-1}| \leq C\lambda^{|i-j|},$$

where:

$$\lambda = \left(\frac{\sqrt{\kappa(A)}-1}{\sqrt{\kappa(A)}+1}\right)^{2/m},$$

and

$$C = \|A^{-1}\| \max\left\{1, \frac{\left(1+\sqrt{\kappa(A)}\right)^2}{2\kappa(A)}\right\}.$$

If A fails to be positive definite but is still m-banded, quasi-centered, bounded, and boundedly invertible:

$$(|A^{-1}|)_{i,j=1}^{n} \leq C_1\lambda_1^{|i-j|},$$

where

$$\lambda_1 = \left(\frac{\kappa(A)-1}{\kappa(A)+1}\right)^{\frac{1}{m}},$$

and

$$C_1 = (m+1)\lambda_1^{-m}\|A^{-1}\|\kappa(A) \max\left\{1, \frac{1}{2}\left[\frac{1+\kappa(A)}{\kappa(A)}\right]^2\right\}.$$

To prove this statement some preliminary work is needed. Let us start with a general complex, separable, Hilbert space H, and let $\mathcal{B}(H)$ denote the Banach algebra of all linear operators on H that are also bounded. Now if $A \in \mathcal{B}(H)$ then A can be represented as a matrix with respect to any complete orthonormal set. In this way, A can be regarded as an element of $B(\ell^2(S))$, where $S = \{1, 2, \ldots, N\}$. In this space the usual matrix product define the action A over the space. A can be considered as a matrix representing a bounded operator in $\mathcal{B}(\ell^2(S))$. Recall that A is *m-banded* if there is an index l such that

$$a_{i,j} = 0, \quad \text{if } j \notin [i-l, i-l+m].$$

A is said to be *centered* and *m-banded* if m is even and the l above can be chosen to be $m/2$. In this case the zero elements of the *centered* and *m-banded* are:

$$a_{i,j} = 0, \quad \text{if } |i-j| > \frac{m}{2}.$$

Remark 3.10. Selfadjoint matrices are naturally centered, i.e., a tridiagonal selfadjoint matrix is centered and 2-banded.

Now let $\mathbb{R}_n$ denote, as usual, the polynomial of degree less than or equal to n. If $K \subseteq \mathbb{C}$ and f is a fixed complex-valued function on K, define the norm

$$\|f\|_K = \sup_{z \in K} |f(z)|$$

and the relative approximation error for the set of polynomial $\mathbb{R}_n$ to an f over the set K as

$$e_n(K) = \inf_{p \in \mathbb{R}_n} \|f - p\|_K.$$

To proceed, a result due to Tchebychev [506], and Bernstein [63], for which a modern presentation is in [386], is needed.

Lemma 3.5. *Let $f(x) = 1/x$ and let $0 < a < b$. Set $r = b/a$ and:*

$$q = q(r) = \frac{\sqrt{r} - 1}{\sqrt{r} + 1}$$

then:

$$e_n([a, b]) = \frac{(1 + \sqrt{r})^2}{2ar} q^{n+1}$$

Proposition 3.7 (Demko et al. [185])**.** *Let A be a positive definite, m-banded, bounded and boundedly invertible matrix in $\ell^2(S)$. Let $[a, b]$ be the smallest interval containing $\sigma(A)$. Set $r = b/a$, $q = q(r)$ as in the Lemma 3.5, and set $C_0 = (1 + \sqrt{r})^2/(2ar)$ and $\lambda = q^{2/m}$. Then,*

$$|A^{-1}| = (|a_{i,j}^{-1}|)_{i,j=1^n} \leq C\lambda^{|i-j|}$$

where:

$$C = C(a, r) = \max\{a^{-1}, C_0\}.$$

Proof. Since A is positive definite and invertible, $0 < a < b$ and A is centered. Thus A^k is centered and km-banded for $k \geq 0$. If p is a polynomial in $\mathbb{R}_k$ then $p(A)$ is km-banded and centered. By the Lemma 3.5, there exists a sequence of polynomials $\{p_n\}_{n \geq 1}$ in $\mathbb{P}_n$ satisfying

$$\left\| \frac{1}{x} - p_n \right\|_{[a,b]} = C_0 q^{n+1}$$

Rewriting it for the matrix:

$$\|A^{-1} - p_n(A)\| = \left\| \frac{1}{x} - p_n \right\|_{\sigma(A)} \leq C_0 q^{n+1}.$$

And now $|i - j| = nm/2 + k$ for $k = 1, 2, \ldots, m/2$ and $i \neq j$. The inequality

$$\frac{2|i-j|}{m} \leq (n+1)$$

holds, and hence

$$|a_{i,j}^{-1}| = |a_{i,j}^{-1} - p_n(a_{i,j})| \leq \|A^{-1} - p_n(A)\| \leq C_0 \lambda^{|i-j|}.$$

In case $i = j$ note that $a^{-1} = \|A^{-1}\|$, and this completes the proof. $\square$

Now following the authors of the result above, there is an extension of the latter for a more generic type of matrix A. Before doing this, the *quasi–centered* matrix definition is needed. A is said to be *quasi–centered* if the central diagonal is contained within the non–zero bands of the matrix, i.e., $A \in \mathcal{B}(\ell^2(S))$ is invertible only if A is quasi–centered. Note also that this is not true for $A \in \ell^2(Z)$.

Proposition 3.8 (Demko et al. [185]). *Let A be m-banded, bounded and boundedly invertible on $\ell^2(S)$. Let $[a, b]$ be the smallest interval containing $\sigma(AA^H)$. Then, setting $r = b/a$, $q = q(r)$ as in Lemma 3.5, and $\lambda_1 = q^{1/m}$, there is a constant C_1 depending on A so that*

$$(|A^{-1}|)_{i,j=1}^{n} = |a_{i,j}^{-1}| \leq C_1 \lambda_1^{|i-j|}.$$

If A is quasi–centered, C_1 can be chosen as $C_1 = (m+1)\|A\|\lambda_1^{-m} C(a, r)$.

Proof. The result follows immediately from the previous proposition by observing that

- $A^{-1} = A^H (AA^H)^{-1}$;
- $\|A\| = \|A^H\|$. □

The proof of Theorem 3.4 is given by two previous propositions. Observe that Theorem 3.4 does not apply to the sequence of matrix from Example 3.7, since their condition number grows like $O(n^2)$, i.e., the spectral condition number is unbounded. Being familiar with the physics of the problem, this has to be expected, since we are trying to approximate the Green's function for d^2/dx^2. Therefore, even if the matrix satisfies a bound of the kind of Theorem 3.4, it deteriorates as $n \to +\infty$.

In many cases understanding the behavior of the decay is important not only from the main diagonal but also when more complicated patterns appear. A useful result in this sense is given in Canuto et al. [127] for Kronecker sums of symmetric and positive definite banded matrices, i.e., $S = I_n \otimes M + M \otimes I_n$, where I_n is the identity matrix in $\mathbb{R}^{n \times n}$, a structure very frequently encountered when dealing with the discretizations of partial differential equations on a Cartesian grids, see, e.g., [351], or other tensor product structures.

Theorem 3.5 (Canuto et al. [127]). *Let $M \in \mathbb{R}^{n \times n}$ be a symmetric and positive definite matrix of bandwidth b. For $k, t \in \{1, 2, \ldots, n^2\}$, let*

$$j = \lfloor {}^{t-1}/_{n} \rfloor + 1, \quad i = t - n\lfloor {}^{t-1}/_{n} \rfloor,$$

and l, m such that

$$m = \lfloor {}^{k-1}/_{n} \rfloor + 1, \quad l = k - n\lfloor {}^{k-1}/_{n} \rfloor.$$

If λ_{min} and λ_{max} are the extreme eigenvalues of M, and $\lambda_1 = \lambda_{min} + \imath\omega$, $\lambda_2 = \lambda_{max} + \imath\omega$, $R = \alpha + \sqrt{\alpha^2 - 1}$ with $\alpha = {}^{|\lambda_1| + |\lambda_2|}/_{\lambda_2 - \lambda_1}$ and $\beta = |\lambda_{max} - \lambda_{min}|$, then the following results hold

1. *If $i \neq l$ and $j \neq m$, then*

$$\left|\left(S^{-1}\right)_{k,t}\right| \leq \frac{1}{2\pi}\frac{64}{\beta^2}\int_{-\infty}^{+\infty}\left(\frac{R^2}{\left(R^2-1\right)^2}\right)^2\left(\frac{1}{R}\right)^{\frac{|i-l|}{b}+\frac{|j-m|}{b}-2} d\omega;$$

2. *If either $i = l$ or $j = m$, then*

$$\left|\left(S^{-1}\right)_{k,t}\right| \leq \frac{1}{2\pi}\frac{8}{\beta}\int_{-\infty}^{+\infty}\frac{R^2}{\left(R^2-1\right)^2\sqrt{\lambda_{min}^2+\omega^2}}\left(\frac{1}{R}\right)^{\frac{|i-l|}{b}+\frac{|j-m|}{b}-1} d\omega;$$

3. *If both $i = l$ and $j = m$, then*

$$\left|\left(S^{-1}\right)_{k,t}\right| \leq \frac{1}{2\pi}\int_{-\infty}^{+\infty}\frac{1}{\lambda_{min}^2+\omega^2}d\omega = \frac{1}{2\lambda_{min}}.$$

Another result that can be useful is the following.

Theorem 3.6 (Jaffard [317]). *The sets of invertible matrix $(A)_{h,k} \in \mathcal{B}(\ell^2(\mathbb{K}))$, $\mathbb{K} = \mathbb{Z}, \mathbb{N}$, such that either*

$$|a_{h,k}| \leq C(1+|h-k|)^{-s},$$

or

$$|a_{h,k}| \leq C\exp(-\gamma|h-k|)$$

are two algebras, respectively, $\mathcal{Q}_s$ and $\mathcal{E}_\gamma$, i.e., their inverses have the same decay behavior.

Remark 3.11. In Theorem 3.6 we used the hypothesis $A \in \mathcal{B}(\ell^2(\mathbb{K}))$; we stress that the fact that the single matrix A of finite dimension is invertible does not ensure that the operator over $\ell^2(\mathbb{K})$ is invertible with bounded inverse.

This kind of feature is not easily proven in general and requires the use of tools from C^*–algebras. For an introduction to these problems and techniques refer to the book by Hagen et al. [288]. A characterization of ℓ^2 invertibility is given for a particular class of matrices/operator in Chapter 4 Remark 4.1.

Decay conditions for general matrices remain an open problem.

Not all of the entries of such inverses, though non–zero, have the same weight. The same principle adopted for the ILU factorizations can be applied when investigating the strategies that generate the approximate inverses with some degree of dropping.

3.5.2 Residual norm minimization

In order to preserve the original notation, define $G = A^{-1}$. Let us find an approximate inverse G through the minimization of the function

$$\begin{aligned} F(G) =& \|I - AG\|_F^2 = \sum_{i=1}^{n} \|\mathbf{e}_i - A\mathbf{g}_i\|_2^2, \text{ (right preconditioning)}, \\ F(G) =& \|I - GA\|_F^2 = \|I - A^T G^T\|_F^2, \text{ (left preconditioning)}, \end{aligned} \tag{3.21}$$

over a certain set S of sparse matrices. A set of sparse matrices with a fixed pattern or with a fixed number of entries per row is considered; namely a pattern is prescribed as a subset $\mathcal{G} \subseteq \{(i,j)\,|\,1 \leq i,j \leq n\}$ of the set of the indexes, for which we define the set S of the underlying matrices as

$$S = \{G \in \mathbb{R}^{n\times n}\,|\,g_{i,j} = 0 \text{ if } (i,j) \notin \mathcal{G}\}.$$

Denoting by $\mathbf{g}_{:,j}$ the j-th column of M and by $\mathcal{J}$ the set $\mathcal{J} = \{i\,|\,(i,j) \in \mathcal{G}\}$, consider the set of column of A that enters in the definition of the $\mathbf{g}_{:,j}$ as the sub–matrix $A_{:,\mathcal{J}}$ of the column indexed by $\mathcal{J}$. Then, consider the set of indexes $\mathcal{I}$ defined as $\mathcal{I} = \{i\,|\,\mathbf{a}_{i,:} \neq \mathbf{0} \text{ and } \mathbf{a}_{i,:} \in \mathcal{J}\}$, and then restrict the attention to the sub–matrix $\hat{A} = A_{\mathcal{I},\mathcal{J}}$, to the unknown vector $\hat{\mathbf{g}}_j = g_j(\mathcal{J})$ and to the right-hand side $\hat{\mathbf{e}}_j = \mathbf{e}_j(\mathcal{I})$. The unconstrained[9] least square problem has to be solved

$$\min \|\hat{\mathbf{e}}_j - \hat{A}\hat{\mathbf{g}}_{:,j}\|_2$$

a problem that can be faced, e.g., by a QR factorization of $\hat{A}$. See, e.g., [271] for details on the QR factorization. The advantages of this approach are that each column $\mathbf{g}_{:,j}$ can be computed, at least in principle, independently from the other columns of G. Due to the sparsity of A, the sub–matrix $\hat{A}$ contains only a few non–zero rows and columns, so each least squares problem has smaller size and often can be solved efficiently by dense matrix techniques.

A heuristic way to determine the sparse pattern, based on the Neumann series expansion of A^{-1}, is choosing the sparse pattern of A^k where $k \geq 1$. Other strategies for the sparse pattern determination of the inverse can be found in [131, 159, 277, 284, 311]. Nevertheless, this approach has a cost that grows with k and there is no guarantee that a good preconditioner would be obtained. Finally, an adaptive step can also be used. It consists of starting with a simple sparse pattern, for example a diagonal pattern, and then proceeding to augment it until a relaxation criterion is satisfied such as:

$$\|\mathbf{e}_j - A\mathbf{g}_{:,j}\|_2 < \varepsilon_j, \;\; \boldsymbol{\varepsilon} = (\varepsilon_1, \ldots, \varepsilon_j, \ldots, \varepsilon_n),$$

combined, or substituted, until a condition on the maximum number of non–zeros in $\mathbf{a}_{:,j}$ is satisfied. Observe also that the solution of each new least squares

[9]The minimization can be considered unconstrained because the constraints are explicit through the $\mathcal{I}$, $\mathcal{J}$.

problem can be computed, with a low computational effort, by updating the QR factorization used at each previous step; see the construction of the GMRES method in § 2.2.7. The standard SPAI algorithm is provided in this way; see Grote and Huckle [284], whose outline is reported in Algorithm 3.12.

Algorithm 3.12: SPAI Algorithm

Input: A sparse $A \in \mathbb{R}^{n \times n}$, sparsity pattern $\mathcal{G}$, drop tolerances $\varepsilon = (\varepsilon_1, \ldots, \varepsilon_2)$, max number of new non–zeros ℓ.

Output: The approximate inverse G.

for $j = 1, 2, \ldots, n$ **do**
- Choose the initial sparsity pattern $\mathcal{J}$ from $\mathcal{G}$;
- Compute the row indexes $\mathcal{I} = \{i \mid \mathbf{a}_{i,:} \neq \mathbf{0} \text{ and } \mathbf{a}_{i,:} \in \mathcal{J}\}$;
- Compute the restriction $\hat{A} = A_{\mathcal{I},\mathcal{J}}$;
- Solve for $\hat{\mathbf{g}}_{:,j} = \arg\min \|\hat{\mathbf{e}}_j - \hat{A}\hat{\mathbf{g}}_{:,j}\|$;
- Compute the residual $\mathbf{r} = \hat{\mathbf{e}}_j - \hat{A}\hat{\mathbf{g}}_{:,j}$;
- **while** $\|\mathbf{r}\|_2 > \varepsilon_j$ **do**
  - $\mathcal{L} = \{l \in \mathcal{I} \mid r_l \neq 0\}$;
  - $\bar{\mathcal{J}} = \{j \mid j \in \mathcal{L} \wedge j \notin \mathcal{J}\}$;
  - **for** $k \in \bar{\mathcal{J}}$ **do**
    - $\rho_k^2 = \|\mathbf{r}\|_2^2 - (\mathbf{r}^T A\mathbf{e}_k)^2 / \|A\mathbf{e}_k\|_2^2$;
  - Remove from $\bar{\mathcal{J}}$ the indexes exceeding ℓ of entries greater than ρ;
  - Compute the new row indexes $\bar{\mathcal{I}} = \{i \mid \mathbf{a}_{i,:} \neq \mathbf{0} \text{ and } \mathbf{a}_{i,:} \in \bar{\mathcal{J}}\}$;
  - $\mathcal{I} = \mathcal{I} \cup \bar{\mathcal{I}}$;
  - $\mathcal{J} = \mathcal{J} \cup \bar{\mathcal{J}}$;
  - Compute the restriction $\hat{A} = A_{\mathcal{I},\mathcal{J}}$;
  - Solve for $\hat{\mathbf{g}}_{:,j} = \arg\min \|\hat{\mathbf{e}}_j - \hat{A}\hat{\mathbf{g}}_{:,j}\|$;
  - Compute the residual $\mathbf{r} = \hat{\mathbf{e}}_j - \hat{A}\hat{\mathbf{g}}_{:,j}$;

A disadvantage of this approach is that the symmetry of the matrix A is not exploited, namely if A is symmetric and positive definite, the sparse approximation of the inverse provided as above is not symmetric in general. Note that this is the case even if the sparsity pattern imposed is symmetric. There is also no guarantee of keeping positive definitness. The algorithm can be implemented without the explicit calculation of the Cholesky factorization (Theorem 3.2 of [332]). For this problem, compute G as $G = G_L^T G_L$ with G_L lower triangular matrix, as in [332] and [333], in which the preconditioned matrix is $G_L A G_L^T$. In this way the matrix is still symmetric and positive definite, and the *Conjugate Gradient* iterative method can be applied. The way of computing G_L is applying one of the above strategies to the Cholesky factor.

On the other hand, if A is not symmetric, a symmetrized version of this preconditioner can be implemented; see [312, 313]. If A is supposed non symmetric, but still positive definite, Algorithm 3.12 computes a non symmetric G. To use it as preconditioner, both matrices

$$G + G^T, \; G + G^T - G^T AG$$

should be symmetrized. The corresponding residual is

$$(I - G^T A)(I - GA) = I - (G + G^T - G^T AG)A,$$

leading to the preconditioner

$$G + G^T - G^T AG.$$

However the above can be not positive definite, in general. Therefore, also in this case, the factored sparse approximate inverses $G = G_L^T G_L$ by considering the Cholesky factor of A can be used as in [332]. On the other hand, the approach in [312, 313], that is closely related to the algorithm from [335], is based on a modification of the functional to optimize in equation (3.21).

The choice is the Kaporin [324] functional, substituting G_L with L to simplify the notation:

$$L = \arg\min K = \arg\min \frac{{}^1\!/\!_n \operatorname{tr}(L^T AL)}{\det(L^T AL)^{1/n}},$$

where the minimization is constrained to a given sparsity pattern for L. To derive this method it is preferable to start with the diagonally scaled version of the matrix A, i.e., $\operatorname{diag}(A)^{-1}A$, to use in the computation a matrix A with ones on the diagonal. The algorithm is built as follows. Call $\mathcal{J}$ the set of allowed indexes for the column $\mathbf{l}_{:,k}$ and set $\tilde{\mathcal{J}} = \tilde{\mathcal{J}} \setminus \{k\}$. The functional K can be rewritten as

$$nK = \frac{\sum_{k=1}^n l_{k,k}^2 a_{k,k} + 2l_{k,k}\langle \mathbf{l}_{\tilde{\mathcal{J}},k}, \mathbf{a}_{\tilde{\mathcal{J}},k}\rangle + \mathbf{l}_{\tilde{\mathcal{J}},k}^T A_{\tilde{\mathcal{J}},\tilde{\mathcal{J}}} \mathbf{l}_{\tilde{\mathcal{J}},k}}{\det(A)^{1/n} \left(l_{1,1} l_{2,2} \cdot \ldots \cdot l_{n,n}\right)^{2/n}}$$

whose *unknowns* are $\mathbf{l}_{\tilde{\mathcal{J}},k}$ and $l_{k,k}$. To find the minimum, derive it and obtain the two conditions:

$$\begin{aligned} \mathbf{l}_{\tilde{\mathcal{J}},k} &= -\, l_{k,k} A_{\tilde{\mathcal{J}},\tilde{\mathcal{J}}}^{-1} \mathbf{a}_{\tilde{\mathcal{J}},k}, \\ l_{k,k}^2 &= \left[a_{k,k} - \mathbf{a}_{\tilde{\mathcal{J}},k}^T A_{\tilde{\mathcal{J}},\tilde{\mathcal{J}}}^{-1} \mathbf{a}_{\tilde{\mathcal{J}},k}\right]^{-1}. \end{aligned}$$

Now, the solution L can be computed columnwise in parallel by solving for each column $\mathbf{l}_{:,k}$ the two smaller linear system related to the $\tilde{\mathcal{J}}$ indexes. To improve this procedure, similarly to what done for SPAI algorithm, improve $\mathbf{l}_{:,k}$ by building an auxiliary 1D–minimization problem, i.e., solve, for $j > k$

$$\lambda_j = \arg\min \frac{({}^1\!/\!_n) \operatorname{tr}\left(L^T + \lambda_j \mathbf{e}_k \mathbf{e}_j^T\right) A \left(L + \lambda_j \mathbf{e}_j \mathbf{e}_k^T\right)}{\det\left(L^T + \lambda_j \mathbf{e}_k \mathbf{e}_j^T\right) A \left(L + \lambda_j \mathbf{e}_j \mathbf{e}_k^T\right)^{1/n}}, \; \mathbf{l}_{:,k}^{(j)} = \mathbf{l}_{:,k} + \lambda_j \mathbf{e}_j,$$

for which a minimum is in

$$\lambda_j = -\frac{\mathbf{a}_{:,j}^T \mathbf{l}_{:,k}}{a_{j,j}} = -\mathbf{a}_{:,j}^T \mathbf{l}_{:,k}.$$

In this way the reduction factor for the functional is found by multiplying for

$$\tau_j = \frac{(a_{j,\mathcal{J}} l_{\mathcal{J},k})^2}{a_{j,j}} = (a_{j,\mathcal{J}} l_{\mathcal{J},k})^2.$$

The analogue of the `while` cycle of algorithm 3.12 can be built looking at the graph representing the sparse structure of A to refine the vector. The outline of this procedure is reported in Algorithm 3.13, the *FSPAI* algorithm.

Algorithm 3.13: FSPAI Algorithm

Input: A sparse matrix $A \in \mathbb{R}^{n \times n}$, sparsity patterns $\mathcal{G}$, drop tolerances $\boldsymbol{\varepsilon} = (\varepsilon_1, \ldots, \varepsilon_2)$, max number of new non–zeros ℓ.

Output: The approximate inverse factor L.

for $k = 1, 2, \ldots, n$ **do**
- Choose the initial sparsity pattern $\mathcal{J}$ from $\mathcal{G}$;
- $\tilde{\mathcal{J}} = \mathcal{J} \setminus \{k\}$;
- Solve $A_{\tilde{\mathcal{J}},\tilde{\mathcal{J}}}\mathbf{y} = \mathbf{a}_{\tilde{\mathcal{J}},k}$;
- Compute $l_{k,k}^2 = \left[a_{k,k} - \mathbf{a}_{\tilde{\mathcal{J}},k}^T \mathbf{y}\right]^{-1}$;
- Compute $\mathbf{l}_{\tilde{\mathcal{J}},k} = -l_{k,k}\mathbf{y}$;
- Compute $\rho = \left(\mathbf{a}_{\tilde{\mathcal{J}},j}^T \mathbf{l}_{\tilde{\mathcal{J}},k}\right)^2$;
- **while** $\rho > \varepsilon_j \vee \text{nnz}(\mathbf{l}_{:,k}) < \ell$ **do**
  - $\bar{\mathcal{J}} = \{j \,|\, \mathbf{l}_{j,k} \neq 0\}$;
  - $\hat{\mathcal{J}} = \{j \,|\, \mathbf{a}_{:,j}^T \mathbf{l}_{:,k} \neq 0 \wedge j \notin \bar{\mathcal{J}}\}$;
  - **for** $j \in \hat{\mathcal{J}}$ **do**
    - $\tau_j = (\mathbf{a}_{:,j}^T \mathbf{l}_{:,k})^2$;
  - Compute $\rho = \max_{j \in \hat{\mathcal{J}}} \tau_j$ and let i be the index of the maximum;
  - $\mathcal{J} = \mathcal{J} \cup \{i\}$;
  - $\tilde{\mathcal{J}} = \mathcal{J} \setminus \{k\}$;
  - Solve $A_{\tilde{\mathcal{J}},\tilde{\mathcal{J}}}\mathbf{y} = \mathbf{a}_{\tilde{\mathcal{J}},k}$;
  - Compute $l_{k,k}^2 = \left[a_{k,k} - \mathbf{a}_{\tilde{\mathcal{J}},k}^T \mathbf{y}\right]^{-1}$;
  - Compute $\mathbf{l}_{\tilde{\mathcal{J}},k} = -l_{k,k}\mathbf{y}$;

Extension to the nonsymmetric case is straightforward but the solution of the auxiliary linear system is no more guaranteed. To obtain this feature, and therefore avoid breakdowns in the algorithm, it is enough to require that all the determinants of principal minors of A are non zero.

Both algorithms 3.12 and 3.13 can be, and have been, implemented in parallel on classical parallel architectures, i.e., standard multi-core architectures, as in [35, 160, 188, 228, 552, 553], and on architecture with hardware accelerators, like the GPU, as in [7, 183, 464]. Few details are shown in § 3.7. An interesting issue is the possibility of separating the workload for the single columns of the matrix, exploiting a feature that, under other points of view, can be disadvantageous, that is the insensitivity of this algorithm to reorderings of A. As have been shown in [57, 207, 238], reorderings can have great effects for many *approximate inverses* strategies, but this is not the case for these algorithms. Reorderings and their effects are also considered in § 3.5.4.

Note that, in general, these algorithms require a high number of parameters in input. Therefore, they are not easy to use and need some manual problem–related tuning.

Global iteration techniques

The computational cost of the SPAI algorithms 3.12 and 3.13 preconditioner can be very high both in space and in time (with the sequential implementation). For trying a mitigation of these issues, Chow and Saad in [162] propose to use a few steps of an iterative method, based only on sparse–sparse operations, to perform the residuals reduction corresponding to each column of the approximate inverse.

Let us start from the G in matrix in equation (3.21), considering it as an unknown matrix for a descent algorithm. The inner product to put in $\mathbb{R}^{n\times n}$ to have the $\|\cdot\|_F^2$ induced is the product

$$< X, Y >_F = \operatorname{tr}\left(Y^T X\right).$$

In the underlying descent algorithm, generate a new iterate $G^{(k+1)}$ from a previous one $G^{(k)}$, taking a step in the direction D with amplitude α, namely:

$$G^{(k+1)} = G^{(k)} + \alpha D,$$

in which α is selected to minimize the objective function in $F(G^{(k+1)})$. The goal is to find α such that

$$\min_{\alpha\in\mathbb{R}} F(G^{(k+1)}) = \min_{\alpha\in\mathbb{R}} F(G^{(k)} + \alpha D) = \min_{\alpha\in\mathbb{R}} \|I - AG^{(k)} - \alpha AD\|_F^2.$$

α should be is such that

$$< I - AG^{(k)} - \alpha AD, AD >_F = 0,$$

and, imposing $R = I - AG^{(k)}$, α should satisfy

$$\alpha = \frac{< R, AD >}{< AD, AD >} = \frac{\operatorname{tr}\left(R^T AD\right)}{\operatorname{tr}\left((AD)^T AD\right)} = \frac{\operatorname{tr}\left(R^T AD\right)}{\|AD\|_F^2}.$$

The simplest choice of a descent direction is represented by $D = R = I - AG^{(k)}$. However, with this choice the matrix $\text{nnz}(G^{(k+1)}) \geq \text{nnz}(G^{(k)})$ $\forall k$, so it is essential to apply some numerical dropping to the algorithm. The choices are essentially two: applying the dropping to $G^{(k)}$, resulting in the possible loss of the certainty of descent or applying the dropping to D, and making it more difficult to control the fill–in of the matrix $G^{(k)}$. This strategy is reported in Algorithm 3.14.

Algorithm 3.14: Global Minimal Residual descent algorithm

Input: Sparse matrix $A \in \mathbb{R}^{n\times n}$, initial guess sparse $G \in \mathbb{R}^{n\times n}$.
Output: Approximate inverse G.

```
while Convergence non reached do
    D ← I − AG;
    α ← tr(D^T AD) / ||AD||_F^2;
    G ← G + αD;
    Apply numerical dropping to G;
```

Another choice is taking D as the direction of the steepest descent, namely the direction opposed to the gradient of the function in (3.21), so D should satisfy the equality

$$F(X + E) = F(X) + < D, E > +O(\|E\|^2).$$

Proposition 3.9 (Steepest Descent). *The array representation of the gradient of F with respect to G is the matrix:*

$$D = -2A^T(I - AM) = -2A^T R.$$

Proof. For any matrix E:

$$\begin{aligned} F(G+E) - F(G) =& \operatorname{tr}\left[(I - A(G-E))^T(I - A(G-E))\right] + \\ & - \operatorname{tr}\left[(I - AG)^T(I - AG)\right] \\ =& \operatorname{tr}\left[(R - AE)^T(R - AE) - R^T R\right] \\ =& - \operatorname{tr}\left[(AE)^T R + R^T AE - (AE)^T(AE)\right] \\ =& - 2\operatorname{tr}(R^T AE) + \operatorname{tr}\left[(AE)^T(AE)\right] \\ =& - 2 < A^T R, E >_F + < AE, AE >_F . \end{aligned}$$

Therefore, the gradient is given by $D = -2A^T R$. □

The Algorithm with this modification is 3.14 in which $D = 2A^T R$.

Other issues such as reformulation of the algorithm above in a column-oriented way, applying self preconditioning, and refining the drop strategies in more effective ways can be found in [162].

To proceed, it would be important to give a condition for the nonsingularity of the computed G.

Proposition 3.10. *Let A be non singular and the residual R of the approximate inverse G satisfies*
$$\|I - AG\| < 1$$
for any consistent matrix norm $\|\cdot\|$, then G is nonsingular.

Proof. By Neumann series :

$$AG = I - (I - AG) = I - N \overset{\|N\|<1}{\Longrightarrow} \det(I - N) \neq 0. \qquad \square$$

This result is clearly not very effective. Indeed, to have such a strict condition on the norm implies having a good balancing of the product AM, and the dropping strategy does not give this property for free. To obtain such a strict value, a decrease of drop tolerance is needed, and therefore a possible increase of the building time. A more useful result can be the following.

Corollary 3.2. *Let A be nonsingular. If there exist two nonsingular diagonal matrices D_1, D_2 such that*

$$\|I - D_1 AGD_2\| < 1,$$

for any consistent matrix norm $\|\cdot\|$, then G is nonsingular.

Proof. It is enough to apply Proposition 3.10 to $A' = D_1A$ and $G' = GD_2$, then G' and G are nonsingular. $\square$

To obtain some better results, involving the spectral properties of A, a columnwise formulation of the algorithms is needed. For these results we refer again to [162].

The following Theorem proposed in [171] shows that there always exists a sparse matrix A for which the hypotheses of Proposition 3.10 can give a dense preconditioner.

Theorem 3.7 (Cosgrove et al. [171])**.** *Given any irreducible sparsity pattern, for each column index i, there exists a matrix A such that*

$$\|A\mathbf{x} - \mathbf{e}_i\|_1 < 1$$

implies that the vector $\mathbf{x}$ is dense.

Saad in [454] shows that this kind of preconditioner can solve problems of fluid dynamics (CFD) sometimes better than the ILU techniques in § 3.4.2. Moreover, Algorithm 3.14 can be used to refine the factored preconditioner computed with algorithms 3.12 and 3.13, i.e., using the preconditioner computed with the latter as a starting guess.

3.5.3 Inversion and sparsification or INVS

The inversion and sparsification of an existing ILU decomposition is a strategy used often to get approximate inverse preconditioners. Three immediate pros are that the latter are produced in factored form, the availability of reliable packages producing ILU factorizations, and the inherent parallel potentialities after the ILU decomposition; see, e.g., van Duin [529] and [73].

Let us focus on the approach by van Duin [529] as reconsidered in [73] and here called *INVS* (from INVersion and Sparsification) for brevity. The strategy is based on performing a sparse inversion technique on the triangular factors of an existing incomplete factorization in the form $M = LDU$, where D is a diagonal matrix and L and U are lower and upper triangular with ones on the main diagonal, respectively. The latter factorization can be obtained with a slight modification of the ILU techniques seen in § 3.4.2: let $M = LU_1$ be a ILU preconditioner and let D be the main diagonal of U_1. D is nonsingular otherwise M is a singular preconditioner. Then, by posing $U = D^{-1}U_1$,

$$M = LDU \rightarrow M^{-1} = U^{-1}D^{-1}L^{-1} = ZD^{-1}W^T$$

are the factorizations for M and for its inverse. However, as proved in Theorem 3.1, Z and W can be dense lower triangular matrices. In order to get a sparse incomplete factorization for A^{-1} and therefore of M^{-1}, a sparsification process for Z and W can be based on threshold and position or both, similar to that seen for ILU decompositions, generating the sparse lower triangular matrices $\tilde{Z}$ and $\tilde{W}$. After having obtained sparse approximations $\tilde{Z}$, $\tilde{W}^T$ for the matrices L^{-1} and U^{-1}, use them to get the *explicit* preconditioner[10] for A^{-1} of the form

$$\tilde{M}^{-1} = \tilde{U}^{-1}D^{-1}\tilde{L}^{-1} = \tilde{Z}D^{-1}\tilde{W}^T,$$

the *INVS*.

To produce the underlying inversion, start writing U as[11]

$$U = I + \sum_{i=1}^{n-1} \mathbf{e}_i \mathbf{u}_{i,:}^T.$$

By observing that $\forall\, j \leq k$, we have $\mathbf{e}_k \mathbf{u}_k^T \mathbf{e}_j \mathbf{u}_j^T = 0$, since the j-th entry of $\mathbf{u}_k$ is zero $\forall\, j \leq k$, rewrite U as

$$U = \prod_{i=n-1}^{1} \left(I + \mathbf{e}_i \mathbf{u}_i^T\right). \tag{3.22}$$

[10]Explicit preconditioner: no linear systems should be solved to apply it but, if used with a Krylov subspace method, just matrix-vector multiplications with $\tilde{Z}$, $\tilde{W}^T$ and D^{-1} are required.

[11]As usual we are using $\mathbf{e}_i$ notation for the vectors of the canonical basis, while $\mathbf{u}_i$ is the i-th row of the matrix U with the element $u_i(j) = 0$ for $j \leq i$.

The inverse of the factors in (3.22) are straightforward[12]:

$$\left(I + \mathbf{e}_i \mathbf{u}_i^T\right)^{-1} = I - \mathbf{e}_i \mathbf{u}_i^T,$$

and then

$$U^{-1} = \prod_{i=1}^{n-1} \left(I - \mathbf{e}_i \mathbf{u}_i^T\right). \tag{3.23}$$

Now, since U^{-1} is also an upper triangular matrix, the above expression can be rewritten as a sum

$$U^{-1} = I + \sum_{i=1}^{n-1} \mathbf{e}_i \hat{\mathbf{u}}_i^T,$$

where $\hat{\mathbf{u}}_i^T$, the strict upper triangular part of the i-th row of U^{-1}, is obtained as

$$\hat{\mathbf{u}}_i^T = -\mathbf{u}_i^T \prod_{j=i+1}^{n-1} \left(I - \mathbf{e}_i \mathbf{u}_i^T\right). \tag{3.24}$$

The expression for L^{-1} can be obtained similarly.

Remark 3.12. From (3.24) observe that no $\hat{\mathbf{u}}_j$ is needed for the calculation of $\hat{\mathbf{u}}_i$ for $i \neq j$, so the whole inversion process can be executed in parallel on a distributed memory machine.

Algorithm 3.15: Sparse product algorithm.

Input: $U \in \mathbb{R}^{n \times n}$ strict upper triangular matrix

for $i = 1, \dots, n-1$ **do**
- $\hat{\mathbf{u}}_i^T \leftarrow -\mathbf{u}_i^T$;
- $j \leftarrow$ first non-zero position in $\hat{\mathbf{u}}_i^T$;
- **while** $j < n$ **do**
  - $\alpha \leftarrow -\hat{\mathbf{u}}_i^T \mathbf{e}_j$;
  - $\hat{\mathbf{u}}_i^T = \hat{\mathbf{u}}_i^T + \alpha \mathbf{u}_j^T$; `// As a sparse operation.`
  - $j \leftarrow$ next non-zero position in $\hat{\mathbf{u}}_i^T$;

A straightforward implementation of the formula (3.24) is in the Algorithm 3.15, that has the flaw of generating dense matrices, recall Theorem 3.1. To sparsify using some dropping strategy, similarly to what done for the incomplete LU factorization, see the modifications needed in Algorithm 3.15.

Pattern drop. A fixed pattern S for the matrix is given, so $\hat{\mathbf{u}}_i^T(k)$ is only calculated when $(i, k) \in S$.

[12]The inverse of the factors in (3.22) are computed by using the Sherman–Morrison formula for the inversion of $(A + \mathbf{u}\mathbf{v}^T)$; see Sherman and Morrison in [477] for details.

Neumann drop. [13] We start from rewriting formula (3.23), namely the Neumann series expansion for the formula (3.22):

$$\begin{aligned} U^{-1} =& I - \sum_{j_1=1}^{n-1} \mathbf{e}_{j_i}\mathbf{u}_{j_1}^T + \sum_{j_2=1}^{n-2}\left(\mathbf{e}_{j_2}\mathbf{u}_{j_2}^T \sum_{j_1=j_2+1}^{n-1} \mathbf{e}_{j_1}\mathbf{u}_{j,1}^T\right) + \\ & - \sum_{j_3=1}^{n-3}\left(\mathbf{e}_{j_3}\mathbf{u}_{j_3}^T \sum_{j_2=j_3+1}^{n-2}\left(\mathbf{e}_{j_2}\mathbf{u}_{j_2}^T \sum_{j_1=j_2+1}^{n-1} \mathbf{e}_{j_1}\mathbf{u}_{j,1}^T\right)\right) + \ldots \end{aligned} \tag{3.25}$$

by truncating this expression at a number of extra term m we obtain the dropping $\hat{U}_m$. The main issue of this approach is that the update $\mathbf{u}_k^T$ can be computed m times in the worst case for $\hat{\mathbf{u}}_i^T$.

Positional fill level. Similarly to $ILU(P)$, define a level of fill initialized for U as:

$$\text{lev}_{i,j} = \begin{cases} 0 & \text{if } \mathbf{u}_i^T(j) \neq 0 \\ +\infty & \text{if } \mathbf{u}_i^T(j) = 0, \end{cases}$$

and the function to update the fill levels is

$$\text{lev}_{i,k} = \min(\text{lev}_{i,j} + 1, \text{lev}_{i,k}).$$

In this way, Algorithm 3.15 becomes Algorithm 3.16.

Algorithm 3.16: Positional fill level inversion of a sparse triangular matrix or *INVK*

Input: $U \in \mathbb{R}^{n\times n}$ strict upper triangular matrix, initial pattern of the matrix $\text{lev}_{i,j}$.

- **for** $j = 1, \ldots, n-1$ **do**
  - $\hat{\mathbf{u}}_i^T \leftarrow -\mathbf{u}_i^T$;
  - $j \leftarrow$ first non-zero position in $\hat{\mathbf{u}}_i^T$;
  - **while** $j < n$ **do**
    - **if** $\text{lev}_{i,j} \leq p$ **then**
      - $\alpha \leftarrow -\hat{\mathbf{u}}_i^T \mathbf{e}_j$;
      - $\hat{\mathbf{u}}_i^T \leftarrow \hat{\mathbf{u}}_i^T + \alpha \hat{\mathbf{u}}_j^T$;
      - $\text{lev}_{i,k} = \min(\text{lev}_{i,j} + 1, \text{lev}_{i,k})$;
    - **else**
      - $\hat{\mathbf{u}}_i^T(j) \leftarrow 0$;
    - $j \leftarrow$ next non-zero position in $\hat{\mathbf{u}}_i^T$;

Positional fill level II. instead of using the level of fill of the approximate

[13] Note that the first two term are available without cost.

inverse matrix, one can choose the level of fill of the original sparse triangular factor. This choice changes only the initialization step in Algorithm 3.16:

$$\text{lev}_{i,j} = \begin{cases} \text{lev}^U_{i,j}, & \text{if } \mathbf{u}_i^T(j) \neq 0, \\ +\infty, & \text{if } \mathbf{u}_i^T(j) = 0. \end{cases}$$

Threshold drop. Algorithm 3.15 can be implemented with the same elementary operators of the incomplete LU factorization with thresholding from Algorithm 3.10:

- in the copy–in phase, step 2, we initialize the set of nonzero entries for the current row $\hat{\mathbf{u}}_i$;
- in the update phase in step 7 we also insert the relevant indices into the set to ensure that the retrieval of the next nonzero at step 10 is performed efficiently;
- at the end of the inner loop, we perform a copy–out operation bringing the row $\hat{\mathbf{u}}_i$ into its desired final state, copying the largest entries up to the maximum allowed number of nonzeros.

Algorithm 3.17: Inversion of triangular matrices with numerical drop or *INVT*

Input: $U \in \mathbb{R}^{n\times n}$ strict upper triangular matrix
for $j = 1, \ldots, n-1$ **do**
    $\hat{\mathbf{u}}_i^T \leftarrow -u_i^T$;
    j location of first nonzero in $\hat{\mathbf{u}}_i^T$;
    while $j < n$ **do**
        $\alpha \leftarrow -\hat{\mathbf{u}}_i^T e_j = -\hat{\mathbf{u}}_i^T(j)$;
        if $|\alpha| > \epsilon$ **then**
            $\hat{\mathbf{u}}_i^T \leftarrow \hat{\mathbf{u}}_i^T + \alpha u_j^T$;
        else
            $\hat{\mathbf{u}}_i^T(j) \leftarrow 0$;
        j location of next nonzero in $\hat{\mathbf{u}}_i^T$;
    Drop elements in $\hat{\mathbf{u}}_i$ as necessary to achieve the desired number of nonzeros.;

From the above discussion, in Algorithm 3.17 we are looking again to implement efficiently the following two operations on a set with an order relation:

1. select and remove the lowest ranked element from a set;
2. add an element to the set.

One possible and efficient solution relies on the *Partially Ordered Set Abstract Data Type* [307]. This guarantees that both insertion of a new element and deletion of the lowest ranked element can be performed with a cost $O(\log(|S|))$, where $|S|$ is the cardinality of the set S.

The copy–out operations in both factorization and inversion can be again implemented by making use either of a partially ordered set or keeping the p largest entries of the current row.

A parallel implementation, exploiting the GPU architecture of this algorithm, is proposed in [73].

With the appropriate implementation for this data structure, we are now in a position to estimate the cost of building an approximate inverse as in Algorithm 3.17.

Theorem 3.8. *Let* $\mathrm{nz}_{\mathbf{u}}$ *be the average number of nonzeros per row for* $\mathbf{u}$, $\mathrm{nz}_{\hat{\mathbf{u}}}$ *for* $\hat{\mathbf{u}}$ *and that the bounds*

$$|S| \leq \gamma\,\mathrm{nz}_{\mathbf{u}}, \tag{3.26}$$

$$\mathrm{nz}_{\hat{\mathbf{u}}} \leq \beta\,\mathrm{nz}_{\mathbf{u}}, \tag{3.27}$$

hold true, where $|S|$ *is the maximum size of the set of entries in any of the* $\hat{\mathbf{u}}_i$ *before the application of the drop rule 11. Then, the computational cost of Algorithm 3.17 is given by*

$$O(\gamma\beta n \cdot \mathrm{nz}_{\mathbf{u}}^2(1 + \log(\gamma\,\mathrm{nz}_{\mathbf{u}}))). \tag{3.28}$$

Proof. Given the bound (3.26), the term $\gamma n \cdot \mathrm{nz}_{\mathbf{u}}$ follows easily nesting the two outer loops. For the remaining factor $\beta\,\mathrm{nz}_{\mathbf{u}}(1+\log(\gamma\,\mathrm{nz}_{\mathbf{u}}))$, consider statement 7. The latter is executed whenever the nonzero in $\hat{\mathbf{u}}_i$ is above the threshold, and this can be expected a number of times within a (moderate) factor times the size of u_i. On each execution, statement 7 requires $\mathrm{nz}_{\mathbf{u}}$ floating-point operations to execute the sparse matrix-vector product and update of the underlying vector, plus $\mathrm{nz}_{\mathbf{u}}\log(|S|)$ operations to update the set S with the (possibly new) nonzero entries. Using again bound (3.26) gives the desired result. □

The above result relies on two crucial assumptions about the size of the sets involved, and specifically that both β and γ are small constants. Assumption (3.27) is easier to justify: it expresses the fact that in many applications we would normally like to have a preconditioner that has a number of nonzeros of the same order as the coefficient matrix A, hence $\beta \approx 1$. Since the number of nonzeros can be enforced at step 11, the above assumptions are reasonable.

Assumption (3.26) is a bit more complex: it relies on the interaction between the profile of $\mathbf{u}$ and $\hat{\mathbf{u}}$. In particular, the hypothesis (3.26) is considered plausible, at least for important classes of problems, see [51, 55]. The latter gives an exponential decaying argument for the entries of the inverse of the

Cholesky factor. The application of the dropping rules at 6 and 11 in Algorithm 3.17 is also acting to keep the number of elements in the set S under control.

The cost of the sparse inversion of sparse factors INVK and INVT has also been analyzed in the paper where they were proposed [529]. The approximate inversion for the upper factor is estimated at

$$C_{\text{invrt}} = O\left(\text{nz}_{\hat{U}} \frac{\text{nz}_U}{n}\right)$$

where nz_U is the number of nonzeros above the main diagonal in U and likewise for the other quantities.

Note that the upper bound for the first term $\text{nz}_{\hat{U}}$ is given by the product $n\beta\text{nz}_{\mathbf{u}}$ while the second term is $\text{nz}_{\mathbf{u}}$. This estimate is then equivalent to our estimate (3.28) under the mild assumption that $\log(\gamma\,\text{nz}_{\mathbf{u}})$ is bounded by a small constant.

Exercise 3.4. Obtain the analogue of formula (3.24) for the lower triangular matrix L.

3.5.4 Incomplete biconjugation: AINV

Let us focus on another approach for preconditioning, based on efficient approximations to the inverse of the underlying matrix. In particular, approaches that do not necessitate a-priori information on the sparsity pattern of A^{-1} are considered here.

Factored approximate inverse preconditioners for general sparse matrices can be efficiently constructed by means of a generalization of the Gram–Schmidt process known as *biconjugation.*

An approximate inverse preconditioner in factored form, or *AINV* for short, was proposed by Benzi et al. in 1996, see [56] and later extended in [58] and in [51]. The main idea comes from an algorithm first proposed in 1948 in [243], a variation of the root–free Cholesky decomposition of A.

AINV strategy is based on the observation that if a matrix $A \in \mathbb{R}^{n\times n}$ is nonsingular, and if two vector sequences $\{\mathbf{z}_i, i = 1\ldots n\}$ and $\{\mathbf{w}_i, i = 1\ldots n\}$ A–biconjugate are given, i.e., $\mathbf{w}_i^T A\mathbf{z}_j = 0$ if and only if $i \neq j$, the biconjugation relation can be expressed as follows:

$$W^T AZ = D = \begin{pmatrix} p_1 & 0 & \ldots & 0 \\ 0 & p_2 & \ldots & 0 \\ \vdots & \vdots & \ddots & \vdots \\ 0 & 0 & \ldots & p_n \end{pmatrix} \tag{3.29}$$

where $p_i = \mathbf{w}_i^T A\mathbf{z}_i \neq 0$. Thus, W and Z must be nonsingular, since D is nonsingular. Therefore, in matrix form,

$$A = W^{-T} D Z^{-1}$$

from which it readily follows that

$$A^{-1} = ZD^{-1}W^T. \tag{3.30}$$

If W and Z are triangular, then they are the inverses of the triangular factors in the familiar LDU decomposition of A (see, e.g., [271]), as can be easily seen by comparing the two expressions

$$A = LDU$$

and

$$A = W^{-T}DZ^{-1}.$$

Algorithm 3.18: Biconjugation

$w_i^{(0)} \leftarrow z_i^{(0)} \leftarrow e_i \qquad 1 \leq i \leq n;$

for $i = 1, \ldots, n$ **do**

  for $j = i, i+1, \ldots, n$ **do**

    $p_j^{(i-1)} \leftarrow \mathbf{a}_{i,:}^T \mathbf{z}_j^{(i-1)}; \qquad q_j^{(i-1)} \leftarrow \mathbf{a}_{:,i}^T \mathbf{w}_j^{(i-1)};$

  for $j = i+1, \ldots, n$ **do**

$$\mathbf{z}_j^{(i)} \leftarrow \mathbf{z}_j^{(i-1)} - \left(\frac{p_j^{(i-1)}}{p_i^{(i-1)}} \right) \mathbf{z}_i^{(i-1)};$$

$$\mathbf{w}_j^{(i)} \leftarrow \mathbf{w}_j^{(i-1)} - \left(\frac{q_j^{(i-1)}}{q_i^{(i-1)}} \right) \mathbf{w}_i^{(i-1)};$$

$\mathbf{w}_i \leftarrow \mathbf{w}_i^{(i-1)}, \ \mathbf{z}_i \leftarrow \mathbf{z}_i^{(i-1)}, \ \mathbf{p}_i \leftarrow \mathbf{p}_i^{(i-1)}, \ 1 \leq i \leq n;$

There are infinitely many biconjugate sequences $\{w\}$ and $\{z\}$ that obey the above relations. To find one of them, it is enough to apply a biconjugation procedure to an appropriate pair of nonsingular matrices $W^{(0)}, Z^{(0)} \in \mathbb{R}^{n \times n}$. From a computational point of view, one can start with $W^{(0)} = Z^{(0)} = I$, thus obtaining Algorithm 3.18, where $\mathbf{a}_{i,:}^T$ is the ith row of A and $\mathbf{a}_{:,i}^T$ is the ith column of A, i.e., the ith row of A^T. If the procedure reaches completion without breakdowns, i.e., all the diagonal element are nonzero, then the resulting matrices W and Z will be triangular. Thus, for symmetric positive definite matrices, the process does not break down. Another interesting feature of Algorithm 3.18 is that the process for building W can proceed independently of Z.

To turn Algorithm 3.18 into a practical approximation procedure, and therefore for a possible preconditioner, we need to "sparsify" the resulting W and Z by dropping elements in the vectors $\mathbf{w}_i$ and $\mathbf{z}_i$. In principle this could be done at the end of Algorithm 3.18, but this would mean storing the (dense) matrices W and Z until the end. In practice, the sparsification is done at each update for the vectors $\mathbf{w}$ and $\mathbf{z}$.

Similar to the case of the incomplete factorization of ILU type, it is possible to prove [56] that the incomplete inverse factorization exists (in exact arithmetic) when A is an H–matrix.

Proposition 3.11 (Benzi et al. [56]). *Let A be an H–matrix and let $\hat{A}$ be the associated M–matrix. If p_i and $\hat{p}_i$ denote the pivots computed by the inverse factorization algorithm 3.18 applied to A and to $\hat{A}$, respectively, then $p_i \geq \hat{p}_i$. Furthermore, if $\hat{p}_i$ denote the pivots computed by the incomplete inverse factorization algorithm applied to A, then $\overline{p}_i \geq \hat{p}_i$.*

We stress that, despite the many similarities, there is a noticeable difference with the case of incomplete factorizations. It is well known that if A is an M-matrix, then the incomplete factorization induces a regular splitting $A = \hat{L}\hat{U} - R$, i.e., $\rho(I - \hat{U}^{-1}\hat{L}^{-1}A) < 1$, while this is not necessarily true for the incomplete inverse factors produced by biconjugation; see [58].

Example 3.8. Consider the symmetric matrix HB/bcsstk07 from the Harwell–Boeing collection [202]. Producing $M^{-1} = ZD^{-1}Z^T$ with a drop tolerance of $\varepsilon = 0.1$, the estimated spectral radius for this splitting is $\rho(I - M^{-1}A) = 1.44 > 1$, and so the splitting is *not* convergent. ✓

In theory, AINV can suffer from breakdown when the coefficient matrix is not an H-matrix. But the process as modified in [51] will not break down for symmetric positive matrices. The modified method was called *Stabilized AINV*, or *SAINV*.

Exercise 3.5. State and prove Proposition 3.11 for the case of A a non singular M–matrix, Definitions 3.3, 3.4, 3.5, 3.6, and 3.7.

Algorithmic variants

The procedure in Algorithm 3.18 is a *right–looking* variant, i.e., when a vector z_i is finalized, it is used to update all the vectors $z_j, j > i$. An alternative formulation is the *left–looking* variant, i.e., all the updates to z_i involving z_j, $j < i$, are performed in a single iteration of the outer loop. The relevant procedure is shown for Z in Algorithm 3.19 and W can be handled in the same way. While the numerical behavior of the two algorithms is the same (in exact arithmetic), the distribution of work in the two variants is quite different. The left-looking variant groups together all the updates to a given column. It tends to perform more (sparse) dot products, using the "true" z_i (i.e., before sparsification). These features have the following interesting properties that can be beneficial from a numerical point of view.

1. The dot products at 5 and 7 in Algorithm 3.19 are computed with the full vector $\mathbf{z}_i$, before the application of the dropping tolerance.

2. The dropping rule on $\mathbf{z}_i$ entries is applied only at the end of the update loop, whereas in the right-looking version it is applied at each update.

Algorithm 3.19: Left Looking Biconjugation for Z

1 $\mathbf{z}_1^{(0)} \leftarrow \mathbf{e}_1; \quad p_1^{(0)} \leftarrow a_{1,1};$
2 **for** $i = 2, \ldots, n$ **do**
3 $\quad \mathbf{z}_i^{(0)} \leftarrow \mathbf{e}_i;$
4 $\quad$ **for** $j = 1, \ldots, i-1$ **do**
5 $\quad\quad p_i^{(j-1)} \leftarrow \mathbf{a}_{j,:}^T \mathbf{z}_i^{(j-1)};$
6 $\quad\quad \mathbf{z}_i^{(j)} \leftarrow \mathbf{z}_i^{(j-1)} - \left(\frac{p_i^{(j-1)}}{p_j^{(j-1)}}\right) \mathbf{z}_j^{(j-1)};$
7 $\quad p_i^{(i-1)} \leftarrow \mathbf{a}_{i,:}^T \mathbf{z}_i^{(i-1)};$

From the experiments in [73], the left–looking variant seems to suffer less from pivot breakdown.

Recall that even if A is sparse, there is no guarantee that Z is sparse too; see also the discussions made at the beginning on the decay of the entries of the inverse of a matrix.

Example 3.9. As an example of the fill-in for the Z matrix in the AINV algorithm, we consider the application to the HB/sherman1[14] matrix (Figure 3.8) of the Harwell-Boeing Collection for various drop tolerances. ✓

Example 3.10. Consider the solution of the linear systems $A\mathbf{x} = \mathbf{b}$ for some matrices A in the Harwell–Boeing collection that are symmetric and positive definite, $\mathbf{b}$ being the vector of all ones. Performances of the SAINV algorithm and the Incomplete Cholesky factorization from § 3.4.1 are compared. The thresholds for the dropping are set to $1e-1$ and to $1e-2$ for both preconditioners, respectively. Fill-in of the matrices, time needed for the solution and number of matrix vector products are all reported in Table (3.4). The PCG and the CG stop when the relative residual is less than $1e-9$ or the number of iterations reaches the size of the matrix. A † is reported if the convergence is not reached within the above criteria. ✓

Approximate Biconjugation Implementation

Consider again Algorithm 3.19. In an actual implementation, the vector $\mathbf{z}_i$ could be stored in full format during the execution of loop 4, and there can be two different ways to apply the dropping rule:

1. at statement 6 the update of $\mathbf{z}_i$ is only performed for a sufficiently large value of p_i/p_j;
2. after the statement 7 a dropping rule is applied to $\mathbf{z}_i$ thereby sparsifying it for final storage.

[14]Information about this matrix can be found in [177].

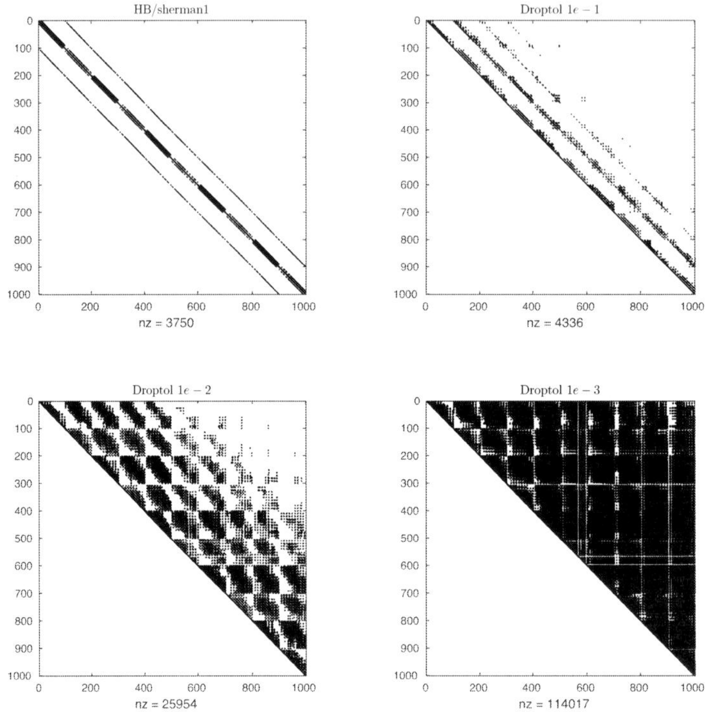

Figure 3.8: $AINV(A, \varepsilon)$ for the HB/sherman1 matrix at various ε.

Note that the application of the first dropping rule based on p_i/p_j was discouraged in the paper that introduces right-looking AINV for symmetric systems [56]. In the experiments in [73] with the left-looking variant, the dropping rule is applied, without adverse numerical effects while at the same time providing a performance advantage. Moreover, in the dropping rule applied to $\mathbf{z}_i$, the usual threshold comparisons are applied there, but also enforcing a limitation on the maximum number of nonzeros allowed, similarly to what happens in the ILUT(P) Algorithm mentioned in § 3.4.

A key observation is that the execution of statement 5 in Algorithm 3.19 computes the dot product between $\mathbf{a}_{j,:}$ and $\mathbf{z}_i$ even if in most cases this product can be exactly zero because of the (mis)match between the position of nonzeros in $\mathbf{a}_{j,:}$ and $\mathbf{z}_i$. Any of the latter operation, even if does not contribute to the result, produces a quadratic cost. We can instead ensure that the loop 4 is executed only whenever necessary. This is equivalent to letting each step j be the lowest index among those not processed yet such that row $\mathbf{a}_{j,:}$ has at least one nonzero element in a column corresponding to a nonzero entry in $\mathbf{z}_i^{(j-1)}$. To achieve this goal, we keep an extra copy of the pattern of $\mathbf{a}$ in a column–oriented format, and, as suggested in [73], do the following:

1. at the beginning of the loop 4, $z_i \leftarrow e_i$. Therefore, the set of indexes

Table 3.4: Solution of linear systems with SAINV and IC preconditioners

PCG		–		IC(1e-1)			IC(1e-2)		
Problem	Size	IT	T(s)	IT	T(s)	Fill %	IT	T(s)	Fill %
HB/nos3	960	230	0.0083	114	0.0090	0.25	30	0.0034	2.06
HB/nos5	468	428	0.0090	65	0.0038	0.65	†	†	†
HB/nos6	675	667	0.0138	29	0.0021	0.32	15	0.0014	0.45
HB/nos7	729	697	0.0156	27	0.0025	0.67	18	0.0017	1.06
HB/494_bus	494	487	0.0089	74	0.0044	0.59	30	0.0024	1.32
HB/662_bus	662	531	0.0101	65	0.0037	0.51	23	0.0017	1.12
HB/1138_bus	1138	1102	0.0287	140	0.0103	0.25	71	0.0059	0.51
HB/gr_30_30	900	34	0.0013	47	0.0032	0.12	12	0.0013	1.36

Preconditioner

PCG		–		SAINV(1e-1)			SAINV(1e-2)		
Problem	Size	IT	T(s)	IT	T(s)	Fill %	IT	T(s)	Fill %
HB/nos3	960			80	0.0078	1.17	34	0.0057	11.54
HB/nos5	468			45	0.0034	2.18	21	0.0022	11.86
HB/nos6	675			31	0.0024	0.89	15	0.0017	4.98
HB/nos7	729			39	0.0027	0.97	16	0.0023	8.47
HB/494_bus	494			42	0.0028	1.83	16	0.0021	12.02
HB/662_bus	662			45	0.0029	1.49	15	0.0023	12.21
HB/1138_bus	1138			70	0.0063	0.76	25	0.0040	3.61
HB/gr_30_30	900			28	0.0024	0.96	13	0.0021	6.89

$\{j\}$ is initialized with $R_{\mathbf{a}_{:,j}} = \{i : \mathbf{a}_{i,j} \neq 0\}$, the set of row indexes of nonzero entries in column i of matrix A;

2. at each iteration in loop 4, choose j to be the smallest index in the set that is greater than the last one visited;

3. at step 6, whenever an entry $\mathbf{z}_i(k)$ becomes nonzero, add the row indexes $R_{\mathbf{a}_{:,k}}$ corresponding to the nonzeros of column k in matrix A to the set of indexes to be visited.

To ease the implementation of the algorithm, one can keep copies of the input matrix A both in a row–oriented and column–oriented storage. Having an extra copy of A in column-oriented format allows to build Z and W at the same time, sharing the same outer loop: when dealing with W we need to access rows and columns of A^T, but these are accessible as the columns/rows (respectively) of A. The inner loop is in any case separate between Z and W, as it runs on a subset of indexes specific to the given triangular factor. The result is Algorithm 3.20; see again [73]. The implementation makes use of a dense work vector zw to compute z_i and w_i. The indexes of the non-zero entries are kept in a heap hp. Another heap rhp is used to hold the indexes of the rows with at least one nonzero in a column matching a nonzero entry in $\mathbf{zw}$, thus giving the set of rows for which we have to compute the scalar products.

Algorithm 3.20: Practical left-looking biconjugation

/* For a sparse matrix A let $C_{\mathbf{a}_{i,:}} = \{j : a_{i,j} \neq 0\}$ the set of column indexes in row i, and similarly let $R_{\mathbf{a}_{:,j}} = \{i : a_{i,j} \neq 0\}$ */

/* For a set S with an order relation $\leq$, let first(S) be the operator returning the smallest element in S */

$\mathbf{z}_1^{(0)} \leftarrow \mathbf{e}_1;\quad p_1^{(0)} \leftarrow a_{1,1};$

for $i = 2, \ldots, n$ **do**

- /* Inner loop over Z_j */
- $\mathbf{zw} \leftarrow \mathbf{e}_i;\quad S \leftarrow R_{\mathbf{a}_{:,i}};$
- **while** $S \neq \emptyset$ **do**
  - $j \leftarrow \text{first}(S);\quad S \leftarrow S - \{j\};$
  - $p(i) \leftarrow \mathbf{a}_{j,:}\mathbf{zw};\quad \alpha \leftarrow (p(i)/p(j))\ ;$
  - **if** $|\alpha| > \epsilon$ *(drop rule)* **then**
    - $\mathbf{zw} \leftarrow \mathbf{zw} - \alpha \mathbf{z}_{:,j};$
    - **for** $k \in R_{\mathbf{z}_{:,j}}$ **do**
      - $S \leftarrow S \cup \{l \in R_{\mathbf{a}_{:,k}} : j < l < i\}\ ;$
- $p(i) \leftarrow \mathbf{a}_{i,:}\mathbf{zw};$
- Apply a drop rule to $\mathbf{zw}$;
- $\mathbf{z}_{:,i} \leftarrow \mathbf{zw};$
- /* Inner loop over W_j */
- $\mathbf{zw} \leftarrow \mathbf{e}_i;\quad S \leftarrow C_{\mathbf{a}_{i,:}};$
- **while** $S \neq \emptyset$ **do**
  - $j \leftarrow \text{first}(S);\quad S \leftarrow S - \{j\};$
  - $q(i) \leftarrow \mathbf{a}_{:,j}^T\mathbf{zw};\quad \alpha \leftarrow (q(i)/q(j))\ ;$
  - **if** $|\alpha| > \epsilon$ *(drop rule)* **then**
    - $\mathbf{zw} \leftarrow \mathbf{zw} - \alpha \mathbf{w}_{:,j}\ ;$
    - **for** $k \in R_{\mathbf{w}_{:,j}}$ **do**
      - $S \leftarrow S \cup \{l \in C_{\mathbf{a}_{k,:}} : j < l < i\}\ ;$
- $q(i) \leftarrow (\mathbf{a}_{:,i})^T\mathbf{zw};$
- Apply a drop rule to $\mathbf{zw}$;
- $\mathbf{w}_{:,i} \leftarrow \mathbf{zw};$

The computational cost is estimated in the following result proposed in [73].

Theorem 3.9. *Let* nz_a *be the average number of nonzeros per row in* A *and* nz_z *for* z*; let the bounds*

$$|S| \leq \gamma\,\mathrm{nz}_a, \tag{3.31}$$

$$\mathrm{nz}_z \leq \beta\,\mathrm{nz}_a, \tag{3.32}$$

hold true, where $|S|$ *is the maximum cardinality of the sets of entries in any of the* z_i *before the application of the drop rule 12. Then the computational cost of Algorithm 3.20 is*

$$O(\gamma n\,\mathrm{nz}_a^2(1+\beta(1+\log(\gamma\,\mathrm{nz}_a)))). \tag{3.33}$$

Proof. Given the bound (3.31), the term $\gamma n\,\mathrm{nz}_a$ follows easily nesting of the two outer loops. The bodies of loops 4 and 15 in Algorithm 3.20 contain the following terms:

- The dot products at 6 and 17 add a term nz_a.
- The cost of 8 and 19 is given by $\beta\,\mathrm{nz}_a$, by assumption (3.32).
- The updates of the set S at 10 and 21 add $\beta\,\mathrm{nz}_a\log(\gamma\,\mathrm{nz}_a))$.

The result follows by considering that the statements 8-10 and 19-21 are executed at most as many times as statements 6 and 17 respectively, because of the dropping rule at 7 and 18. □

The situation is thus analogous to that of Theorem 3.8. To be more precise, note that the bound β in (3.32) refers to the ratio between the size of the rows in the upper triangle Z and the rows in matrix A. When enforcing that the number of nonzeros in the preconditioner is comparable to that in the matrix A, the actual value of β will be approximately one half of (3.27). Indeed, in that case we are comparing the upper triangle of the inverse to the upper triangle of the incomplete factorization. On the other hand, the biconjugation process is applied twice, for both Z and W, so that the ratio of the number of nonzeros in the preconditioner with respect to the number of nonzeros in A is again β just as in the case of INVT.

The application of the dropping rules in statements 7, 12 18 and 23 of Algorithm 3.20 have the effect of enforcing a control over the size of the set S, thereby improving the factor γ and the overall timings needed for preconditioner construction. A key element here is the fact that with dropping rules 12 and 23 we limit the number of accepted nonzeros.

The original AINV proposed in [56] can suffer from pivot breakdown when applied to matrices that are not H-matrices. In [51] a more robust version, called SAINV, is proposed. It computes the pivots p_i via the formula

$$p_i \leftarrow \mathbf{z}_i^T A\mathbf{z}_i,$$

Algorithm 3.21: Practical left-looking biconjugation stabilized

/* For a sparse matrix A let $C_{\mathbf{a}_{i,:}} = \{j : a_{ij} \neq 0\}$ the set of column indexes in row i, and similarly let $R_{\mathbf{a}_{:,j}} = \{i : a_{ij} \neq 0\}$ */

/* For a set S with an order relation $\leq$, let $\text{first}(S)$ be the operator returning the smallest element in S */

$\mathbf{z}_1^{(0)} \leftarrow \mathbf{e}_1$; $p_1^{(0)} \leftarrow a_{1,1}$;
for $i = 2, \ldots, n$ **do**
    /* Inner loop over Z_j */
    $\mathbf{zw} \leftarrow \mathbf{e}_i$; $S \leftarrow R_{\mathbf{a}_{:,i}}$;
    while $S \neq \emptyset$ **do**
        $j \leftarrow \text{first}(S); S \leftarrow S - \{j\}$;
        $p(i) \leftarrow ((\mathbf{w}_{:,j})^T A)\mathbf{zw}$; $\alpha \leftarrow (p(i)/p(j))$;
        if $|\alpha| > \epsilon$ *(drop rule)* **then**
            $\mathbf{zw} \leftarrow \mathbf{zw} - \alpha \mathbf{z}_{:,j}$;
            for $k \in R_{\mathbf{z}_{:,j}}$ **do**
                $S \leftarrow S \cup \{l \in R_{\mathbf{a}_{:,k}} : j < l < i\}$;
    Apply a drop rule to $\mathbf{zw}$;
    $\mathbf{z}_{:,i} \leftarrow \mathbf{zw}$;
    /* Inner loop over W_j */
    $\mathbf{zw} \leftarrow \mathbf{e}_i$; $S \leftarrow C_{\mathbf{a}_{i,:}}$;
    while $S \neq \emptyset$ **do**
        $j \leftarrow \text{first}(S)$; $S \leftarrow S - \{j\}$;;
        $q(i) \leftarrow (AZ_j)^T \mathbf{zw}$; $\alpha \leftarrow (q(i)/q(j))$; ;
        if $|\alpha| > \epsilon$ *(drop rule)* **then**
            $\mathbf{zw} \leftarrow \mathbf{zw} - \alpha \mathbf{w}_{:,j}$;
            for $k \in R_{\mathbf{w}_{:,j}}$ **do**
                $S \leftarrow S \cup \{l \in C_{\mathbf{a}_{k,:}} : j < l < i\}$;
    Apply a drop rule to $\mathbf{zw}$;
    $\mathbf{w}_{:,i} \leftarrow \mathbf{zw}$;
    $p(i) \leftarrow q(i) \leftarrow (\mathbf{w}_{:,i})^T A \mathbf{z}_{:,i}$;

instead of the simplified formula

$$p_i \leftarrow \mathbf{a}_{i,:}\mathbf{z}_i,$$

and thus if A is symmetric and positive definite, the pivots cannot be zero. The same kind of reasoning can be applied to nonsymmetric matrices, but in general, if A is nonsymmetric, we do not necessarily define a scalar product. On the other hand, there are important cases where pivot breakdown cannot occur for SAINV also in the nonsymmetric case, in particular, if the symmetric part of the matrix if positive definite [441].

If we apply the full formula to the left-looking algorithm we obtain Algorithm 3.21. The product with A is applied at steps 6, 16 and 23. Note that the two triangular matrices W and Z are no longer independent of each other. Indeed, the computation of the p_i and q_i must be performed at step 23 where the relevant elements of both W and Z are available. On the other hand, stabilization is not always beneficial. As an example, in all the tests performed in [73], there are no significant advantages in using Algorithm 3.21 over 3.20.

Diagonal shifting for SPD matrices

Miming the strategy used to avoid the non–positive pivots for the incomplete Cholesky factorization, one can be tempted to try applying to this case a diagonal shifting technique. If the pivots become near to the machine precision or even negative, one could in principle try to compute the AINV preconditioner on a matrix $A_\alpha = A + \alpha\,\text{diag}(A)$, instead of calculating it onto the matrix A. An α for which the preconditioner can be computed surely exists, i.e., in the worst case a sufficiently large α can be taken to make A_α diagonally dominant. Observe that nothing guarantees that this preconditioner for A_α is also a good preconditioner for A. The other question arising from this strategy is determining the optimal α. Numerical experiments have shown that the optimal α is usual less than the α that makes the A_α diagonally dominant. Finally, this is a high cost strategy viable only as a last resort for difficult problems for which all the other choices are ineffective.

3.5.4.1 Reordering strategies

Approximate inverse preconditioners in factored form obtained through biconjugation § 3.5.4 or through the inversion and sparsification procedure § 3.5.3 are sensitive to the ordering of the unknowns. Indeed, if P_1 and P_2 are permutation matrices we can always think of replacing the linear system $A\mathbf{x} = \mathbf{b}$ with,

$$P_1 A P_2 \mathbf{y} = P_1 \mathbf{b}, \quad \mathbf{x} = P_2 \mathbf{y},$$

thus computing an incomplete factorization of the permuted matrix $P_1 A P_2$. As usual, if A is symmetric or Hermitian, this should be preserved after any manipulation.

Reordering algorithms can be crucial in certain cases, but their treatise is technical. In [54], some reordering strategies for performing the AINV factorization over a generic matrix $A \in \mathbb{R}^{n\times n}$ are proposed. Further details on reordering algorithms can be found in [205] and in [206].

An implementation of reordering algorithms can be found in the HSL package, a collection of Fortran codes for large scale scientific computation that can be found in URL `http://www.hsl.rl.ac.uk/`. Numerical experiments relative to the combination of these algorithms with the AINV preconditioner can be found in [54].

3.6 Preconditioning sequences of linear systems

The solution of sequences of linear systems as

$$A^{(k)}\mathbf{x}^{(k)} = \mathbf{b}^{(k)},\ k \geq 0, \{A^{(k)}\}_k \subset \mathbb{R}^{n\times n},\ \{\mathbf{x}^{(k)}\}_k, \{\mathbf{b}^{(k)}\}_k \subset \mathbb{R}^n, \qquad (3.34)$$

is an important problem that arises in several contexts in scientific computing. Note that we use the term *sequences of linear systems* whenever at least the $A^{(k)}$ share something and/or do not differ much in some sense. The most frequent and natural example is the solution of time dependent partial differential equations by implicit methods.

Example 3.11. Time-dependent PDEs. Consider for instance a partial differential equation of the form

$$\frac{\partial u}{\partial t} = \mathcal{L}u + g, \qquad (3.35)$$

on a region with Dirichlet boundary conditions and an initial condition $u(x, 0) = u_0(x)$. An implicit time discretization with time step τ_m and a finite difference/element approach in space for $\mathcal{L}$ with stepsize h gives a sequence of linear systems if $\mathcal{L} = A$ is, e.g., linear

$$(I - \frac{\tau_m}{h^2}A)u^{m+1} = u^m + \tau_m\Phi(h, g^{m+1}, u_m, u_{m-1}, \ldots). \quad m = 0, 1, 2, \ldots \quad (3.36)$$

If $\mathcal{L}$ is non linear, by using a quasi-Newton step, we can get again linear systems as in (3.36). Typically the time step τ_m will not be constant, but change adaptively from time to time. ✓

Example 3.12. *U*nconstrained optimization. [15]. Consider the Newton method for finding a zero $\mathbf{x}^*$ of the function $F(\mathbf{x}) = 0$, $F : \mathbb{R}^n \to \mathbb{R}^n$ at least of class $\mathcal{C}^1(\mathbb{R}^n)$. Let us apply a Krylov subspace method for solving the linear systems

$$J\left(\mathbf{x}^{(k)}\right)\mathbf{s}^{(k)} = -F\left(\mathbf{x}^{(k)}\right),\quad k = 1, 2, \ldots, \qquad (3.37)$$

[15] An account of the Newton and inexact Newton method can be found in [187], specific information on the Newton–Krylov methods can be found in [330].

where $\mathbf{x}^{(k)}$ is the current candidate solution, J is the Jacobian matrix, or at least an approximated version of the operator providing the action of the Jacobian operator on a vector. Then, a candidate approximation for $\mathbf{x}^*$ can be found in the sequence $\{\mathbf{x}^{(k)}\}$ generated starting from an initial guess $\mathbf{x}^{(0)}$ as usual:

$$\mathbf{x}^{(k+1)} = \mathbf{x}^{(k)} + \mathbf{s}^{(k)}.$$

So also in this case (3.37) is a sequence of linear systems. If a Krylov solver is used to solve these linear systems, those methods are usually called *Newton–Krylov methods.* ✓

More generally, sequences of systems of the form (3.34) occur in the numerical solution of discretized nonlinear systems of ordinary and partial differential equations with implicit methods; see, e.g., [64, 351, 533] and the references therein. Sequences of linear systems also occur in other contexts, such as regularization of linear and non linear ill–posed least squares problems, see [291] and, e.g., [126, 305], selective nonlinear diffusion models for image processing, see [17], trust region methods in nonlinear optimization, and elsewhere. See [78] for an application to a nonlinear PDE model in image restoration.

The linear systems of the underlying sequences may need a suitable preconditioner. Reusing the same preconditioner for all linear systems in the sequence each time often leads to slow convergence, whereas recomputing a preconditioner each time is both costly and wasteful. Clearly, there is a broad range of possibilities within these two extremes. It should be possible to modify an existing preconditioner at a cost much lower than recomputing a preconditioner from scratch; even if the resulting preconditioner can be expected to be less effective than a brand new one in terms of iteration count, the overall cost should be reduced.

Due to these motivations, there is intense research on the update of preconditioners, starting from the seminal paper by Meurant [389]. For brevity, in the sequel we present only a couple of fairly general strategies for updating the sequence of generic preconditioners based on incomplete factorizations. Some other interesting approaches can be found in [42–44, 86, 125, 209, 507, 508]. Other strategies, mainly based on block Krylov methods and not considered here, can be devised if the linear systems (3.34) share the same right–hand side.

3.6.1 Updating incomplete or approximate inverse factorizations

Given the sequence (3.34), here we give some recipes and some sufficient conditions for updating sequences of both incomplete factorizations and approximate inverse factorizations for $\{A^{(k)}\}$ efficiently and reliably, i.e.,

$$M^{(k)} = \tilde{L}(\tilde{D} + \tilde{E}_k)^{-1}\tilde{U}^T, \tag{3.38}$$

for approximating $\{A^{(k)}\}$ and

$$P^{(k)} = \tilde{Z}(\tilde{D} + \tilde{E}_k)^{-1}\tilde{W}^T, \tag{3.39}$$

for approximating $\{\left(A^{(k)}\right)^{-1}\}$. The aim is to use one (of both) of $M^{(k)}$ and $P^{(k)}$ as preconditioners.

In order to generate the above sequences of updates, let us start from $A^{(0)}$, the first matrix of the sequence, also called here the *seed* matrix. Suppose that the inverse of $A^{(0)}$ can be decomposed formally (i.e., we NEVER produce a factorization for either the inverses or any of the underlying matrices $A^{(k)}$ that can have full factors even if $A^{(k)}$ is sparse) as

$$(A^{(0)})^{-1} = ZD^{-1}W^T,$$

approximated by a preconditioner in factored form

$$P^{(0)} = \tilde{Z}\tilde{D}^{-1}\tilde{W}^T, \tag{3.40}$$

where $\tilde{Z}$, $\tilde{W}$ are sparse unit lower triangular matrices, while $\tilde{D}$ is a nonsingular diagonal matrix. To generate the factorization (3.40) we can start from the preconditioners discussed in § 3.5 and, in particular, the INVT and AINV preconditioners; see §§ 3.5.3 and 3.5.4.

By assuming that

$$A^{(k)} = A^{(0)} + \Delta_k, \tag{3.41}$$

we can express formally the inverse of $A^{(k)}$, the kth element of the sequence, as

$$\begin{aligned} (A^{(k)})^{-1} =& (A_0 + \Delta_k)^{-1} = (W^{-T}DZ^{-1} + W^{-T}E_kZ^{-1})^{-1} \\ =& Z(D + E_k)^{-1}W^T, \text{ where } E_k = W^T\Delta_k Z. \end{aligned} \tag{3.42}$$

Therefore, a candidate updated preconditioner for $\left(A^{(k)}\right)^{-1}$ is

$$P^{(k)} = \tilde{Z}(\tilde{D} + \tilde{E}_k)^{-1}\tilde{W}^T, \tag{3.43}$$

where

$$\tilde{E}_k = g(E_k) = g(W^T\Delta_k Z),$$

also called the *updating matrix*, that can be built by applying a sparsification function g to the correction factor $\tilde{W}^T\Delta_k\tilde{Z}$.

The updating function g

The function $g(X)$ serves to generate a sparse or structured matrix from X. As an example, if X is a full matrix, $g(X) = \tilde{X}$, where $\tilde{X}$ is a sparse matrix made by selecting only certain entries of X. In practice, often even the extraction of a banded submatrix (e.g., the main diagonal), gives surprisingly good results; see [41, 44, 49, 67, 71].

On the other hand, if the entries of $(A^{(0)})^{-1}$ decay fast away from the main diagonal, we can consider the matrix function $g = g_m$ such that

$$g_m : \mathbb{C}^{n\times n} \to \mathbb{C}^{n\times n},$$

that acts by extracting m upper and lower bands (with respect to the main diagonal, which is the 0-diagonal) of its matrix argument generating an (m,m)–*banded matrix*. A substantial saving can be made by approximating

$$g_m(\tilde{W}^T \Delta_k \tilde{Z}) \text{ with } g_m(\tilde{W}^T)\, \Delta_k\, g_m(\tilde{Z}), \quad m > 0$$

with a reasonable quality of the approximation, i.e., under suitable conditions and provided $m > 0$, if the relative error

$$\frac{||g_m(\tilde{W}^T\, \Delta_k\, \tilde{Z}) - g_m(\tilde{W}^T)\Delta_k g_m(\tilde{Z})||}{||\tilde{W}^T\, \Delta_k\, \tilde{Z}||}$$

is moderate.

Let us state a couple of results on the decay of the entries of the inverse of the seed matrix that can be derived as corollaries of Theorem 3.4.

Theorem 3.10. *Let $A \in \mathbb{C}^{n\times n}$ be a nonsingular (m,m)–banded matrix, with $A \in \mathcal{B}(l^2(S))$ and condition number $\kappa_2(A) \geq 2$. Then, by denoting with $b_{i,j}$ the i,j–entry of A^{-1} and with*

$$\beta = \left(\frac{\kappa_2(A)-1}{\kappa_2(A)+1}\right)^{1/2m},$$

for all $\tilde{\beta} > \beta$, $\tilde{\beta} < 1$, there exists a constant $c = c(\tilde{\beta}, A) > 0$ such that

$$|b_{i,j}| \leq c\,\tilde{\beta}^{|i-j|},$$

with

$$c \leq (2m+1)\frac{\kappa_2(A)+1}{\kappa_2(A)-1}\|A^{-1}\|\kappa_2(A) \leq 3(2m+1)\|A^{-1}\|\kappa_2(A).$$

Proof. The desired result follows by Theorem 3.4 [185, Theorem 2.4] for a finite dimensional Hilbert space, using a (m,m)-banded matrix $A \in \mathcal{B}(l^2(S))$, i.e., a $2m$–banded matrix, and $\kappa_2(A) \geq 2$. Indeed, by direct application of the second part of Theorem 3.4 we obtain:

$$\begin{aligned}|b_{i,j}| \leq &(2m+1)\lambda_1^{-2m}\|A^{-1}\|\kappa_2(A)\cdot \\ &\cdot \max\left\{1, \frac{1}{2}\left[\frac{1+\kappa_2(A)}{\kappa_2(A)}\right]^2\right\}\left(\frac{\kappa_2(A)-1}{\kappa_2(A)+1}\right)^{\frac{|i-j|}{2m}}.\end{aligned}$$

Since $\lambda_1 = \left(\frac{\kappa_2(A)-1}{\kappa_2(A)+1}\right)^{\frac{1}{2m}}$, and $\kappa_2(A) \geq 2$, the maximum assumes the value 1, i.e., $\max\left\{1, \frac{1}{2}\left[\frac{1+\kappa_2(A)}{\kappa_2(A)}\right]^2\right\} = 1$, and therefore:

$$\begin{aligned}
|b_{i,j}| \leq & (2m+1)\frac{\kappa_2(A)+1}{\kappa_2(A)-1}\|A^{-1}\|\kappa_2(A)\left(\frac{\kappa_2(A)-1}{\kappa_2(A)+1}\right)^{\frac{|i-j|}{2m}} \\
\leq & 3(2m+1)\|A^{-1}\|\kappa_2(A)\left(\frac{\kappa_2(A)-1}{\kappa_2(A)+1}\right)^{\frac{|i-j|}{2m}} \\
\leq & c\tilde{\beta}^{|i-j|}.
\end{aligned}$$

□

We can note immediately that the results in Theorem 3.10, without suitable further assumptions, can be of very limited use because:

- the decay of the extradiagonal entries can be very slow, in principle arbitrarily slow;
- the constant c in front of the bound depends on the *condition number* of A and we are usually interested in approximations of A^{-1} such that their condition numbers can range from moderate to high;
- the bound is too far to be tight in general. A trivial example is given by a diagonal matrix with entries $a_{j,j} = j$, $j = 1, \ldots, n$. We have $b_{i,j} = 0$, $i \neq j$ but of course $\kappa_2(A) = a_{n,n}/a_{1,1} = n$.
- If we take $m = n$ and n is very large, then $\tilde{\beta}$ must be chosen very near 1 and it is very likely that no decay can be perceptible with the bound in Theorem 3.10.

However, the issues presented here are more properly connected with the decay properties of the matrices Z, W (and therefore of $\tilde{Z}$, $\tilde{W}$).

Corollary 3.3. *Let $A \in \mathbb{C}^{n\times n}$ be invertible, $A \in \mathcal{B}(l^2(S))$, and with its symmetric part positive definite. Then for all i, j with $j > i$, the entries $z_{i,j}$ in $W = L^{-H}$ and $w_{i,j}$ in $Z = U^{-1}$ satisfy the following upper bound:*

$$|z_{i,j}| \leq c_1\,\tilde{\beta}_1^{j-i}, \quad |w_{i,j}| \leq c_2\,\tilde{\beta}_2^{j-i}, \quad j > i$$

(note that $z_{i,j}$, $w_{i,j} = 0$ for $j \leq i$), where

$$0 < \tilde{\beta}_1,\ \tilde{\beta}_2 \leq \tilde{\beta} < 1$$

and c_1, c_2 are positive constants, $c_1, c_2 \leq c_3 \cdot \kappa_2(A)$.

Proof. The desired result follows by Theorem 3.10 and [59, Theorem 4.1]. Since $WD^{-1/2} = L^{-T} = A^{-1}L$ and the fact that $l_{i,j} = 0$ for $i < j$ and

$i - j > m$ we find that $w_{i,j} = \sum_{k=j}^{j+m-1} b_{i,k} l_{k,j}$ for each $i \leq j$. By using Theorem 3.4 we easily get that:

$$|w_{i,j}| \leq \sum_{k=1}^{j+m-1} |b_{i,k}||l_{k,j}| \leq c \sum_{k=1}^{j+m-1} \beta^{i-j} |l_{k,j}|.$$

Without loss of generality, we can assume that $\max_{i=1,\ldots,n} a_{i,i} = 1$ and thus $|l_{i,j}| \leq 1$ (otherwise, to enforce such condition, it is sufficient to divide A by its largest diagonal entry: β remains unchanged and c is replaced by another constant $\tilde{c}$). Therefore, we obtain, by a direct application of Theorem 3.10,

$$|w_{i,j}| \leq \tilde{c} \sum_{k=0}^{m} \tilde{\beta}^{j-i+k} = \tilde{c}\tilde{\beta}_2^{j-i} \sum_{k=0}^{m} \tilde{\beta}_2^k \leq c_2 \tilde{\beta}_2^{j-i}.$$

In conclusion the existence of c_3 and the bound for the $\tilde{\beta}_1$ follows by applying the same procedure to the matrix Z. □

If the underlying seed matrix $A^{(0)}$ is, e.g., diagonally dominant, then the decay of the entries of $(A^{(0)})^{-1}$ and therefore of W, Z ($\tilde{W}$, $\tilde{Z}$) is faster and more evident. This can be very useful for at least two aspects:

- the factors $\tilde{W}$, $\tilde{Z}$ of the underlying approximate inverse in factored form can have a narrow band for drop tolerances even just slightly more than zero;
- banded approximations can be used not only for post–sparsifying $\tilde{W}$, $\tilde{Z}$ in order to get more sparse factors, but also the update process can benefit from the fast decay.

We can use the above properties in the following result to obtain an *a–priori* estimate of the error produced using a bound of the correction factors. Clearly, this is very important for the implementation of the underlying updating strategy. For simplicity, let $\Delta_k = \alpha_k I$ below, i.e., $A^{(k)} = A^{(0)} + \alpha_k I$.

Theorem 3.11. *Let $A^{(0)} \in \mathbb{C}^{n \times n}$ be invertible, $A^{(0)} \in \mathcal{B}(l^2(S))$, and with its symmetric part positive definite. Let $g_m = [\cdot]_m$ be a matrix function extracting the m upper and lower bands of its argument. Then, given the matrices from Corollary 3.3, we have*

$$[\tilde{W}^H \tilde{Z}]_m = [\tilde{W}^H]_m [\tilde{Z}]_m + E(A^{(0)}, m), \quad |(E(A^{(0)}, m))_{i,j}| \leq c_4 \tilde{\beta}^{|i-j|}$$

where $c_4 = c_1 c_2$.

Proof. For a fixed value of the size n, let m be the selected bandwidth, then we have:

$$([\tilde{W}^H]_m [\tilde{Z}]_m)_{i,j} = \sum_{k=\max\{1,\max\{i-m,j-m\}\}}^{\min\{n,\max\{i+m,j+m\}\}} (\tilde{W}^H)_{i,k} (\tilde{Z})_{k,j}$$

$$([\tilde{W}^H \tilde{Z}]_m)_{i,j} = \begin{cases} \sum_{k=1}^{n} (\tilde{W}^H)_{i,k} (\tilde{Z})_{k,j}, & |i-j| \leq k, \\ 0, & \text{otherwise.} \end{cases}$$

Let us assume that $m = 1$ (tridiagonal matrix), in this case the difference $E(A, m)$ can be expressed, by direct inspection, as:

$$\begin{aligned}([\tilde{W}^H \tilde{Z}]_1 - [\tilde{W}^H]_1[\tilde{Z}]_1)_{i,j} = & - \chi_{\max(1,i-1)\leq 1\leq j+1}(\tilde{W}^H)_{i,k}(\tilde{Z})_{k,j} \\ & - \chi_{\max(1,i-1)\leq 2\leq j+1}(\tilde{W}^H)_{i,k}(\tilde{Z})_{k,j} \\ & + n\chi_{i\leq j+1}(\tilde{W}^H)_{i,k}(\tilde{Z})_{k,j},\end{aligned}$$

where χ is a function on the i, j–indices that takes value 1 when i and j satisfy the condition written at its subscript and 0 otherwise. In the same way, if $m = 2$ (pentadiagonal matrix), then we find:

$$\begin{aligned}([\tilde{W}^H \tilde{Z}]_2 - [\tilde{W}^H]_2[\tilde{Z}]_2)_{i,j} = & - \chi_{\max(1,i-2)\leq 1\leq j+2}(\tilde{W}^H)_{i,k}(\tilde{Z})_{k,j} \\ & - \chi_{\max(1,i-2)\leq 2\leq j+2}(\tilde{W}^H)_{i,k}(\tilde{Z})_{k,j} \\ & - \chi_{\max(1,i-2)\leq 3\leq j+2}(\tilde{W}^H)_{i,k}(\tilde{Z})_{k,j} \\ & + n\chi_{i\leq j+2}(\tilde{W}^H)_{i,k}(\tilde{Z})_{k,j}.\end{aligned}$$

Thus we assume, as inductive hypothesis, that for $m < n/2$ we have:

$$\begin{aligned}(E(A^{(0)}, m))_{i,j} = & - \sum_{l=1}^{m+1} \chi_{\max(1,i-m)\leq l\leq j+m}(\tilde{W}^H)_{i,k}(\tilde{Z})_{k,j} \\ & + n\chi_{i\leq j+m}(\tilde{W}^H)_{i,k}(\tilde{Z})_{k,j}.\end{aligned} \tag{3.44}$$

The conclusion follows by induction over n (since the admissible values of m are bounded by $n/2$) and again direct inspection. In conclusion, the bound on the correction term is given by the triangle inequality and an application of Corollary 3.3 to (3.44). □

We see that a fast decay of entries of $(A^{(0)})^{-1}$ guarantees that the essential component of the proposed update matrix, i.e., $\tilde{E} = \tilde{W}^T \Delta_k \tilde{Z}$, can be cheaply, easily and accurately approximated by the product $g_m(\tilde{W}^T)\Delta_k g_m(\tilde{Z})$ (or $g_m(\tilde{W}^T)g_r(\Delta_k)g_m(\tilde{Z})$ if necessary and if, e.g., the extradiagonal entries of Δ_k show a decay), without performing the possibly time and memory consuming matrix-matrix product $\tilde{W}^T \Delta_k \tilde{Z}$.

On the other hand, if the decay of the entries of $(A^{(0)})^{-1}$ is fast and Δ_k is αI, a diagonal approximation of $\tilde{W}^T \operatorname{diag}(\Delta_k)\, \tilde{Z}$ can be accurate enough. In this case, there is no need to apply the approximation in Theorem 3.11. The update matrix $\tilde{E}$ can be produced explicitly by the exact expression of $\alpha\, diag(\tilde{W}^T \tilde{Z})$ we give in the following corollary.

Corollary 3.4. *Let $A \in \mathbb{C}^{n\times n}$ be invertible, $A^{(0)} \in \mathcal{B}(l^2(S))$, and with its symmetric part (m, m)–banded and positive definite, $1 \leq m \leq n$. Then, the diagonal approximation for $\tilde{E}$ generated by the main diagonal of $\tilde{Z}^H \tilde{W}$ is given by*

$$\tilde{E} = \operatorname{diag}(\tilde{Z}^H \tilde{W}) = (d_{i,i}),$$

where

$$d_{i,i} = 1 + \sum_{j=1,\ i-j \leq m}^{i-1} z_{j,i} w_{j,i}, \quad 1 \leq i \leq n.$$

Proof. The claimed thesis follows by induction on i and by the definition of banded matrix, the same technique as the proof of Theorem 3.11 is used, while observing that both $\tilde{Z}^H$ and $\tilde{W}$ have ones on the main diagonals. □

Most of the numerical experiments performed for many applications use the above approximation of $g_m(\tilde{W}^T \Delta_k \tilde{Z})$ proposed in Theorem 3.11 and in Corollary 3.4 without perceptible loss of accuracy; see, e.g., [77]. Other approximations of the updating matrix tailored for the sequences of linear systems arising in quasi-Newton methods can be found in [41].

3.6.1.1 Convergence analysis of the updates*

To analyze the update (3.43) we proceed in a way analogous to [41]. Clearly, as we have seen also with the construction of incomplete factorizations in § 3.4, the quality of the preconditioner $P^{(k)}$ depends on the elements discarded by g. In order to quantify this statement we introduce the operator $O_g(\cdot)$, where "O"stands for offdiagonal entries,

$$O_g(A) = A - g(A),$$

for a generic sparsification function g and a matrix A. Considering that the update itself can be approximated by $\tilde{\Delta}_k = f(\Delta_k)$, for f a sparsification function that can be different from g, we obtain

$$\begin{aligned} \tilde{E}_k =& g(\tilde{W}^T \tilde{\Delta}_k \tilde{Z}) \\ =& \tilde{W}^T \Delta_k \tilde{Z} - O_g(\tilde{W}^T \Delta_k \tilde{Z}) - g(\tilde{W}^T O_f(\Delta_k) \tilde{Z}) \\ =& \tilde{W}^T \Delta_k \tilde{Z} + C_1 + C_2. \end{aligned} \tag{3.45}$$

Let us state a result on the distance of $(\tilde{P}^{(k)})^{-1}$ from $A^{(k)}$.

Theorem 3.12. *Let* $\zeta = \|\tilde{W}\|\|\tilde{Z}\|$ *and* $\nu = \|\tilde{W}^{-T}\|\|\tilde{Z}^{-1}\|$, $\tilde{P}^{(0)}$ *and* $\tilde{P}^{(k)}$ *be defined as in* (3.40) *and* (3.43). *Assume that, for some* $\varepsilon > 0$,

$$\|A^{(0)} - (P^{(0)})^{-1}\| = \varepsilon \|A^{(0)}\|. \tag{3.46}$$

Then,

$$\|A^{(k)} - (P^{(k)})^{-1}\| \leq \varepsilon \|A^{(0)}\| + \nu(\|C_1\| + \|C_2\|). \tag{3.47}$$

Moreover, if

$$\|A^{(k)} - (P^{(k)})^{-1}\| > \varepsilon \|A^{(0)}\|, \tag{3.48}$$

then

$$\|A^{(k)} - (P^{(k)})^{-1}\| \leq \frac{\varepsilon \|A^{(0)}\| + \nu(\|C_1\| + \|C_2\|)}{\|A^{(k)} - A^{(0)}\| - \varepsilon \|A^{(0)}\|} \|A^{(0)} - (P^{(0)})^{-1}\|. \tag{3.49}$$

Proof. From (3.41), (3.42) and (3.45) we have

$$\begin{aligned} A^{(k)} - (P^{(k)})^{-1} =& (A^{(k)} - A^{(0)}) + (A^{(0)} - (P^{(0)})^{-1}) + ((P^{(k)})^{-1} - (P^{(0)})^{-1}) \\ =& \Delta_k + (A_0 - (P^{(0)})^{-1}) - \tilde{W}^{-T}\tilde{E}_k\tilde{Z}^{-1} \\ =& \Delta_k + (A_0 - (P^{(0)})^{-1}) - \Delta_k\tilde{W}^{-T}(C_1 + C_2)\tilde{Z}^{-1}. \end{aligned}$$

Substituting the definition in (3.45), and using the main hypothesis gives the first bound in (3.47). Combining together (3.46) and (3.48) we derive the bound

$$\|A^{(0)} - A^{(k)}\| - \varepsilon\|A^{(0)}\| = \|A^{(0)} - A^{(k)}\| - \|A^{(0)} - (P^{(0)})^{-1}\| \leq \|A^{(k)} - (P^{(0)})^{-1}\|,$$

that together with (3.47) gives (3.49). □

The bound (3.49) states that the distance of $(\tilde{P}^{(k)})^{-1}$ from $A^{(k)}$ depends on the quality of the seed preconditioner (3.40) and on the discarded quantities C_1 and C_2. Indeed, when the norm of the latter are small enough, we can have $\|A^{(k)} - (P^{(k)})^{-1}\| < \|A^{(k)} - (P^{(0)})^{-1}\|$.

As a corollary of results in [41, 49] we get the following result.

Theorem 3.13. *Assume that the seed preconditioner* (3.40) *satisfies*

$$\exists\, \varepsilon > 0 \,:\, \|A^{(0)} - (P^{(0)})^{-1}\| = \varepsilon. \tag{3.50}$$

Then the right preconditioned matrix $A^{(k)}P^{(k)}$ *can be written as*

$$A^{(k)}P^{(k)} = I + R^{(k)}P^{(k)}, \qquad R^{(k)} = A^{(k)} - (P^{(k)})^{-1},$$

with

$$\|R^{(k)}\| \leq \varepsilon + c\|\Delta_k\|$$

and, given ζ *as in Theorem 3.12,*

$$\exists\, c > 0 \,:\, \|R^{(k)}P^{(k)}\| \leq \zeta(\varepsilon + c\|\Delta_k\|)\|(\tilde{D} + \tilde{E}_k)^{-1}\|. \tag{3.51}$$

Proof. From (3.40) the inverse of the updated preconditioner is given by,

$$(P^{(k)})^{-1} = \tilde{W}^{-T}(\tilde{D} + g(\tilde{W}^T\tilde{\Delta}_k\tilde{Z}))\tilde{Z}^{-1}.$$

The residual

$$R^{(k)} = A^{(k)} - (P^{(k)})^{-1} = A_0 + \Delta_k - \tilde{W}^{-T}(\tilde{D} + g(\tilde{W}^T\tilde{\Delta}_k\tilde{Z}))\tilde{Z}^{-1}$$

can be split as

$$R^{(k)} = (A^{(0)} - (P^{(0)})^{-1}) + (\Delta_k - \tilde{W}^{-T}(g(\tilde{W}^T\tilde{\Delta}_k\tilde{Z}))\tilde{Z}^{-1}),$$

giving

$$\|R^{(k)}\| \leq \|A^{(0)} - (P^{(0)})^{-1}\| + \|\Delta_k - \tilde{W}^{-T}(g(\tilde{W}^T\tilde{\Delta}_k\tilde{Z}))\tilde{Z}^{-1}\|.$$

The first term is then bounded by hypothesis (3.50). If we denote the second term of by ϱ we only need to prove that there exists a positive constant c such that $\varrho < c\|\Delta_k\|$. Indeed,

$$
\begin{aligned}
\varrho =&\|\Delta_k - \tilde{W}^{-T}(g(\tilde{W}^T\tilde{\Delta}_k\tilde{Z}))\tilde{Z}^{-1}\| \\
\leq&\|\tilde{W}^{-T}\|\|\tilde{W}^T(\Delta_k - \tilde{\Delta}_k)\tilde{Z} + \tilde{W}^T(\tilde{\Delta}_k)\tilde{Z} - g(\tilde{W}^T\tilde{\Delta}_k\tilde{Z})\|\|\tilde{Z}^{-1}\| \\
=&\|\tilde{W}^{-T}\|\|\tilde{W}^T(\Delta_k - \tilde{\Delta}_k)\tilde{Z} + \tilde{W}^T O_f(\Delta_k)\tilde{Z} + O_g(\tilde{W}^T\tilde{\Delta}_k\tilde{Z})\|\|\tilde{Z}^{-1}\| \\
\leq&\kappa(\tilde{W})\kappa(\tilde{Z})(\|O_f\| + \|O_g\|\|f\|)\|\Delta_k\|.
\end{aligned}
$$

Thus, we proved that $\varrho < c\|\Delta_k\|$ with $c = \kappa(\tilde{W})\kappa(\tilde{Z})(\|O_f\| + \|O_g\|\|f\|)$. Finally, by substituting $\|P^{(k)}\| \leq \|\tilde{W}\|\tilde{Z}\|\|(D + \tilde{E}_k)^{-1}\|$, we find the thesis (3.51). □

We stress that this result highlights the connections of the radius of the cluster of the eigenvalues of the updated preconditioner linear system $A^{(k)}P^{(k)}$ with the accuracy of the seed preconditioner matrix $P^{(0)}$ and with the accuracy of the updates for Δ_k, $\tilde{W}$ and $\tilde{Z}$.

3.6.1.2 Applying the preconditioner updates

Preconditioner updates can be applied successfully in various applications. Among them we recall:

- time-dependent partial differential equations of various type; see [49, 68, 86, 209, 497] for problems generating real matrices and [67] for problems generating complex symmetric matrices;
- unconstrained and constrained optimization [41–43, 164, 209, 557];
- image restoration [78];
- integrodifferential models from biomedical sciences [79];
- model reduction [9];
- fractional partial differential equations [72, 209].

For brevity we resume here the application of preconditioner updates in § 3.6.1 for just some fractional partial differential equation models, which generate sequences of linear systems whose matrices are full. An introduction to fractional partial differential equations and their applications can be found in the classical reference [433].

Example 3.13 (Fractional PDEs [72]). Consider the following partial/fractional differential equation problem, i.e., where we can find also fractional partial derivatives [433]:

$$\begin{cases} \dfrac{\partial u(x,y,t)}{\partial t} = & \nabla \cdot (< (a(x), b(y)), \nabla u >) \quad (x,y) \in \Omega, t > 0, \\ & +\mathcal{L}^{(\alpha,\beta)} u, \\ u(x,y,0) = & u_0(x,y), \\ u(x,y,t) = & 0 \qquad\qquad\qquad\qquad (x,y) \in \partial\Omega, t > 0. \end{cases}, \tag{3.52}$$

where the fractional operator is given by

$$\begin{aligned} \mathcal{L}^{(\alpha,\beta)} \cdot = & \, d_+^{(\alpha)}(x,y) \, {}_{\mathrm{RL}}D^{\alpha}_{a,x} \cdot + d_-^{(\alpha)}(x,y) \, {}_{\mathrm{RL}}D^{\alpha}_{x,b} \cdot \\ & + d_+^{(\beta)}(x,y) \, {}_{\mathrm{RL}}D^{\beta}_{a,y} \cdot + d_-^{(\beta)}(x,y) \, {}_{\mathrm{RL}}D^{\beta}_{y,b} \cdot \end{aligned} \tag{3.53}$$

and ${}_{\mathrm{RL}}D_{,}$ operators, for a given $\alpha > 0$ and $m \in \mathbb{Z}^+$ such that $m - 1 < \alpha \leq m$ are the left–side Riemann-Liouville fractional derivative

$${}_{\mathrm{RL}}D^{\alpha}_{a,x} y(x) = \frac{1}{\Gamma(m-\alpha)} \left(\frac{d}{dx} \right)^m \int_a^x \frac{y(\xi) d\xi}{(x-\xi)^{\alpha-m+1}},$$

and the right–side Riemann-Liouville fractional derivative

$${}_{\mathrm{RL}}D^{\alpha}_{x,b} y(x) = \frac{1}{\Gamma(m-\alpha)} \left(-\frac{d}{dx} \right)^m \int_x^b \frac{y(\xi) d\xi}{(\xi - x)^{\alpha-m+1}}.$$

To discretize the problem, consider the five points discretization for the Laplace operator combined with the shifted Grünwald-Letnikov approximation for the fractional:

$$\begin{aligned} {}_{\mathrm{RL}}D^{\alpha}_{a,x_k} y(x) \approx & \; \tfrac{1}{h^\alpha} \sum_{j=0}^{k} (-1)^j \tbinom{\alpha}{j} y((k-j)x), \quad k = 0, \ldots, N, \\ {}_{\mathrm{RL}}D^{\alpha}_{x_k,b} y(x) \approx & \; \tfrac{1}{h^\alpha} \sum_{j=0}^{k} (-1)^j \tbinom{\alpha}{j} y((k+j)x), \quad k = 0, \ldots, N. \end{aligned}$$

In this way we can define the lower and upper Toeplitz triangular matrices, respectively

$$B_L^{(\alpha)} = \frac{1}{h^\alpha} \begin{bmatrix} \omega_0^{(\alpha)} & \omega_1^{(\alpha)} & 0 & \ldots & 0 \\ \omega_1^{(\alpha)} & \ddots & \ddots & \ddots & \vdots \\ \vdots & \ddots & \ddots & \ddots & \vdots \\ \vdots & \ddots & \ddots & \ddots & \omega_1^{(\alpha)} \\ \omega_N^{(\alpha)} & \omega_{N-1}^{(\alpha)} & \ldots & \omega_1^{(\alpha)} & \omega_0^{(\alpha)} \end{bmatrix}, \; B_U^{(\alpha)} = \left(B_L^{(\alpha)} \right)^T,$$

where the coefficients $\omega_j^{(\alpha)}$ are defined as

$$\omega_j^{(\alpha)} = (-1)^j \binom{\alpha}{j}, \; j = 0, 1, \ldots, N.$$

In [383] it is proved that the Backward Euler method based on the Grünwald-Letnikov formula is consistent and unconditionally stable. Therefore, we use the Backward Euler method here. With this choice, the matrix in (3.36) can be written as

$$\begin{aligned} & I - \tau A^{(m)}, \text{ with} \\ & A^{(m)} = D^{(\alpha)}_{+,x}(B^{(\alpha)}_L \otimes I_{Ny}) + D^{(\alpha)}_{-,x}(B^{(\alpha)}_U \otimes I_{Ny}) + \\ & \quad + D^{(\beta)}_{+,y}(I_{Nx} \otimes B^{(\beta)}_L) + D^{(\beta)}_{-,y}(I_{Nx} \otimes B^{(\beta)}_U) + L, \end{aligned} \tag{3.54}$$

where τ is the (here constant) time step, $D_{\pm,x}$ and $D_{\pm,y}$ are block diagonal matrices made by the evaluation of the functions $d^{(\alpha)}_{\pm,x}$ and $d^{(\beta)}_{\pm,y}$ on the grid points in the x–direction and the y–direction, respectively, and L is the matrix generated by the second order centered differences discretization of $\nabla \cdot (< (a(x), b(y)), \nabla u >)$.

As a first choice for the solution of problem (3.52), consider the coefficients

$$\begin{aligned} a(x) =& 1 + 0.5\sin(3\pi x),\ b(y) = 1 + 0.7\sin(4\pi y) \\ d^{\alpha}_{+}(x,y,t) =& d^{\alpha}_{-}(x,y,t) = e^{4t}x^{4\alpha}y^{4\beta}, \\ d^{\beta}_{+}(x,y,t) =& d^{\beta}_{-}(x,y,t) = e^{4t}(2-x)^{4\alpha}(2-y)^{4\beta} \\ u_0(x,y) =& x^2y^2(2-x)^2(2-y)^2. \end{aligned} \tag{3.55}$$

over the domain $[0,2] \times [0,2]$ and the time interval $t \in [0,2]$. In Table 3.5 we use GMRES(50) algorithm stopped when the relative residual is less than $\varepsilon = 1e{-}6$ or the total iterations are more than `MAXIT` $= 1000$. The approximate

Table 3.5: Example 3.13. Problem in (3.52) with coefficients (3.55). $Nx = Ny = 80$, $Nt = 50$

			GMRES(50)		FIXED PREC.		UPDATED PREC.	
α	β	fill-in	IT	T(s)	IT	T(s)	IT	T(s)
1.6	1.2	5.38 %	†	†	4.45 26.24	7.68e-01	2.04 26.08	**2.95e-01**
1.3	1.8	6.51 %	†	†	4.55 27.57	7.79e-01	2.20 23.71	**3.09e-01**
1.5	1.5	4.41 %	†	†	4.43 28.04	7.61e-01	2.04 26.08	**2.93e-01**

inverses are computed with a drop tolerance $\delta = 0.1$. The update formula (3.43) for the preconditioner is used with only a diagonal update, i.e., the function $g(\cdot)$ extracts only the main diagonal of its matrix argument. Table 3.5 shows the average timings and the number of external/internal iterations. When a "†" is reported, at least one of the iterative solvers does not reach the prescribed tolerance in `MAXIT` iterations. ✓

3.6.1.3 Updating incomplete factorizations by matrix interpolation*

As suggested in [49], the preconditioners update can be generalized by using matrix interpolation. In the style of [71, 209, 210], here we concentrate on quadratic matrix interpolation and therefore start to build our preconditioner from three *seed* preconditioners computed for three appropriately chosen different matrices in the sequence (3.34). However, this paradigm is fairly general and can be applied to a higher degree or spline–like interpolation.

Note that the matrix interpolation approach for updating preconditioners requires that the matrices used for interpolation are known in advance while this is not required in the updating paradigm described in the previous section.

Given the underlying sequence of $n \times n$ matrices $\{A_i\}_{i=0}^{p}$, let us begin with the computation of three reference approximate inverse preconditioners for the matrices A_{i_0}, A_{i_1} and A_{i_2} chosen appropriately from the sequence $\{A_i\}_i$, i.e.,

$$\left\{\text{approximate inverses of } A_i^{-1}\right\} = Z_i D_i^{-1} W_i^T, \quad i = i_0, i_1, i_2. \tag{3.56}$$

The choice of A_{i_0}, A_{i_1} and A_{i_2} is problem-dependent and it is made in order to maximize the probabilities to get a reasonable approximation for all the interpolated preconditioners. In order to build up the underlying interpolated preconditioners, we need to provide preconditioners P_{i_j} for (a few) matrices A_{i_j}, $j = 0, 1, 2, \dots, p$, in approximate inverse form. Given A_{i_j}, $j = 0, 1, 2, \dots, p$, we need to generate $p + 1$ well defined and sparse approximations in factored form

$$P_{i_j} = Z_{i_j}\, D_{i_j}^{-1}\, W_{i_j}^T \tag{3.57}$$

for $A_{i_j}^{-1}$, then we build the preconditioner factors by mean of quadratic interpolation of the couples, called here "points": (α_i, Z_i^T), (α_i, W_i), $i = i_0, i_1, i_2$, where $\{\alpha_i\}_{i=0}^{s}$ is the discretization of a parameter α linked to the problem and to the i indices. For example, if we are dealing with a variable time step integrator, α can be is the time step. Otherwise, some interpolation parameter or some parameter related to the time variable in PDEs with non-constant coefficients. The functions for the interpolation are given by the quadratic polynomial

$$p_2(\alpha; \cdot) = a + b\alpha + c\alpha^2, \tag{3.58}$$

where we obtain, for a generic triplet of matrices M_1, M_2, M_3, the expression for the coefficients a, b, c:

$$\begin{aligned}
a &= \frac{\alpha_0\,(\alpha_0 - \alpha_1)\,\alpha_1 M_3 + \alpha_2\,(\alpha_1\,(\alpha_1 - \alpha_2)\,M_1 + \alpha_0\,(\alpha_2 - \alpha_0)\,M_2)}{(\alpha_0 - \alpha_1)\,(\alpha_0 - \alpha_2)\,(\alpha_1 - \alpha_2)},\\
b &= \frac{\left(\alpha_1^2 - \alpha_0^2\right) M_3 + \left(\alpha_0^2 - \alpha_2^2\right) M_2 + \left(\alpha_2^2 - \alpha_1^2\right) M_1}{(\alpha_0 - \alpha_1)\,(\alpha_0 - \alpha_2)\,(\alpha_1 - \alpha_2)},\\
c &= \frac{(\alpha_0 - \alpha_1)\,M_3 + (\alpha_1 - \alpha_2)\,M_1 + (\alpha_2 - \alpha_0)\,M_2}{(\alpha_0 - \alpha_1)\,(\alpha_0 - \alpha_2)\,(\alpha_1 - \alpha_2)}.
\end{aligned}$$

Therefore, we build the approximations for the Z_α and W_α matrices as functions of the parameter α by using equations (3.58) with the coefficients a, b, c computed for the $\{W_{i_j}\}_{j=0}^2$ and $\{Z_{i_j}\}_{j=0}^2$ matrices coming from the factored approximate inverses of the reference matrices:

$$Z_\alpha = p_2(\alpha; \{Z_{i_j}\}_{j=0}^2) \text{ and } W_\alpha = p_2(\alpha; \{W_{i_j}\}_{j=0}^2).$$

We can take as a preconditioner the approximate inverse of the generic matrix A_i of the sequence given by

$$M_i^{-1} = Z_\alpha (D_i + [W_\alpha^T \, \Delta \, Z_\alpha]_k)^{-1} W_\alpha^T,$$

where D_i is the diagonal matrix coming from one of the reference preconditioners (3.57). The matrix Δ is given by

$$\Delta = A_i - A_{i_j},$$

and the operator $[\cdot]_k$ extracts the k upper and lower diagonals of a matrix ($[\cdot]_0$ gives the main diagonal, i.e., it is a simple realization of a sparsification function g). Therefore, what we do from the computational point of view is working with $[\Delta]_k$ and the k banded approximation of Z_α and W_α. In this way, the matrix Δ is not completely computed. Then, to choose between the different reference matrices, i.e., between the different D_i, taking into account the value assumed by the α parameter for each single linear system, we take the index $i = i_j$ as the one that realizes

$$i^* = \arg \min_{i_j = i_0, i_1, i_2} \|A_i - A_{i_j}\|_F, \quad \Delta = A_i - A_{i^*}. \tag{3.59}$$

Thus, the updated preconditioner in factored form is

$$M_{i,k}^{-1} = Z_\alpha (D_{i^*} + E_k)^{-1} W_\alpha^T, \text{ with } \begin{cases} Z_\alpha = p_2(\alpha; \{Z_{i_j}\}_{j=0}^2), \\ W_\alpha = p_2(\alpha; \{W_{i_j}\}_{j=0}^2), \\ E_k = \left[W_\alpha^T \Delta Z_\alpha\right]_k. \end{cases} \tag{3.60}$$

For the memory requirements, assuming a reasonable worst case with a non cancellation rule, we can estimate approximately at most three times the sum of the nonzero element of the reference matrices:

$$\text{nnz}(p_2(\alpha, M_1, M_2, M_3)) \leq 3 \sum_i \text{nnz}(M_i).$$

Moreover, the asymptotic computational cost for the construction of the preconditioner $M_{i,k}^{-1}$ and its application is the same of linear interpolation and of the preconditioner updates described in the previous sections.

Using the upper bound for the condition numbers of $n \times n$ regular triangular matrices from [349], we can also get a bound for the condition number of both the matrices Z_α and W_α obtained by the quadratic interpolation formula (3.58).

In particular, the condition number of the interpolated matrices is bounded by the condition number of the reference matrices; see [71]. Therefore, reference preconditioners with a reasonable condition number can generate interpolated preconditioners with a reasonable condition number too.

We define the matrix for later use

$$C_\alpha = L_\alpha D_{i^*} U_\alpha - L_i D_i U_i. \tag{3.61}$$

Similarly to what was observed for updated preconditioners, the interpolated preconditioners $\{M_{i,k}^{-1}\}$ in (3.60) can cluster the eigenvalues of the matrices $\{A_i\}$.

Theorem 3.14 (Bertaccini and Durastante [71]). *Let us consider the sequence of linear systems $A_i x = b_i$, $i = 0, 1, \ldots, s$ and the interpolated preconditioners in (3.60). If there exist $\delta \geq 0$ and $t \ll n$ such that*

$$W_\alpha^{-T} E_k Z_\alpha^{-1} - \Delta = U\Sigma V^T, \quad \Sigma = \text{diag}(\sigma_1, \sigma_2, \ldots, \sigma_n),$$

$$\sigma_1 \geq \sigma_2 \geq \ldots \geq \sigma_t \geq \delta > \sigma_{t+1} \geq \ldots \geq \sigma_n, \tag{3.62}$$

and

$$\max_{\alpha \in (\alpha_1, \alpha_2)} \|D_{i^*}^{-1} E_k\| \leq \frac{1}{2}, \tag{3.63}$$

then there exist matrices C_α, Δ, F and a scalar constant c_α such that

$$M_{i,k}^{-1} A_i = I + M_{i,k}^{-1} C_\alpha + \Delta + F,$$

with $\text{Rank}(\Delta) = t \ll n$*, independent from α and*

$$\|F\|_2 \leq \frac{2\delta c_\alpha}{\beta} \frac{\sqrt{n\left((n-1)a_W^2 + 2\right)}\sqrt{n\left((n-1)a_Z^2 + 2\right)}}{2\, n\, b_W b_Z},$$

where

$$\beta = \min\{\max_{r=1,\ldots,n} |d_r|^{-1}, c\}, \; d_r = (D_{i^*})_{r,r} \tag{3.64}$$

with c constant used to avoid zero pivots in the matrix D_{α_1},

$$b_W = \min_{i,j=1,2,\ldots,n} |w_{i,j}^{(i_0)}|^{1/2}, \; b_Z = \min_{i,j=1,2,\ldots,n} |z_{i,j}^{(i_0)}|^{1/2}.$$

and

$$\|M_{i,k}^{-1} C_\alpha\|_2 \leq 2 c_\alpha \kappa_2(L_\alpha) \kappa_2(U_\alpha) \frac{\max_r |d_r|}{\min_r |d_r|} + \|M_{i,k}^{-1} A_{\alpha_1}\|_2.$$

Example 3.14 (Preconditioner updates by matrix interpolation, inspired by a similar example in [71]). Let us consider the finite element approximation of the convex combination ($\alpha \in [0, 1]$) of the following equations

$$\begin{cases} u_t + a_1(x,y)u_x + b_1(x,y)u_y - \nabla \cdot (k_1(x,y)\nabla u) = f_1(x,y), & (x,y) \in \Omega, \\ u_t + a_2(x,y)u_x + b_2(x,y)u_y - \nabla \cdot (k_2(x,y)\nabla u) = f_2(x,y), & (x,y) \in \Omega, \end{cases},$$

and $k_i(x,y), a_i(x,y) b_i(x,y) > 0 \; \forall (x,y) \in \Omega \; i = 1, 2$, with the same boundary conditions on the borders of the domain. The boundary is parametrized by the equations

$$\mathcal{C}_1 \begin{cases} x = b\cos\left(t\left(\frac{a}{b} - 1\right)\right) + t\cos(a-b), & a = 7, b = 1, \\ y = t\sin(a-b) - b\sin\left(t\left(\frac{a}{b} - 1\right)\right), & t \in [0, 2\pi] \end{cases} \backslash \mathcal{C}_2 \begin{cases} x = 0.3\cos(t), \\ y = 0.3\sin(t). \end{cases}$$

The discretization is obtained by applying the Backward Euler method for the time derivative and taking the test functions in the space of the piecewise linear continuous polynomials

$$P1_h = \left\{ v \in H^1(\Omega) \,|\, \forall\, K \in \mathcal{T}_h,\ v|_K \in \mathbb{R}_1[x] \right\}.$$

The coefficient functions are chosen as

$$a_1(x) = 50(1 + 0.9\sin(100\pi x)),\ \ b_1(y) = 50(1 + 0.3\sin(100\pi y)),$$
$$k_1(x, y) = x^2 y^2 \exp(-x^2 - y^2),$$
$$a_2(x, y) = \cos^2(2x + y),\ \ b_2(x, y) = \cos^2(x + 2y),\ \ k_2(x, y) = x^4 + y^4,$$
$$f_1(x, y) = f_2(x, y) = \pi^2(\sin(x) + \cos(y)).$$

and boundary condition as

$$u(x, y, t) = x,\ (x, y) \in \mathcal{C}_1,\ \ u(x, y, t) = y,\ (x, y) \in \mathcal{C}_2,\ \forall t \geq 0.$$

The initial condition is the null function over the entire domain. FreeFem++ software [296] is used. The results of the preconditioning strategies for this problem are reported in Table 3.6 with GMRES. The reference AINV preconditioners are computed with a drop tolerance $\delta = 1e - 2$, and only the main diagonal of matrix Δ is used, i.e., we are using $[\Delta]_k$ with $k = 1$. The values of α used for reference are $\alpha = 0, 0.5, 1$, respectively. From these exper-

AINV fixed		AINV updated		AINV interpolated		AINV interpolated 2	
IT	T(s)	IT	T(s)	IT	T(s)	IT	T(s)
24	0.069517	24	0.045601	24	0.060492	24	0.048347
28	0.057633	27	0.050633	27	**0.055109**	25	0.062979
32	0.074485	31	0.058652	31	**0.060355**	26	0.060886
37	0.078756	35	0.067780	35	0.068347	28	**0.064882**
42	0.096048	40	0.079990	40	0.082364	29	**0.067872**
47	0.118547	44	0.095973	44	0.094332	29	**0.062821**
51	0.122968	48	0.104468	48	0.107474	29	**0.067773**
55	0.127817	53	0.118813	53	0.117902	29	**0.067542**
59	0.142804	58	0.136582	58	0.140420	29	**0.063895**
63	0.156484	63	0.155178	63	0.150704	29	**0.063404**
66	0.166710	69	0.188811	69	0.179450	30	**0.062864**

Table 3.6: Example 3.14. Value of $\alpha = 0 : 1e - 1 : 1$, $\alpha = 0 : 1e - 1 : 1$, size of the matrix $n = 10765$, *GMRES* algorithm. The columns IT and T(s) contains respectively the average number of iterations and the average time needed to perform them for solving the αth linear system.

iments the performance of the fixed ILU preconditioner is omitted because, even if it is convergent, generates matrices close to singular or badly scaled. Also the results with the unpreconditioned GMRES are omitted because they never reach convergence due to stagnation or reach the maximum number of iteration. These results are obtained by choosing the following indices in equation (3.59) $\begin{pmatrix} 1 & 1 & 1 & 2 & 2 & 2 & 2 & 2 & 3 & 3 & 3 \end{pmatrix}$ that fit well with the expected behavior of the interpolation. ✓

3.7 Parallelizing preconditioners and available software

Iterative methods and preconditioners we have discussed so far are usually employed to solve *very large* linear problems. Therefore, as we have hinted in § 1.3, whenever necessary, they are implemented in parallel computing platforms to achieve faster solution speed and to accommodate larger problems.

The incomplete factorization of the matrix of the system, see § 3.4, requires the solution of sparse triangular systems: this operation is intrinsically inefficient on most parallel architectures, see, e.g., [394] for the case of NVIDIA GPUs. The cause of this inefficiency is indeed the intrinsic sequential nature of the solution of a triangular system. Consider, e.g., the well known backward substitution algorithm for a lower triangular system given in Algorithm 3.22. We observe that there exist intrinsic dependencies in Algorithm 3.22 that can

Algorithm 3.22: Solution of a Lower Triangular System

Input: $L \in \mathbb{R}^{n\times n}$ lower triangular, $\mathbf{b} \in \mathbb{R}^n$
Output: $\mathbf{x} \in \mathbb{R}^n$ such that $L\mathbf{x} = \mathbf{b}$

1 $\mathbf{x} \leftarrow \mathbf{b}$;
2 **for** $i = 1, \ldots, n$ **do**
3 $\quad x_i \leftarrow x_i - \mathbf{l}_{i,1:i-1}\mathbf{x}_{1:i-1}$;
4 $\quad x_i \leftarrow x_i / l_{i,i}$;

not be faced by any rearrangement of the workload for a parallel execution. When working with dense matrices, the first possible parallelization relies on the instruction given in line 3. The workload implied by this statement is distributed among the available processors, i.e., the execution of the scalar product. Eventual block partitioning of the matrix can add more workload to the parallel pipeline, see [87, 295, 353, 354]. Obviously, for these strategies to be efficient, the number of operations actually implied by line 3 should be large enough, and this goes against the sparseness of the triangular factor. This is one of the main reasons that the *approximate inverse preconditioners* we have discussed in 3.5 are so appealing. In that case, the triangular factors are used to perform matrix–vector multiplications and no triangular system solves are needed at all. Matrix–vector multiplication is an operation that can be executed efficiently on tightly coupled and highly parallel architecture, such as the GPUs; see, e.g., [73] and references therein.

As we have discussed in § 1.3, parallelizing both the preconditioner and the iterative method relies deeply on the parallelization of the **B**asic **L**inear **A**lgebra **S**ubprograms (BLAS), that represents the standard low–level routines of any linear algebra library. Thus, all the libraries we now very briefly recall, rely upon highly efficient machine–specific implementations of the BLAS, that are available for many of the modern architectures, see Table 3.7. Particularly we want to mention the **A**utomatically **T**uned **L**inear **A**lgebra **S**oftware (ATLAS) project that represents an ongoing research effort focused on providing portable performance for the BLAS.

The original BLAS package was concerned only with vectors and matrices densely stored, see e.g, [347] and the `LINPACK` library [199] whose parallel implementation `HPL` (High Performance Linpack) [428] is used to benchmark and rank supercomputers for the TOP500 list[16]. The evolution of the `LINPACK` libraries is represented by the `LAPACK` project [186], a Fortran 90 library that provides routines for solving linear systems (also in the least–squares sense), eigen– and singular value problems. It also includes the routines needed for computing the matrix factorizations (LU, Cholesky, QR, SVD, Schur, generalized Schur) associated with these classical problems; while *dense* and *banded matrices* are handled, general sparse matrices are not considered. For the application of the strategies we have seen in this chapter in the *large* and *sparse* case, one may refer to the Trilinos' `Tpetra` package. It implements linear algebra objects within the framework of *hybrid parallelism*, see § 1.3. Thus, it uses at least two levels of parallelism, one for distributed–memory parallelism and inside any process any suitable shared–memory parallel model, i.e., it supports both OpenMP, POSIX Threads and Nvidia's CUDA programming models. To use incomplete factorization, relaxation and domain decomposition schemes within the `Tpetra` package, the Trilinos' packages `Ifpack2` and `AztecOO` can be used. Krylov methods and approximate inverses are also implemented in the `PSBLAS` package [230], with a parallelism framework that uses again both distributed– and shared–memory parallelism with GPUs support. Along the line of hybrid parallelism, we can find also the `MAGMA` project [200, 516, 517], aimed to develop a *dense* linear algebra library similar to `LAPACK` but for heterogeneous/hybrid architectures like Multicore+GPU systems. It also contains a sparse package that includes the relative sparse BLAS routines and functions for the solution of sparse linear systems with Krylov iterative methods and incomplete factorization preconditioners. A similar project on the commercial front is the `PARALUTION` library for sparse iterative methods, again with a special focus on hybrid multi–core and accelerator technology. The latter includes Krylov methods, ILU factorizations (with the different thresholding strategies and multilevel approach) and approximate inverses.

We also mention briefly the `cuSOLVER` library, that implements direct meth-

---

[16]The updated TOP500 list is at `https://www.top500.org/`.

Table 3.7: A list of Basic Linear Algebra Subprograms libraries. The architectures are indicated with their respective name and with "CP" if Cross–Platform, $\mathbb{R}$ and $\mathbb{C}$ stand for real and complex matrices, respectively, while D and S for Dense and Sparse respectively. For the codes regarding the licenses see `http://www.gnu.org/licenses/license-list.html`

	Arch.	Type	Licensing	
BLAS	CP	$\mathbb{R}\mathbb{C}$DS	PD	www.netlib.org/blas
ATLAS	CP	$\mathbb{R}\mathbb{C}$DS	BSD-style	sourceforge.net/projects/math-atlas/
SPBLAS	CP	$\mathbb{R}\mathbb{C}S$	PD	math.nist.gov/spblas/
PSBLAS	CP	$\mathbb{R}\mathbb{C}S$	BSD	people.uniroma2.it/salvatore.filippone/psblas/
cuBLAS	NVIDIA GPU	$\mathbb{R}\mathbb{C}D$	NVIDIA Corporation	developer.nvidia.com/cublas
clBLAS	CP	$\mathbb{R}\mathbb{C}D$	Apache License v2.0	github.com/clMathLibraries/clBLAS
clSPARSE	CP	$\mathbb{R}\mathbb{C}S$	Apache License v2.0	clmathlibraries.github.io/clSPARSE/
GotoBLAS	Intel Nehalem/Atom VIA Nanoprocessor AMD Opteron	$\mathbb{R}\mathbb{C}D$	BSD	www.tacc.utexas.edu/research-development... .../tacc-software/gotoblas2
OpenBLAS	x86 x86-64 MIPS ARM ARM64	$\mathbb{R}\mathbb{C}D$	BSD	www.openblas.net
Intel MKL	Intel Pentium, Core Intel Xeon CPUs Intel Xeon Phi	$\mathbb{R}\mathbb{C}SD$	Proprietary, freeware	software.intel.com/en-us/intel-mkl-support/

ods for both dense and sparse linear systems on the NVIDIA GPU architectures. For working with NVIDIA GPUs, the library `CUSP` [172], an open source C++ library, provides a flexible and high–level interface for solving sparse linear systems that implements approximate inverse preconditioners. With similar functions the `CULA` library can also be considered.

Many of the strategies we have seen in this chapter are also included in the latest versions of the MATLAB program, that provides the support for computation with the GPUs. There is also the *SuiteSparse* library, that is a suite of sparse matrix algorithms including many of the ordering methods we have discussed in § 3.5.4.1.

3.8 Which general purpose preconditioner should I use?

After this brief review of general purpose preconditioners for Krylov iterative methods and of Flexible GMRES (FGMRES), we try to give some suggestions and remarks on the preconditioning strategy to use and point the reader to a bibliography for other schemes not discussed here. However, even if precondioning is very often mandatory to use iterative methods, we should remark that here it is much harder to give hints because of the higher problem-dependent nature of any preconditioning strategy. Therefore, here we give just some short hints.

If the underlying linear system is large but not enormous and we do not need to focus too much on a parallel implementation, for generic real and symmetric or Hermitian definite (positive or negative) problems, we can try CG with Incomplete Cholesky decomposition, first with zero fill–in and then increasing the level of fill and/or using a threshold in order to limit the fill–in [329, 385].

For generic Hermitian problems (and therefore also, e.g., symmetric and indefinite) we should use ILU techniques. If the preconditioner does not accelerate iterations enough, we can allow enough fill–in either by an appropriate threshold (very problem–dependent, of course) or by increasing the level of fill or both. In particular, for ill-conditioned problems, we could need to resort to preconditioners sensibly denser than the matrices of the originating problems. For more discussions, see, e.g., [20, 161].

For for nonsymmetric–real and non-Hermitian linear systems, the choice is even more delicate and again does depend on the problem; see [55, 161]. If the matrix has no special properties (otherwise see the next chapter) to enable fast matrix–vector multiplication, try ILU techniques first with zero fill-in and then increasing the level of fill and/or using a threshold in order to limit the fill–in. Moreover, some implementations based on incomplete biconjugation, are more robust than standard incomplete factorizations; see [60].

Approximate inverse preconditioners are very attractive both for real symmetric and for nonsymmetric linear systems, in particular in the factored form; see [47] and § 3.5.4. They have a higher setup cost than ILU-like, in general but their implementation is carried out by sparse matrix by vector multiplications and this can be highly beneficial. In particular, for highly parallel computing architectures such as the GPUs and shared and distributed memory machines, matrix-vector multiplication can be implemented very efficiently. Recall that parallel implementations for triangular systems are much less efficient. See, e.g., [33, 130] for the effects on GPUs. Aside from the potential to exploit parallel hardware, there are also frameworks where sparse approximate inverses are required. An example is the updates of inverse incomplete factorization preconditioners for solving cheaply sequences of large and sparse linear systems discussed in § 3.6.

Chapter 4

Preconditioners for some structured linear systems

This chapter is devoted to the treatment of some classes of structured linear systems. Structure is a fairly general word that along the years has incorporated various and very different classes of matrices. Among them: Toeplitz, banded, semiseparable, saddle point, complex symmetric, etc. In a certain sense, we can also see sparsity and localization (in the sense outlined in [48]) as particular cases of structure. For our purposes, we often investigate the structures that permit the use of matrix–vector products with less than the standard $O(n^2)$.

Note that using the specific structure for each class of the underlying linear system requires different and specialized strategies and algorithms each time. Thus, this chapter is necessarily longer than the previous one and there will be no "Which preconditioner should I use?" section at the end of the chapter.

4.1 Toeplitz and block Toeplitz systems

Toeplitz matrices arise in a wide set of contexts. They appear in the *finite difference discretization* (with uniform mesh) of differential operators [351]; in *integral equations*; in the treatment of *queue* and related problems [155], *image deconvolution, deblurring* and *filtering* [293]; in the *numerical computation of Padé coefficients* [103]; *time series* treatment [223] and many others.

This topic has attracted great interest in recent years. Analysis of this *linear space* of matrices can be performed with instruments coming from functional analysis and with the classical instruments of numerical linear algebra, [93–95, 283], and a great number of results are available on this topic.

A basic knowledge of Fourier series, i.e., harmonic analysis, is assumed here. References for these topics can be found in in [213, 334, 560].

Definition 4.1 (Toeplitz Matrix). A Toeplitz matrix is a matrix of the form

$$T_n = \begin{bmatrix} t_0 & t_{-1} & \dots & t_{2-n} & t_{1-n} \\ t_1 & t_0 & t_{-1} & \dots & t_{2-n} \\ \vdots & t_1 & t_0 & \ddots & \vdots \\ t_{n-2} & \dots & \ddots & \ddots & t_{-1} \\ t_{n-1} & t_{n-2} & \dots & t_1 & t_0 \end{bmatrix}, \tag{4.1}$$

i.e., its entries are constant along the diagonals. A subset of this linear space of matrices is given by the matrices for which exists an $f \in \mathbb{L}^1([-\pi,\pi])$, such that

$$t_k = \frac{1}{2\pi}\int_{-\pi}^{\pi} f(\theta)e^{-ik\theta}d\theta, \;\; k = 0, \pm 1, \pm 2, \dots,$$

the t_k are the Fourier coefficients of f. In this case we write $T_n = T_n(f)$ where f is the *generating function* of the matrix $T_n(f)$.

This class of matrices take its name from *Otto Toeplitz* (1881–1940) in light of his early work on bilinear form related to Laurent series [514]. Here we focus only on the fundamental properties needed for the iterative solution of linear systems

$$T_n(f)\mathbf{x} = \mathbf{b}, \;\; \mathbf{x}, \mathbf{b} \in \mathbf{R}^n.$$

As one may suspect, many properties of a Toeplitz matrix generated by a function $f(x)$, are connected with the properties of f itself.

Proposition 4.1.

1. *The operator* $T_n : \mathbb{L}^1[-\pi,\pi] \to \mathbb{C}^{n\times n}$ *defined by equation* (4.1) *is linear and positive, i.e., if* $f \geq 0$ *then* $T_n(f) = T_n(f)^H$ $\forall n$ *and* $\mathbf{x}^H T_n(f)\mathbf{x} \geq 0$ $\forall \mathbf{x} \in \mathbb{C}^n$.

2. *Given* $f \in \mathbb{L}^1[-\pi,\pi]$ *such that* $m_f = \operatorname{ess\,inf}(f)$ *and* $M_f = \operatorname{ess\,sup}(f)$. *If* $m_f > -\infty$ *then* $m_f \leq \lambda_j(T_n(f))$ $\forall j = 1,\ldots,n$ *and if* $M_f < \infty$ *then* $M_f \geq \lambda_j(T_n(f))$ $\forall j = 1,\ldots,n$. *Moreover if* f *is not identical to a real constant and the above inequalities hold, then*

$$m_f < \lambda_j(T_n(f)) < M_f \ \ \forall j = 1,\ldots,n.$$

Proof.

1. We start proving the linearity property: $\forall\, f, g \in \mathbb{L}^1[-\pi,\pi]$ and $\alpha,\beta \in \mathbb{C}$ so

$$\begin{aligned} t_k =& [\alpha f + \beta t]_k = \frac{1}{2\pi}\int_{-\pi}^{\pi} (\alpha f(t) + \beta g(t)) e^{-ikt} dt = \\ =& \alpha[f]_k + \beta[g]_k = \alpha[T_n(f)]_k + \beta[T_n(g)]_k. \end{aligned}$$

   For the positivity we need to have $f \geq 0$. We first show that $T_n(f) = T_n(f)^H$ $\forall n \geq 0$. Since f is real, $t_k = \overline{t_{-k}}$ because

$$\frac{1}{2\pi}\int_{-\pi}^{\pi} f(t)e^{-ikt}dt = \overline{\frac{1}{2\pi}\int_{-\pi}^{\pi} f(t)e^{ikt}dt}.$$

   We need to prove that $\mathbf{x}^H T_n(f)\mathbf{x} \geq 0$ $\forall\, \mathbf{x} \in \mathbb{C}^n$:

$$\begin{aligned} \mathbf{x}^H T_n(f)\mathbf{x} =& \sum_{j,k=0}^{n-1} \overline{x}_j \left(T_n(f)\right)_{j,k} x_k = \sum_{j,k=0}^{n-1} \overline{x}_j t_{j-k} x_k = \\ =& \sum_{j,k=0}^{n-1} \overline{x}_j x_k \frac{1}{2\pi}\int_{-\pi}^{\pi} f(t) e^{-i(j-k)t} dt = \\ =& \frac{1}{2\pi}\int_{-\pi}^{\pi} f(t) \sum_{j,k=0}^{n-1} \overline{x}_j x_k e^{ikt} e^{-ijt} dt, \end{aligned}$$

   then by defining $q_x(t) = \sum_{l=0}^{n-1} x_l e^{ilt}$,

$$\begin{aligned} \mathbf{x}^H T_n(f)\mathbf{x} =& \frac{1}{2\pi}\int_{-\pi}^{\pi} f(t) q_x(t)\overline{q_x}(t) dt = \\ =& \frac{1}{2\pi}\int_{-\pi}^{\pi} f(t)|q_x(t)|^2 dt \geq 0. \end{aligned}$$

Moreover, if $x > 0$, then $q_x(t)$ is zero only in a finite number of points. Moreover, if f is different from zero in a subset of $[-\pi, \pi]$ with non zero measure we obtain $\mathbf{x}^H T_n(f)\mathbf{x} > 0$.

2. It is a simple consequence of the previous discussions. It is enough to observe that both

$$\begin{aligned} T_n(f - m_f) =& T_n(f) - m_f T_n(1) = T_n(f) - m_f I, \\ T_n(M_f - f) =& M_f T_n(1) - T_n(f) = M_f I - T_n(f), \end{aligned}$$

are positive semidefinite. Moreover, the strict inequalities hold for the observation made on the polynomial $q_x(t)$, i.e., whenever f is not constant. □

Asymptotic distribution of the eigenvalues of $\{T_n(f)\}_n$

Let us recall another tool, useful to study the spectrum of the underlying matrices: the *asymptotic distribution* of the eigenvalues and of the singular values of the matrix sequence $\{T_n(f)\}_n$.

Definition 4.2 (Asymptotic eigenvalue distribution). Given a sequence of matrices $\{X_n\}_n \in \mathbb{C}^{d_n \times d_n}$ with $d_n = \{\dim X_n\}_n \overset{n\to+\infty}{\longrightarrow} \infty$ monotonically and a μ-measurable function $f : D \to \mathbb{R}$, with $\mu(D) \in (0, \infty)$, we say that the sequence $\{X\}_n$ is distributed in the sense of the eigenvalues as the function f and write $\{X_n\}_n \sim_\lambda f$ if and only if,

$$\lim_{n\to\infty} \frac{1}{d_n} \sum_{j=0}^{d_n} F(\lambda_j(X_n)) = \frac{1}{\mu(D)} \int_D F(f(t))dt, \;\; \forall F \in \mathcal{C}_c(D), \tag{4.2}$$

where $\lambda_j(\cdot)$ indicates the j-th eigenvalue.

Definition 4.3 (Asymptotic singular values distribution). Given a sequence of matrices $\{X_n\}_n \in \mathbb{C}^{d_n \times d_n}$ with $d_n = \{\dim X_n\}_n \overset{n\to+\infty}{\longrightarrow} \infty$ monotonically and a μ-measurable function $f : D \to \mathbb{R}$, with $\mu(D) \in (0, \infty)$, we say that the sequence $\{X\}_n$ is distributed in the sense of the singular values as the function f and write $\{X_n\}_n \sim_\sigma f$ if and only if

$$\lim_{n\to\infty} \frac{1}{d_n} \sum_{j=0}^{d_n} F(\sigma_j(X_n)) = \frac{1}{\mu(D)} \int_D F(|f(t)|)dt, \;\; \forall F \in \mathcal{C}_c(D),$$

where $\sigma_j(\cdot)$ is the j-th singular value.

The core result of this class is the Grenander and Szegö Theorem [283], with the related results in [18, 419], that prove the relation (4.2) for the matrices $T_n(f)$ and f the generating symbol with the restriction that f is in the Wiener class, i.e., the Fourier coefficients of f are a sequence in the space ℓ^1. A stronger result, i.e., a result with weaker assumption, is the one obtained by Tyrtyshnikov [522].

Theorem 4.1 (Eigenvalue and Singular Value Distribution). *Given the generating function f, $T_n(f)$ is distributed in the sense of the eigenvalues (Definition 4.2) as f, written also as $T_n(f) \sim_\lambda f$, if one of the following conditions hold:*

1. *Grenander and Szegö [283]: f is real valued and $f \in \mathbb{L}^\infty$,*
2. *Tyrtyshnikov [522]: f is real valued and $f \in \mathbb{L}^2$.*

Moreover, $T_n(f)$ is distributed in the sense of the singular values (Definition 4.3) as f, written also as $T_n(f) \sim_\sigma f$, if one of the following conditions hold:

1. *Avram [18], Parter [419]: $f \in \mathbb{L}^\infty$,*
2. *Tyrtyshnikov [522]: $f \in \mathbb{L}^2$.*

Example 4.1. As an example of the application of Theorem 4.1, let us compute the eigenvalues of the matrix $T_n(2 - 2\cos(\theta))$, that generates the symmetric tridiagonal matrix with first row $\mathbf{r} = [2, -1, 0, \ldots, 0]$. In Figure 4.1 both f and the computed eigenvalues for $n = 100$ are reported. ✓

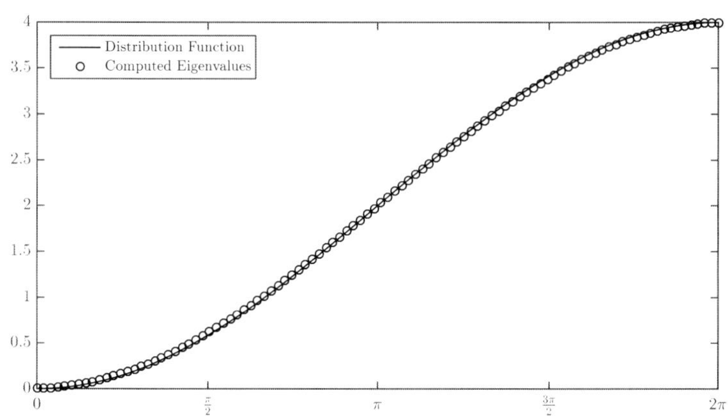

Figure 4.1: Example of eigenvalue distribution for Toeplitz Matrix.

Remark 4.1 ($\ell^2(\mathbb{Z}^+)$ invertibility of Toeplitz operators and matrices). As we have discussed in § 3.5 one could be interested, in some cases, in investigating the invertibility of the Toeplitz operator $T(f)$ over $\ell^2(\mathbb{Z}^+)$ to derive some properties of its finite sections, i.e., Toeplitz matrices $T_n(f)$. In general, a linear bounded operator on $\ell^2(\mathbb{Z}^+)$ is boundedly invertible if and only if its kernel (as operator) is $\{0\}$ and its range is all of $\ell^2(\mathbb{Z}^+)$. This property can be expressed in terms of having a zero difference between the dimension of the kernel and of the cokernel, $\ell^2 \setminus \overline{\text{range}(T(f))}$. For Toeplitz operators, with continuous symbols, this condition is reduced in checking that, given the counterclockwise orientation of $\mathbb{T} = [0, 2\pi]$, $0 \notin f(\mathbb{T})$, i.e., that the symbol is invertible, and that the *winding number* of the curve $f(\mathbb{T})$ around the origin is exactly 0, i.e.,

$$\nu(f,0) = \oint_{f(\mathbb{T})} \frac{dz}{z} = 0.$$

Differential operators with a zero in the symbol are not invertible as an operator on $\ell^2(\mathbb{Z}^+)$, even if *every* finite section of it is an invertible matrix. For further details on these issues, like the extension to the case of discontinuous symbol, see [94, 95, 288].

We now discuss briefly the computational complexity of the operation $T_n\mathbf{v}$ for a generic vector $\mathbf{v} \in \mathbb{R}^n$. In principle, a matrix–vector product, being the matrix T_n dense in general, costs $O(n^2)$ flops. However, T_n depends only on $2n-1$ parameters. Therefore, we expect that the cost of the operation can be sensibly reduced. In the sequel, we show that the particular Toeplitz structure allows to reduce the cost of this operation to $O(n\log(n))$ operations. To obtain this, we need to introduce another algebra of matrices that we will use also as a preconditioner, the *circulant* matrices.

Definition 4.4 (Circulant Matrix). A **circulant matrix** $C_n \in \mathbb{R}^{n\times n}$ is a Toeplitz matrix in which each row is a cyclic shift of the row above it, i.e., $(C_n)_{i,j} = c_{(j-i) \mod n}$:

$$C_n = \begin{bmatrix} c_0 & c_1 & c_2 & \cdots & \cdots & c_{n-1} \\ c_{n-1} & c_0 & c_1 & \ddots & & \vdots \\ c_{n-2} & c_{n-1} & c_0 & c_1 & \ddots & \vdots \\ \vdots & \ddots & \ddots & \ddots & \ddots & c_2 \\ \vdots & & \ddots & \ddots & c_0 & c_1 \\ c_1 & \cdots & \cdots & c_{n-2} & c_{n-1} & c_0 \end{bmatrix}.$$

Proposition 4.2 (Characterization of circulant matrices). *The matrices in the circulant algebra $\mathcal{C}$ are characterized by being simultaneously diagonalized by the unitary matrix F_n*

$$(F_n)_{j,k} = \frac{1}{\sqrt{n}} e^{2\pi i\, j\, k/n}, \quad j,k = 1,\ldots,n.$$

Therefore,

$$\mathcal{C} = \left\{ C_n \in \mathbb{C}^{n\times n} \mid C_n = F^H D F \;:\; D = \operatorname{diag}(d_0, d_1, \ldots, d_{n-1}) \right\}.$$

Proof. Let us suppose that $C = C_n$ is a circulant matrix. We need to solve the eigenvalue equation

$$C\mathbf{y} = \lambda\mathbf{y},$$

from which we obtain

$$\sum_{k=0}^{m-1} c_{n-m+k} y_k + \sum_{k=m}^{n-1} c_{k-m} y_k = \lambda y_m, \quad m = 0, 1, \ldots, n-1,$$
$$\sum_{k=0}^{n-1-m} c_k y_{k+m} + \sum_{k=n-m}^{n-1} c_k y_{k-(n-m)} = \lambda y_m, \quad m = 0, 1, \ldots, n-1.$$

That is a linear difference equation with constant coefficients, for which a solution is of the form $y_k = \zeta^k$. Therefore, we find

$$\sum_{k=0}^{n-1-m} c_k \zeta^k + \frac{1}{\zeta^n} \sum_{k=n-m}^{n-1} c_k \zeta^k = \lambda, \quad m = 0, 1, \ldots, n-1. \tag{4.3}$$

To obtain an *eigenvalue* and an *eigenvector* from equation (4.3) we can simply pose $\zeta^n = 1$, i.e., one of the nth distinct roots of unity, in this way

$$\lambda = \sum_{k=0}^{n-1} c_k \rho^k, \ \mathbf{y} = (1, \zeta, \zeta^2, \ldots, \zeta^{n-1})^T.$$

By normalizing the eigenvector basis and choosing the canonical expression of the nth root of the unity we obtain that if C is circulant then $C \in \mathcal{C}$. To prove now that if $C_n \in \mathcal{C}$ then C_n satisfies Definition 4.4 and is sufficient to perform the product $F^H D F$ and look explicitly at its entries. □

The unitary matrix F_n in Proposition 4.2 is indeed the Fourier matrix F_n. Therefore, the product $C_n\mathbf{y}$ can be formed with Algorithm 4.1. The cost

---

Algorithm 4.1: Circulant matrix–vector product

Input: First row of the circulant matrix $\mathbf{c} = [c_0, c_1, \ldots, c_{n-1}]$, $\mathbf{y}$
Output: $\mathbf{x} = C\mathbf{y}$

$\mathbf{f} \leftarrow F_n \mathbf{y}$;
$\boldsymbol{\lambda} \leftarrow \sqrt{n} F_n \mathbf{c}$; `// We are computing the eigenvalues of` C_n
$\mathbf{z}^T \leftarrow [f_1\lambda_1, f_2\lambda_2, \ldots, f_n\lambda_n]$;
$\mathbf{x} \leftarrow F_n^H \mathbf{z}$;

---

of computing the matrix–vector product is reduced to the cost of computing matrix–vector products with the Fourier matrix. This can be achieved by using the implementation of the DFT (Discrete Fourier Transform) Algorithm[1]

[1] There exist many low cost implementation of the DFT algorithms based on different approaches, like the one based on *divide et impera* like the Cooley and Tukey [170], Bruun [114] and Guo et al. [285], the one by Winograd [547] based on the factorization of the polynomial $x^n - 1$ or the Good–Thomas [272, 509] working on the Chinese remainder Theorem. For an up–to–date implementation of the DFT refer to the FFTW library [251] at `http://www.fftw.org/`.

by Cooley and Tukey [170], also known as Fast Fourier Transform, or *FFT*, that performs the computation in $O(n\log(n))$ operations with an accuracy of $O(\varepsilon\log(n))$.

The good news is that we can compute the product $T_n(f)\mathbf{v}$ in $O(n\log(n))$ operations simply by embedding the $T_n(f)$ matrix in a circulant matrix of size $2n$ in the following way:

$$C_{2n}\begin{bmatrix}\mathbf{v}\\ \mathbf{0}_n\end{bmatrix} = \begin{bmatrix}T_n(f) & E_n\\ E_n & T_n(f)\end{bmatrix}\begin{bmatrix}\mathbf{v}\\ \mathbf{0}_n\end{bmatrix} = \begin{bmatrix}T_n(f)\mathbf{v}\\ E_n\mathbf{v}\end{bmatrix}$$

where E_n is defined to make C_{2n} circulant, and therefore is

$$E_n = \begin{bmatrix} 0 & t_{n-1} & \dots & t_2 & t_1\\ t_{1-n} & 0 & t_{n-1} & \dots & t_2\\ \vdots & t_{1-n} & 0 & \ddots & \vdots\\ t_{-2} & \dots & \ddots & \ddots & t_{n-1}\\ t_{-1} & t_{-2} & \dots & t_{1-n} & 0 \end{bmatrix},$$

and applying Algorithm 4.1 with C_{2n} at the cost of $O(n\log(n))$ operations.

Remark 4.2. Observe that Algorithm 4.1 is efficient only if we have to compute a single product with $T_n(f)$ or C_n. If the product has to be computed repeatedly, as is exactly our case, we should modify it by using as input directly the eigenvalues of C_{2n} or C_n and gain a FFT application for each iteration.

Direct Toeplitz solvers

Direct Toeplitz solvers for linear systems have been built using a strong property characterizing the inverse of a Toeplitz matrix, i.e., that the inverse can be decomposed into the sum of two matrices that are both products of two Toeplitz matrices. This results descends from the formula of Gohberg and Semencul [267] and it has given rise to many direct algorithms, see Table 4.1 for the references of some of them and [115, 297] for a general discussion and information on the stability issues. We will not devote other space to direct methods, focusing instead on the use of iterative methods.

Table 4.1: Toeplitz Direct Solvers

Algorithm	Complexity	Algorithm	Complexity
Levinson [352]	$O(n^2)$	Ammar and Gragg [8]	$O(n\log^2(n))$
Trench [518]	$O(n^2)$	Bitmead and Anderson [89]	$O(n\log^2(n))$
Zohar [559]	$O(n^2)$	de Hoog [180]	$O(n\log^2(n))$
Chan and Hansen [151]	$O(n^2)$	Brent et al. [103]	$O(n\log^2(n))$

Multilevel and quasi–Toeplitz matrices

Multilevel operators of this kind appear naturally from multidimensional problems discretized with uniform meshes. Among the most popular examples we can find discretizations of the systems of ODEs, see, e.g., [64, 76], of systems of PDEs, see e.g., [122, 134, 360]; regularization problems, see, e.g., [83, 292, 323].

Let us start with two particular examples of a *multilevel operator*:

BTTB the block Toeplitz matrix of size nm with Toeplitz blocks of size m, i.e., $T_{n,m}(f) \in \mathbb{C}^{nm \times nm}$,

$$T_{n,m}(f) = \begin{bmatrix} T_m^{(0)} & T_m^{(-1)} & \dots & T_m^{(2-n)} & T_m^{(1-n)} \\ T_m^{(1)} & T_m^{(0)} & \ddots & \dots & \vdots \\ \vdots & \ddots & T_m^{(0)} & \ddots & \vdots \\ \vdots & \dots & \ddots & \ddots & T_m^{(-1)} \\ T_m^{(n-1)} & T_m^{(n-2)} & \dots & T_m^{(1)} & T_m^{(0)} \end{bmatrix}, \tag{4.4}$$

that has an overall Toeplitz structure and in which each block $T_m^{(j)}$ for $j = 1-n, \dots, 0, \dots, n-1$ is a Toeplitz matrix of size m itself,

BCCB the block Circulant matrix of size nm with Circulant blocks of size m, i.e., $C_{n,m} \in \mathbb{C}^{nm \times nm}$,

$$C_{n,m} = \begin{bmatrix} C_m^{(0)} & C_m^{(1)} & C_m^{(2)} & \dots & \dots & C_m^{(n-1)} \\ C_m^{(n-1)} & C_m^{(0)} & C_m^{(1)} & C_m^{(2)} & & \vdots \\ C_m^{(n-2)} & C_m^{(n-1)} & C_m^{(0)} & C_m^{(1)} & \ddots & \vdots \\ \vdots & \ddots & \ddots & \ddots & \ddots & C_m^{(2)} \\ \vdots & & \ddots & \ddots & C_m^{(0)} & C_m^{(1)} \\ C_m^{(1)} & \dots & \dots & C_m^{(n-2)} & C_m^{(n-1)} & C_m^{(0)} \end{bmatrix}, \tag{4.5}$$

that has an overall Circulant structure and in which each block $C_m^{(j)}$ for $j = 0, \dots, n-1$ is a Circulant matrix of size m itself.

In the general case we can give the following definition,

Definition 4.5 (Multilevel Toeplitz and Circulant Matrix). Given a number of levels $m \in \mathbb{N}$ we define the set of n–admissible multiindices $k = (k_1, \dots, k_m)$ and $l = (l_1, \dots, l_m)$ where $n = (n_1, \dots, n_m)$ and the following inequalities hold

$$0 \leq k_j l_j \leq n_j - 1, \quad j = 1, 2, \dots, m.$$

The m–level Toeplitz matrix $T_n \in \mathbb{C}^{N(n) \times N(n)}$, with $N(n) = \prod_{j=1}^{m} n_j$ is therefore defined as the matrix

$$T_N = [t_{k-l}],$$

while the m–level Circulant matrix C_N is obtained whenever

$$t_{k-l} = t_{(k-l) \mod n}$$

where, by definition,

$$k \mod n = (k_1 \mod n_1, \ldots, k_m \mod n_m).$$

We can again associate the m–level Toeplitz matrix with a complex–valued **generator function** f of m real variables which is 2π–periodic in each of them.

Definition 4.6 (m–level generator function). Given $f : Q_m = [0, 2\pi]^m \to \mathbb{C}$ that is 2π–periodic in each variable. If $f \in \mathbb{L}^1$ we can consider the multiindices $k = (k_1, \ldots, k_m)$ and associate it to its Fourier series,

$$f(\mathbf{x}) \approxeq \sum_{k \in \mathbb{Z}^m} t_k e^{i \cdot k_1 x_1} \cdot \ldots \cdot e^{i \cdot k_m x_m}$$

In this way we have the coefficients t_k needed for the construction of the multilevel matrices of Definition 4.5.

We can now formulate again the same question about *asymptotic distribution* of m–level Toeplitz matrices we answered for the 1–level case with Theorem 4.1. The case with $f \in \mathbb{L}^\infty$ can be recovered from the original work of Grenander and Szegö [283], for which a detailed account is in the work of Sakrison [461]. The case of $\mathbb{L}^2$ and $\mathbb{L}^1$ generator functions is obtained in Tyrtyshnikov [522], Tyrtyshnikov and Zamarashkin [524]. We return on these results in § 4.1.1 when we deal with circulant preconditioners.

For the multilevel case no structured decomposition of the inverse is known, i.e., we do not have any generalization of the Gohberg and Semencul [267] formula and, in general, we need to resort to iterative methods. Therefore, from here on we concentrate on some examples of appropriate preconditioners for the iterative methods in Chapter 2 to solve Toeplitz, multilevel Toeplitz and Toeplitz-like systems.

Note that it is even more important to reduce the computational cost of matrix-vector products with multilevel Toeplitz matrices. For the BTTB case we can apply the same embedding strategies we have used for the $T_n(f)$ matrix, i.e., we embed the $T_{n,m}(f)$ into a BCCB matrix $C_{2n,2m}$ and extend the vector $\mathbf{v}$ we want to multiply to a vector of size $4nm$. We can diagonalize the BCCB matrix $C_{2n,2m}$ with the multilevel Fourier matrix $F_{2n,2m} = F_{2n} \otimes F_{2m}$ and compute the product. This procedure has the cost of $O(nm \log(nm))$. Some other approaches can be found in [37, 273, 326].

Finally, let us consider the so called *perturbed* Toeplitz systems, or *quasi*–Toeplitz that are systems in which the matrix is given by $A = T_n(f) + E$. In this class of problems we can find cases where E has a structure or not. Let

E be a Hankel matrix, i.e., is of the form

$$E = \begin{bmatrix} a_0 & a_1 & a_2 & \cdots & \cdots & a_{n-1} \\ a_1 & a_2 & \iddots & \iddots & \iddots & \vdots \\ a_2 & \iddots & \iddots & \iddots & \iddots & \vdots \\ \vdots & \iddots & \iddots & \iddots & \iddots & a_{2n-4} \\ \vdots & \iddots & \iddots & \iddots & a_{2n-4} & a_{2n-3} \\ a_{n-1} & \cdots & \cdots & a_{2n-4} & a_{2n-3} & a_{2n-2} \end{bmatrix}, \ (E)_{i,j} = e_{i+j-2}.$$

This is an example of the so–called Toeplitz plus Hankel system, for which various techniques exist, see, e.g., [266, 298, 299, 338, 388, 396], and arises often in the context of regularization problem, e.g., [365, 400]. Another relevant case of this form is the one in which E is a *banded* matrix, the so–called *Toeplitz–plus–band* systems, and, more specifically, a *diagonal* matrix [139]. Problems of this kind arise in the solution of Fredholm integro-differential equations [184] and signal processing [129]. More in general *quasi*–Toeplitz and multilevel *quasi*–Toeplitz systems arise, e.g., in the treatment of partial differential models because of the boundary and/or interface conditions.

Another generalization of the spectral theory of Toeplitz matrices is represented by the theory of *Generalized Locally Toeplitz* (GLT) sequences. It stems from the seminal work by Tilli [510] on Locally Toeplitz (LT) sequences. An account of this theory can be found in the book by Garoni and Serra-Capizzano [256].

In the following sections we discuss some preconditioners for Toeplitz, *quasi*-Toeplitz and multilevel *quasi*-Toeplitz systems. Multigrid preconditioners, briefly recalled in § 4.2, are also a powerful tool that can be used for various Toeplitz systems.

4.1.1 Circulant preconditioners

The first approach, proposed simultaneously and independently by Strang [493] and Olkin [411], is based on the use of the Conjugate Gradient method (see § 2.2.2) for symmetric and positive definite Toeplitz linear systems with a circulant preconditioner P_n, i.e., a preconditioner based on a circulant matrix. As discussed in § 2.2.2, the cost of the preconditioned Conjugate Gradient iterations depends mainly on the cost of the matrix–vector products performed:

$$T_n\mathbf{v}, \quad P_n^{-1}\mathbf{v}, \quad \mathbf{v} \in \mathbb{R}^n.$$

Therefore a naïve implementation results in a cost per iteration of $O(2n\log(2n))$ for the $T_n\mathbf{v}$ product and, choosing P_n as a circulant matrix C_n, of $O(2n\log(n))$ for the solution of $C_n^{-1}\mathbf{v}$. In this way the cost-per-iteration is augmented by 50%.

However, a more efficient implementation is possible. In order to discuss it, let us introduce the ω–circulant matrices.

Definition 4.7 (ω–Circulant Matrix). Let $\omega = \exp(i\theta)$ for $\theta \in [-\pi, \pi]$. A matrix W_n of size n is said to be an ω–**circulant matrix** if it has the spectral decomposition[2]

$$W_n = \Omega_n^H F_n^H \Lambda_n F_n \Omega_n,$$

where F_n is the Fourier matrix, $\Omega_n = \text{diag}(1, \omega^{-1/n}, \ldots, \omega^{-(n-1)/n})$ and Λ_n is the diagonal matrix of the eigenvalues. In particular 1–circulant matrices are circulant matrices while -1–circulant matrices are the skew–circulant matrices.

To reduce the computational complexity of the Toeplitz matrix-vector product, it is sufficient to exploit the circulant/skew–circulant decomposition of a Toeplitz matrix discovered by Pustyl'nikov [438]:

$$T_n = U_n + V_n,\ U_n = F_n^H \Lambda_n^{(1)} F_n,\ V_n = \Omega_n^H F_n^H \Lambda_n^{(2)} F_n \Omega_n \tag{4.6}$$

where

$$\begin{aligned} \mathbf{e}_1^T U_n &= {}^1\!/_2 \left[t_0, t_{-1} + t_{n-1}, \ldots, t_{-(n-1)+t_1}\right], \\ W_n \mathbf{e}_1 &= {}^1\!/_2 \left[t_0, -(t_{n-1} - t_{-1}), \ldots, -(t_{-1} - t_{n-1})\right]^T. \end{aligned}$$

By Definition 4.7 we have

$$\begin{aligned} C_n^{-1} T_n &= C_n^{-1}(U_n + V_n) = C_n^{-1}\left(F_n^H \Lambda_n^{(1)} F_n + \Omega_n^H F_n^H \Lambda_n^{(2)} F_n \Omega_n\right) \\ &= F_n^H \Lambda_n^{-1} F_n \left(F_n^H \Lambda_n^{(1)} F_n + \Omega_n^H F_n^H \Lambda_n^{(2)} F_n \Omega_n\right) \\ &= F_n^H \left[\Lambda_n^{-1}\left(\Lambda_n^{(1)} + F_n \Omega_n^H F_n^H \Lambda_n^{(2)} F_n \Omega_n F_n^H\right)\right] F_n. \end{aligned}$$

Therefore, we solve $C_n^{-1} T_n \mathbf{x} = C_n^{-1} \mathbf{b}$ as

$$\Lambda_n^{-1}\left(\Lambda_n^{(1)} + F_n \Omega_n^H F_n^H \Lambda_n^{(2)} F_n \Omega_n F_n^H\right) \underbrace{F_n \mathbf{x}}_{=\tilde{\mathbf{x}}} = \underbrace{\Lambda_n^{-1} F_n \mathbf{b}}_{=\tilde{\mathbf{b}}},$$

with 4 FFTs per iteration. The overall cost per iteration is then again $O(n \log(n))$, asymptotically the same as the case without preconditioning.

The use of circulant preconditioners is therefore good from the point of view of computational costs. However, to be considered as an effective strategy, we need also to check that this approach also reduces the number of Conjugate Gradient iterations. In the real symmetric case, a sufficient condition is a cluster in the spectrum of the preconditioned system. To move in this direction we should distinguish two cases. One in which we have information on the symbol, or the generating function, of our Toeplitz matrix sequence, and the other in which we have only information on the entries of the matrix, i.e., we are aware only of a truncated trigonometric polynomial.

[2]Pay attention to the definition: in some cases instead of the Ω_n value we have taken its complex conjugate, see, e.g., [236], thus the decomposition becomes $W_n = \Omega_n F_n^H \Lambda_n F_n \Omega_n^H$.

Circulant preconditioners from the kernels

In the first case, i.e., if we have information on the symbol, the circulant preconditioner can be constructed starting from the *kernel* functions, see [146]. Under appropriate hypotheses, some useful theorems guarantee that the underlying preconditioners work. Providing preconditioners from the kernel comes directly from classical Fourier analysis and needs the convolution of two functions. The definition is proposed for scalar functions but can be easily generalized.

Definition 4.8 (Convolution product). Given two scalar functions f and g in the Schwartz space, i.e., $f, g \in \mathcal{C}^{\infty}(\mathbb{R})$ such that $\exists\, C^{(f)}_{\alpha,\beta}, C^{(g)}_{\alpha',\beta'} \in \mathbb{R}$ with $\|x^{\alpha}\partial_{\beta} f(x)\|_{\infty} \leq C^{\alpha\beta}$ and $\|x^{\alpha'}\partial_{\beta'} g(x)\|_{\infty} \leq C^{\alpha'\beta'}$, α, β, α', β' scalar indices, we define the **convolution operation**, "$*$", as

$$[f * g](t) = \int_{-\infty}^{+\infty} f(\tau)g(t-\tau)d\,\tau = \int_{-\infty}^{+\infty} g(\tau)f(t-\tau)d\,\tau.$$

Therefore, if we take two arbitrary 2π–periodic continuous functions,

$$f(\theta) = \sum_{k=-\infty}^{+\infty} t_k e^{ik\theta} \text{ and } g = \sum_{k=-\infty}^{+\infty} s_k e^{ik\theta}$$

their convolution product, in the sense of Definition 4.8, is given by

$$[f * g](\theta) = \sum_{k=-\infty}^{+\infty} s_k t_k e^{ik\theta}.$$

Now, given a kernel $\mathcal{K}_n(\theta)$ defined on $[0, 2\pi]$ and a generating function f for a Toeplitz sequence $T_n(f)$, we consider the circulant matrix C_n with eigenvalues given by

$$\lambda_j(C_n) = [\mathcal{K}_n * f]\left(\frac{2\pi j}{n}\right), 0 \leq j \leq n, \tag{4.7}$$

from which we obtain easily the first column of the matrix C_n from proposition 4.2. Some popular choices for the kernels are reported in Table 4.2, together with the first column of their associate circulant preconditioners. These convolution products, namely $\mathcal{K}_n * f$, are nothing but smooth approximations of the generating function f itself.

By this approach we can switch the problem of finding an appropriate preconditioner to the problem of approximating the generating function of the underlying Toeplitz matrix. To clarify this point, let us consider the following two results from [134, 146]. The first one is the characterization of the performance of the T. Chan preconditioner, the so–called *optimal* (circulant) *preconditioner*, that is the minimizer of $\|C_n - T_n\|_F$ over the algebra of all $n \times n$ circulant matrices $\mathcal{C}$ [150], i.e.,

$$c(T_n) = \arg\min_{C_n \in \mathcal{C}} \|C_n - T_n\|_F. \tag{4.8}$$

Table 4.2: Kernels $\mathcal{K}_n(\theta)$ and first column for some classical circulant preconditioners C_n for Toeplitz matrices $T_n(f)$. The main diagonal of C_n is always equal to t_0. For Huckle's preconditioner eigenvalues for $k = 0, \ldots, n-1$ are given instead.

Kernel	$\mathcal{K}_n(\theta)$	$C_n(\theta)$	Ref.
Dirichlet	$\mathcal{D}_n(\theta) = \dfrac{\sin((n + {}^1\!/_2)\theta)}{\sin({}^\theta\!/_2)}$	$\begin{cases} t_k, & 0 < k \leq \lfloor {}^n\!/_2 \rfloor, \\ t_{k-n}, & \lfloor {}^n\!/_2 \rfloor < j < n, \\ c_{n+k}, & 0 < -k < n. \end{cases}$	[493]
R. Chan	$\mathcal{D}_{n-1}(\theta)$	$t_k + \bar{t}_{n-k},\ 0 < k \leq n-1$	[132]
Fejér	$\mathcal{F}_n(\theta) = \dfrac{1}{n}\left(\dfrac{\sin({}^{n\theta}\!/_2)}{\sin({}^\theta\!/_2)}\right)^2$	$\begin{cases} \dfrac{(n-j)t_j + j t_{j-n}}{n}, & 0 \leq j < n, \\ c_{n+j}, & -n < j < 0. \end{cases}$	[150]
Huckle	$\mathcal{F}_p(\theta),\ 1 \leq p < n$	$\lambda_k = \displaystyle\sum_{j=-p+1}^{p+1} t_j \left(1 - \frac{\lvert j \rvert}{p}\right) e^{\frac{2\pi i j k}{n}}$	[309]
Modified Dirichlet	${}^1\!/_2\,(\mathcal{D}_{n-1}(\theta) + \mathcal{D}_{n-2}(\theta))$	$\begin{cases} t_1 + {}^1\!/_2\bar{t}_{n-1}, & k = 1, \\ t_k + t_{n-k}, & 2 \leq k \leq n-2, \\ {}^1\!/_2 t_{n-1} + \bar{t}_1, & k = n-1. \end{cases}$	[146]
De la Vallèe Poussin	$2\mathcal{F}_{2m}(\theta) - \mathcal{F}_m(\theta) \quad (m = \lfloor {}^n\!/_2 \rfloor)$	$\begin{cases} t_k + {}^k\!/_m \bar{t}_{2m-k}, & 1 \leq k \leq m, \\ {}^{2m-k}\!/_m t_k + \bar{t}_{2m-k}, & m < k \leq 2m, \\ 0, & k = 2m. \end{cases}$	[560]
von Hann	${}^1\!/_4\,(\mathcal{D}_n(\theta - {}^\pi\!/_n) + 2\mathcal{D}_n(\theta) + \mathcal{D}_n(\theta + {}^\pi\!/_n))$	$\cos^2({}^{\pi k}\!/_{2n}) t_k + \cos^2({}^{\pi(n-k)}\!/_{2n}) \bar{t}_{n-k}$	[146]
Hamming	$0.23\,(\mathcal{D}_n(\theta - {}^\pi\!/_n) + \mathcal{D}_n(\theta)) + 0.54\mathcal{D}_n(\theta + {}^\pi\!/_n)$	$(0.54 + 0.46\cos({}^{\pi k}\!/_n)) t_k + \ldots$ $\ldots + (0.54 + 0.46\cos({}^{\pi(n-k)}\!/_n)) \bar{t}_{n-k}$	[289]
Bernstein	${}^1\!/_2\,(\mathcal{D}_{n-1}(\theta) + \mathcal{D}_{n-1}(\theta + {}^\pi\!/_n))$	$\frac{1}{2}\left[(1 + \exp({}^{i\pi k}\!/_n))\, t_k + (1 - \exp({}^{i\pi k}\!/_n)) \bar{t}_{n-k}\right]$	[146]

By simple manipulations and using the characterization given by Proposition 4.2 we have

$$\lambda_j(c(T_n)) = [\mathcal{F}_n * f]\left(\frac{2\pi j}{n}\right), \quad 0 \leq j < n,$$

where $\mathcal{F}_n(\theta)$ is the Fejèr kernel (see Table 4.2) and

$$\lambda_j(c(T_n)) = \frac{1}{n}\left(f_0 + f_1 + \ldots + f_{n-1}\right)\left(\frac{2\pi j}{n}\right), \quad 0 \leq j < n,$$

that is the Cesáro summation process of order 1 for the Fourier series of f. From standard Fourier analysis it is well known that $\mathcal{F}_n * f$ converges uniformly to f whenever f is a 2π–periodic function. Therefore, we can reasonably suppose that $c(T_n)$ works well as a preconditioner for the sequence $T_n(f)$ when f is a 2π–periodic function. With the following proposition we establish exactly this.

Proposition 4.3 (Chan and Yeung [147], Tyrtyshnikov [521]). *Let f be a 2π–periodic continuous positive function and $c(T_n(f))$ the T. Chan's circulant preconditioner for $T_n(f)$. Then the spectrum of the eigenvalues of the matrices of the sequence $\{c(T_n(f))^{-1}T_n(f)\}_n$ is clustered around 1.*

Proof. As a first step, we need to prove that both $\|c(T_n(f))\|_2$ and $\|c(T_n(f))^{-1}\|_2$ are bounded. We drop f from the definition of $T_n(f)$ to simplify the notation.

To prove the underlying boundedness result, let us prove that

$$\lambda_{\min}(T_n) \leq \lambda_{\min}(c(T_n)) \leq \lambda_{\max}(c(T_n)) \leq \lambda_{\max}(T_n), \tag{4.9}$$

by exploiting the fact that both T_n and $c(T_n)$ are Hermitian, the first one by Proposition 4.1 while the second one from (4.8), because $c(T_n) = F_n \operatorname{diag}(F_n T_n F_n^H) F_n^H$. Now, by using the fact that $c(T_n)$ is a circulant matrix, we obtain the relation (4.9) as

$$\begin{aligned}
\lambda_{\max}(c(T_n)) =& \lambda_k = \frac{\mathbf{e}_k^H F_n T_n F_n^H \mathbf{e}_k}{\mathbf{e}_k^H \mathbf{e}_k} \leq \max_{\mathbf{x}\neq 0} \frac{\mathbf{x}^H F_n T_n F_n^H \mathbf{x}}{\mathbf{x}^H \mathbf{x}} \\
=& \max_{\mathbf{x}\neq 0} \frac{\mathbf{x}^H T_n \mathbf{x}}{\mathbf{x}^H \mathbf{x}} = \lambda_{\max}(T_n); \\
\lambda_{\min}(T_n) =& \min_{\mathbf{x}\neq 0} \frac{\mathbf{x}^H F_n T_n F_n^H \mathbf{x}}{\mathbf{x}^H \mathbf{x}} \leq \frac{\mathbf{e}_j^H F_n T_n F_n^H \mathbf{e}_j}{\mathbf{e}_j^H \mathbf{e}_j} = \lambda_j = \lambda_{\min}(c(T_n)).
\end{aligned}$$

Therefore, we have bounded both $\|c(T_n(f))\|_2$ and $\|c(T_n(f))^{-1}\|_2$ in terms of the maximum and the minimum of f, see again Proposition 4.1. To proceed, we need a result from theory of function approximation: the Weierstrass approximation Theorem in its trigonometric form. By this result we have

$f : [-\pi, \pi] \to \mathbb{R}^+$ that is continuous and 2π–periodic. Then, $\forall \varepsilon > 0$ there exist $M > 0$ and a trigonometric polynomial $p_M(\theta)$

$$p_M(\theta) = \sum_{k=-M}^{M} a_k e^{ik\theta}, \text{ with } \bar{a}_k = a_{-k}$$

such that $\|f - p_M\|_\infty < \varepsilon$. For all $n > 2M$,

$$\begin{aligned} c(T_n(f)) - T_n(f) = & c(T_n(f)) - c(T_n(p_M)) - T_n(f) + T_n(p_M) + \\ & + c(T_n(p_M)) - T_n(p_M). \end{aligned}$$

and then

$$\begin{aligned} \|c(T_n(f)) - c(T_n(p_M)) - T_n(f) + T_n(p_M)\|_2 \leq & \|c(T_n(f)) - c(T_n(p_M))\|_2 + \\ & + \|T_n(f) + T_n(p_M)\|_2 \\ \leq & \|c\|_2 \|f - p_M\|_\infty + \\ & + \|f - p_M\|_\infty \leq 2\varepsilon, \end{aligned}$$

because $\|c\|_2 = 1$. Indeed, for any A it is true that $\|c(A)\|_2^2 = \rho(c(A)) = \rho(A) = \|A\|_2^2$ and $\|c(I)\|_2^2 = \|I\|_2^2 = 1$.

It remains to show that $D_n = c(T_n(p_M)) - T_n(p_M)$ has a clustered spectrum. Decompose $D_n = N_n' + R_n$, where

$$N_n = c(T_n(f)) - c(T_n(p_M)) - T_n(f) + T_n(p_M) + N_n', \; \|N_n\|_2 \leq 3\varepsilon \; \forall n \geq N,$$

and both N_n and R_n are Hermitian Toeplitz matrices. Now if we choose the first row of N_n as

$$\mathbf{e}_1^T N_n' = (0, -{}^{a_{-1}}/_{n}, \ldots, -{}^{Ma_{-M}}/_{n}, 0, \ldots, 0)\,,$$

$$\begin{aligned} \|N_n'\|_2 \leq & \|N_n'\|_\infty = 2\sum_{k=1}^{M} \frac{i}{n}|b_k| = 2\sum_{k=1}^{M} \frac{i}{n}\left|\frac{1}{2\pi}\int_{-\pi}^{\pi} p_M(t)e^{-ikt}dt\right| \\ \leq & 2\sum_{k=1}^{M} \frac{i}{n}\left(\left|\frac{1}{2\pi}\int_{-\pi}^{\pi} (p_M(t) - f(t))e^{-ikt}dt\right| + \left|\frac{1}{2\pi}\int_{-\pi}^{\pi} f(t)e^{-ikt}dt\right|\right) \\ \leq & 2\sum_{k=1}^{M}\left(\|f - p_m\|_\infty + \max_{x\in[-\pi,\pi]} f\right) = \frac{2}{n}M(M+1)\left(\varepsilon + \max_{x\in[-\pi,\pi]} f\right). \end{aligned}$$

Therefore, there exists an $N > 0$ such that $\forall n > N$ $\|N_n'\|_2 \leq \varepsilon$. R_n is the Hermitian Toeplitz matrix with first row

$$\mathbf{e}_1^T R_n = (0, \ldots, 0, {}^{n-M}/_{n} a_M, \ldots, {}^{n-1}/_{M} a_1)\,,$$

whose rank is such that $\text{rank}(R_N) \leq 2M$ and $c(T_n) - T_n$ can be seen as the sum of a matrix of "small norm" and a matrix of "small rank" R_N, $\text{rank}(R_N) < 2M$. By the Cauchy interlacing theorem (see appendix A), the matrix $c(T_n) - T_n$ has at most $2M$ eigenvalues with absolute values larger than 3ε for $n \geq N$. Therefore, we have existence of a cluster at 1 in the sense of Definition 3.1. □

We recall (here without proof) a more general result that holds for circulant preconditioners whose kernel are such that their convolution product with f converge uniformly.

Theorem 4.2 (Chan and Yeung [146]). *Lef f be a 2π–periodic continuous positive function. Let $\mathcal{K}_n(\theta)$ be a kernel such that $\mathcal{K}_n * f \overset{n\to+\infty}{\longrightarrow} f$ uniformly on $[-\pi, \pi]$. If C_n is the sequence of circulant matrices with eigenvalues given by (4.7), then the sequence $\{C_n^{-1}T_n(f)\}_n$ has a cluster of eigenvalues around 1 in the sense of Definition 3.1.*

By knowing the generating function f, we can choose the "best" preconditioner, reducing the problem to the approximation of the symbol f with a suitable convolution kernel. Consider in this regard the following example from [399].

Example 4.2. Consider the sequence $\{T_n(f)\}$ with f the 2π–periodic extension of $f(\theta) = \theta^4 + 1$ and $\theta \in [-\pi, \pi]$. This is a continuous function with Fourier coefficients given by

$$t_k = \begin{cases} \frac{1}{5}\left(5 + \pi^4\right), & k = 0, \\ \frac{4(-1)^k\left(\pi^2 k^2 - 6\right)}{k^4}, & k \neq 0. \end{cases}.$$

A sample of the spectrum of the preconditioned system for the choices of preconditioner in Table 4.2 is shown in Figure 4.2 for $n = 64$. ✓

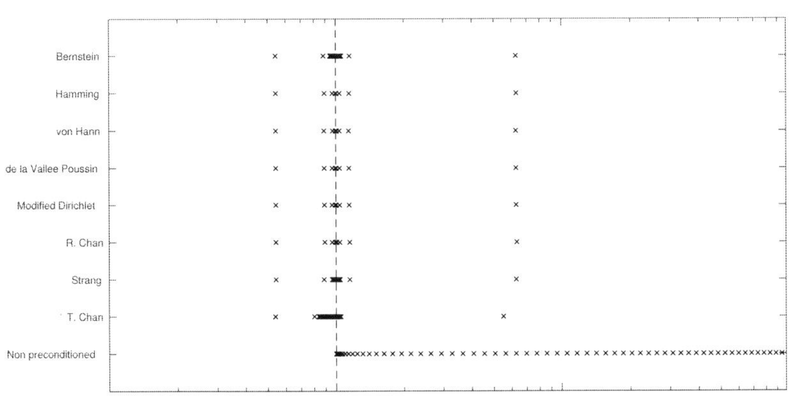

Figure 4.2: The spectra of the matrices $C_n^{-1}T_n(f)$ for the various choices of C_n from Table 4.2 is given, $n = 64$.

Another approach based on circulant matrices is the so–called *superoptimal preconditioner.* The word superoptimal was chosen because the former was born to be a competitor for T. Chan's *optimal* preconditioner. Recall that the *optimal* preconditioner minimizes of $\|C_n - T_n\|_F$ over the circulant matrices. The superoptimal preconditioner is defined to minimize the following quantity

$$\arg \min_{\substack{C_n \in \mathcal{C}, \\ \det C_n \neq 0}} \|I - C_n^{-1} T_n\|_F, \tag{4.10}$$

i.e., such that $\{C_n^{-1}T_n\}_n$ is as close as possible to the identity I_n. This preconditioner has been proposed independently by Tismenetsky [511] and by Tyrtyshnikov [521].

We can define the superoptimal preconditioner for a generic $A_n \in \mathbb{C}^{n\times n}$.

Theorem 4.3 (Chan et al. [138], Tyrtyshnikov [521]). *Let $A_n \in \mathbb{C}^{n\times n}$ be such that both A_n and $c(A_n)$ are nonsingular. Then the superoptimal circulant preconditioner for A_n exists and is given by $c(A_nA_n^H)c(A_n^H)^{-1}$.*

Observe that, if A_n is not necessarily a Toeplitz matrix, computing the *superoptimal* circulant preconditioner costs $n + 2$ FFTs plus $2n^2$ additions and $2n + n^2$ multiplication, while for a Toeplitz matrix T_n the cost is sensibly reduced to 6 FFTs, $8n$ additions and $14n$ multiplications, see [137]. Moreover, the following clustering result holds.

Theorem 4.4 (Chan et al. [138]). *Let $\{T_n(f)\}_n$ be a sequence of Hermitian Toeplitz matrices generated by a positive function in the Wiener class. Then the sequence $\{c(T_n^2)c(T_n)^{-1}T_n\}$ has a cluster of eigenvalues at 1.*

We can also develop an ω–circulant extension of the preconditioner of T. Chan. Following [236, 310], we aim to minimize the quantity

$$\|W_n - T_n(f)\|_F$$

over all the ω–circulant matrices W_n, see Definition 4.7. By using the decomposition for a generic ω–circulant matrix,

$$\|W_n - T_n(f)\|_F = \|\Omega_n^H F_n^H \Lambda_n F_n \Omega_n - T_n(f)\|_F = \|\Omega_n^H C_n \Omega_n - T_n(f)\|_F.$$

Therefore, our optimal ω–circulant preconditioner is obtained by taking the circulant matrix C_n such that

$$\arg\min \|C_n - \Omega_n T_n(f) \Omega_n^H\|_F = \arg\min \|C_n - T_n^{\omega}(f)\|_F,$$

and this is simply $C_n = c(T_n^{\omega}(f))$. Therefore, we have a clear way to build the preconditioner:

$$W_n = \Omega_n^H c(T_n^{\omega}(f))\Omega_n = c_{\omega}(T_n(f)). \tag{4.11}$$

Now, from equation (4.11) we find the optimal ω–circulant preconditioner for a given matrix $T_n(f)$, but what about ω? Obviously we aim to the ω giving the optimal result, i.e.,

$$\omega = \arg\min_{\omega} \|c(T_n^{\omega}(f)) - T_n^{\omega}(f)\|_F.$$

Thus,

$$\|c(T_n^{\omega}(f)) - T_n^{\omega}(f)\|_F^2 = \frac{1}{n}\sum_{j=1}^{n-1}(n-j)j|t_{-j}|^2 +$$

$$+\frac{1}{n}\sum_{j=1}^{n-1}(n-j)j|t_j|^2 - \frac{2}{n}\Re\left(\omega\sum_{j=1}^{n-1}(n-j)j\overline{t_{-j}}t_{n-j}\right).$$

And so we can compute the optimal $\omega = \exp(i\theta)$ as the solution of the following one–dimensional real optimization problem in θ:

$$\theta = -\arg\left(\sum_{j=1}^{n-1}(n-j)j\overline{t_j}t_{j-n}\right) + 2k\pi, \quad k \in \mathbb{Z}. \tag{4.12}$$

Concerning the clustering properties of this preconditioner, results similar to Proposition 4.3 can be established.

Proposition 4.4 (Fischer and Huckle [236]). *Lef f be a 2π–periodic continuous positive function with the associated sequence of Toeplitz matrices $\{T_n(f)\}_n$. If W_n is the optimal ω–circulant preconditioner for $T_n(f)$, then the spectra of $\{W_n^{-1}T_n(f)\}_n$ is clustered around* 1.

In practice, to be a sensible improvement over the optimal T. Chan preconditioner $c(T_n)$, the optimal ω–circulant preconditioner W_n from equation (4.11), needs non-banded Toeplitz matrices. Indeed, as we see from (4.12), if $T_n(f)$ is banded with a bandwidth that is less then $n/2$, then $\omega = 1$ and then the two preconditioners behave the same way at most. On the other hand, if the underlying Toeplitz matrix is closely related to a skew–circulant matrix, the use of a skew–circulant preconditioner minimizes the Frobenius norm and leads to a faster convergence.

Observe that it is possible to develop a ω–circulant version of the Circulant Strang preconditioner (see Table 4.2). This extension is more interesting to avoid singular preconditioning matrices (or component matrices for a hybrid preconditioner) than minimizing a residual norm; see, e.g, [65, 66, 75] for theory and applications to various differential problems. Another extension of the Strang preconditioner used for least squares problems is in [141].

The ill–conditioned case

Let us suppose now that we have located the analytic expression of f, one or more zeros of the function f.

Definition 4.9. The point x^* is called a zero of order $q \in \mathbb{N}$ of f if

$$\exists\, q > 0,\ f^{(q)}(x^*) = 0,\ f^{(q+1)}(x^*) \neq 0.$$

The preconditioners devised from the kernel functions of Table 4.2 are not appropriate for this case and we need to focus on other strategies.

The first strategy on which to focus our attention is the one by Potts and Steidl [434] that uses ω–circulant matrices as preconditioners. Observe that this strategy *depends* on the knowledge of the location of the zeros of f. Choose a uniform grid $\{x_k\}_k$ such that

$$x_k = \omega_n + \frac{2\pi k}{n},\ \omega_n \in [-\pi, -\pi + {}^{2\pi}/_{m}),\ :\ f(x_k) \neq 0,\ k = 0, \dots, n-1. \quad (4.13)$$

Then, define the ω–circulant preconditioner W_n as the ω_n-circulant matrix

$$W_n(f) = \Omega_n^H F_n^H \Lambda_n F_n \Omega_n,\ \Lambda_n = \text{diag}(f(x_0), \dots, f(x_{n-1})). \quad (4.14)$$

As we have observed in the definition of ω–circulant matrices (Definition 4.7), the cost of the application of the preconditioner, once the diagonal matrix Λ_n is computed, is $O(n \log n)$. Moreover, the computation of the Λ_n can be computed by an FFT sweep and so it costs another $O(n \log n)$ flops. W_n is also positive definite by definition. The effectiveness of this preconditioner is established by the following clustering result.

Theorem 4.5 (Potts and Steidl [434]). *Let f be a continuous nonnegative 2π–periodic function with a zero of order $2s$ in $x = 0$. Let $T_n(f)$ be the corresponding Toeplitz matrix with preconditioner $W_n(f)$ defined by position* (4.14). *Then, the sequence of matrices $\{W_n(f)^{-1}T_n(f)\}_n$ have a cluster at* 1.

The above clustering result can be easily extended to the case in which f is a non–negative 2π–periodic function with several zeros of even order, i.e.,

$$f(x) = \prod_{i=1}^{m} (x - x_i)^{2s_i} \tilde{f}(x),\ \tilde{f} > 0.$$

On the other hand, this approach for zeros of odd order seems not to be so effective, see again [434].

The other approach we want to discuss now is again in the terms of a kernel approach devised in Chan et al. [143, 149], Potts and Steidl [435] that does not require the explicit knowledge[3] of the generating function f. The kernel we are now introducing can be seen as a generalization of either the Fejér or the Jackson kernel from Table 4.2, indeed it is the so–called *generalized Jackson kernel*, and is defined as follows.

[3]In the proof by Potts and Steidl [435] the position of the zeros of the f was originally required while in the one by [143, 149] this requirement was not present. Indeed the preconditioner defined is the same; what is different is how its properties are proved.

Definition 4.10 (Generalized Jackson Kernel). Given $\theta \in [-\pi, \pi]$, $\mathbb{N} \ni r \geq 1$ and $\mathbb{N} \ni m > 0$ such that $r(m-1) < n \leq rm$, i.e., $m = \lceil n/r \rceil$, the generalized Jackson kernel function is defined as,

$$\mathcal{K}_{m,2r}(\theta) = \frac{k_{m,2r}}{m^{2r-1}} \left(\frac{\sin(\frac{m\theta}{2})}{\sin(\frac{\theta}{2})} \right)^{2r},$$

where $k_{m,2r}$ is a normalization constant such that

$$\frac{1}{2\pi} \int_{-\pi}^{\pi} \mathcal{K}_{m,2r}(\theta) d\theta = 1.$$

Theorem 4.6 (Chan et al. [143]). *Let f be a nonnegative 2π–periodic continuous function with a zero of order 2ν at θ_0. Let $r > \nu$ and $m = \lceil n/r \rceil$. Then there exists numbers α, β independent from n and such that the spectrum of $J^{-1}_{n,m,r} T_n(f)$ is clustered in $[\alpha, \beta]$ and at most $2\nu + 1$ eigenvalues are not in $[\alpha, \beta]$ for n sufficiently large.*

In order to check that the underlying preconditioner is well-defined, we need to check that the preconditioned system has eigenvalues that are bounded away from zero. Again, this is analogue to what we did for the ω-circulant preconditioners in Theorem 4.5 and thus the proof is omitted.

Theorem 4.7. *Let f be a nonnegative 2π–periodic continuous function with a zero of order 2ν at θ_0. Let $r > \nu$ and $m = \lceil n/r \rceil$. Then, there exists a constant C independent from n and a value n_0 such that for all $n > n_0$ all eigenvalues of $J^{-1}_{n,m,r} T_n(f)$ are larger than C.*

These two results guarantee the convergence of the PCG in $O(1)$ iterations, i.e., not depending on the size of the system. Similarly to what was done for the ω–circulant preconditioners, we can go from the presence of a single zero to the one of multiple by decomposing the generating function in factors. By this simple extension, one can prove the following generalization of the previous results.

Theorem 4.8 (Chan et al. [143]). *Let f be a non–negative 2π–periodic continuous function with s zeros of order $2\nu_j$ at θ_j for $j = 1, 2, \ldots, s$. Let $r > \max \nu_j$, $m = \lceil n/r \rceil$. Then, there exist positive numbers $\alpha < \beta$, independent of n, such that at most $2\nu + s$ eigenvalues of $J^{-1}_{n,m,r} T_n(f)$ are outside the interval $[\alpha, \beta]$. Moreover, all the eigenvalues of the preconditioned matrix are larger than a positive constant.*

Remark 4.3. Finding good preconditioners for ill–conditioned Toeplitz systems depends strongly on the matching of the zeros of f. This idea will be exploited also in the next section (§ 4.1.2) for preconditioners based on non-circulant matrix algebras.

The non–Hermitian and the multilevel cases

Consider now a complex–valued generating function f. Then, in general, the matrices $\{T_n(f)\}$ are complex-valued non-Hermitian. It is well known, see also Chapter 2, that in this case the Conjugate Gradient method is no more viable and we need to use other appropriate Krylov subspace methods like, e.g., GMRES, BiCGstab and their appropriate variants.

In some cases, solvers for the normal equations like CGLS [90], CGS or one of the variants of the LSQR method, can be interesting, at least theoretically. Similar to what is observed for the Hermitian case, by using Krylov methods for the normal equations and proving the existence of a cluster of singular values for the preconditioned system, one can speed up the convergence. On the other hand, the simple existence of clusters of eigenvalues for the underlying sequence of preconditioned systems can not be enough for fast convergence. Recall that, in the non-Hermitian case, the convergence estimates based only on the eigenvalue analysis can fail and one needs to consider other factors to study the convergence of iterations; see also §§ 2.2.8 and 2.3.4.

Let us start with the case of skew–Hermitian Toeplitz matrices. Recall that skew–Hermitian Toeplitz matrices are matrices of the form

$$T_n(f) = \begin{bmatrix} t_0 & t_1 & \ldots & t_{n-2} & t_{n-1} \\ -\bar{t}_1 & t_0 & t_1 & \ldots & t_{n-2} \\ -\bar{t}_2 & \ddots & \ddots & \ddots & \vdots \\ \vdots & \ddots & \ddots & \ddots & t_1 \\ -\bar{t}_{n-1} & \ldots & -\bar{t}_2 & -\bar{t}_1 & t_0 \end{bmatrix} = t_0 I_n + iH_n, \quad H_n = H_n^H.$$

The generating function of $T_n(f)$ is of the form $f = t_0 + ig$, g being a real valued positive function. We can use for $T_n(f)$ any of the circulant preconditioner C_n from Table 4.2 computed for the generating function f. Indeed, $T_n^H T_n = t_0^2 I_n + H_n^H H_n$, thus the singular values of T_n are bounded from below by $|t_0| > 0$. Hence, we can bound the singular values of $C_n^{-1}T_n$ by using results like Theorem 4.2. As an example, results of this form are in [136] for g in the Wiener class and C_n as the Strang preconditioner.

In general, i.e., for f a generic complex valued function, we can again distinguish between the case in which f has one or more zeros or none at all. In the latter case, we can try again to choose the *optimal circulant preconditioner*, i.e., the projection on the algebra of the circulant matrices of $T_n(f)$. By using the same notation of the definition of the T. Chan preconditioner in (4.8), we can consider the solution of $c(T_n(f))^{-1}T_n(f)$. For this choice, the following cluster result has been proved in [148] obtaining the analogue of Proposition 4.3.

Proposition 4.5 (Chan and Yeung [148]). *Let f be a 2π–periodic continuous complex–valued function. Let $\{T_n(f)\}_n$ be the associated sequence of Toeplitz matrices, and $\{c(T_n(f))\}_n$ the related sequence of the optimal circulant preconditioners. Then,*

1. *the spectra of* $\{c(T_n(f)) - T_n(f)\}_n$ *is clustered around zero;*

2. *if* f *has no zeros, both* $\|c(T_n(f))\|_2$ *and* $\|c(T_n(f))^{-1}\|_2$ *are uniformly bounded and the sequence* $\{c(T_n(f))^{-1}T_n(f)\}$ *has a cluster of eigenvalues in* 1.

In order to have superlinear convergence, we need also to know that all the singular values of the preconditioned system are uniformly bounded from below. While in the Hermitian case the absence of zeros of f was sufficient to prove superlinear convergence, this is no more true in general. If f has no zeros, we need just a lower bound on the singular values of $T_n(f)$ because $||c(T_n(f))||_2$ and $||c(T_n(f))^{-1}||_2$ are uniformly bounded. As an example, $f(\theta) = e^{i\theta}$ has no zero in $[pi, \pi]$ but $T_n(f)$ is singular for all n. Serra-Capizzano and Tilli proved [474] that such a lower bound for the singular values exists if f is *weakly sectorial.*

Definition 4.11 (Weakly Sectorial). A function $f : [-\pi, \pi]^q \to \mathbb{C}$ with $f \in \mathbb{L}^1$ is called weakly sectorial if there exists a straight line r in the complex plane such that the origin does not belong to the line and the range of f is included in one side of the line r.

If f is weakly sectorial, the lower bound on the singular values of $T_n(f)$ is the distance of that straight line from the origin, and we are done.

By using the tools above, we can obtain again similar asymptotic convergence result of the case without zeros for the optimal circulant preconditioner, a number of iterations that is an $O(\alpha \log n + 1)$ for matrices whose condition number $\kappa_2(T_n(f))$ is $O(n^\alpha)$ for $\alpha \geq 0$.

Some specialization of this kind of approach exists for particular classes of generating functions f. In particular, for rational functions $f = p/q$, p and q polynomials, see [144, 339].

Another way to face non-Hermitian Toeplitz linear systems that has received a lot of attention in recent years is based on a multi-iterative method obtained through a stationary Hermitian–skew–Hermitian splitting and Krylov subspace methods; see the seminal paper [25] and also § 2.1.1 and § 3.3. The splitting we are interested in here is the one that decomposes a Toeplitz matrix T_n in its circulant and skew–circulant part from (4.6). Given our non–Hermitian Toeplitz matrix $T_n(f)$, we decompose it in $T_n = C_n + S_n$ with C_n a circulant matrix and S_n a skew–circulant matrix, i.e., S_n is a (-1)–circulant matrix. In this way we can define the two step CSCS stationary iterative method as

$$\begin{cases} (\alpha I_n + C_n)\mathbf{x}^{(k+1/2)} = (\alpha I_n - S_n)\mathbf{x}^{(k)} + \mathbf{b}, \\ (\alpha I_n + S_n)\mathbf{x}^{(k+1)} = (\alpha I_n - C_n)\mathbf{x}^{(k+1/2)} + \mathbf{b}, \end{cases} \tag{4.15}$$

with $\mathbf{x}^{(0)}$ a given starting vector, $k \geq 0$, $\alpha > 0$. Each step of the SCSC method costs $O(n \log n)$ flops. The method is convergent whenever C and S are positive definite.

Theorem 4.9 (Bai et al. [26], Ng [398]). *Given $T_n(f)$ let C_n and S_n be its circulant and skew–circulant matrix decomposition, and let $\mathbb{R} \ni \alpha > 0$. If C_n and S_n are positive definite then the iteration matrix M_α of the method* (4.15) *is given by*

$$M_\alpha = (\alpha I_n + S_n)^{-1}(\alpha I_n - C_n)(\alpha I_n + C_n)^{-1}(\alpha I_n - S_n)$$

for which the following bound holds

$$\rho(M_\alpha) \leq \max_j \frac{|\alpha - \lambda_j(C_n)|}{|\alpha + \lambda_j(C_n)|} \cdot \max_j \frac{|\alpha - \lambda_j(S_n)|}{|\alpha + \lambda_j(S_n)|} < 1, \ \forall \alpha > 0.$$

Moreover, if the spectra of the matrices C_n and S_n is contained in $\Omega = [\gamma_{min}, \gamma_{max}] \times i[\eta_{min}, \eta_{max}]$, then the optimal value of α is given by

$$\alpha^* = \begin{cases} (\gamma_{min}\gamma_{max} - \eta_{max})^{1/2}, & \eta_{max} < (\gamma_{min}\gamma_{max})^{1/2}, \\ (\gamma_{min}^2 + \eta_{max}^2)^{1/2}, & \eta_{max} \geq (\gamma_{min}\gamma_{max})^{1/2}. \end{cases}$$

We can use the iteration matrix M_α defined implicitly by the above stationary iterations as a preconditioner for our Krylov subspace method with a fixed amount of iterations; see again the discussion in § 3.3.

Stationary iterative methods based on Hermitian–skew–Hermitian splittings, where the Hermitian and the skew-Hermitian linear systems are solved with Krylov projection methods and preconditioned with various strategies can be an interesting (and sometimes optimal) alternative for the solution of nonsymmetric Toeplitz systems under certain conditions; see, e.g., [74, 197], where the real part of the matrix is used.

Remark 4.4. In previous sections, we observed superlinear convergence of some preconditioned Krylov subspace method. Unfortunately, this is not true anymore in the multilevel case, i.e., using block preconditioners; see [475], in particular, even if the clustering and the boundedness from zero of the eigenvalues is proved.

Therefore, block preconditioning strategies will not give superlinear convergence for block–Toeplitz systems. Hovever, some interesting generalizations of the previously discussed approaches exist but are not considered here.

Remark 4.5. In general we can use all the other circulant preconditioners we have developed in this section as point–preconditioners, i.e., acting on each block level, for the underlying multilevel matrices. As an example, the Strang preconditioner for block symmetric Toeplitz matrices with symmetric Toeplitz blocks. Nevertheless, the severe limitation that the convergence cannot be superlinear for the preconditioned iterations by using $\{C_n^{-1}T_n(f)\}_n$ still holds.

Like the CSCS method that we discussed above, a preconditioned HSS stationary iterative method based on the use of multilevel symbol of the sequence of matrices can be built to handle linear systems whose matrices are multilevel Toeplitz. An approach of this type is proposed in [196].

4.1.2 Unitary transform–based preconditioners

Let us give some brief notes to non–circulant preconditioners for Toeplitz and Toeplitz-like linear systems. As seen in § 4.1.1, circulant matrices are those that can be diagonalized by Fourier transform, thus allowing the execution of both a fast construction and application of this matrices as preconditioners. Therefore, the alternative to the circulant choice is looking for an algebra of matrices such that

$$\mathcal{L} = \left\{ M_n \in \mathbb{C}^{n\times n} \;:\; M_n = Q_n^H \Lambda_n Q_n,\; Q_n^H Q_n = I_n,\; \Lambda_n \text{ diagonal matrix} \right\}.$$

In this way, we can reproduce the definition of both the *optimal* preconditioner over this algebra, see (4.8), and the *super–optimal* preconditioner, see (4.10), as

$$P_n = \arg \min_{\substack{M_n\in\mathcal{L}\\ \det(M_n)\neq 0}} \|M_n - T_n\|_F \text{ or } P_n^{-1} = \arg \min_{\substack{M_n\in\mathcal{L}\\ \det(M_n)\neq 0}} \|I_n - M_n^{-1}T_n\|_F.$$

First, in analogy of what we have seen with the discrete Fourier transform, the matrix–vector product for a matrix in $\mathcal{L}$ should be done cheaply than the standard one. Nevertheless, this is non sufficient for making the underlying an efficient strategy. The other important needed characteristic is: given the transformation Q_n for the algebra $\mathcal{L}$, we need to find a *sparse basis* for all the space.In this way the construction of the minimizer P_n can be done effectively by

$$(\Lambda_n)_{i,i} = \frac{(Q_n T_n \mathbf{e}_1)_i}{(Q_n \mathbf{e}_1)_i},\; i = 1, 2, \ldots, n,$$

and then its application costs only 2 fast transformations Q_n and $O(n)$ operations for the inversion of the diagonal matrix Λ_n. A general discussion for this kind of optimal approximations in matrix algebra is in [191].

Among the most interesting classes of algebra, i.e., of transformation Q_n, we mention

- the algebra of **trigonometric transforms**; see [92, 96, 142, 189, 190];
- the algebra of **Hartley transforms**; see [85, 319].

Other classes of preconditioners are based on the Wavelet transform; see [294, 359].

In the next section we devote some attention to the case of Toeplitz and band–Toeplitz preconditioners for Toeplitz linear systems.

4.1.3 Toeplitz and band–Toeplitz preconditioners

We discussed in § 4.1 the inverse structure of a Toeplitz matrix by observing that in general this is not a Toeplitz matrix both for one level and multilevel matrices. Nonetheless, since the matrix–vector product with a Toeplitz matrix costs $O(n \log n)$, determining a suitable generating function g to use the sequence of matrices $T_n(g)$ as preconditioners for the sequence $T_n(f)$ for a given f is an appealing idea. For this reason, several strategies of this form have been devised, e.g., [133, 140, 145, 198, 290, 300, 320, 467, 468, 495].

Historically the first approach in this sense was the Toeplitz preconditioner in Chan [133]. This is a suitable choice for real generating functions f that are $f \geq 0$, so we come back to it after considering the well conditioned case.

If $f > 0$, a somewhat natural choice is taking the Toeplitz matrix generated by $g = 1/f$ as a preconditioner for $T_n(f)$. Nevertheless, $T_n(1/f)$ can be quite difficult to assemble. Indeed, computing explicitly the Fourier coefficients of $1/f$ can be an impossible task. A way to overcome this problem is proposed in Chan and Ng [140] by approximating the integral defining the Fourier coefficients of $1/f$ in the following way:

$$t_k = \frac{1}{2\pi}\int_0^{2\pi} \frac{1}{f(\theta)} e^{-i\,k\theta} d\,\theta \approx z_k^{(s)} = \frac{1}{sn}\sum_{j=0}^{sn-1} \frac{1}{[\mathcal{K} * f]\left(\frac{2\pi j}{sn}\right)} e^{-2\pi i\, jk/sn},\ k \in \mathbb{Z},$$

where the interval $[0, 2\pi]$ is subdivided into $sn - 1$ subintervals of equal length, and the convolution Kernel $\mathcal{K}$ is chosen as either one of the kernels from Table 4.2 or the δ function. Observe that in this case we need only $[\mathcal{K} * f]\left(\frac{2\pi j}{sn}\right) \neq 0$ in each $0 \leq j < sn$ for the $z_k^{(s)}$ to be defined. The cost for obtaining the $\{z_k^{(s)}\}_{k=0}^{\pm(n-1)}$ coefficients is only 2 FFTs for sn–dimensional vectors, thus $O(2sn \log sn)$ flops. Moreover, for $s = 1$ the Toeplitz matrix $P_n^{(s)}(\mathcal{K}, f)$ assembled with the $z_k^{(1)}$ is simply a circulant preconditioner, that reduces to Strang's or T. Chan's preconditioner for $\mathcal{K}$ respectively the Dirichlet or the Fejér kernel, see again Table 4.2. While for $s > 1$ we have a Toeplitz matrix, the matrices $T_n^{(s)}$ are the sum of ω–circulant matrices for

$$P_n^{(s)}(\mathcal{K}, f) = \sum_{t=0}^{s-1} W_n^{(t)},\ \ W_n^{(t)} = \Omega_n^H F_n^H \Lambda_n F_n \Omega_n,\ \omega = e^{-\frac{2\pi i t}{s}}. \tag{4.16}$$

That is coherent, for $s = 2$, with the decomposition by Pustyl'nikov [438] discussed at the beginning of § 4.1.1. Now, since $P_n^{(s)}$ is a Toeplitz matrix, we can apply the underlying preconditioner in $O(2n \log(2n))$ flops. For this preconditioner, analogously to Theorem 4.2, the following clustering properties hold.

Theorem 4.10 (Chan and Ng [140]). *Let f be a 2π–periodic continuous function and $s \geq 1$. Let $\mathcal{K}$ be a kernel such that $\mathcal{K} * f \overset{unif.}{\longrightarrow} f$ on $[0, 2\pi]$ and $P_n^{(s)}(\mathcal{K}, f)$ be the preconditioner defined in (4.16). Then, for all $\varepsilon > 0$, there exist positive integers N, M such that, for all $n > N$, the matrices in the sequence $\{\left(P_n^{(s)}(\mathcal{K}, f)\right)^{-1} T_n(f)\}_n$ have clustered spectra at 1 with at most M eigenvalues with distance greater than ε from 1.*

Another preconditioner we mention is the one by Hanke and Nagy [290] for banded Toeplitz matrices. Let us start from the case of T_n β–banded Hermitian positive definite Toeplitz matrix and restrict it to the case in which $\beta < n/2$:

$$T_n = \begin{bmatrix} t_0 & t_1 & \dots & t_\beta & & & & & \\ \bar{t}_1 & \ddots & \ddots & \ddots & t_\beta & & & & \\ \vdots & \ddots & \ddots & \ddots & \ddots & \ddots & & & \\ \bar{t}_\beta & \dots & \bar{t}_1 & t_0 & t_1 & \dots & t_\beta & & \\ & \ddots & \ddots & \ddots & \ddots & \ddots & \ddots & \ddots & \\ & & \bar{t}_\beta & \dots & \bar{t}_1 & t_0 & t_1 & \dots & t_\beta \\ & & & \ddots & \ddots & \ddots & \ddots & \ddots & \vdots \\ & & & & \ddots & \ddots & \ddots & \ddots & t_1 \\ & & & & & \bar{t}_\beta & \dots & \bar{t}_1 & t_0 \end{bmatrix},$$

that can be embedded into a circulant matrix C defined as

$$C_{n+\beta} = \begin{bmatrix} T_n & T_{1,2} \\ T_{2,1} & T_{2,2} \end{bmatrix}, T_{2,1} = [L, 0, U], \quad T_{1,2} = T_{2,1}^H, \tag{4.17}$$

with

$$L = \begin{bmatrix} t_\beta & & \\ \vdots & \ddots & \\ t_1 & \dots & t_\beta \end{bmatrix}, \; U = \begin{bmatrix} \bar{t}_\beta & \dots & \bar{t}_1 \\ & \ddots & \vdots \\ & & \bar{t}_\beta \end{bmatrix}, \; T_{2,2} = \begin{bmatrix} t_0 & t_1 & \dots & t_{\beta-1} \\ \bar{t}_1 & \ddots & \ddots & \vdots \\ \vdots & \ddots & \ddots & t_1 \\ \bar{t}_{\beta-1} & \dots & \bar{t}_1 & t_0 \end{bmatrix}.$$

With a FFT sweep we can compute the matrix Λ of the eigenvalues of $C_{n+\beta}$. We assemble its inverse as $C_{n+\beta}^{-1} = F^H \Lambda^{-1} F$ and partition it in the same block form of $C_{n+\beta}$. If C is positive definite,

$$C_{n+\beta}^{-1} = \begin{bmatrix} M & M_{1,2} \\ M_{2,1} & M_{2,2} \end{bmatrix},$$

and M is a Hermitian positive definite Toeplitz matrix, we can use the matrix $P = M^{-1}$ as a preconditioner for the banded matrix T_n. Observe that neither M nor $C_{n+\beta}^{-1}$ need to be computed. We can obtain $\mathbf{z} = P^{-1}\mathbf{r} = M\mathbf{r}$ as the first n components of

$$\mathbf{z} = P^{-1}\mathbf{r} = M\mathbf{r} = C^{-1} \begin{bmatrix} \mathbf{r} \\ \mathbf{0} \end{bmatrix} = F^H \left(\operatorname{diag}(\Lambda^{-1}) \circ F \begin{bmatrix} \mathbf{r} \\ \mathbf{0} \end{bmatrix} \right). \tag{4.18}$$

If C fails to be positive definite, at least one of the diagonal elements of Λ is nonpositive. Nevertheless, the number of eigenvalues for which this could occur is limited by β, since, by construction, the principal $n \times n$ leading submatrix of C is positive definite, i.e., T_n is positive definite. Therefore, we can define Λ^{-1} also for this case by setting to zero all the non positive entries of Λ. In this way the following clustering results can be established.

Theorem 4.11 (Hanke and Nagy [290]). *Let T_n be a β–banded Hermitian, positive definite Toeplitz matrix and let $P = M^{-1}$ be the preconditioner defined from the leading principal submatrix of $C_{n+\beta}^{-1}$ of size n. If ν is the number of non positive eigenvalues of C we have*

$$P^{-1}T = MT = I + R, \quad \mathrm{rank}(R) \leq \beta + \nu \leq 2\beta.$$

This construction can be replicated to also generate a preconditioner for the non–Hermitian case to be applied to the normal equations. All the procedure is repeated in the same way by taking the matrix $T_n^H T_n$ as a starting point. Observe also that in this case we have to focus only on the possibility of having exactly zeros eigenvalues of C. In this way, the rank of the small rank matrix in the clustering results is between 2β and 3β for the case of a singular C. Moreover, the scheme is suitable for a natural extension to the block case. If $T_{n,m}$ is a BTTB block banded with banded blocks matrix and we assume that the maximal band structure in either dimension is β, then we can embed the matrix in an opportune BCCB matrix by appending β rows and columns in each block and other β additional rows and columns of blocks at the whole matrix. Then, the remaining part of the building procedure is similar and the results can be easily extended. In principle, if the matrix T_n is not banded, we could not apply this strategy. Nevertheless, if there is some degree of decay of the extradiagonal coefficients of T_n, then we could discard the diagonals exceeding β and replicate the construction. Some details on this approach are in Strohmer [495].

A modification of this approach has been presented in [362]. Repeating the construction for the circulant extension of T_n we observe that

$$C_{n+\beta}^{-1} C_{n+\beta} = \begin{bmatrix} M & M_{1,2} \\ M_{2,1} & M_{2,2} \end{bmatrix} \begin{bmatrix} T_n & T_{1,2} \\ T_{2,1} & T_{2,2} \end{bmatrix} = \begin{bmatrix} I_n & \\ & I_\beta \end{bmatrix}$$

Thus one can choose the Schur complement

$$S = M - M_{1,2} M_{2,2}^{-1} M_{2,1},$$

as a preconditioner for T_n. In this form S is not a Toeplitz matrix, but since $C_{n+\beta}^{-1}$ can be diagonalized by an FFT of size $n + \beta$, we can compute the products $\mathbf{z} = S\mathbf{r}$ by using 3 FFTs of size $n + \beta$ and solving a Toeplitz system with matrix $M_{2,2}$ of size $\beta \times \beta$.

This construction can be replicated embedding $T_n(f)$ into an ω–circulant matrix $W_{n+\beta}(f)$, as proposed in [236]. In analogy to the construction in (4.17), we perform the embedding in the following way

$$W_{n+\beta} = \begin{bmatrix} T_n & T_{2,1}^H \\ T_{2,1} & T_{2,2} \end{bmatrix}, \quad T_{2,1} = \begin{bmatrix} \omega t_\beta & & & 0 & \dots & 0 & \overline{t_\beta} & \dots & \overline{t_1} \\ \vdots & \ddots & & \vdots & \ddots & \vdots & & \ddots & \vdots \\ \omega t_1 & \dots & \omega t_\beta & 0 & \dots & 0 & & & \overline{t_\beta} \end{bmatrix}$$

where $\omega = \exp(i\theta)$ with $\theta \in [-\pi, \pi]$. In this way, the diagonal matrix $\Lambda_{n+\beta}$ containing the eigenvalues of $W_{n+\beta}$ is obtained as

$$\Lambda_{n+\beta} = F_{n+\beta}\Omega_{n+\beta}W_{n+\beta}\Omega_{n+\beta}^H F_{n+\beta}^H = F_{n+\beta}C_{n+\beta}F_{n+\beta}^H.$$

Once the eigenvalues have been obtained in this way, the inverse of the ω–circulant matrix is obtained as

$$W_{n+\beta}^{-1} = \begin{bmatrix} P & P_{1,2} \\ P_{1,2} & P_{2,2} \end{bmatrix} = \Omega_{n+\beta}^H F_{n+\beta}^H \Lambda_{n+\beta}^{-1} F_{n+\beta}\Omega_{n+\beta}, \tag{4.19}$$

taking note of the position of the non–positive entries of $\Lambda_{n+\beta}$ and putting a zero in the same position of $\Lambda_{n+\beta}^{-1}$, as we have done for its circulant counterpart. We can now state the analogue of Theorem 4.11 for this new approximate inverse preconditioner.

Theorem 4.12 (Fischer and Huckle [236]). *Let T_n be a β–banded Hermitian, positive definite Toeplitz matrix, and let P be the preconditioner defined from the leading principal submatrix of $W_{n+\beta}^{-1}$ of size n in equation* (4.19).

1. *If $W_{n+\beta}$ is positive definite then M_n is positive definite and $PT_n = I_n + R_n$, where* $\text{rank}(R_n) \leq \beta$*,*

2. *If $W_{n+\beta}$ has ν non–positive eigenvalues then $PT_n = I_n + R_n$, where* $\text{rank}(R_n) \leq \beta + \nu \leq 2\beta$.

To choose the "optimal" value of ω, i.e., the value of $\theta = \arg(\omega)$, we observe that for each non–positive eigenvalue, the estimate given by Theorem 4.12 deteriorates. Thus, one can try to choose θ such that as many eigenvalues as possible are positive.

Example 4.3. Consider both $P_n^{(s)}(\mathcal{K}, f)$ the preconditioner from (4.16) and the preconditioner P from the circulant embedding in (4.18). For both these preconditioners use PCG method with a tolerance of $\varepsilon = 10^{-7}$. Let us start from the the $P_n^{(s)}(\mathcal{K}, f)$ for the solution of the linear system with coefficient matrix $T_n(f)$, $f(\theta) = \theta^4 + 1$ with right–hand side the vectors of all 1. As we see from Table 4.3, $P_n^{(1)}(\mathcal{K}, f)$ for $\mathcal{K}$ gives back the T. Chan *optimal* preconditioner, while the performance for the same Kernel are increased for $s = 3$.

The other part of the example is devoted to the application of the preconditioner by Hanke and Nagy [290] to the Hermitian banded problem for $t_0 = 1$ and $t_1 = t_6 = -1/4$. As a comparison we consider again the T. Chan's *optimal* preconditioner. In Figure 4.3 we report the clustering obtained for this

Table 4.3: Example 4.3. Convergence of the PCG method with $P_n^{(s)}(\mathcal{K}, f)$ preconditioner for $\mathcal{K}$ the Fejér Kernel and $s = 1, 3$. Also the T. Chan's *optimal* and no preconditioner are used for comparisons.

n	$s = 1$ Fejér IT	$s = 3$ Fejér IT	T. Chan IT	Unpreconditioned IT
64	7	5	7	36
128	6	5	6	55
256	6	4	6	66
512	6	4	6	70
1024	5	4	5	71

preconditioner with $n = 1024$. As we have discussed in the analysis, this preconditioner gives a sharper cluster than the one based on circulant matrices. ✓

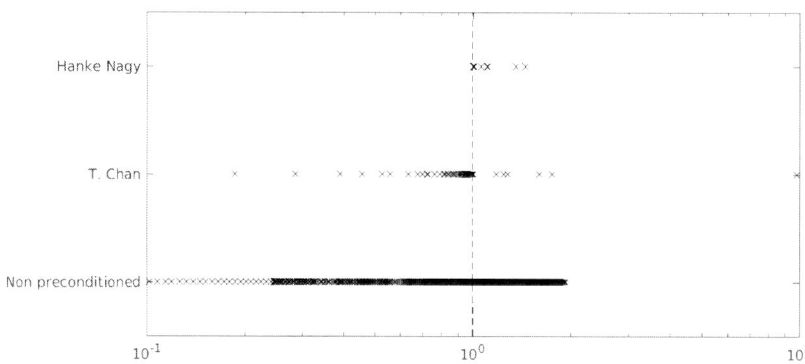

Figure 4.3: Example 4.3. Clustering of the eigenvalues for the Hanke and Nagy [290] preconditioners. Comparison with the clustering obtained with the T. Chan preconditioner and the spectrum of the unpreconditioned system.

We can now treat the case of nonnegative generating functions, i.e., the case of Toeplitz preconditioners for ill–conditioned Toeplitz systems. As promised we start from the preconditioner introduced in Chan [133] that, in a certain sense, contains the main idea. As usual, we deal with generating functions that have one or more zeros $\{\theta_i\}_i$ of various order $\{p_i\}_i$. We build a trigonometric polynomial $s(\theta)$ that matches the zeros of f with the same order and use the Toeplitz matrix $T_n(s)$ as a preconditioner for $T_n(f)$. Let us start from the case of $f \geq 0$, $T_n(f)$ Hermitian and semi–definite with a unique zeros in $\theta = 0$ of order $p = 2\ell$. Select the trigonometric polynomial

$$s_\ell(\theta) = (2 - 2\cos(\theta))^\ell = \left[2\sin\left(\frac{\theta}{2}\right)\right]^{2\ell} = \sum_{k=-\ell}^{\ell} (-1)^k \binom{2\ell}{\ell+k} e^{i\,k\theta}. \quad (4.20)$$

Thus, $T_n(s_\ell)$ can be computed easily by the last equality, resulting in a Hermitian matrix of bandwidth $2\ell + 1$. Hence, for any vector $\mathbf{x}$, the product $T_n^{-1}(s_\ell)\mathbf{x}$ can be computed in $O(\ell^2 n)$ operations. Thus, we can show that this preconditioner enhances the preconditioned system condition number to a $O(1)$.

Theorem 4.13 (Chan [133]). *Let f be a nonnegative 2π–periodic continuous function having a zero of order $p = 2\ell$ at 0. Then, the condition number $\kappa_2(T_n^{-1}(s_\ell)T_n(f))$ is uniformly bounded for all $n > 0$.*

Proof. By the continuity assumption, there exists $\delta > 0$ such that f is continuous in $N_\delta(0)$, $N_\delta(0) = (-\delta, \delta)$. We define the function $g(\theta) = f(\theta)/s_\ell(\theta)$. Clearly g is positive and continuous for each $\theta \in N_\delta(0) \setminus \{0\}$. Since $\theta = 0$ is a zero of order $p = 2\ell$ for f we have

$$\lim_{\theta \to 0} g(\theta) = \frac{f^{(2\ell)}(0)}{(2\ell)!} > 0,$$

and hence g is continuous and positive in $N_\delta(0)$. g shows a positive minimum over $[-\pi, \pi]$ and therefore there exist $c_1, c_2 > 0$ such that $c_1 \leq f(\theta)/s_\ell(\theta) \leq c_2$. By using the usual relationship between Toeplitz matrices and generating functions (Proposition 4.1) we have

$$c_1 \leq \frac{\mathbf{x}^H T_n(f)\mathbf{x}}{\mathbf{x}^H T_n(s_\ell)\mathbf{x}} \leq c_2, \quad \forall \mathbf{x} \in \mathbb{R}^n.$$

Thus, we have the bound $\kappa_2(T_n^{-1}(s_\ell)T_n(f)) \leq c_2/c_1$. □

Example 4.4. Let us look at the clustering for the preconditioners $P(s_1)$ and $P(s_2)$ for the Toeplitz matrix generated, respectively, by the functions $f(\theta) = \theta^2$ and $f(\theta) = \theta^4$. As we observe from Figure 4.4 the spectrum is bounded away from zeros coherently with Theorem 4.13. Nevertheless, there is no cluster around 1 for the eigenvalues of the preconditioned system. So, both the circulant preconditioners described in the ill conditioned case have better performance that is reflected in the number of iterations for the solution with the PCG for the systems $T_n(f)$ with $\mathbf{b} = [1, 1, \ldots, 1]^T$, stopped when the relative residual is less than $\varepsilon = 10^{-7}$ or the maximum number of iterations reaches 1000. ✓

The next point to discuss is what happens when the zero θ_0 is not in the origin. In this case, we can simply shift back the zero to the origin by

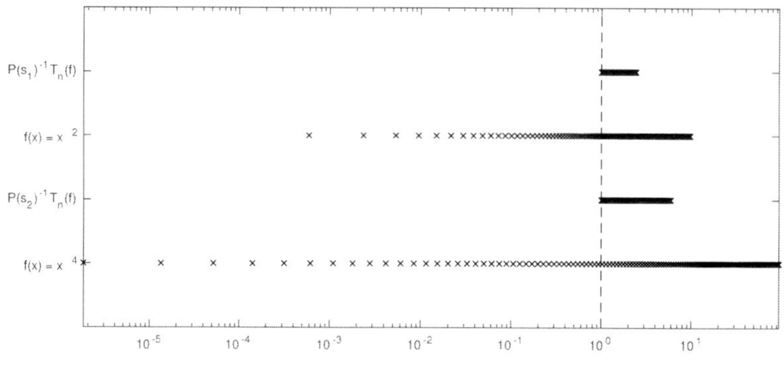

Figure 4.4: Example 4.4. Clustering of the eigenvalues for the band–Toeplitz preconditioners $P(s_1)$ and $P(s_2)$ for two ill–conditioned Toeplitz matrix.

modifying the generating function $f(\theta)$ as $f(\theta-\theta_0)$. Then, in order to be sure that we are solving an equivalent linear system, we modify it as

$$T_n(f(\theta-\theta_0))\Omega_n^H\mathbf{x}=\Omega_n^H\mathbf{b},\ \ \Omega_n=\operatorname{diag}\left(1,e^{i\theta_0},\ldots,e^{i2\theta_0},\ldots,e^{i(n-1)\theta_0}\right)$$

that has the same solution as the starting linear system since $T_n(f(\theta-\theta_0))=\Omega_n^H T_n(f)\Omega_n$.

Now we are ready to treat the case of multiple zeros with any order. Let $\{\theta_i\}_{i=1}^m$ be the zero of order $\{p_i=2\ell_i\}_{i=1}^m$ of the nonnegative continuous function f defined in $[-\pi,\pi]$. If $\ell=\sum_{i=1}^m \ell_i$, we can define the trigonometric polynomial

$$s_\ell(\theta)=\prod_{i=1}^{m}\left(2-2\cos(\theta-\theta_j)\right)^{\ell_j},$$

and then use $T_n(s_\ell)$ as the preconditioner for $T_n(f)$. The coefficients for building $T_n(s_\ell)$ can be obtained by simple trigonometric manipulation. Consider the example $\{\theta_1=0,\theta_2=\pi\}$, where

$$p_1=p_2=2\ell_1=2\ell_2=4 \quad\Rightarrow\quad \ell=\ell_1+\ell_2=4.$$

Table 4.4: Example 4.4. PCG method used for solving the ill–conditioned Toeplitz system. The $P(s_\ell)$ and circulant preconditioners built with the Jackson kernel with $r = 2, 3, 4$

			$f(\theta) = \theta^4$		
Preconditioner	-	$P(s_2)$	$J_{n,m,2}$	$J_{n,m,3}$	$J_{n,m,4}$
n	IT	IT	IT	IT	IT
64	103	15	11	10	11
128	397	17	12	10	10
256	863	18	14	11	12
512	981	19	16	11	13
1024	952	19	19	10	10
2048	982	19	23	9	11
			$f(\theta) = \theta^2$		
Preconditioner	-	$P(s_1)$	$J_{n,m,2}$	$J_{n,m,3}$	$J_{n,m,4}$
n	IT	IT	IT	IT	IT
64	38	10	7	7	7
128	84	11	7	8	8
256	178	11	8	8	8
512	373	11	8	8	8
1024	768	11	8	8	8
2048	995	11	8	8	8

We need to compute the coefficients of the polynomial

$$\begin{aligned}
s_4(\theta) =& \prod_{i=1}^{2} (2 - 2\cos(\theta - \theta_j))^{\ell_j} = \prod_{i=1}^{2} \left[\left(1 - e^{i(\theta-\theta_j)}\right)\left(1 - e^{i(\theta-\theta_j)}\right)\theta_j)\right]^{\ell_j} \\
=& \prod_{j=1}^{2} \left[\left(1 - ze^{-i\theta_j}\right)\left(1 - z^{-1}e^{i\theta_j}\right)\right]^{\ell_j} \\
=& \frac{(-1)^4}{z^4} \prod_{j=1}^{2} \left(e^{i\theta_j} - 2z + z^2 e^{-i\theta_j}\right)^{\ell_j} = \frac{\left(-z^2 - 2z - 1\right)^2 \left(z^2 - 2z + 1\right)^2}{z^4} \\
=& \frac{1}{z^4} - \frac{4}{z^2} + 6 - 4z^2 + z^4 = \sum_{j=-4}^{4} b_j z^j, \quad z = e^{i\theta}.
\end{aligned}$$

In general we have

$$s_\ell(\theta) = \frac{(-1)^\ell}{z^\ell} \prod_{j=1}^{m} \left(e^{i\theta_j} - 2z + z^2 e^{-i\theta_j}\right)^{\ell_j} = \sum_{j=-\ell}^{\ell} b_j z^j, \quad z = e^{i\theta}.$$

matrix $T_n(s_\ell)$. Observe that the matrix $T_n(s_\ell)$ is a band matrix of band at most $2\ell+1$. The generalization of Theorem 4.13 for this case follows immediately yielding similar results.

Remark 4.6. Observe that in the case in which f is non–negative and has some zeros in $[-\pi, \pi]$ we have built band–Toeplitz preconditioners only for the cases in which these zeros are of even order. This is completely natural, because any non–negative trigonometric polynomial can have only zeros of even order. Thus the spectrum of the preconditioned system can not be contained in an interval if f has a zero of odd order.

To improve the convergence rate of the band–Toeplitz preconditioners (recall the absence of clustering results, see also Example 4.4), several approaches have been developed. We start by considering briefly Chan and Tang [145] based on the use of the Remez algorithm [444, 502] used for finding simple best approximations in the uniform norm to functions with basis functions that are in a Chebyschev space. That is, we are looking for the trigonometric polynomial p that minimizes the norm $\|f - p/f\|_\infty$ for f the generating function of our Toeplitz matrix. If we want a Toeplitz preconditioner with band ν, we need to find the coefficients of the polynomial

$$p(x) = b_0 + \sum_{j=1}^{\nu-1} 2b_j \cos(jx),$$

that minimizes

$$\min_{b_0,\ldots,b_{\nu-1}} \left\| \frac{f-p}{f} \right\|_\infty \overset{\text{if } f>0}{=} \min_{p_0,\ldots,p_{\nu-1}} \|1 - P(x)\|_\infty,$$

with $P(x) \sum_{j=0}^{\nu-1} p_j \phi_j(x)$ $\phi_0(x) = 1/f(x)$ and $\phi_j(x) = 2\cos(jx)/f(x)$ for $j > 0$. This can be achieved with the standard Remez algorithm if $f > 0$ and with its modification by Tang [502] if $f \geq 0$. The minimization properties enable not only the possibility of matching the zeros of f, as also the $T_n(s_\ell)$ preconditioner did, but also finding the optimal band–Toeplitz preconditioner of fixed band. This means that the spectrum of $T_n^{-1}(p)T_n(f)$ lies in (r, R) for any dimension n with the minimal achievable ratio R/r, for a fixed band.

Theorem 4.14 (Optimal band–Toeplitz preconditioner, Chan and Tang [145], Serra-Capizzano [467]). *Let $f \geq 0$ be a continuous function and p a be a polynomial of degree ν such that the relative error ϵ is*

$$\min_{b_0,\ldots,b_{\nu-1}} \left\| \frac{f-p}{f} \right\|_\infty = \epsilon < 1,$$

then

1. *the condition number of the preconditioned matrix is a $O(1)$,*

$$\kappa_2(T_n^{-1/2}(p)T_n(f)T_n^{-1/2}(p)) < \frac{1+\epsilon}{1-\epsilon}, \quad \forall n \geq 1$$

   and $\kappa_2(T_n^{-1/2}(p)T_n(f)T_n^{-1/2}(p)) \to {}^{1+\epsilon}/_{1-\epsilon}$ for $n \to +\infty$;

2. *We have no cluster in* $({}^1\!/_{1+\epsilon}, {}^1\!/_{1-\epsilon})$ *because the set*

$$\bigcup_{n\in\mathbb{N}}\bigcup_{i\leq n}\lambda_i\left(T_n^{-1}(p)T_n(f)\right),$$

is dense in $[{}^1\!/_{1+\epsilon}, {}^1\!/_{1-\epsilon}]$*;*

3. *the number of PCG iteration* N *for convergence within a tolerance* ε *for a given* ϵ *is*

$$N = \frac{1+\epsilon}{2(1-\epsilon)}\log\frac{1}{\varepsilon}+1.$$

Since the Remez Algorithm can be computationally expensive, in particular when the bandwidth is relatively large, two techniques suggested in Serra-Capizzano [467] can give an approximate minimization of $\|(f-p)/f\|_\infty$. Let us define the polynomial s_k as the one of minimum degree k that matches f on all zeros with the same order, see (4.20). Then, define polynomial p of degree ν as the polynomial

$$p = s_k g_{l-k}, \quad \text{with degree of } p_{l-k} \text{ equal to } l \geq k, \tag{4.21}$$

and g_{l-k} a trigonometric polynomial of degree $l-k$ that can be chosen in one of the two following way:

1. g_{l-k} is the best Chebyschev approximation of $\hat{f} = f/s_k$. This means using again the Remez Algorithm, even if for a lower number of coefficients. We call $p^{(1)}$ the polynomial obtained with this choice in (4.21).

2. g_{l-k} is the trigonometric polynomial of degree $l-k$ interpolating $\hat{f}$ at the $l-k+1$ zeros of the $(l-k+1)$-th Chebyschev polynomial of the first kind, that can be computed with few FFTs of order $l-k$. We call $p^{(2)}$ the polynomial obtained with this choice in (4.21).

Results such as Theorem 4.13 can be stated also for these choices of the generating functions. Moreover, both the $T_n(p)$ and $T_n(p^{(1)})$ preconditioners generate clusters around 1, while the $T_n(p^{(2)})$ is only quasi–optimal: the number of outliers increases as $O(\log(\log(n)))$, thus the deterioration becomes sensible only when $\hat{f}$ is quite "irregular", see again [467] for details. Concerning the computational cost, if we want to maintain the optimality of $O(n\log n)$ operations while choosing a classical band solver, the maximum (half)bandwidth we can admit for our preconditioners is $\log^{1/2} n$. From this point of view better results can be obtained by using multigrid methods for the application of the preconditioner, see § 4.2. Other modifications of this strategy for the positive definite case are in [408], while extensions to the multilevel cases can be found in [361, 391, 397, 407, 409, 470, 472].

4.2 Notes on multigrid preconditioning

Multigrid (MG) methods are classes of algorithms for solving linear systems using a hierarchy. They are an example of a class of techniques called multiresolution methods, very useful in problems exhibiting multiple scales of behavior. For example, many basic relaxation methods exhibit different rates of convergence for short- and long-wavelength components, suggesting these different scales be treated differently, as in a Fourier analysis approach to multigrid. MG methods can be used as solvers as well as preconditioners.

The main idea of multigrid is to accelerate the convergence of a basic iterative method, which generally reduces short-wavelength error when used for differential operators, by a correction of the fine grid solution approximation, accomplished by solving a coarse problem. The coarse problem, while cheaper to solve, is similar to the fine grid one because it also has short- and long-wavelength errors. It can also be solved by a combination of relaxations and appeal to still coarser grids. This recursive process is repeated until a grid is reached where the cost of direct solution is negligible compared to the cost of one relaxation sweep on the fine grid. This multigrid cycle typically reduces all error components by a fixed amount bounded well below one, independent of the fine grid mesh size. The typical application for multigrid is in the numerical solution of some classes of PDEs. In particular, they have been shown quite effective for elliptic partial differential equations in multiple dimensions.

Here we focus on some notes on the latter case for the so called *Algebraic Multigrid*, or AMG for short. This approach faces the linear system without considering its underlying geometry.

For an extensive treatment of the Multigrid *solvers* we suggest the books by Briggs et al. [111], Ruge and Stüben [448], Trottenberg et al. [519] and the references therein.

To understand better the main principles of AMG, *error smoothing* and *coarse grid correction*, let us start from a simple elliptic PDE model problem.

Example 4.5 (1D – Poisson Equation). We have already used other and more complete versions of this model problem to benchmark our linear system solvers (see, e.g, Example 2.2). Consider the following boundary value problem

$$\begin{cases} -u_{xx}(x) = f(x), & x \in (0,1), \\ u(0) = u(1) = 0. \end{cases}, \quad f \in \mathcal{C}^1([0,1]).$$

and proceed with the standard centered finite difference discretization on the grid $\Omega_h = \{x_k\}_{k=0}^{n+2} = \{kh\}$ and $h = 1/(n+2)$, that gives rise to the linear system

$$\frac{K}{/}h^2 T_n(2 - 2\cos(\theta))\mathbf{u} = \mathbf{f},$$

where $\mathbf{f}$ contains the values of the forcing term on the nodes $\{x_k\}_{k=1}^{n+1}$, i.e., $f_k =$

$f(x_k)$. Similarly, the vector $\mathbf{u}$ contains the unknowns $u_k = u(x_k)$. By denoting $A_n = T_n(2-2\cos(\theta))$, from the analysis we made for Toeplitz matrices in § 4.1, we know that

$$A_n \mathbf{v}^{(i)} = \lambda_i \mathbf{v}^{(i)},$$

with eigenvalues and eigenvectors given by

$$\lambda_i = 2 - 2\cos\left(\frac{i\pi}{n+1}\right), \quad v_j^{(i)} = \sin\left(\frac{ij\pi}{n+1}\right), \; i,j = 1,\ldots,n.$$

In § 2.1.1 we suggested, introducing the *local Fourier analysis*, that splitting methods, regardless of the goodness of the convergence ratio, can *smooth* the high Fourier frequencies of the residual. Let us consider the sampling of the eigenvector we have found on the grid space given by the discrete boundary value problem in Figure 4.5. Thus, since the eigenvectors are the samplings of

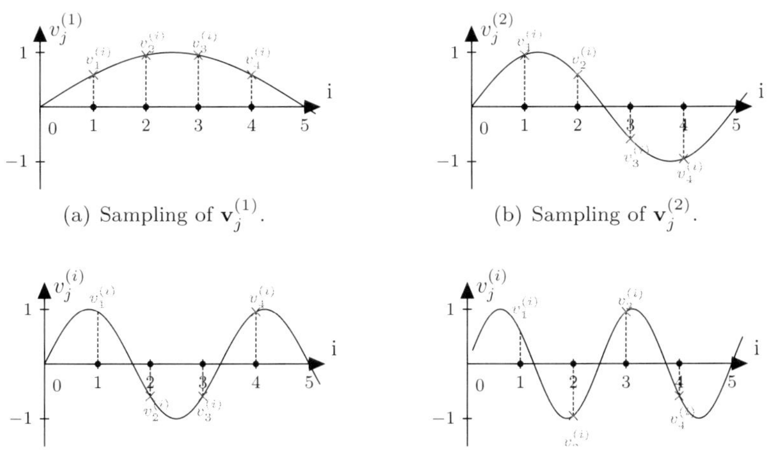

(a) Sampling of $\mathbf{v}_j^{(1)}$. (b) Sampling of $\mathbf{v}_j^{(2)}$. (c) Sampling of $\mathbf{v}_j^{(3)}$. (d) Sampling of $\mathbf{v}_j^{(4)}$.

Figure 4.5: Sampling of the eigenvectors of $T_n(2-2\cos(\theta))$.

the functions $\sin(ix)$ for $i = 1,\ldots,n$ on the grid, they represent exactly the *frequency* in the span of the grid function Fourier series. So, we can divide this set of frequencies in two subsets

$$\begin{array}{ll} \text{Low frequencies} & \left\{\mathbf{v}^{(i)} = \sin(i\mathbf{y}) : \mathbf{y} = \frac{j\pi}{n+1},\, j = 1,\ldots,n \; i = 1,\ldots,n/2-1\right\}, \\ \text{High frequencies} & \left\{\mathbf{v}^{(i)} = \sin(i\mathbf{y}) : \mathbf{y} = \frac{j\pi}{n+1},\, j = 1,\ldots,n \; i = n/2,\ldots,1\right\}. \end{array}$$

If we try to solve our liner system $A_n\mathbf{u} = \mathbf{f}$ by the Jacobi method (2.5), we find

$$\begin{aligned}\mathbf{u}^{(k+1)} =&\mathbf{u}^{(k)} + D_n^{-1}(\mathbf{f} - A_n\mathbf{u}^{(k)}) = \\ =&\left(I_n - \frac{1}{2}A_n\right)\mathbf{u}^{(k)} + \frac{1}{2}\mathbf{f}, \\ D_n =& \operatorname{diag}(A_n) = 2I_n\end{aligned} \tag{4.22}$$

where the iteration matrix is $J_n = I_n - A_n/2$ such that $\rho(J_n) \to 1$ for $n \to +\infty$, thus implying a very slow rate of convergence. Nevertheless, in this phase we are more interested in what the method does on the *frequencies* of the error. Therefore, we try also to consider the use of the *weighted*, or *damped*, Jacobi, see Table 2.1, that is

$$\mathbf{u}^{(k+1)} = \left[(1-\omega)I_n + \omega\left(I_n - \frac{1}{2}A_n\right)\right]\mathbf{u}^{(k)} + \frac{1}{2}\omega\mathbf{f}, \tag{4.23}$$

The iteration matrix is now $J_n^\omega = I_n - \frac{\omega}{2}A_n$. Observe that the best spectral conditioning for the eigenvalues $\mu_1^\omega = \lambda_{\max}(J_n^\omega)$ is obtained for $\mu_1^1 < \mu_1^\omega$ $\forall\omega \in (0,1)$. Therefore, it seems that we have only made the convergence worse. For making sense of this choice let us now decompose the initial error $\mathbf{e}^{(0)}$ and the iteration matrix J_n^ω in the eigenvectors basis $V = [\mathbf{v}^{(1)}, \ldots, \mathbf{v}^{(n)}]$:

$$\mathbf{e}^{(0)} = \sum_{i=1}^{n} \alpha_i \mathbf{v}^{(i)}, \quad J_n^\omega = V\Lambda_n^\omega V^T \text{ where } \Lambda_n^\omega = \operatorname{diag}(\mu_i^\omega),$$

in this way the error at the kth step, calling the true solution $\overline{\mathbf{u}} = A_n^{-1}\mathbf{f}$, is given by:

$$\begin{aligned}\overline{\mathbf{u}} - \mathbf{u}^{(k)} = \mathbf{e}^{(k)} =&(J_n^\omega)^k\mathbf{e}^{(0)} = V(\Lambda_n^\omega)^k V^T\mathbf{e}^{(0)} = V(\Lambda_n^\omega)^k\alpha \\ =&\sum_{i=1}^{n}\left(1 - 2\omega\sin^2\left(\frac{i\pi}{2(n+1)}\right)\right)^k \alpha_i v^{(i)}.\end{aligned}$$

Thus, the ith entry of the error $\mathbf{e}^{(k)}$ is defined in terms of the ith eigenvalues of the iteration matrix J_k^ω:

$$\beta_i = \left(1 - 2\omega\sin^2\left(\frac{i\pi}{2(n+1)}\right)\right)^k \alpha_i \approx \left(1 - \omega\frac{\pi^2}{2}h^2\right)^k \alpha_i.$$

Exploiting this information, we can choose an *optimal* ω that minimizes the absolute values of the β_i in the high frequencies. Therefore, we need to choose a ω such that $|\mu_{n/2}^\omega| = |\mu_n^\omega|$, i.e., such that $1 - \omega = 2\omega - 1$, thus $\omega_{\text{opt}} = 2/3$. If we select the method for $\omega = \omega_{\text{opt}}$, we obtain $|\beta_i| \approx \frac{1}{3^k}|\alpha_i|$. Thus, in the domain of the high frequencies, for whatever value of k, we find that the error has become a *smooth function*, as has been depicted in Figure 4.6.

Nevertheless, this situation does not alter the convergence properties of the

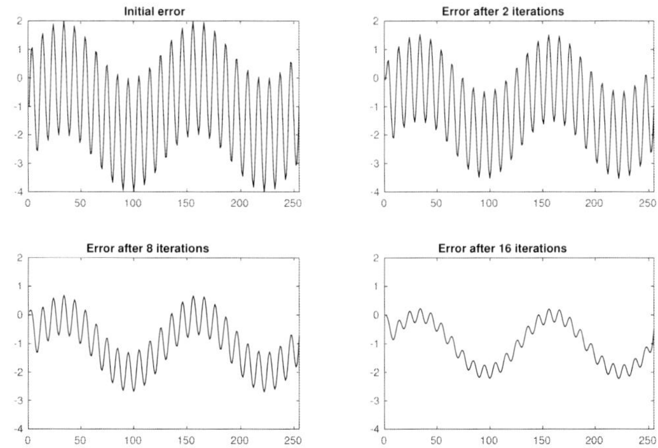

Figure 4.6: Smoothing of the error by damped Jacobi method

damped Jacobi method, because it is influenced by the spectral radius that still goes asymptotically to one. Let us now introduce the second ingredient of the multigrid algorithm, the *iterative refinement procedure.*

Let us now suppose having an approximation $\tilde{\mathbf{u}}$ of the solution $\overline{\mathbf{u}}$ obtained through some iterations of (4.23). We can define the error as $\mathbf{e} = \overline{\mathbf{u}} - \tilde{\mathbf{u}}$. Observe that if we compute in some way the error $\mathbf{e}$ then the solution of the linear system can be obtained as $\overline{\mathbf{u}} = \tilde{\mathbf{u}} + \mathbf{e}$. To compute it we solve the linear system

$$A_n\mathbf{e} = A_n\overline{\mathbf{u}} - A_n\tilde{\mathbf{u}} = \mathbf{f} - A_n\tilde{\mathbf{u}} = \mathbf{r}.$$

Thus, the error satisfies the so called *error equation*: $A\mathbf{e} = \mathbf{r}$. At a first view, it may seem that we make no gain, since we are back to solve a linear system with the same coefficient matrix. Nevertheless, having swapped from the need of computing $\mathbf{u}$ to $\mathbf{e}$ gives us the chance to exploit the information that the error has been smoothed. Going back to the grid, start the discretization of the underlying problem on a grid Ω_h and consider a grid with a doubled step–size Ω_{2h}: observe that only the *low frequency* components are meaningful grid functions on this grid, i.e., high frequencies do not interfere on the Ω_{2h}. Thus, we can summarize our aims in the following way: we want to exploit our smoothing strategy only where it works, i.e., on the domain of the high frequencies. Let us be more specific and restrict ourselves to the case of odd n. In this case,

$$\begin{aligned}
\Omega_h &= \left\{\frac{i\pi}{n+1} \; : \; i = 1, \ldots, n\right\},\\
\Omega_{2h} &= \left\{\frac{2i\pi}{n+1} \; : \; i = 1, \ldots, \frac{n-1}{2}\right\} = \left\{\frac{i\pi}{\frac{n-1}{2}+1} \; : \; i = 1, \ldots, \frac{n-1}{2}\right\},
\end{aligned}$$

so that we can use the grid Ω_{2h} to compute an approximation for the grid of step–size h, i.e., we can restrict both the matrices of our system and the residual vector on the coarse grid to solve there the error equation. Thus,

$$\Omega_h \rightsquigarrow A_n \mathbf{u} = \mathbf{f}, \quad \mathbf{u}, \mathbf{f} \in \mathbb{R}^n,$$
$$\Omega_{2h} \rightsquigarrow A_{\frac{n-1}{2}} \tilde{\mathbf{e}} = \tilde{\mathbf{r}}, \quad \tilde{\mathbf{e}}, \tilde{\mathbf{r}} \in \mathbb{R}^{\frac{n-1}{2}}.$$

Having computed a solution of the error equation on the coarser grid we now need a way to extend it on the finer one, i.e., a map that sends $\tilde{\mathbf{e}}$ from Ω_{2h} to Ω_h to update our solution $\tilde{\mathbf{u}}$. As have been shown in Figure 4.7, this procedure

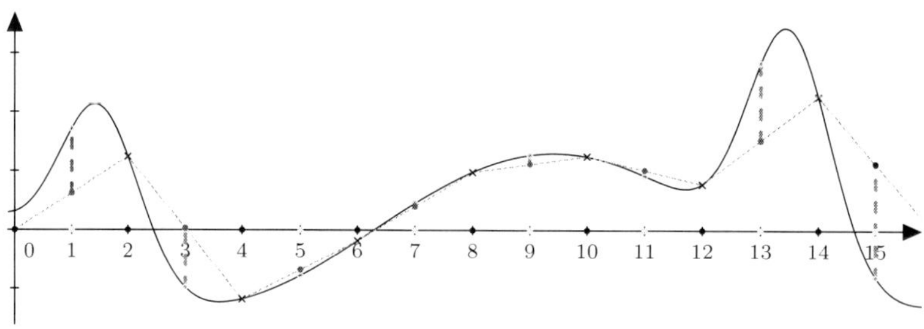

Figure 4.7: Mapping from Ω_{2h} to Ω_h.

can have some difficulties in extending $\mathbf{u}$, e.g., in the case of an oscillating function. Nevertheless, if we can guarantee that our extension procedure works on reasonably smooth data, like the error after some steps of a smoother, the results can be reliable.

To formalize this idea, we need to define the second main ingredient of our multigrid architecture, i.e., a *coarsening strategy*:

1. a restriction operator $I_h^{2h} : \Omega_h \to \Omega_{2h}$,
2. a prolongation operator $I_{2h}^h : \Omega_{2h} \to \Omega_h$,
3. the discretization matrix at the lower level, i.e., $A_{\frac{n-1}{2}}$.

We can make additional distinctions between high and low frequencies for the error equation with respect to the actual grid Ω_{2h} and a coarser grid Ω_{4h} to iterate our coupling of smoothing iterations and iterative refinement by coarsening. Observe that we need to do something peculiar on the coarsest grid in which we face a very small linear system, possibly a single linear scalar equation, that can be solved efficiently by a direct method. ✓

Since not all the linear systems come from elliptic PDEs, let us work on Example 4.5 to produce a prototype of the multigrid Algorithm based on the selection of an opportune smoothing and coarsening strategy by keeping in mind the need of working in a general setting. To proceed in a way that is analogue of what we have done in the Example 4.5 we continue to speak about *grids*, even if this can be intended simply as an index set, not corresponding to any geometrical grid. Let us consider a sequence of coarser grids $\{\Omega_k\}_k$

$$\Omega_l \supset \ldots \supset \Omega_k \supset \Omega_{k-1} \supset \ldots \supset \Omega_0, \tag{4.24}$$

where the coarsest grid is characterized by the index 0 while l represents the finest grid. For each level of the grid, i.e., for each index k consider the linear operators

1. $S_k^{(1)}, S_k^{(2)} : \mathbb{R}^{n_k \times n_k} \to \mathbb{R}^{n_k}$ iteration matrices of the smoothers, they can be either equal or different (one for the restriction phase and one for the prolongation),
2. $A_k : \mathbb{R}^{n_k \times n_k} \to \mathbb{R}^{n_k}$ restriction of the matrix of the linear system,
3. $I_k^{k-1} : \Omega_k \to \Omega_{k-1}$, restriction operator,
4. $I_{k-1}^k : \Omega_{k-1} \to \Omega_k$, prolongation operator,

where n_k is the number of unknowns of the system at level k. With the above notation, the original linear system $A\mathbf{x} = \mathbf{b}$ reads as $A_l\mathbf{x}_l = \mathbf{f}_l$. Let us also introduce a parameter γ indicating what type of cycle is to be employed on the coarser level and the number of cycles that needs to be carried out on the current level. The underlying idea of this parameter is making the cycle perform more work on the coarser level in principle doing more work in reduced dimension. When $\gamma = 1$ the cycle is usually called a V–cycle, while for $\gamma > 1$ we talk about a W–cycle, see Figure 4.8.

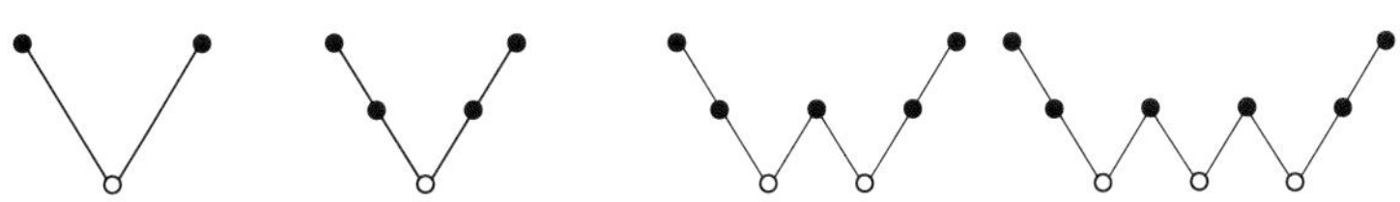

(a) $l = 2, \gamma = 1$ (b) $l = 3, \gamma = 1$ (c) $l = 3, \gamma = 2$ (d) $l = 3, \gamma = 3$

Figure 4.8: Multigrid V–cycle and W–cycle.

With this notation, we can express the outline on one cycle of the multigrid, see Algorithm 4.2, and describe it graphically in the V–cycle case in Figure 4.9. While in geometric multigrid the restriction matrix can be the discretization of the underlying differential problem on the coarser grids, in a generic context this can be no more possible, i.e., an automatic way to restrict the operator is needed. Usually, the *Galerkin condition* is used. In order to use the latter, define the sequences of restriction matrices as

$$I_{k-1}^k = (I_k^{k-1})^T, \quad A_{k-1} = I_k^{k-1} A_k I_{k-1}^k, \; \forall\, k = l-1, \ldots, 0, \tag{4.25}$$

assuming it to have *full rank* restriction and prolongation operators. These choices are useful in many ways, e.g., if the starting matrix A is symmetric

Algorithm 4.2: Multigrid Cycle (MGM)

Data: Multigrid structure $\{A_k\}_{k=l}^0$, l, $\{S_k^{(1)}\}_{k=l}^0$, $\{S_k^{(2)}\}_{k=l}^0$, $\{I_k^{k-1}\}_{k=l}^0$ and $\{I_{k-1}^k\}_{k=l}^0$, ${\mathbf{b}_k}_{k=l}^0$, initial guess $\mathbf{u}^{(j)}$.

Output: Approximation $\mathbf{u}^{(j+1)}$ to the solution of $\mathbf{x}_l$.

Input: $\mathbf{u}_k^{(j+1)} = \text{MGM}(A_k, \mathbf{b}_k, \mathbf{x}_k^{(j)}, k, \nu_1, \nu_2, \gamma)$

`// Presmoothing`

ν_1 steps of *presmoother* $S_k^{(1)}$ applied to $A_k\tilde{x}_k^{(j)} = \mathbf{b}_k$;

`// Coarse Grid Correction`

Compute the residual $\mathbf{r}_k^{(j)} = \mathbf{b}_k - A_k\tilde{x}_k^{(j)}$;

Restrict the residual $\mathbf{r}_{k-1}^{(j)} = I_k^{k-1}\mathbf{r}_k^{(j)}$;

if $k = 1$ **then**

  Direct solver for $A_{k-1}\mathbf{e}_{k-1}^{(j)}$;

else

  for $i = 1, \ldots, \gamma$ **do**

    $\mathbf{e}_{k-1}^{(j)} = \text{MGM}(A_{k-1}, \mathbf{r}_{k-1}, 0, k-1, \nu_1, \nu_2, \gamma)$

Prolong the error $\mathbf{e}_k^{(j)} = I_{k-1}^k \mathbf{e}_{k-1}^{(j)}$;

Update the approximation $\mathbf{x}_k^{(j)} = \tilde{\mathbf{x}}_k^{(j)} + \mathbf{e}_k^{(j)}$;

`// Postsmoothing`

ν_2 steps of *postsmoother* $S_k^{(2)}$ applied to $A_k\tilde{x}_k^{(j+1)} = \mathbf{b}_k$ with initial guess $\mathbf{x}_k^{(j)}$;

---

and positive definite, then all its restrictions are also symmetric and positive definite. Moreover, all intermediate coarse grid correction operators

$$\Pi_k = I - I_{k+1}^k A_{k+1}^{-1} I_k^{k+1} A_k, \tag{4.26}$$

become automatically *orthogonal projectors*. As it stands, we can express one application of the multigrid cycle as an iteration with a classical stationary scheme, i.e., we can express an iteration matrix M_l that fully describes one sweep of multigrid recursively as

$$\begin{cases} M_0 = 0, & k = 0, \\ M_k = (S_k^{(1)})^{\nu_1}\left(I_k - I_{k-1}^k(I_{k-1} - M_{k-1}^{\gamma})A_{k-1}^{-1}I_k^{k-1}A_k\right)(S_k^{(2)})^{\nu_2} & k \geq 1. \end{cases}.$$

Therefore, a *multigrid method* is simply the fixed point iteration with matrix M_l. We can characterize the convergence by means of Theorem 2.1, i.e., we need to have $\rho(M_l) < 1$. We start investigating the convergence of this method after the selection of an opportune *smoothing* and *coarsening* strategy for the underlying class of matrices of interest.

Theorem 4.15 (Mandel [373], McCormick [382]). *Let $A \in \mathbb{R}^{n\times n}$ be symmetric and positive definite. Assume that the prolongation operators I_{k-1}^k have*

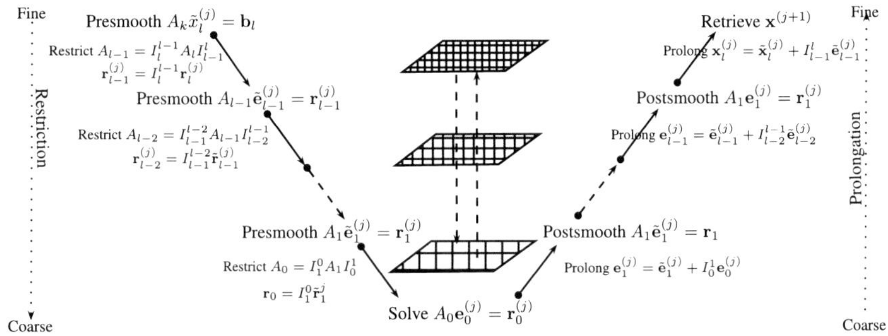

Figure 4.9: One full iteration of a multigrid V–cycle with pre– and post–smoother and Galerkin condition.

full rank and that the Galerkin conditions (4.25) *hold. Furthermore, given the orthogonal projector* Π_k *from* (4.26), *if*

$$\forall\, \mathbf{e}_k \,\exists\, \delta_1 > 0 \;:\; \|S_k^{(2)}\mathbf{e}_k\|_A^2 \leq \|\mathbf{e}_k\|_A^2 - \delta_1\|\Pi_k\mathbf{e}_k\|_A^2, \tag{4.27}$$

independently of $\mathbf{e}_k$ *and* k*, then the multigrid method based on the* V*–cycle in Algorithm 4.2, with* $\gamma = 1$*,* $\nu_1 = 0$ *and* $\nu_2 \geq 1$ *(no* pre–smoother*), has a converge factor bounded above by* $\sqrt{1-\delta_1}$ *with* $\delta_1 \leq 1$*. Otherwise, if the following condition holds*

$$\forall\, \mathbf{e}_k \,\exists\, \delta_2 > 0 \;:\; \|S_k^{(1)}\mathbf{e}_k\|_A^2 \leq \|\mathbf{e}_k\|_A^2 - \delta_2\|\Pi_k S_k^{(1)}\mathbf{e}_k\|_A^2, \tag{4.28}$$

independently of $\mathbf{e}_k$ *and* k*, then the multigrid method based on the* V*–cycle in Algorithm 4.2, with* $\gamma = 1$*,* $\nu_1 \geq 1$ *and* $\nu_2 = 0$ *(no* post–smoother*), has a converge factor bounded above by* $1/\sqrt{1-\delta_2}$*,* $\delta_2 \leq 1$*.*

Remark 4.7. If we have a multigrid cycle (Algorithm 4.2) with both a pre– and post–smoother satisfying the relations (4.27) and (4.28), we can give an estimate of the convergence factor as $\sqrt{1-\delta_1/1-\delta_2}$.

Theorem 4.15 states that it is the balance between the *smoothing* and the *coarsening* that makes the convergence possible. As we suggested in Example 4.5, in the case of PDEs, the interplay between these two components can be interpreted in terms of the frequencies of the error, e.g., condition (4.27) means that the error $\mathbf{e}_k$ that can not be efficiently reduced by the orthogonal projector Π_k needs to be uniformly reduced by $S_k^{(1)}$. On the other hand, on the components on which Π_k is effective, i.e., that are in the range of I_{k-1}^k, the smoothing operator is allowed to be ineffective. Therefore, while the geometric multigrid relies on the availability of a smoother that is tuned on the problem (see again the construction of the damped Jacobi in Example 4.5), for the algebraic multigrid (or AMG for short) we need a suitable matrix dependent coarsening strategy. More specifically, in geometric solvers, all coarser levels are predefined and we select the smoothing so that this coupling is effective. In the AMG case we fix a simple smoothing procedure and explicitly construct an operator–dependent coarsening strategy.

Exercise 4.1. Prove (with details) Theorem 4.15.

Let us stress that the hypotheses (4.27) and (4.28) are not so easy to check since they mix together properties of the coarsening strategy, through the orthogonal projector Π_k, and of the smoother. It is preferable to have some easier to verify *sufficient conditions*, to imply hypotheses (4.27). To obtain them, define the following scalar product

$$< \mathbf{x}, \mathbf{y} >_{A^2} = < D^{-1}A\mathbf{x}, \mathbf{y} >, \text{ with } A, D \text{ SPD},$$

where we assume $D = \text{diag}(A)$. Then, the following two conditions can be used

$$\begin{aligned} &\exists\, \alpha > 0: \quad \|S_k^{(2)}\mathbf{e}_k\|_A^2 \leq \|\mathbf{e}_k\|_A^2 - \alpha\|\mathbf{e}_k\|_{A^2}^2, && \text{(smoothing condition)}\\ &\exists\, \beta > 0: \quad \|\Pi_k\mathbf{e}_k\|_A^2 \leq \beta\|\mathbf{e}_k\|_{A^2}^2, && \text{(approximation property)} \end{aligned}$$

with $\delta_1 = \alpha/\beta$. Similarly, for hypothesis (4.28) we have again the conditions

$$\begin{aligned} &\exists\, \alpha > 0: \quad \|S_k^{(1)}\mathbf{e}_k\|_A^2 \leq \|S_k^{(1)}\mathbf{e}_k\|_A^2 - \alpha\|\mathbf{e}_k\|_{A^2}^2, && \text{(smoothing condition)}\\ &\exists\, \beta > 0: \quad \|\Pi_k\mathbf{e}_k\|_A^2 \leq \beta\|\mathbf{e}_k\|_{A^2}^2. && \text{(approximation property)} \end{aligned}$$

with $\delta_2 = \alpha/\beta$. Smoothing conditions are usually the easiest part to be obtained, while, on the other hand, the approximation property can range from trickier to almost impossible. A way to further relax the approximation property is by reducing the hypotheses in Theorem 4.15 to have a proof only for the convergence of the two–grid method, i.e., the case with $l = 2$, $\gamma = 1$ from Figure 4.8. Moreover, since the two relations can be obtained in the same way, we restrict to the case in which smoothing is performed only after each coarse grid correction step, i.e., we consider only the smoothing matrix $S_k^{(2)}$.

Theorem 4.16 (Brandt [102]). *Let A_k be positive definite matrix and let us consider a Multigrid Cycle (Algorithm 4.2) with $\gamma = 1$ and $l = 2$ in which we identify the two level with k and K, respectively and $S_k^{(1)} = 0$, i.e., no pre–smoother is used. If $S_k^{(2)}$ satisfies the smoothing property, the interpolation operator I_K^k has full rank and for each $\mathbf{e}_k$, we have*

$$\exists\, \beta > 0: \min_{\mathbf{e}_k} \|\mathbf{e}_k - I_K^k\mathbf{e}_K\|_D^2 \leq \beta\|\mathbf{e}_k\|_A^2,$$

with β independent from $\mathbf{e}_k$, then, $\beta \geq \alpha$ and the two–grid method converges with a convergence factor $\sqrt{1 - \alpha/\beta}$.

Proof. Since the range of Π_k is orthogonal to the range of I_K^k in the $< \cdot, \cdot >_A$ scalar product,

$$\|\mathbf{e}_k\|_A^2 = < A_k\mathbf{e}_k, \mathbf{e}_k - I_K^k\mathbf{e}^K >,$$

for all the $\mathbf{e}_k$ in the range of Π_k and all the $\mathbf{e}^K$. Then, by means of usual Cauchy–Schwarz inequality,

$$\begin{aligned}\|\mathbf{e}_1^2\|_A^2 =& < A^{1/2}(D^{-1}A)^{1/2}\mathbf{e}_k, A^{1/2}(D^{-1}A)^{-1/2}(\mathbf{e}_k - I_K^k\mathbf{e}_K) > \\ \leq& \|A^{1/2}(D^{-1}A)^{1/2}\mathbf{e}_k\| \|A^{1/2}(D^{-1}A)^{-1/2}(\mathbf{e}_k - I_K^k\mathbf{e}_K)\| \\ =& \|\mathbf{e}_k\|_{A^2}\|\mathbf{e}_k - I_K^k\mathbf{e}_K\|_D.\end{aligned}$$

Then, by using the hypothesis on the left side of the previous equation, we obtain the approximation property $\exists\, \beta > 0 : \|\Pi_k\mathbf{e}_k\|_A^2 \leq \beta\|\mathbf{e}_k\|_{A^2}^2$, and then, by means of the relative smoothing property,

$$\begin{aligned}\|S_k^{(2)}\Pi_k\mathbf{e}_k\|_A^2 \leq& \|\Pi_k\mathbf{e}_k\|_A^2 - \alpha\|\Pi_k\mathbf{e}_k\|_{A^2}^2 \leq \|\Pi_k\mathbf{e}_k\|_A^2 - \frac{\alpha}{\beta}\|\Pi_k\mathbf{e}_k\|_A^2 \\ =& \left(1 - \frac{\alpha}{\beta}\right)\|\Pi_k\mathbf{e}_k\|_A^2 \leq \left(1 - \frac{\alpha}{\beta}\right)\|\mathbf{e}_k\|_A^2.\end{aligned}$$

□

Extending the above convergence properties to the full V–cycle case is virtually impossible by using merely algebraic arguments, i.e., without recurring when possible to underlying geometric properties. Nevertheless, it often turns out that two level convergence can be extended, at least approximatively, to the V–cycle in a way independent of the problem size. However, here we consider mostly using AMG as a preconditioner. For further discussions on these issues see, e.g., [448, 496, 519]. There exist several proofs of convergence for multigrid methods applied to a wide number of elliptic partial differential equations; consider, e.g., [16, 31, 98, 201, 287]. On the remaining part of the section we consider properties for some class of matrices. For proving the convergence of the two–grid method in the symmetric positive definite case other ways are indeed possible; see e.g., [279], in which a general class of stationary iterative methods is introduced for solving SPD systems.

4.2.1 Smoothers and smoothing properties

Differently from what we have seen in Example 4.5, see Figure 4.6, in the algebraic multigrid approaches, smoothing in the geometric/intuitive sense can be meaningless. Instead, we need to define the error $\mathbf{e}$ to be *algebraically* smooth if it converge slowly with respect to $S_k^{(1)}$, that is, if $S_k^{(1)}\mathbf{e} \approx \mathbf{e}$. In other words, it is smooth when we need a coarser level to further reduce it and speed up the convergence.

In the following we discuss the *smoothing property* (see discussion at the end of § 2.1.1) for the *Gauss–Seidel* (2.6) and the *damped Jacobi* (Table 2.1) methods.

Lemma 4.1 (Mandel [373], Mandel et al. [374]). *Given any SPD matrix A, the smoothing properties* Gauss–Seidel *and for* damped Jacobi *methods are* *equivalent, respectively, to:*

$$\begin{aligned}\alpha M_{(2)}^T D^{-1} M_{(2)} &\leq M_{(2)} + M_{(2)}^T - A, \\ \alpha(A - M_{(1)}^T) D^{-1}(A - M_{(1)}) &\leq M_{(1)} + M_{(1)}^T - A\end{aligned}$$

where $M_{(i)}$, *for* $i = 1, 2$, *is the matrix defining the splitting* $S^{(i)} = I - M_{(i)}^{-1} A$.

Theorem 4.17 (Brandt [102], Ruge and Stüben [448]). *Given any SPD matrix A and any positive vector* $\mathbf{v}$ *we define,*

$$\gamma_- = \max_{i=1,\ldots,n} \left(\frac{1}{v_i a_{i,i}} \sum_{j<i} v_j |a_{i,j}| \right), \quad \gamma_+ = \max_{i=1,\ldots,n} \left(\frac{1}{v_i a_{i,i}} \sum_{j>i} v_j |a_{i,j}| \right).$$

Then the Gauss–Seidel method (2.6) *satisfies the smoothing properties if,* $\alpha \leq (1 + \gamma_-)^{-1}(1 + \gamma_+)^{-1}$ *and* $\alpha \leq (\gamma_- \gamma_+)^{-1}$, *respectively.*

Proof. From Table 2.1 we observe that $M_{(1)} = M_{(2)}$ the strictly lower triangular part of A, therefore $M_{(i)} + M_{(i)}^T - A = D$ for $i = 1, 2$. Thus, by Lemma 4.1, the smoothing properties are equivalent to

$$\alpha \leq \rho(D^{-1} M_{(2)} D^{-1} M_{(2)}^T)^{-1} \text{ and } \alpha \leq \rho(D^{-1}(A - M_{(1)}) D^{-1}(A - M_{(1)}^T))^{-1},$$

respectively. By selecting the induced matrix norm given by,

$$\|A\|_{\mathbf{v}} = \max_{i=1,\ldots,n} \left(\frac{1}{v_i} \sum_{j<i} v_j |a_{i,j}| \right), \tag{4.29}$$

and using the hypothesis we prove the Theorem, since the spectral radius $\rho(\cdot)$ is bounded by any *consistent* matrix norm. □

A similar result can be stated for the Jacobi method.

Theorem 4.18 (Ruge and Stüben [448]). *Consider a SPD matrix A and* $\gamma_0 \geq \rho(D^{-1}A)$. *Then, the Jacobi method* (2.5) *with* $0 < \omega < 2/\gamma_0$ *satisfies the smoothing property if* $\alpha \leq \omega(2 - \omega\gamma_0)$ *and* $\alpha \leq \omega \min\{2, (2-\omega\gamma_0)/(1-\omega\gamma_0)^2\}$, *respectively.*

Proof. Again by Lemma 4.1, we can restate the the smoothing inequalities as

$$\rho\left(D^{-1}A + \frac{\alpha}{\omega^2} I\right) \leq \frac{2}{\omega} \text{ and } \rho\left(D^{-1}A + \alpha\left(D^{-1}A - \frac{1}{\omega} I\right)^2\right) \leq \frac{2}{\omega},$$

that can be obtained by

$$\gamma_0 + \frac{\alpha}{\omega^2} \text{ and } \max_{0 \leq \lambda \leq \gamma_0} \left\{ \lambda + \alpha \left(\lambda - \frac{1}{\omega} \right)^2 \right\} \leq \frac{2}{\omega},$$

respectively, from which the statement follows. □

We can conclude that the Gauss–Seidel method satisfies the smoothing property uniformly for the class of symmetric M–matrices (Definition 3.4). By [466, Theorem 2.6] we know that if M is an M–matrix then there exist vectors $\mathbf{v} > 0$ (element by element) such that $M\mathbf{v} > 0$ (element by element). Therefore, we can take this $\mathbf{v}$ in (4.29) to get

$$\gamma_- = \max_{i=1,\ldots,n} \left(\frac{1}{v_i a_{i,i}} \sum_{j<i} v_j a_{i,j} \right) = \max_{i=1,\ldots,n} \left(1 - \frac{1}{v_i a_{i,i}} \sum_{j\leq i} v_j a_{i,j} \right) < 1,$$

$$\gamma_+ = \max_{i=1,\ldots,n} \left(\frac{1}{v_i a_{i,i}} \sum_{j>i} v_j a_{i,j} \right) = \max_{i=1,\ldots,n} \left(1 - \frac{1}{v_i a_{i,i}} \sum_{j\geq i} v_j a_{i,j} \right) < 1,$$

and thus $\alpha = 1/4$ and $\alpha = 1$ for the two smoothing properties. Similarly, this relationship can be extended to the case of SPD matrices that are obtained by symmetrically flipping the signs of the off–diagonal elements (even of all of them). Another case in which the norm (4.29) can be used is the one of ℓ–banded SPD matrices A, i.e., with at most ℓ non–vanishing elements per row. By selecting $\mathbf{v}$ such that $v_i = a_{i,i}^{-1/2}$, we obtain $\gamma_-, \gamma_+ < \ell - 1$. Therefore, $\alpha = \ell^{-2}$ and $\alpha = (\ell-1)^{-2}$, respectively. On the other hand, for the damped Jacobi method used for systems with symmetric M–matrices we can choose $\gamma_0 = 2$ in Theorem 4.18. Then, as we did in Example 4.5, choose the optimal values for ω in terms of γ_0, thus obtaining the largest values of α. These two values are easily computed as, respectively, $\omega = \gamma_0^{-1}$ and $\omega = 3/(2\gamma_0)$, for which the smoothing inequalities are satisfied for $\alpha = 1/\gamma_0$ and $\alpha = 3/\gamma_0$ giving $\alpha = 1/2$ and $\alpha = 3/2$ if $\gamma_0 = 2$.

Another class that satisfies the smoothing property is given by Toeplitz matrices (see § 4.1) generated by an even function f, for which we know that $\rho(A) \leq \max_{\theta\in[-\pi,\pi]} f(\theta)$ (Proposition 4.1) and, moreover, that $D = \text{diag}(A)$ is just a constant multiple of the identity matrix. Therefore, the damped Jacobi method with the suitable smoothing constant can be easily obtained also in this case. Observe that if f is not known, we can compute an $O(n)$ estimate of $\rho(A)$ by either Frobenius or ∞–norm of A.

Another smoother that is particularly useful in the (possibly multilevel) Toeplitz case is the *Richardson method* § 2.1.3.4, as proposed in [231–233, 469]. These works are driven by the close relations between Toeplitz matrices and matrices from trigonometric algebras we have already seen in § 4.1.2.

Remark 4.8. Krylov subspace and multigrid methods have been coupled in several ways. It is possible to use a Krylov subspace method as pre– or post–smoother, i.e., substituting the stationary methods $S_k^{(i)}$ with some iterations of GMRES/GMRES(m) or BiCGstab, say.

If we do not use a stationary method, the convergence analysis obtained through Theorem 4.15 cannot be used anymore and ad–hoc techniques should be considered instead; see, e.g., [4, 218, 412].

Another possibility is using the so–called K–cycle in which some Krylov subspace iterations are used instead of a direct solver at line 5 of Algorithm 4.2, see, e.g., [405, 406].

In general, these kinds of approaches seem to be numerically promising. Nevertheless, they present many issues in the analysis which are mainly due to the fact that these methods are nonlinear.

Remark 4.9. Other methods we have seen in Chapter 3 that can be used as smoother for Multigrid method are the *domain decomposition methods* [152] and the *incomplete factorizations*; see Section 3.4. Consider the case of the Additive Schwarz preconditioners from [118, 119, 123, 124, 152, 216]. At the kth level, we divide Ω_k into m_k subsets $\Omega_{k,i}$ of size $n_{k,i}$, possibly overlapping. Then, for each i, the restriction operator $R_i^{(k)}\, \mathbb{R}^{n_k} \to \mathbb{R}^{n_{k,i}}$, maps a vector $\mathbf{x}^{(k)}$ to the vector $\mathbf{x}_i^{(k)}$ made of the components of $\mathbf{x}^{(k)}$ with indices in $\Omega_i^{(k)}$, and the prolongation operator is defined simply as its transpose. Then, the classical preconditioner is defined as

$$P_{\text{AS}}^{-1} = \sum_{i=1}^{m} R_i^{(\delta)^T} A_i^{(\delta)^{-1}} R_i^{(\delta)},$$

and is used as the matrix of the smoother at the kth level. Many variants of these multilevel strategies have been investigated, see, e.g., [216] for some comparisons. Similarly, both approximate inverse preconditioners [112, 113, 503, 556] and incomplete LU factorizations [3, 490, 538, 548, 556] have been used as smoothers inside multigrid methods.

Exercise 4.2. Given a symmetric Toeplitz matrix $T_n(f)$ with a generating function $f > 0$, provide an estimate for the smoothing constant α obtained with the Richardson method § 2.1.3.4, i.e., an expression of α, and the parameter ω in terms of f.

4.2.2 Coarsening strategies and approximation properties

As we have seen with equation (4.24), we can regard the coarse–level variables as a subset of the ones of the fine–level. Therefore, for every Ω_k we can devise a splitting in two disjoint subsets C_k and F_k, where C_k contains the variables for the coarser level, $\Omega_{k-1} = C_k$, and F_k its complement, i.e., $\Omega_k = C_k \cup F_k$, $C_k \cap F_k = \emptyset$. As done in the previous section, we restrict to the two–grid setting, denoting by k, K the fine and coarse level respectively and call *C/F–splitting* the above construction, see Figure 4.10. Consider *interpolation* operators of the form

$$(\mathbf{e}_k)_i = \left(I_K^k \mathbf{e}_K\right)_i = \begin{cases} (\mathbf{e}_K)_i, & i \in C_k, \\ \sum\limits_{j \in P_{k,i}} w_{i,j}^k (\mathbf{e}^K)_j, & i \in F_k, \end{cases} \tag{4.30}$$

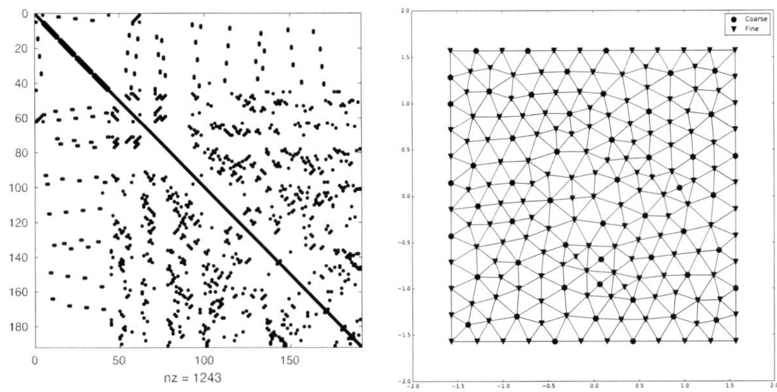

(a) Pattern of the matrix (b) C/F–splitting

Figure 4.10: Pattern and C/F–splitting for the finite element discretization of an airfoil problem, example made with the `PyAMG` library [40].

where $P_{k,i} \subset C_k$ is the set of *interpolatory variables* and the $w_{i,j}^k$ are the weights of the interpolation formula. The approximation property of the scheme depends on both these choices and needs to be weighted by considering the construction and the application costs of the operators.

Let us use the Gauss–Seidel method as the smoother for a symmetric M–matrix. We can observe that, by the Cauchy–Schwarz inequality, the following relation between the errors holds:

$$\|\mathbf{e}\|_A^2 =< A\mathbf{e}, \mathbf{e} >=< D^{-1/2}A\mathbf{e}, D^{1/2}\mathbf{e} >\leq \|\mathbf{e}\|_D\|\mathbf{e}\|_{A^2}.$$

Therefore, $\|\mathbf{e}\|_{A^2} \ll \|\mathbf{e}\|_A \ll \|\mathbf{e}\|_D$, that is:

$$< A\mathbf{e}, \mathbf{e} >= \frac{1}{2}\sum_{i,j}(-a_{i,j})(e_i - e_j)^2 + \sum_i \left(\sum_j a_{i,j}\right) e_i^2 \ll \sum_i a_{i,i}e_i^2. \quad (4.31)$$

The most interesting case is the one in which we are far from diagonal dominant case, in which the multilevel approach is not useful, that is $\sum_{j\neq i} |a_{i,j}| \approx a_{i,i}$. This gives

$$\sum_{j\neq i} \frac{|a_{i,j}|}{a_{i,i}} \frac{(e_i - e_j)^2}{e_1^2} \ll 1.$$

For each node i, a smooth error varies slowly in the direction of strong $i \to j$ connection, i.e., when $|a_{i,j}|/a_{i,i}$ is relatively large. To formalize this concept, we can define the *connections* between grid points. That is, for the level k we define a point $i \in \Omega_k$ to be *directly connected* to a point $j \in \Omega_k$ if $a_{i,j} \neq 0$. The set of direct neighbors of the point $i \in \Omega_k$ is then defined as

$$N_i^k = \{j \in \Omega_k \ : \ j \neq i, \ a_{i,j} \neq 0\}.$$

Therefore, if more elements of N_i^k with strong connections are contained in $P_{k,i}$, our algebraically smooth error satisfies better

$$\frac{1}{\sum\limits_{j\in P_{i,k}} a_{i,j}^k} \sum_{j\in P_{i,k}} a_{i,j}^k e_j \approx \frac{1}{\sum\limits_{j\in N_i^k} a_{i,j}^k} \sum_{j\in N_i^k} a_{i,j}^k e_j, \quad i \in \Omega_k.$$

Inserting this in (4.30) gives the interpolation weights:

$$w_{i,j}^k = -\frac{\sum_{j\in N_i^k} a_{i,j}}{\sum_{l\in P_{i,k}} a_{i,l}} \frac{a_{i,j}}{a_{i,i}} > 0. \tag{4.32}$$

To prove the *approximation property* and thus the convergence of the two grid methods with these weights we use Theorem 4.16 stated on the orthogonal projector (4.26).

Theorem 4.19. *Let A be a weakly diagonally dominant symmetric M–matrix with $s_i = \sum_j a_{i,j} \geq 0$. With a fixed value of $\beta \geq 1$ we select a splitting $\Omega_k = C_k \cup F_k$ so that for each $i \in F_k$ there is a set $P_{i,k} \subseteq C_k \cap N_i^k$ such that*

$$\sum_{j\in P_{i,k}} |a_{i,j}| \geq \frac{1}{\beta} \sum_{j\in N_i^k} |a_{i,j}|. \tag{4.33}$$

Then, the interpolation (4.30) *with weights* (4.32) *satisfies the hypothesis of Theorem 4.16.*

Proof. Without loss of generalities we can exclude from the analysis the variables that are not coupled to any other variable, i.e., matrix rows with $\sigma_i = a_{i,i}$. They will always be in F_k not requiring any interpolation whatsoever. Therefore, we are assuming that $P_{i,k} \neq \emptyset$ for all $i \in F_k$. By (4.31) we can give the following estimate for the error $\mathbf{e}_k$

$$\|\mathbf{e}_k\|_A^2 =< A\mathbf{e}_k, \mathbf{e}_k >\geq \sum_{i\in F} \left[\sum_{j\in P_{i,k}} (-a_{i,j})((\mathbf{e}_k)_i - (\mathbf{e}_k)_j)^2 + s_i(\mathbf{e}_k)_i^2 \right].$$

Moreover, by means of the Cauchy–Schwarz inequality, we have

$$\begin{aligned}
\|\mathbf{e}_k - I_K^k \mathbf{e}_K\|_D^2 =& \sum_{i\in F} a_{i,i}\left((\mathbf{e}_k)_i - \sum_{j\in P_{i,k}} w_{i,j}^k(\mathbf{e}_k)_j\right)^2 \\
=& \sum_{i\in F} a_{i,i}\left(\sum_{j\in P_{i,k}} w_{i,j}^k((\mathbf{e}_k)_i - (\mathbf{e}_k)_j)\right. \\
&\left. + \left(1 - \sum_{j\in P_{i,k}} w_{i,j}^k\right)(\mathbf{e}_k)_i\right)^2 \\
\leq& \sum_{i\in F} a_{i,i}\left(\sum_{j\in P_{i,k}} w_{i,j}^k((\mathbf{e}_k)_i - (\mathbf{e}_k)_j) + \right. \\
&\left. + \left(1 - \sum_{j\in P_{i,k}} w_{i,j}^k\right)(\mathbf{e}_k)_i\right).
\end{aligned}$$

Observe that, by our interpolation formula, the row sum is given by

$$s_i = a_{i,i}(1 - \sum_{j\in P_{i,k}} w_{i,j}^k).$$

Therefore, we have the thesis if

$$a_{i,i} w_{i,j}^k \leq \beta |a_{i,j}| \quad \forall i \in F_k, \quad \forall j \in P_{i,k}.$$

By Definition 4.32, we observe that this is equivalent to the property

$$a_{i,i} w_{i,j}^k = a_{i,i}\left(-\frac{\sum_{j\in N_i^k} a_{i,j}}{\sum_{l\in P_{i,k}} a_{i,l}} \frac{a_{i,j}}{a_{i,i}}\right) = -\frac{\sum_{j\in N_i^k} a_{i,j}}{\sum_{l\in P_{i,k}} a_{i,l}} a_{i,j} \leq \beta |a_{i,j}|,$$

that is equivalent to (4.33), from which the thesis follows. □

Observe that the above interpolation formula can be formally applied to any M–matrix and to any splitting such that $C_k \cap N_i^k \neq \emptyset$ for each $i \in F_k$, while we restricted ourselves to the class of weakly diagonally dominant M–matrices, in which the approximation property holds uniformly. Indeed, this requirement is sufficient but not necessary.

Theorem 4.20. *Let A be a symmetric M–matrix with $s_i = \sum_j a_{i,j} \geq -c$ for some $c \geq 0$ and such that there exists $\varepsilon > 0$ for which $< A\mathbf{x}, \mathbf{x} > \geq \varepsilon < \mathbf{x}, \mathbf{x} >$ for all $\mathbf{x}$. With a fixed value of $\beta \geq 1$ we select a splitting $\Omega_k = C_k \cup F_k$ so that for each $i \in F_k$ there is a set $P_{i,k} \subseteq C_k \cap N_i^k$ such that*

$$\sum_{j\in P_{i,k}} |a_{i,j}| \geq \frac{1}{\beta} \sum_{j\in N_i^k} |a_{i,j}|.$$

Then the interpolation (4.30) *with weights* (4.32) *satisfies the hypothesis of Theorem 4.16 with* $\tilde{\beta} = \tilde{\beta}(\varepsilon, c, \beta)$. *Moreover,* $\tilde{\beta} \to +\infty$ *if either* $c \to +\infty$ *or* $\varepsilon \to 0$.

The generalization to the class of M–matrices whose row sums are uniformly bounded from below and whose eigenvalues are uniformly bounded away from zero implies that we do not have anymore an approximation property that holds in the class of symmetric M–matrices, differently from what happens with the smoothing property. We need to choose β by accommodating two conflicting requests. From one side, large values of β make the coarsening rapid and aggressive. From the other, large values of β can make the convergence factor tend to 1, and the convergence can become very slow. A somewhat reasonable compromise is selecting $\beta = 2$ which means that we use about half of the information on the finer level.

Now we treated the case of *negative connections* in the context of M–matrices, nevertheless, there is also the possibility of facing matrices with *positive connections*, i.e., positive off–diagonal elements. Unfortunately, the treatment of these connections in interpolation is, in general, more difficult than that of negative ones. We observe that the error $(\mathbf{e}_k)_j$, that corresponds to $a_{i,j} > 0$, changes slowly among each other. If we assume that some $i \in F_k$ has both negative and positive connections, we can identify two nonempty sets of direct neighbors of these points $i \in \Omega_k$ as

$$N_i^{k,+} = \{j \in \Omega_k \,:\, j \neq i,\, a_{i,j} > 0\}, \quad N_i^{k,-} = \{j \in \Omega_k \,:\, j \neq i,\, a_{i,j} < 0\}.$$

In this case our decomposition of Ω_k needs to be such that at least one connection of either sign is contained in C_k and then provides two sets of interpolation points such that

$$\emptyset \neq P_{i,k}^- \subseteq C_k \cap N_i^{k,-}, \quad \emptyset \neq P_{i,k}^+ \subseteq C_+ \cap N_i^{k,+}, \quad P_{i,k} = P_{i,k}^- \cup P_{i,k}^+.$$

Then, to compute the weights for the interpolation formula (4.30), we assume that

$$a_{i,i}(\mathbf{e}_k)_i + \eta_i \sum_{j \in P_{i,k}} a_{i,j}^-(\mathbf{e}_k)_j + \mu_i \sum_{j \in P_{i,k}} a_{i,j}^+(\mathbf{e}_k)_j = 0$$

with

$$\eta_i = \sum_{j \in N_i^k} a_{i,j}^- / \sum_{j \in P_{i,k}} a_{i,j}^-, \qquad \mu_i = \sum_{j \in N_i^k} a_{i,j}^+ / \sum_{j \in P_{i,k}} a_{i,j}^+,$$

where $a_{i,j}^-$ is $a_{i,j}$ if $a_{i,j} < 0$ and 0 otherwise and, similarly, $a_{i,j}^+$ is $a_{i,j}$ if the latter is positive and zero otherwise. Thus the interpolation weights are given by

$$w_{i,j}^k = \begin{cases} -\eta_i \frac{a_{i,j}}{a_{i,i}} > 0, & j \in P_{i,k}^-, \\ -\mu_i \frac{a_{i,j}}{a_{i,i}} < 0, & j \in P_{i,k}^+. \end{cases} \tag{4.34}$$

which gives an identical row–sum back again

$$s_i = a_{i,i}(1 - \sum_{j \in P_{i,k}} w_{i,j}^k)$$

providing to us an extension of Theorem 4.19 for the class of weakly diagonally dominant matrices.

Theorem 4.21. *Let A be an SPD matrix with $t_i = a_{i,i} - \sum_{j \in N_{i,k}} |a_{i,j}| \geq 0$. Given a fixed value of $\beta \geq 1$, select a splitting $\Omega_k = C_k \cup F_k$ so that for each $i \in F_k$, if $N_i^{k,-} \neq \emptyset$, there is a set $P_{i,k}^{-} \subseteq C_k \cap N_i^{k,-}$ such that*

$$\sum_{j \in P_{i,k}^{-}} |a_{i,j}| \geq \frac{1}{\beta} \sum_{j \in N_i^{k,-}} |a_{i,j}|,$$

and, if $N_i^{k,+} \neq \emptyset$, there is a set $P_{i,k}^{+} \subseteq C_k \cap N_i^{k,+}$ such that

$$\sum_{j \in P_{i,k}^{+}} a_{i,j} \geq \frac{1}{\beta} \sum_{j \in N_i^{k,+}} a_{i,j}.$$

Then, the interpolation (4.30) *with weights* (4.34) *satisfies the hypothesis of Theorem 4.16.*

The extension to the case in which $t_i \geq -c$ for some $c \geq 0$, analogue to Theorem 4.20, is straightforward. Moreover, observe that if no highly oscillatory behavior occurs on the positive correction, i.e., the strength of the positive correction is negligible, we can use the standard weights (4.32) ignoring the positive connections, i.e., we can simply proceed by not interpolating from positive connections while adding them to the diagonal terms (to preserve row sums), i.e.,

$$\tilde{a}_{i,i}(\mathbf{e}_k)_i + \eta_i \sum_{j \in P_{i,k}} a_{i,j}^{-}(\mathbf{e}_k)_j = 0, \quad \tilde{a}_{i,i} = a_{i,i} + \sum_{j \in N_i^k} a_{i,j}^{+}, \quad \eta_i = \frac{\sum_{j \in N_i^k} a_{i,j}^{-}}{\sum_{j \in P_{i,k}} a_{i,j}^{-}}$$

that yields the interpolation weights for $i \in F$ and $j \in P_{i,k}$ given by $w_{i,j}^k = -\eta_i \tilde{a}_{i,j}/\tilde{a}_{i,i}$ that satisfy the modified row sum $\tilde{a}_{i,i}(1 - \sum_{j \in P_{i,k}} w_{i,j}^k) = s_i$. Within this framework it is possible to prove the analogues of Theorem 4.19 in the case of essentially positive matrices; see [102] for details.

We presented the classical AMG approach to coarsening and, in a way, the worst–case scenario for bounds and estimates. For example, instead of dealing directly with the non interpolatory part of the error estimates, we can first eliminate all the strong connections that point to the one at hand. This procedure from [519, Appendix A] is called *standard interpolation*. Other possible improvements are the *Jacobi smoothing* from [336] that modifies the previous approaches trying to enforce the approximation property, through the usual variational formulation, in the case in which a presmoother is used. Whenever some geometrical information on the problem is available, one could use it to select which points should be included in the various $P_{i,k}$ by selecting them also considering the discretization stencil used. In this case, one should arrange the splitting in such a way that the coarse variables are a maximal set with respect to the properties of being not strongly coupled among each other and surrounding the suitable variables of the fine level.

Observe that the changes in the neighborhood where we interpolate should be done carefully. If the radius of this set is increased too much, we can reduce the sparsity of the resulting Galerkin operator, obtaining potentially more computationally expensive methods both in time and space. Whether or not this pays, depends on the considered problem. For discussions and details on these issues, see [448, 519], in which an automatic algorithm for producing some of these coarsening strategies is also proposed.

Consider briefly an interpolation approach based on the opposite point of view. Instead of working on increasingly complex interpolation approaches based on the couplings, take any fine variable as an interpolation of a single coarser variable. This choices is nothing else than a piecewise constant interpolation from the coarse–level variables to the selected node or, more generally, to a fixed subset of nodes that in this case is called an *aggregate*, see Figure 4.11. Development of this very simple interpolation strategy is know as the *Ag-*

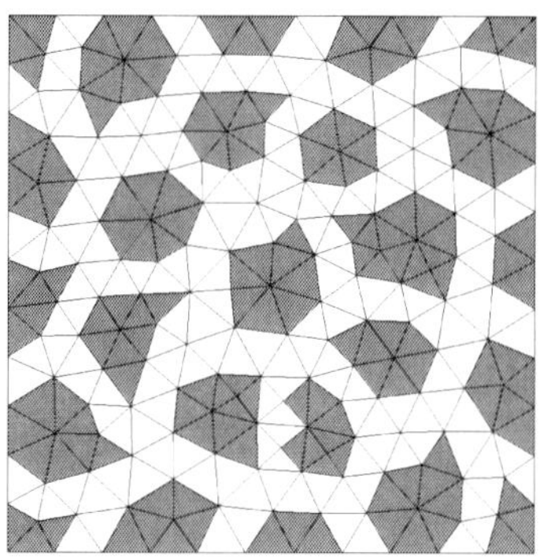

Figure 4.11: Aggregates for a two level smoothed aggregation coarsening for the problem considered in Figure 4.10.

gregation Multigrid, see again the review [496] and [97, 102, 405, 519, 530, 531]. There exist several automatic algorithms implementing the coarsening strategy for the algebraic multigrid approach we have described and, indeed, many that we left out of our discussion. A review of this material can be found in [5], where the behavior of these algorithms is studied in detail. In particular, we mention the approach by Cleary et al. [165, 166], that uses a heuristic approach to enforce the properties we have discussed.

Example 4.6 (Coarsening strategies). Let us focus first on the construction of the algebraic multigrid hierarchy for a matrix A discretizing the 2D Laplacian with zero boundary conditions on a 500×500 grid with centered differences

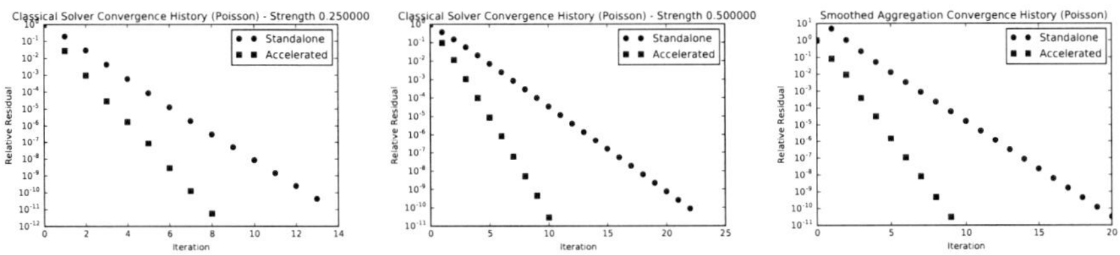

(a) Classical Strength 0.25 (b) Classical Strength 0.50 (c) Smoothed Aggregation

Figure 4.12: Example 4.6. Convergence history for the AMG with different coarsening strategy (Standalone) and for the PCG with the relative AMG preconditioner (Accelerated).

of second order. In Table 4.5 we consider the construction of the levels with the classical coarsening strategy, enforcing a different strength requirement on the connections and comparing it with the coarsening obtained with the smoothed aggregation. To a greater level of the strength parameter on the

Table 4.5: Example 4.6. Construction of the coarser levels with different strength of connection and with smoothed aggregation

	Strength: 0.25		Strength: 0.50		Smoothed Aggregation	
level	variables	nonzeros	variables	nonzeros	variables	nonzeros
0	250000	1248000	250000	1248000	250000	1248000
1	125000	1121002	125000	1121002	41750	373416
2	31252	280662	31252	280662	4704	41554
3	7825	70657	7851	71399	532	4526
4	1937	17971	1990	18946	65	509
5	483	4725	555	5803	9	65
6	124	1352	148	1704		
7	29	293	46	548		
8	7	41	17	171		
9			5	23		

connections correspond coarser levels with more variables and thus more levels, while the smoothed aggregation shows a more aggressive coarsening strategy. In Figure 4.11 we observe how this is reflected on the convergence history of the algorithm, both as a solver and as a preconditioner for the Conjugate Gradient algorithm. ✓

Multigrid methods for Toeplitz matrices $T_n(f)$, with both $f > 0$ and $f \geq 0$, were first proposed by Fiorentino and Serra-Capizzano [231, 232, 233]. Other approaches, extending these initial efforts in various directions and accommodating the issues of ill–conditioned and multilevel systems discussed in § 4.1, can be found in Chan et al. [135], Donatelli [194], Huckle and Staudacher

[314], Serra-Capizzano [469], Sun et al. [499], with extensions taking into account also circulant systems in [471, 473].

4.2.3 Multigrid preconditioners, parallelization and available software

Multigrid preconditioners have been diffusely applied, mostly for elliptic PDEs with both structured and unstructured grids. We do not focus explicitly on any of these application considering instead the common ground.

Note that the multilevel method we have described, i.e., Algorithm 4.2, provides an *approximate inverse* of A by suitably combining some *approximate inverses* of the hierarchy of matrices that represent A in increasingly coarser spaces from (4.24). We can therefore devise two distinct phases of our preconditioning routine based on the general multigrid algorithm: a *setup phase*, summarized in Algorithm 4.3, and an *application phase*, in which the preconditioner is applied to the residual vectors of our Krylov subspace method as $\mathbf{z} = M_l^{-1}\mathbf{r}$, see § 3.2.

Algorithm 4.3: Setup phase of MGM preconditioner

Input: $A \in \mathbb{R}^{n\times n}$, Set of indices/grid Ω

$A_l \leftarrow A$, $\Omega_l \leftarrow \Omega$;

Set up $S_k^{(1)}$ and $S_k^{(2)}$;

for $k = l, l-1, \ldots, 1$ **do**

  Generate Ω_{k-1} from Ω_k;

  Define I_k^{k-1} and put $I_{k-1}^{k} = (I_k^{k-1})^T$;

  Compute $A_{k-1} = I_k^{k-1} A_k I_{k-1}^{k}$;

  Set up $S_{k-1}^{(1)}$ and $S_{k-1}^{(2)}$;

For this second phase we can identify two basic approaches, the *additive* and the *multiplicative* multilevel:

additive: at each level, the smoother and the coarse–space correction are applied to the restriction of the vector $\mathbf{r}$ to that level, and the resulting vectors are added;

multiplicative: at each level, the smoother and the coarse–space correction are applied in turn, the former on a vector resulting from the application of the latter and/or vice versa.

An example of multiplicative preconditioner, where the smoother (or possibly the smoothers) is applied before and after the coarse–space correction, is the symmetrized multiplicative preconditioner or, more generally, the V–cycle preconditioner summarized in Algorithm 4.4.

Application of this kind of preconditioners to the PDEs setting are, e.g.,

Algorithm 4.4: V–cycle preconditioner

Input: Vector $\mathbf{v} \in \mathbb{R}^{n \times n}$, MGM preconditioner in Algorithm 4.3
Output: Vector $\mathbf{w} = M_l^{-1}\mathbf{v}$

$\mathbf{v}_l \leftarrow \mathbf{v}$;
`/* Recursively traverse levels downward: smoothing then residual restriction */`
for $k = l, l-1, \ldots, 2$ **do**
  $\mathbf{y}_k \leftarrow S_k^{(1)}\mathbf{v}_k$, ($\nu_1$ times);
  $\mathbf{r}_k \leftarrow \mathbf{v}_k - A_k\mathbf{y}_k$;
  $\mathbf{v}_{k-1} \leftarrow I_k^{k-1}\mathbf{r}_k$;
$\mathbf{y}_0 \leftarrow S_0^{(1)}\mathbf{v}_0$, ($\nu_1$ times);
`/* Recursively traverse levels upward: residual prolongation then smoothing */`
for $k = 1, \ldots, l$ **do**
  $\mathbf{y}_k \leftarrow \mathbf{y}_k + I_{k-1}^k\mathbf{y}_{k-1}$;
  $\mathbf{r}_k \leftarrow \mathbf{v}_k - A_k\mathbf{y}_k$;
  $\mathbf{r}_k \leftarrow S_k^{(2)}\mathbf{r}_k$, ($\nu_2$ times);
  $\mathbf{y}_k \leftarrow \mathbf{y}_k + \mathbf{r}_k$;
$\mathbf{w} \leftarrow \mathbf{y}_l$;

in [70, 154, 192, 195, 209, 403, 431, 442, 534] and many others. We refer again to the review [496].

There exist many libraries implementing algebraic multigrid methods, as a solver or as a preconditioner, possibly for Krylov subspace methods. We briefly mention some of them.

Let us start mentioning the `PyAMG` library by Bell et al. [40], in which AMG solvers are implemented with a convenient Python interface. With this tool, test cases can be implemented in a simple way with several coarsening strategies we discussed and hinted at in § 4.2.2. For large scale computation, we suggest the `MLD2P4` library [174] by D'Ambra et al. [173], i.e., the *Multi–Level Domain Decomposition Parallel Preconditioners Package*, based on the `PSBLAS` library [230]. This is a package of parallel[4] algebraic multi–level preconditioners, in which several versions of one–level additive and of multi–level additive and hybrid Schwarz algorithms are implemented. We stress that, for the multi–level case, some of the purely algebraic approaches we have discussed are applied to obtain the relative coarsening strategy, i.e., no geometric background on the problem from which the matrix is obtained is needed to build the preconditioner. Moreover, parallel versions of the domain decomposition techniques are also included, see [118, 119]. The other library, with

[4]The parallel implementation is based on a *Single Program Multiple Data* (SPMD) paradigm for distributed memory architectures.

a long development history, started in the late 1990s, is the `HYPRE` linear solvers library [225], developed at the Lawrence Livermore National Laboratory and that includes the classical algebraic multigrid preconditioners known as `BoomerAMG` [550]. This library is mostly focused on scalable solvers for large–scale scientific simulation based on parallel multigrid methods of both geometric and algebraic type, i.e., for both structured and unstructured grid problems. From the `Trilinos` project [342], the packages `ML` and `MueLu` should be considered. `ML` is made of a variety of parallel multigrid schemes for both preconditioning and solving systems arising (primarily) from elliptic PDE discretizations and is mostly based on some smoothed–aggregation strategy Noticeably, it also contains an example of a matrix-free algebraic multigrid. `MueLu` otherwise provides an architecture for algebraic multigrid methods for symmetric and nonsymmetric systems based on smoothed aggregation and does not implement any smoother itself that should be added manually (or linked through other `Trilinos` packages).

We conclude this section with an example of application of the multigrid strategy.

Example 4.7 (Hypre/BoomerAMG). Consider again the solution of the 2D Laplacian problem with zero boundary conditions on an $n \times n$ grid; thus the number of the unknowns is $N = n^2$. The second order centered finite difference discretization with a 5-point stencil is used. We solve only for the interior nodes. The algorithm is stopped when the relative residual $\tau = \|\mathbf{r}\|_2/\|\mathbf{b}\|_2$ is less than $\tau = 1e-7$. The selected (*pre* and *post*) smoother is one sweep

Table 4.6: Example 4.7. Solution of the 2D Laplacian problem with AMG approach, both as a solver and as a preconditioner

	AMG		CG		PCG + AMG		FGMRES+AMG	
n	IT	τ	IT	τ	IT	τ	IT	τ
33	7	6.31e-09	58	8.64e-08	5	2.85e-08	5	3.08e-09
66	8	3.55e-08	114	9.26e-08	6	9.62e-08	5	5.18e-08
99	8	2.34e-08	169	8.45e-08	6	9.68e-09	5	7.30e-08
132	10	5.70e-08	224	9.88e-08	7	6.54e-09	6	2.57e-08
165	9	2.40e-08	281	9.58e-08	6	8.30e-08	6	8.87e-09
250	10	4.34e-08	427	9.55e-08	7	6.22e-09	6	2.73e-08

of the Gauss–Seidel or SOR method, while the coarsening strategy is the standard coarsening strategy we have described in § 4.2.2 with a maximum number of level $l = 20$. We used the parallel implementation from the library `Hypre/BoomerAMG` [550]. This is indeed one of the cases in which the multigrid algorithms are effective and easy to apply. ✓

4.3 Complex symmetric systems

Complex linear systems arise naturally in a wide context of problems, like quantum mechanics [321], electromagnetism [67, 167, 168], optical tomography [12], etc.

Most of the preconditioned iterative methods we considered in previous pages belonging to the class of Krylov solvers, like GMRES, PCG, etc., can be applied to both real- and complex-valued linear systems, see the review [483] for some details, and multigrid methods as well, see, e.g., [370, 525]. Nevertheless, much large scale software is especially tuned for working with real–valued systems. This issue can be solved simply by recasting the complex problem in a real formulation. Observe that, in many cases, the complex entries of the matrix are localized, e.g., on the main diagonal, thus many preconditioning strategies applied directly to the complex linear system will have the undesirable effect of spreading nonreal entries in the preconditioning matrix. Consider, e.g., incomplete factorization in the case of a matrix that has its only complex entries on the main diagonal. In other cases, direct preconditioning of the complex system can be a viable approach; see, e.g., [348].

As suggested in [178, 246], there exist several mathematically equivalent formulations of a complex system

$$C\mathbf{w} = \mathbf{d}, \quad C \in \mathbb{C}^{n\times n}, \quad \mathbf{w}, \mathbf{b} \in \mathbb{C}^n, \tag{4.35}$$

that can be obtained by writing explicitly its real and imaginary terms:

$$(A + \imath B)(\mathbf{x} + \imath \mathbf{y}) = \mathbf{b} + \imath \mathbf{c}. \tag{4.36}$$

Then, by equating the real and imaginary parts in (4.36), consider the following four augmented 2×2 block formulations detailed in Table 4.7:

$$\mathcal{M}_j\mathbf{z} = \mathbf{f}, \quad \mathcal{M} \in \mathbb{R}^{2n\times 2n}, \quad \mathbf{z}, \mathbf{f} \in \mathbb{R}^{2n}, \tag{4.37}$$

We focus on finding preconditioning strategies for one of the systems (4.37), i.e., a real preconditioner. Observe that, while every preconditioner for the complex form (4.35) has indeed a real equivalent, not all real preconditioners for (4.37) come from a complex one.

Let us start from the spectral properties of the real equivalent approximations for the complex systems in Table 4.7.

Theorem 4.22 (Freund [246]). *Let $J = X^{-1}CX$ be the Jordan canonical form of C in (4.35). Then, the matrix $\mathcal{M}_1$ in Table 4.7 has the Jordan normal form*

$$\begin{bmatrix} J & 0 \\ 0 & \bar{J} \end{bmatrix} = X_1^{-1}\mathcal{M}_1 X_1, \qquad \text{where } X_1 = \begin{bmatrix} X & -\imath\bar{X} \\ -\imath X & \bar{X}. \end{bmatrix}$$

In particular, $\sigma(\mathcal{M}_1) = \sigma(A) \cup \overline{\sigma(A)}$. Moreover, the matrices $\mathcal{M}_2$ and $-\mathcal{M}_2$ are similar and $\sigma(\mathcal{M}_2) = \{\lambda \in \mathbb{C} : \lambda^2 \in \sigma(\bar{C}C)\}$.

Table 4.7: Equivalent 2×2 block formulations for a complex linear system

j	$\mathcal{M}_j$	$\mathbf{z}$	$\mathbf{f}$	Spectral Properties
1	$\begin{bmatrix} A & -B \\ B & A \end{bmatrix}$	$\begin{bmatrix} \mathbf{x} \\ \mathbf{y} \end{bmatrix}$	$\begin{bmatrix} \mathbf{b} \\ \mathbf{c} \end{bmatrix}$	If $\lambda \in \sigma(C)$, then $\lambda, \bar{\lambda} \in \sigma(\mathcal{M}_1)$; If C is Hermitian (positive definite), then $\mathcal{M}_1$ is symmetric (positive definite).
2	$\begin{bmatrix} A & B \\ B & -A \end{bmatrix}$	$\begin{bmatrix} \mathbf{x} \\ -\mathbf{y} \end{bmatrix}$	$\begin{bmatrix} \mathbf{b} \\ \mathbf{c} \end{bmatrix}$	If $\lambda \in \sigma(\mathcal{M}_2)$, then $-\lambda, \pm\bar{\lambda} \in \sigma(\mathcal{M}_2)$; If C is symmetric, then $\mathcal{M}_2$ is symmetric.
3	$\begin{bmatrix} B & A \\ A & -B \end{bmatrix}$	$\begin{bmatrix} \mathbf{x} \\ \mathbf{y} \end{bmatrix}$	$\begin{bmatrix} \mathbf{c} \\ \mathbf{b} \end{bmatrix}$	If $\lambda \in \sigma(\mathcal{M}_3)$, then $-\lambda, \pm\bar{\lambda} \in \sigma(\mathcal{M}_3)$; If C is symmetric, then $\mathcal{M}_3$ is symmetric; $\sigma(\mathcal{M}_3) = \sigma(\mathcal{M}_2)$.
4	$\begin{bmatrix} B & -A \\ A & B \end{bmatrix}$	$\begin{bmatrix} \mathbf{x} \\ -\mathbf{y} \end{bmatrix}$	$\begin{bmatrix} \mathbf{c} \\ \mathbf{b} \end{bmatrix}$	If $\lambda \in \sigma(C)$, then $-\imath\lambda, \imath\bar{\lambda} \in \sigma(\mathcal{M}_4)$; If C is Hermitian (positive definite), then $\mathcal{M}_4$ is skew symmetric (with eigenvalues that have positive imaginary parts).

The spectral properties of $\mathcal{M}_3$ and $\mathcal{M}_4$ can be readily obtained with results similar to the above. Note that we can expect a worse convergence rate for a Krylov iterative method applied to an equivalent real formulation with respect to the method directly applied to the originating complex system. Indeed, the above spectral properties suggest that the spectrum of $\mathcal{M}_i$ is, in general, not contained in a half plane and that many eigenvalues can surround the origin see, e.g., Figure 4.13. Nevertheless, there are cases in which this problem is indeed less severe. In [178] it is observed that the convergence rates of iterative methods in the equivalent and original formulations are more or less comparable for some very representative classes of model problems, e.g., in the case of Hermitian C, the equivalent formulation preserves the same convergence rate of the original complex formulation. However, preconditioning of the iterative method on the real equivalent system is almost always mandatory. In the following, we focus on general purpose preconditioning techniques, more than referring to specialized solvers for some specific problem. On this ground, one can consider the approach based on either stationary iteration methods or matrix splitting from [24, 235, 308, 356, 370, 381, 549], polynomial preconditioners [245] or on incomplete factorizations [50, 67, 178, 308].

Example 4.8. Let us consider the solution of a complex linear system (4.35) by rewriting it in the real formulation from Table 4.7 and using the GMRES method. To precondition the resulting equivalent system

$$\mathcal{M}_i\mathbf{x} = \mathbf{f}, \ i = 1, 2, 3, 4.$$

Consider, as in [178], the use of some ILU factorizations, see § 3.4. In particular, we use the ILUT preconditioner with pivoting and the row–MILU approach in § 3.4.3. In Table 4.8 we report the number of iterations (for GM-

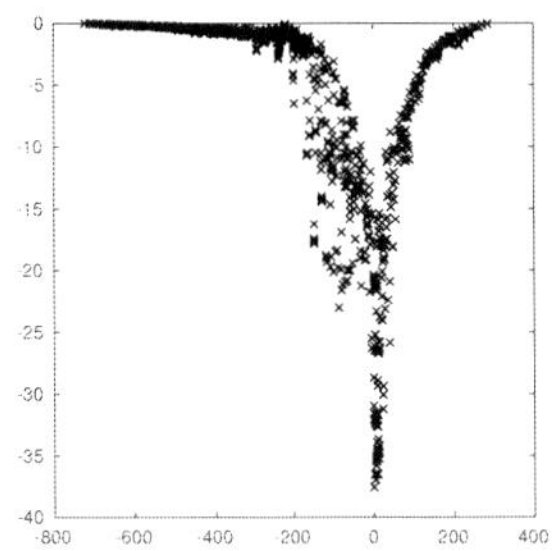

(a) $\sigma(C)$

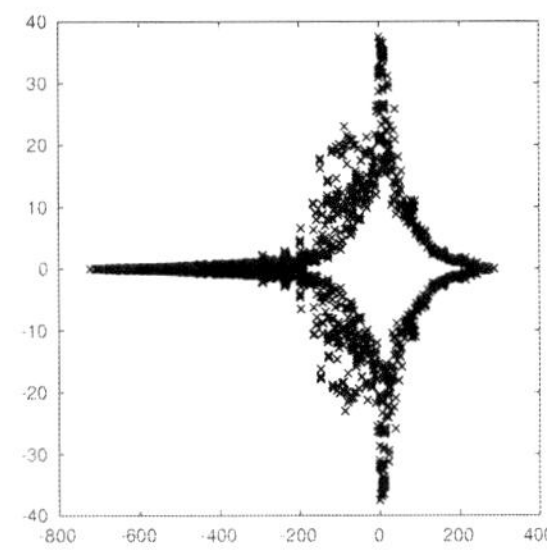

(b) $\sigma(\mathcal{M}_1)$

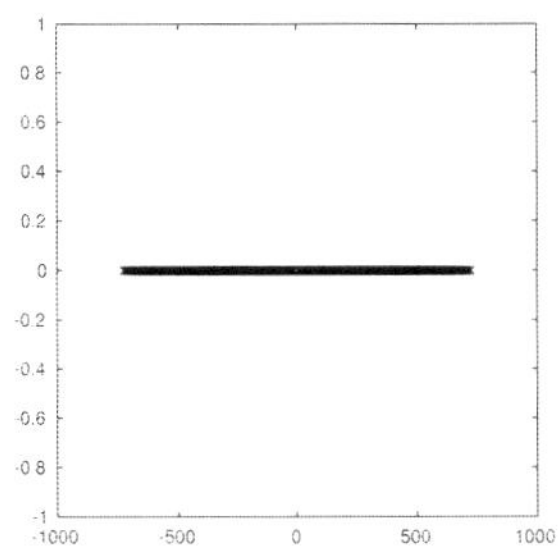

(c) $\sigma(\mathcal{M}_2)$

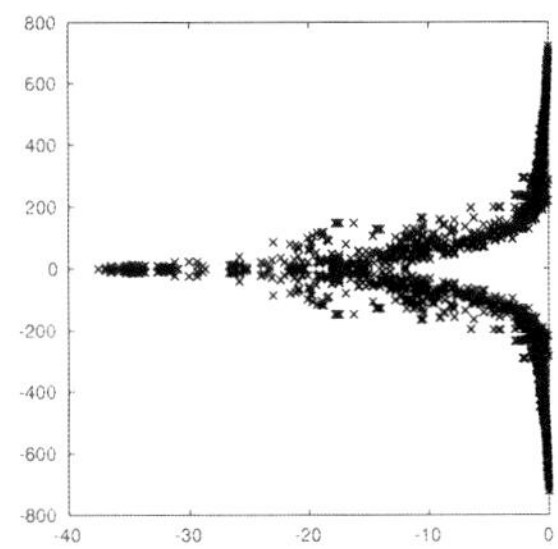

(d) $\sigma(\mathcal{M}_4)$

Figure 4.13: Complex matrix YOUNG2C that arise in the modeling of acoustic scattering phenomena. Discretization of the Helmholtz equation with Dirichlet boundary conditionsnd varying k. Grid with 29×29 interior points.

RES case equal to the number of matrix–vector products performed) for some test matrices from the University of Florida sparse matrix collection [177] with the right–hand side given by $\mathbf{d} = (1+\imath)C\boldsymbol{e}$, where $\mathbf{e} = (1, 1, \ldots, 1)^T$ and iteration stopped when the relative residual is less than $\varepsilon = 1e - 6$. For the test matrix `DWG961A` we reduced the problem size to cut off the null subspace and replaced some zero pivots that appeared during the computation applying the drop tolerance. We can observe that preconditioning is really necessary and the strategy gives acceptable performances. However, the incomplete factors are $2n \times 2n$ matrices, and reasonably, the amount of storage required for the incomplete L, U factors is usually expected to be higher than for the (incomplete) factors of the original $n \times n$ complex matrix. Moreover, since the matrix–vector products with the coefficient matrices from Table 4.7 can be implemented using only a single copy of both A and B, finding a preconditioner that requires only the *approximate* solution of the underlying sparse real linear systems of order n can be preferable. We discuss these kind of preconditioners based on the 2×2–blocks augmented matrices $\mathcal{M}_j$ in the following § 4.4 discussing the topic of *saddle point matrices*; see also [50]. ✓

Table 4.8: Example 4.8. Performances using equivalent real formulations for some complex linear systems

		$\mathcal{M}_1$		$\mathcal{M}_2$	
Matrix	Size	GMRES	ILUT(1e-1)	GMRES	ILUT(1e-1)
YOUNG1C	841	1248	1	601	9
YOUNG2C	841	1248	1	601	9
DWG961A	705	1036	1	1094	†
			ILUT(1e-2)		ILUT(1e-2)
DWG961B	961	1556	296	1689	40
		$\mathcal{M}_3$		$\mathcal{M}_4$	
Matrix	Size	GMRES	ILUT(1e-1)	GMRES	ILUT(1e-1)
YOUNG1C	841	591	4	514	4
YOUNG2C	841	591	4	514	4
DWG961A	705	1094	2	1144	2
			ILUT(1e-2)		ILUT(1e-2)
DWG961B	961	1702	31	1627	213

4.3.1 Sequences of parametric complex symmetric linear systems

Consider the sequence of complex symmetric linear systems (see also the discussion in § 3.6.1)

$$A_j\mathbf{x}_j = \mathbf{b}_j, \quad A_j = A + \alpha_j\, E_j, \quad j = 0, \ldots, s, \tag{4.38}$$

A Hermitian, $E_0, \ldots, E_s$ complex diagonal matrices and $\alpha_0, \ldots, \alpha_s$ scalar complex parameters. In order to approximate the solutions of (4.38), we use a Krylov iterative method. Following [67], let us focus on a preconditioning strategy based on matrices P_j approximating A_j where

$$P_j = \tilde{A} + \alpha_j K_j, \quad j = 0, \ldots, s, \tag{4.39}$$

and $\tilde{A}$ is a (sparse and/or structured or localized) matrix which approximates A in (4.38) and K_j, $j = 0, \ldots, s$, are suitable matrix corrections. The preconditioner is based on updated incomplete factorization in the style of those discussed in § 3.6.1. As usual, the correction matrices K_j are computed by balancing the main requirements of having cheap–to–compute updated preconditioners (4.39), $||P_j^{-1}||$ not too large and $\|A_j - P_j\|$ small.

From here on, the sequence of linear systems (4.38) with A Hermitian and large, sparse, possibly ill–conditioned matrices can be given simultaneously or sequentially. The latter case occurs, for instance, when the right-hand side $\mathbf{b}_j$ depends on the previous solution $\mathbf{x}_{j-1}$, as in the case of time-dependent PDEs.

Since we assumed that A is Hermitian and positive definite, we can reasonably assume that there exists an incomplete LDL^H factorization, see § 3.4, defined as usual as

$$P = \tilde{L}\tilde{D}\tilde{L}^H \tag{4.40}$$

such that

$$A = LDL^H \simeq \tilde{L}\tilde{D}\tilde{L}^H, \tag{4.41}$$

where L, $\tilde{L}$ are unit lower triangular and D, $\tilde{D}$ are diagonal matrices. Note that if we take the (exact) inverse factorization of A we have:

$$A^{-1} = ZD^{-1}Z^H \Rightarrow A = Z^{-H}DZ^{-1} \Rightarrow L = Z^{-H},\ L^H = Z^{-1}. \tag{4.42}$$

Let us suppose that αE has diagonal entries with positive real part. The proposed preconditioner based on (4.41) for $A_{\alpha,E} = A + \alpha E$ can be written as

$$P_{\alpha,E} = \tilde{L}\left(\tilde{D} + \alpha B\right)\tilde{L}^H. \tag{4.43}$$

Analogously, with a slight abuse of notation, we can define the preconditioner for (4.38) that is based on the approximate inverse preconditioner for A, see § 3.5.4:

$$Q \equiv P^{-1} = \tilde{Z}\tilde{D}^{-1}\tilde{Z}^H \tag{4.44}$$

as

$$Q_{\alpha,E} \equiv P_{\alpha,B}^{-1} = \tilde{Z}\left(\tilde{D} + \alpha B\right)^{-1}\tilde{Z}^H, \tag{4.45}$$

where the component matrices in (4.44) can be computed by Algorithm 3.20, which is robust if A is a Hermitian positive (or negative) definite matrix. Of course, $\tilde{D} + \alpha B$ must be invertible.

Proposition 4.6 (Bertaccini [67]). *Let us consider the exact LDL^H-factorization of A. By defining $P_{\alpha,E} = L\left(D + \alpha B\right)L^H$ and choosing $B = Z^H EZ$, we have $P_{\alpha,E} \equiv A_{\alpha,E}$, independently of α.*

Proof. Defining $P_{\alpha,E} = L\left(D + \alpha B\right)L^H$, i.e., using the exact LDL^H factorization of A, we have:

$$\begin{aligned} P_{\alpha,E} - A_{\alpha,E} =& Z^{-H}\left(D + \alpha B\right)Z^{-1} - \left(Z^{-H}DZ^{-1} + \alpha E\right) \\ =& \alpha\left(LBL^H - E\right), \end{aligned} \tag{4.46}$$

and therefore

$$B = L^{-1}EL^{-H} = Z^H EZ.$$

□

Unfortunately, we cannot use the form of B suggested in Proposition 4.6 in practice. Indeed, in general, only an incomplete LDL^H-factorization of A (or an incomplete $ZD^{-1}Z^H$-factorization of A^{-1}) is available, i.e., we don't have L to generate $L^{-H} = Z$ or directly Z but only (sparse) approximations. On the other hand, even if the exact Z is available, to apply the underlying preconditioner, we would have to solve, at each iteration step of the Krylov subspace solver, a linear system whose matrix $D + \alpha Z^H EZ$ is usually full.

Similar to what was done in the real case in § 3.6.1, for $k \geq 1$, we define the *order k modified preconditioner* (or the *order k updated preconditioner*) as

$$P_{\alpha,E}^{(k)} = \tilde{L}(\tilde{D} + \alpha B_k)\tilde{L}^H \tag{4.47}$$

with the same notation as in (4.43) and where B_k is the symmetric positive definite band matrix given by

$$B_k = \tilde{Z}_k^T \, E \, \tilde{Z}_k. \tag{4.48}$$

$\tilde{Z}_k$ is obtained by extracting the $k-1$ upper diagonals from $\tilde{Z}$ if $k > 1$ or $B_1 = \text{diag}(\tilde{Z}^H E \tilde{Z})$ if $k = 1$. Similarly, we define the *order k (inverse) modified preconditioner* (or the *order k (inverse) updated preconditioner*) as

$$Q_{\alpha,E}^{(k)} = \tilde{Z}(\tilde{D} + \alpha B_k)^{-1}\tilde{Z}^H. \tag{4.49}$$

Note that:

- B_k is always nonsingular since $\tilde{Z}_k$ is a unit upper triangular matrix and therefore nonsingular and αE is diagonal whose entries have a positive real part.
- From the previous item, $\tilde{D}+\alpha B_k$ is nonsingular. Indeed, $\alpha E = \Re(\alpha E) + \imath\Im(\alpha E) = E_R + \imath E_I$, where E_R is a diagonal matrix whose real entries are non negative by hypothesis. Therefore, we can write

  $$\tilde{D} + \alpha B_k = \left(\tilde{D} + \tilde{Z}_k^H E_R \tilde{Z}_k\right) + \imath \tilde{Z}_k^H E_I \tilde{Z}_k.$$

  Let us consider the matrix in brackets. Recalling that $\tilde{Z}_k$ is nonsingular, we have $\tilde{Z}_k^H E_R \tilde{Z}_k$ is positive semidefinite and $\tilde{D}$ has real positive entries. By using Gershgorin circles, we have the thesis.
- The *order k* preconditioners $P_{\alpha,E}^{(k)}$ based on incomplete LDL^H-threshold factorization with $k \geq 1$ require the computation of the matrix $\tilde{Z}$ or of an approximation of Z by using $\tilde{L}$. Note that the computation of an approximation of Z for computing B_k can be done just once for all scalars α and matrices E in (4.38). The Algorithm 3.20 generates both $\tilde{L}$ and $\tilde{Z}$ at the same time.

Now, let $\Re(M) = (M + \overline{M})/2$ and $\Im(M) = (M - \overline{M})/(2\imath)$.

Proposition 4.7 (Bertaccini [67]). *Let the incomplete factorization $\tilde{L}\tilde{D}\tilde{L}^H$ in* (4.40) *(the approximate inverse factorization $\tilde{Z}\tilde{D}^{-1}\tilde{Z}^H$ in* (4.44)*) be positive definite. If E (E_j in* (4.38)*) is such that $\Re(\alpha E)$ is positive definite, then $P_{\alpha,E}^{(k)}$ in* (4.47) *($Q_{\alpha,E}^{(k)}$ in* (4.49)*) is nonsingular. Moreover, $\Re(P_{\alpha,E}^{(k)})$ ($\Re(Q_{\alpha,E}^{(k)})$) is positive definite.*

Proof. The result follows from (4.47), (4.49) by observing that $\Re(\tilde{D} + \alpha B_k)$ is positive definite for $k \geq 0$, the incomplete factorization (4.41) for the seed matrix A (the preconditioner based on the approximate inverse preconditioner (4.44)) is positive definite and therefore $P^{(k)}_{\alpha,E}$ ($Q^{(k)}_{\alpha,E}$) is nonsingular. □

To conclude the analysis, we can proceed in the style of § 3.6.1 by considering the (exact) LDL^H factorization of A and the (exact) $ZD^{-1}Z^H$ factorization of the inverse for the matrix A in (4.38)), where $Z = L^{-H}$. The results below extend those formerly given in § 3.6.1 for symmetric positive definite matrices, α_0,... α_s real and positive and $E_j = I$ for all j.

Theorem 4.23 (Bertaccini [67]). *Let us consider the sequence of algebraic linear systems in* (4.38). *Let A be Hermitian positive definite, $\alpha \in \mathbb{C}$. Moreover, let $\delta > 0$ be a constant such that the singular values of the matrix $E - LB_kL^H$, B_k as in* (4.48), *are as follows*

$$\sigma_1 \geq \sigma_2 \geq \ldots \geq \sigma_t \geq \delta \geq \sigma_{t+1} \geq \ldots \geq \sigma_n \geq 0,$$

and $t \ll n$. Then, if

$$\max_{\alpha \in \{\alpha_0,\ldots,\alpha_s\}} |\alpha| \cdot \|D^{-1}Z_k^H E Z_k\|_2 \leq 1/2, \tag{4.50}$$

there exist matrices F, Δ and a constant c_α such that

$$\left(P^{(k)}_{\alpha,E}\right)^{-1}(A + \alpha\, E) = I + F + \Delta, \tag{4.51}$$

$$\|F\|_2 \leq \frac{2\max_{\alpha \in \{\alpha_0,\ldots,\alpha_s\}} |\alpha| c_\alpha \delta}{\lambda_{\min}(A)} \left(\frac{\|Z\|_2}{\min_i \|z_i\|_2}\right)^2, \quad Z = [z_1 \ \cdots \ z_n], \ z_i \in \mathbb{C}^n,$$

rank$(\Delta) \leq t \ll n$, *the rank of Δ does not depend on α, c_α is a constant such that* $\lim_{|\alpha|\to 0} c_\alpha = 1$, *$c_\alpha$ of the order of unity. The same properties hold true for*

$$Q^{(k)}_{\alpha,E} \cdot (A + \alpha E),$$

with $Q^{(k)}_{\alpha,E}$ defined as in (4.49).

The results of Theorem 4.23, without further assumptions, have a limited role in practice in order to explain the convergence of preconditioned iterations. For example, the norm of $\tilde{Z}$ and of $\tilde{L}^{-1}$ ($Z = L^{-H}$) can be large and therefore the spectrum of the preconditioned matrix cannot be considered clustered at all. Note that $||\tilde{Z}||$, $||Z||$ (and therefore $||L^{-1}||$, $||\tilde{L}^{-1}||$) can be large if, e.g., the entries of A^{-1} do not decay or decay very slowly away from the main diagonal, see again the discussion we made at the beginning of § 3.5. To this end, we give a result on the decay of the entries of Z.

Theorem 4.24 (Bertaccini [67]). *Let A be Hermitian, positive (or negative) definite and normalized, $A^{-1} = ZD^{-1}Z^H$, $D = \text{diag}(d_1, \ldots, d_n)$, $Z = (z_{i,j})$. Then, for $j > i$, we have:*

$$|z_{i,j}| \leq \sqrt{|d_j|} \frac{1-t^n}{1-t} \max \left\{ |\lambda_{\min}^{-1}(A)|, \frac{(1+\sqrt{|\lambda_{\max}(A)|/|\lambda_{\min}(A)|})^2}{2|\lambda_{\max}(A)|} \right\} t^{j-i}, \tag{4.52}$$

where $t = \left((\sqrt{|\lambda_{\max}(A)|/|\lambda_{\min}(A)|} - 1)/(\sqrt{|\lambda_{\max}(A)|/|\lambda_{\min}(A)|} + 1) \right)^{1/n}$, *for* $1 \leq i < j \leq n$.

When the bandwidth and the condition number of A are not too large, the entries of Z (i.e., of L^{-H}) are bounded in a rapidly decaying fashion away from the main diagonal along rows. In this case, we can find a constant such that the inverse of L (and thus $P_{\alpha,E}^{(k)}$)) has uniformly bounded norm. By theorems 4.23 and 4.24, we have the following consequence.

Corollary 4.1 (Bertaccini [67]). *Let A be a normalized symmetric positive definite diagonally dominant matrix, and let αE, $\alpha \in \mathbb{C}^+ = \{z \in \mathbb{C} : \Re(z) \geq 0\}$, be a diagonal matrix whose entries have positive real part. Then, $P_{\alpha,E}^{-1} A_{\alpha,E}$ ($Q_{\alpha,E} A_{\alpha,E}$) has a clustered spectrum.*

We recall that if $\kappa_2(V_{\alpha,E})$ is moderate, the eigenvalue distribution is relevant for the convergence of GMRES, see § 2.2.8. The hypotheses in Theorem 4.23 and Corollary 4.1 may look quite restrictive for several classes of problems. However, practical experience shows that preconditioned iterations can converge fast even when A is quite far from being diagonally dominant and there is no strong structural decay of the entries of $Z = L^{-H}$ away from the main diagonal.

Example 4.9 (Helmholtz equation). An example of a problem whose discretization produces complex symmetric linear systems is the Helmholtz equation with complex wave numbers

$$-\nabla \cdot (c\nabla u) + \sigma_1(j)u + \imath\sigma_2(j)u = f_j, \quad j = 0, \ldots, s, \tag{4.53}$$

where $\sigma_1(j)$, $\sigma_2(j)$ are real coefficient functions and c is the diffusion coefficient. The above equation describes the propagation of damped time-harmonic waves. We consider (4.53) on the domain $\mathcal{D} = [0, 1] \times [0, 1]$ with σ_1 constant, σ_2 a real coefficient function, $c(x, y) = e^{-x-y}$, u satisfies Dirichlet boundary conditions in $\mathcal{D}$. We discretize the problem with finite differences on a $n \times n$ grid, $N = n^2$, and mesh size $h = 1/(n + 1)$. We obtain the $s + 1$ linear systems ($j = 0, \ldots, s$):

$$A_j \mathbf{x}_j = \mathbf{b}_j, \; A_j = H + h^2\sigma_1(j)I + \imath h^2 D_j, \; D_j = \text{diag}(d_1, \ldots, d_N), \tag{4.54}$$

where H is the discretization of $-\nabla \cdot (c\nabla u)$ by means of centered differences. The $d_r = d_r(j)$, $r = 1, \ldots, N$, $j = 0, \ldots, s$, are the values of $\sigma_2(j)$ at the grid points. We consider the solution of the related 2D model problem in the unit square by using GMRES without preconditioning and the order 1 (diagonal), 2 (tridiagonal) and 3 (pentadiagonal) updated approximate inverse preconditioner (4.49). Results are collected in Table 4.9. Observe that the updates of

Table 4.9: Example 4.9. Solution with the GMRES method and updated AINV preconditioner for the Helmholtz equation with complex wave numbers. Average number of iterations and average times is given for the solution with updated preconditioner $Q_{\alpha,E}$ where k represents the order of the correction

GMRES	I		$Q_{\alpha,E}$ $k=1$		$Q_{\alpha,E}$ $k=2$		$Q_{\alpha,E}$ $k=3$	
σ_1	IT	T(s)	IT	T(s)	IT	T(s)	IT	T(s)
50	46	4.56e-02	28	**2.92e-02**	27	3.18e-02	27	3.18e-02
100	46	4.63e-02	28	**2.83e-02**	27	3.11e-02	27	3.11e-02
200	45	4.33e-02	28	**2.79e-02**	27	3.05e-02	27	3.05e-02
400	46	4.54e-02	28	**2.88e-02**	27	3.14e-02	27	3.14e-02
800	45	4.42e-02	29	**2.84e-02**	27	3.13e-02	27	3.13e-02

order k greater than one are sometimes not competitive for this problem. If σ_1 is positive and not negligible with respect to the entries of H, say, the inverse of (4.54) has entries which exhibit a fast decay away from the main diagonal; see § 3.5. Therefore, it is reasonable that diagonal corrections give good approximations with the minimum computational cost. Moreover, order $k > 1$ approximations require the solution of a k–banded linear system in complex arithmetic per iteration, which can represent a not negligible computational cost since the preconditioned iterative method converges in almost the same number of iterations with the underlying diagonal corrections. ✓

4.4 Saddle point systems

Let us now focus on the iterative solution of the ubiquitous class of *saddle point problems*, that in block 2×2 form can be written as

$$\begin{bmatrix} A & B_1^T \\ B_2 & -C \end{bmatrix} \begin{bmatrix} \mathbf{x} \\ \mathbf{y} \end{bmatrix} = \begin{bmatrix} \mathbf{f} \\ \mathbf{g} \end{bmatrix}, \quad \begin{array}{l} A \in \mathbb{R}^{n\times n}, \\ B_1, B_2 \in \mathbb{R}^{m\times n}, \\ C \in \mathbb{R}^{m\times m}, \end{array} \quad n \geq m. \tag{4.55}$$

where the blocks A, B_1, B_2 and C satisfy one or more of the following conditions:

S1 A is symmetric, i.e., $A = A^T$;

S2 the symmetric part of A, $H = (A + A^T)/2$ is positive semidefinite;

S3 $B_1 = B_2 = B$;

S4 C is symmetric, $C = C^T$, and positive semidefinite;

S5 C is the zero matrix, i.e., $C = 0$.

If all the above conditions are satisfied, A is symmetric positive semidefinite and we have symmetric linear systems of the form (4.58). Systems of this form arise, e.g., in the first order optimality conditions for the equality–constrained quadratic programming problem; see Example 4.10 below.

Example 4.10. Among the many fields in which *large* and *sparse* saddle point problems occur, the first one that we point out is *numerical optimization* [401] and, specifically, the *equality–constrained quadratic programming problem*

$$\begin{cases} \min_{\mathbf{x}} & q(\mathbf{x}) = \frac{1}{2}\mathbf{x}^T G\mathbf{x} + \mathbf{x}^T\mathbf{c}, \\ \text{subject to:} & A\mathbf{x} = \mathbf{b}. \end{cases}, \; G = G^T, \; G \geq 0 \tag{4.56}$$

whose first–order optimality conditions for $\tilde{\mathbf{x}}$ give rise to the 2×2 block system

$$\begin{bmatrix} G & -A^T \\ A & 0 \end{bmatrix} \begin{bmatrix} \tilde{\mathbf{x}} \\ \mathbf{y} \end{bmatrix} = \begin{bmatrix} -\mathbf{c} \\ \mathbf{b} \end{bmatrix}, \tag{4.57}$$

The variable $\mathbf{y}$ contains the Lagrange multipliers. Any solution $(\tilde{\mathbf{x}}, \mathbf{y})$ of (4.57) is a saddle point for the Lagrangian

$$\mathcal{L}(x, y) = \frac{1}{2}\mathbf{x}^T A\mathbf{x} - \mathbf{f}^T\mathbf{x} + (B\mathbf{x} - \mathbf{g})^T\mathbf{y},$$

hence the name *saddle point problem* which is popular for the block linear systems (4.57). Recall that $(\tilde{\mathbf{x}}, \tilde{\mathbf{y}}) \in \mathbb{R}^{n+m}$ is a *saddle point* if

$$\mathcal{L}(\tilde{\mathbf{x}}, \mathbf{y}) \leq \mathcal{L}(\tilde{\mathbf{x}}, \tilde{\mathbf{y}}) \leq \mathcal{L}(\mathbf{x}, \tilde{\mathbf{y}}), \forall x \in \mathbb{R}^n, \; \forall y \in \mathbb{R}^m.$$

By expressing $\tilde{\mathbf{x}} = \mathbf{x} + \mathbf{p}$ and rearranging the equations we obtain the *saddle point problem*

$$\begin{bmatrix} G & A^T \\ A & 0 \end{bmatrix} \begin{bmatrix} -\mathbf{p} \\ \mathbf{y} \end{bmatrix} = \begin{bmatrix} \mathbf{g} \\ \mathbf{h} \end{bmatrix}, \; \mathbf{h} = A\mathbf{x} - \mathbf{b}, \; \mathbf{g} = \mathbf{c} + G\mathbf{x}. \tag{4.58}$$

In this form (4.58) is usually called *Karush–Kuhn–Tucker system.* ✓

Systems of the form (4.57) arise also in nonlinearly constrained optimization (sequential quadratic programming and interior point methods), in fluid dynamics (Stokes problem), incompressible elasticity, circuit analysis, structural analysis, etc.

Another important case is when conditions from **S1** to **S4** are satisfied, but not **S5**. In this case we have a block linear system of the form

$$\begin{bmatrix} A & B^T \\ B & -C \end{bmatrix} \begin{bmatrix} \mathbf{x} \\ \mathbf{y} \end{bmatrix} = \begin{bmatrix} \mathbf{f} \\ \mathbf{g} \end{bmatrix}, \quad \begin{matrix} A \in \mathbb{R}^{n \times n}, \\ B \in \mathbb{R}^{m \times n}, \\ C \in \mathbb{R}^{m \times m}. \end{matrix} \quad n \geq m. \tag{4.59}$$

Problems of this kind arise in a wide number of contexts. Some examples come from stabilized mixed finite element methods, where stabilization is used to comply with the Ladyzhenskaya–Babušska–Brezzi (or inf-sup) condition to the discretization of models for slightly compressible fluids or solids, usually leading to nonzero matrices C. Others are in optimization, in which the solution of systems of the above form arises in regularization, least squares and in several interior point algorithms. As noted in [53], the phrase "generalized saddle point problem has been used in the literature primarily to allow for the possibility of a nonsymmetric coefficient matrix A in (4.55)". In such problems it is usually either $A \neq A^T$ (with condition **S2** satisfied), or $B_1 \neq B_2$ and sometimes both. Among these cases, one of the recurring problem is represented by the solution of linearized Navier–Stokes equations by some technique (often Newton or Picard iterations).

Linear systems of saddle point type appear usually with real coefficients. Thus, we restrict ourselves to this case. On the other hand, complex coefficient matrices do sometimes arise. Nevertheless, most of the results and algorithms we discuss can be extended to cover these cases.

Here we follow the review [53] noting that the definition of generalized saddle point problem as a linear system of the form (4.55), where the blocks A, B_1, B_2 and C satisfy one or more conditions **S1**–**S5**, is the most general.

In order to discuss iterative methods and preconditioners for the underlying problems, we need some preliminaries, covered in sections 4.4.1 and 4.4.2.

4.4.1 Invertibility, block factorizations and the Schur complement

To solve saddle point linear systems one could use several direct algorithms like those based on factorizations of the underlying system, on the use of the Schur–complement, *etc.* These algorithms can be appropriate under certain conditions. Nevertheless, as we have seen in Chapter 3, direct algorithms can be successfully adapted and used as preconditioners for iterative methods for sparse saddle point systems.

Let us begin by investigating some properties of the underlying matrices useful for the iterative algorithms and preconditioners for saddle point problems we consider in the next sections.

If the matrix A in (4.55) is invertible, then the saddle point matrix admits the following block triangular factorization.

$$\mathcal{M} = \begin{bmatrix} A & B_1^T \\ B_2 & -C \end{bmatrix} = \begin{bmatrix} I & 0 \\ B_2 A^{-1} & I \end{bmatrix} \begin{bmatrix} A & 0 \\ 0 & -(C + B_2 A^{-1} B_1^T) \end{bmatrix} \begin{bmatrix} I & A^{-1} B_1^T \\ 0 & I \end{bmatrix},$$

where $S = -(C + B_2 A^{-1} B_1^T)$ is the *Schur complement* of A in $\mathcal{M}$, or equivalently as

$$\mathcal{M} = \begin{bmatrix} A & 0 \\ B_2 & S \end{bmatrix} \begin{bmatrix} I & A^{-1}B_1^T \\ 0 & I \end{bmatrix} = \begin{bmatrix} I & 0 \\ B_2 A^{-1} & I \end{bmatrix} \begin{bmatrix} A & B_1^T \\ 0 & S \end{bmatrix}.$$

Then, by assuming the nonsingularity of A, whenever S is nonsingular, then $\mathcal{M}$ is nonsingular. To obtain the nonsingularity of the Schur complement we impose some restrictions on the matrices A, B_1, B_2 and C. Let us consider the following results from [53].

Theorem 4.25. *Assume A and C symmetric positive definite, $B_1 = B_2 = B$. If $\ker(C) \cap \ker(B^T) = \{0\}$, then the saddle point matrix $\mathcal{M}$ is nonsingular. In particular, $\mathcal{M}$ is invertible if B has full rank.*

Proof. Let us start with $C = 0$. In this case, the Schur complement reduces to $S = -BA^{-1}B^T$, a symmetric negative semidefinite matrix. Therefore, S and thus $\mathcal{M}$, is invertible if and only if $\operatorname{rank}(B) = m$. On the other hand, if A is symmetric and positive definite, $B_1 = B_2 = B$ and $C \neq 0$ is symmetric positive semidefinite, then $S = -(C + BA^{-1}B^T)$ is symmetric negative semidefinite. By imposing that $\ker(C) \cap \ker(B^T) = \{0\}$, we get that S is symmetric and negative definite, hence invertible and $\mathcal{M}$ is invertible. □

Theorem 4.26. *Assume that A is symmetric positive semidefinite, $B_1 = B_2 = B$ has full rank, and $C = 0$. Then $\mathcal{M}$ is nonsingular if and only if $\ker(A) \cap \ker(B) = \{0\}$.*

Proof. Let $\mathbf{u} = [\mathbf{x}^T, \mathbf{y}^T]^T \in \ker(\mathcal{M})$, hence $A\mathbf{x} + B^T\mathbf{y} = 0$ and $B\mathbf{x} = 0$. Then,

$$\mathbf{x}^T A\mathbf{x} = -\mathbf{x}^T B^T \mathbf{y} = (B\mathbf{x})^T \mathbf{y} = 0,$$

but $\mathbf{x}^T A\mathbf{x} = 0$ implies $A\mathbf{x} = 0$ since A is symmetric and positive definite. Therefore $\mathbf{x} \in \ker(A) \cap \ker(B)$, thus $\mathbf{x} = 0$. Also, $\mathbf{y} = 0$ since $B^T\mathbf{y} = 0$ and B^T has full column rank. Therefore, $\mathbf{u} = 0$, and $\mathcal{M}$ is nonsingular. For the necessary part we assume $\ker(A) \cap \ker(B) \neq \{0\}$, thus $\mathbf{x} \in \ker(A) \cap \ker(B)$ and $\mathbf{x} \neq 0$. Letting $\mathbf{u} = [\mathbf{x}^T, 0]^T$ we have also $\mathcal{M}\mathbf{u} = 0$, thus $\mathcal{M}$ is singular. □

Example 4.11. Let us focus on the definiteness hypothesis on the matrix A, so consider the following matrix $\mathcal{M}$

$$\mathcal{M} = \left[\begin{array}{cc|c} 1 & 0 & -1 \\ 0 & -1 & 1 \\ \hline -1 & 1 & 0 \end{array}\right] = \begin{bmatrix} A & B^T \\ B & 0 \end{bmatrix},$$

clearly this $\mathcal{M}$ is singular. A is indefinite both in general and on the $\ker(B)$. On the other hand, if we consider the slight modification of the previous matrix given by

$$\mathcal{M} = \left[\begin{array}{cc|c} 1 & 0 & 3 \\ 0 & 0 & 2 \\ \hline 3 & 2 & 0 \end{array}\right] = \begin{bmatrix} A & B^T \\ B & 0 \end{bmatrix},$$

we fall under the hypothesis of Theorem 4.26 and obtain the non singularity. Observe also that $\mathcal{M}$ is invertible also if A is positive definite on $\ker(B)$. ✓

What happens if B is rank deficient? If we look at it in the optimization context of Example 4.10, this means that we have some redundant constraints. Thus, we can think of eliminating them directly from the problem. The other interesting case is the one in which B has exactly one zero eigenvalue. In general we can use tools to restrict the problem on the non singular part, but, since the global linear system is consistent by construction, we can also overcome these difficulties by using one of Krylov subspace methods with a zero vector as the initial guess. In this way, the null solution is never recovered and the singularity is not a problem; see, e.g., the discussion on the GMRES in § 2.2.7. So, let us look at the converse of these results in a more general case.

Theorem 4.27 (Gansterer et al. [255]). *Let $C = 0$. If the matrix $\mathcal{M}$ is nonsingular, then* $\operatorname{rank}(B_1) = m$ *and* $\operatorname{rank}([A^T B_2^T]^T) = n$.

Proof. By contradiction, if $\operatorname{rank}(B_1) < m$ then there exists a nonzero vector $\mathbf{y} \in \mathbb{R}^m$ with $B_1^T\mathbf{y} = 0$. Therefore letting $\mathbf{u} = [0, \mathbf{y}^T]^T$, we find $\mathcal{M}\mathbf{u} = 0$. Again, by contradiction, if $\operatorname{rank}([A^T B_2^T]^T) < n$ then there exists a nonzero vector $\mathbf{x} \in \mathbb{R}^n$ such that $[A^T B_2^T]^T\mathbf{x} = 0$. Letting $\mathbf{u} = [\mathbf{x}^T, 0]$, we obtain $\mathcal{M}\mathbf{u} = 0$. □

Nevertheless, these conditions can not be reversed. That is, in the general case with $C = 0$ the rank conditions are not sufficient to ensure that $\mathcal{M}$ is invertible.

Theorem 4.28 (Benzi and Golub [52]). *Let the symmetric part of A, $H = (A + A^T)/2$ be positive semidefinite, $B_1 = B_2 = B$ have full rank and C is symmetric positive semidefinite (possibly zero). Then:*

1. $\ker(H) \cap \ker(B) = \{0\}$ *implies $\mathcal{M}$ invertible,*
2. *$\mathcal{M}$ invertible implies* $\ker(A) \cap \ker(B) = \{0\}$.

The converses of 1 and 2 do not hold in general.

Exercise 4.3. Prove Theorem 4.28.

In general the converse of the results in Theorem 4.28 are false; see the following Example.

Example 4.12. Consider the matrix

$$\mathcal{M} = \left[\begin{array}{ccc|c} 1 & -1 & 0 & 0 \\ 1 & 0 & 0 & 0 \\ 0 & 0 & 0 & 1 \\ \hline 0 & 0 & 1 & 0 \end{array}\right] = \begin{bmatrix} A & B^T \\ B & 0 \end{bmatrix}.$$

We have $\ker(H) \cap \ker(B) = \operatorname{span}([0, 1, 0]^T) \neq \{0\}$, H is the symmetric part of A, and yet $\mathcal{M}$ is invertible, so the converse of (1) does not hold. Similarly, let us consider the singular matrix

$$\mathcal{M} = \left[\begin{array}{cc|c} 0 & -1 & 0 \\ 1 & 0 & 1 \\ \hline 0 & 1 & 0 \end{array}\right] = \begin{bmatrix} A & B^T \\ B & 0 \end{bmatrix},$$

for which A is nonsingular, thus $\ker(A) \cap \ker(B) = \{0\}$ hence also the converse of (2) is false in general. ✓

Since we know under what conditions the matrix $\mathcal{M}$ is non singular, we can now focus our attention on finding some useful expressions for the inverse of the saddle point matrix. We know very well that computing the inverse of a *large* and *sparse* matrix is often not a great idea; see also the beginning of Chapter 3. Nevertheless, some formulations of the inverse of the saddle point matrix can be used to develop useful preconditioning strategies and the related theoretical analysis; see the end of this chapter. The characterization of the inverses we discuss are the ones from [367]. From what we have seen, when $\mathcal{M}$ is non singular, the Schur complement S is nonsingular. Therefore, we can write the following decomposition for the inverse of $\mathcal{M}$.

$$\begin{aligned} \mathcal{M}^{-1} &= \begin{bmatrix} A & B_1^T \\ B_2 & -C \end{bmatrix}^{-1} \\ &= \begin{bmatrix} A^{-1} + A^{-1}B_1^T S^{-1} B_2 A^{-1} & -A^{-1}B_1^T S^{-1} \\ -S^{-1}B_2A^{-1} & S^{-1} \end{bmatrix}. \end{aligned} \tag{4.60}$$

We can also have the case in which A is singular while C is not, for which the previous expression is of limited interest in the numerical solution of saddle point problems.

A particular case is the one in which **S1**, **S3** and **S5** hold, thus the Schur complement reduces to $S = -BA^{-1}B^T$, and $\mathbf{g} = 0$ in equation (4.55). Then, by the explicit expression of $\mathcal{M}^{-1}$, we find the solution.

$$\begin{bmatrix} \tilde{\mathbf{x}} \\ \tilde{\mathbf{y}} \end{bmatrix} = \begin{bmatrix} (I + A^{-1}B^T S^{-1} B)A^{-}\mathbf{f} \\ S^{-1}BA^{-1}\mathbf{f} \end{bmatrix}. \tag{4.61}$$

The matrix $P = -A^{-1}B^T S^{-1} B = A^{-1}B^T (BA^{-1}B^T)^{-1} B$ is a *projector*, i.e., is such that $P^2 = P$. Moreover, $P\mathbf{v} \in \operatorname{Im}(A^{-1}B^T)$ and $\mathbf{v} - P\mathbf{v}$ is orthogonal to $\operatorname{Im}(B^T)$ for all $\mathbf{v} \in \mathbb{R}^n$. Although this can appear as a mere curiosity, it opens the possibility of decoupling the problem and solving it with the (preconditioned) LSQR method [45]. To proceed in this direction, we need to observe that $\tilde{\mathbf{x}}$ in equation (4.61) can be written as $\tilde{\mathbf{x}} = (I - P)\hat{\mathbf{x}}$, $\hat{\mathbf{x}} = A^{-1}\mathbf{f}$, the solution of the unconstrained problem, and $\tilde{\mathbf{x}}$ is orthogonal to the space spanned by the columns of B^T. Furthermore, $\hat{\mathbf{x}} = P\hat{\mathbf{x}} + \tilde{\mathbf{x}}$, that is the decomposition into a part in the range of $A^{-1}B^T$ and one orthogonal to the range of B^T. Observe that, in general, the projection made by P will be oblique, the condition under which it is orthogonal is clearly when the range of

$A^{-1}B^T$ is equal to the range of B^T. Next we need to note that $\mathbf{f} - B^T\tilde{\mathbf{y}} = A\tilde{\mathbf{x}}$ and $B\tilde{\mathbf{x}} = 0$, thus,

$$0 = B\tilde{\mathbf{x}} = (BA^{-1})(A\tilde{\mathbf{x}}) = (A^{-1}B^T)^T(\mathbf{f} - B^T\tilde{\mathbf{y}}).$$

Therefore, the vector $\mathbf{f} - B^T\tilde{\mathbf{y}}$ is A^{-1}–orthogonal to the range of B^T. So, $\tilde{\mathbf{y}}$ is the solution of the generalized least squares problem with respect to the A^{-1} norm:

$$\|\mathbf{f} - B^T\tilde{\mathbf{y}}\|_{A^{-1}} = \min_{\mathbf{u}} \|\mathbf{f} - B^T\mathbf{u}\|_{A^{-1}}. \tag{4.62}$$

Observe that this is not the simple solution with the LSQR of the problem $B^T\mathbf{u} \approx \mathbf{f}$. A modification of the algorithm to let it work with the $\|\cdot\|_{A^{-1}}$ norm is necessary. Indeed, if we have the Cholesky factorization of the matrix $A = LL^T$, we can restate the generalized problem in $\|\cdot\|_{A^{-1}}$

$$\|L^{-1}(\mathbf{f} - B^T\tilde{\mathbf{y}})\| = \min_{\mathbf{u}} \|L^{-1}(\mathbf{f} - B^T\mathbf{u})\|, \tag{4.63}$$

and then use the standard LSQR to obtain the solution, e.g., see Paige [415]. What we want to do, instead, is follow [45] to produce a suitable modification of the LSQR method that works directly in the $\|\cdot\|_{A^{-1}}$ norm. To introduce this we need to follow the same strategies of Chapter 2 to get iterative methods from the Lanczos tridiagonlization (Algorithm 2.5). Let us start from the simple case in which we apply it to the block matrix

$$\begin{bmatrix} I & B \\ B^T & 0 \end{bmatrix} \begin{bmatrix} \mathbf{x} \\ \mathbf{y} \end{bmatrix} = \begin{bmatrix} \mathbf{b} \\ 0 \end{bmatrix}, \tag{4.64}$$

by doing this we generate a bidiagonal matrix E_k from the usual tridiagonal matrix T_k that shares the same singular values of the matrix B, that is

$$E_k = \begin{bmatrix} \alpha_1 & & \\ \beta_2 & \ddots & \\ & \ddots & \alpha_k \\ & & \beta_{k+1} \end{bmatrix}, \quad T_k = \begin{bmatrix} \hat{\alpha}_1 & \hat{\beta}_2 & & \\ \hat{\beta}_2 & \hat{\alpha}_2 & \ddots & \\ & \ddots & \ddots & \hat{\beta}_k \\ & & \hat{\beta}_k & \hat{\alpha}_k \end{bmatrix}.$$

Moreover, let us suppose we started the process with the vector $\hat{\mathbf{v}} = [\mathbf{b}^T, 0]^T$. Then, the set of vectors we obtain are of the following form:

$$\hat{\mathbf{v}}^{(i)} = \begin{cases} \begin{bmatrix} \mathbf{u}^{(j)} \\ 0 \end{bmatrix}, & i = 2j-1, \quad j \in \mathbb{N}, \\ \\ \begin{bmatrix} 0 \\ \mathbf{v}^{(j)} \end{bmatrix}, & i = 2j, \qquad\quad j \in \mathbb{N}. \end{cases},$$

and

$$\hat{\alpha}_i = \begin{cases} 1, & i = 2j-1, \quad j \in \mathbb{N}, \\ 0, & i = 2j, \qquad\quad j \in \mathbb{N}. \end{cases}.$$

Since $\{\hat{\mathbf{v}}^{(i)}\}_{i=1}^{k}$ is an orthonormal basis, then both the $\{\mathbf{u}^{(j)}\}_{j=1}^{\lceil k+1/2\rceil}$ and the $\{\mathbf{v}^{(j)}\}_{j=1}^{\lceil k/2\rceil}$ are orthonormal sets. Then we only need to rename the $\hat{\beta}_i$ as

$$\hat{\beta}_i = \begin{cases} \beta_j, & i = 2j-1, \quad j \in \mathbb{N}, \\ \alpha_j, & i = 2j, \qquad\;\; j \in \mathbb{N}. \end{cases}.$$

In this way, we can rewrite the Lanczos tridiagonalization algorithm as the Golub–Kahan lower–bidiagonalization Algorithm 4.5.

Algorithm 4.5: Golub–Kahan lower–bidiagonalization (GKLB)

Input: Matrix B, starting vector $\hat{\mathbf{v}} = [\mathbf{b}^T, 0]^T$, k number of iteration.
Output: $U_j = [\mathbf{u}^{(j)}, \ldots, \mathbf{u}^{(j)}]$, $V_j = [\mathbf{v}^{(j)}, \ldots, \mathbf{v}^{(k)}]$, E_k.

$\beta_1 \leftarrow \|\mathbf{b}\|$, $\mathbf{u}^{(1)} \leftarrow \mathbf{b}/\beta_1$;
$\mathbf{v}^{(1)} \leftarrow B^T\mathbf{u}^{(1)}$, $\alpha_1 \leftarrow \|\mathbf{v}^{(1)}\|$, $\mathbf{v}^{(1)} \leftarrow \mathbf{v}^{(1)}/\alpha_1$;
for $j = 1, 2, \ldots, k$ **do**
  $\mathbf{u}^{(j+1)} \leftarrow B\mathbf{v}^{(j)} - \alpha_j\mathbf{u}^{(j)}$, $\beta_{j+1} \leftarrow \|\mathbf{u}^{(j+1)}\|$;
  if $\beta_{j+1} \neq 0$ **then**
    $\mathbf{u}^{(j+1)} \leftarrow \mathbf{u}^{(j+1)}/\beta_{j+1}$
  else
    break;
  $\mathbf{v}^{(j+1)} \leftarrow B^T\mathbf{u}^{(j+1)} - \beta_{j+1}\mathbf{v}^{(j)}$, $\alpha_{j+1} \leftarrow \|\mathbf{v}^{(j+1)}\|$;
  if $\alpha_{j+1} \neq 0$ **then**
    $\mathbf{v}^{(j+1)} \leftarrow \mathbf{v}^{(j+1)}/\alpha_{j+1}$
  else
    break;

As usual we can also express this algorithm in matrix form obtaining the following recurrence relations:

$$\begin{aligned} U_{k+1}\left(\beta\mathbf{e}_1\right) =& \mathbf{b}, \\ BV_k =& U_{k+1}E_k, \\ B^T U_{k+1} =& V_k E_k^T + \alpha_{k+1}\mathbf{v}_{k+1}\mathbf{e}_{k+1}^T. \end{aligned}$$

The lower bidiagonal matrix E_k represents the restriction of B to the space spanned by the column of B_k with respect to the basis formed by the columns of U_{k+1}; see [271]. The solution of the linear system (4.64) can be now obtained by this algorithm. It is sufficient to permute the rows and the columns of the tridiagonal matrix T_k of the auxiliary subproblem, i.e., the projected problem

$$T_k\hat{\mathbf{p}}_k = \hat{\beta}_1\mathbf{e}_1,$$

with $k = 2l + 1$ that corresponds to

$$\begin{bmatrix} I & E_k \\ E_k^T & 0 \end{bmatrix} \begin{bmatrix} \mathbf{s}_{k+1} \\ \mathbf{t}_k \end{bmatrix} = \beta_1 \begin{bmatrix} \mathbf{e}_1 \\ 0 \end{bmatrix},$$

where $[\mathbf{s}_{k+1}^T \mathbf{t}_k^T]^T$ is the permutation of $\hat{\mathbf{p}}_{2k+1}$. Hence, we need to compute the solution of the least square problem

$$\mathbf{t}_k = \arg\min_{\mathbf{t}\in\mathbb{R}^k} \|E_k \mathbf{t} - \beta_1 \mathbf{e}_1\|, \; \mathbf{y}_k = V_k \mathbf{t}_k, \; \mathbf{x}_k = U_{k+1}\mathbf{s}_{k+1}.$$

Moreover, as we did for the construction of the GMRES method, one can solve the least squares problem through (a cheap update of) the QR factorization, see [416].

We can now solve the problem (4.63) without using explicitly the Cholesky factor of $A = LL^T$. From what we did with algorithm 4.5, we now need to apply the GKLB to the operator $L^{-1}B$ and with the transformed vectors $\mathbf{w}^{(i)} = M\mathbf{u}^{(i)}$ for an opportune matrix M. Since the vectors $\{\mathbf{u}^{(j)}\}_{j=1}^{\lceil k+1/2 \rceil}$ form an orthonormal set,

$$\mathbf{w}^{(i)^T}(MM^T)^{-1}\mathbf{w}^{(j)} = \delta_{i,j}.$$

Therefore it is sufficient to choose $M = L$ to cancel all the applications of L^{-1} from the algorithm and transform all the occurrences of $L^{-T}M^{-1}$ into applications of A^{-1}. With these choices, by repeating the construction of Algorithm 4.5, we obtain the GLKB(A^{-1}) procedure, Algorithm 4.6. We can

---

Algorithm 4.6: Golub–Kahan lower–bidiagonalization for A^{-1} (GLKB(A^{-1}))

Input: Matrices A, B, starting vector $\hat{\mathbf{v}} = [\mathbf{b}^T, 0]^T$, k number of iteration.

Output: $U_j = [\mathbf{u}^{(j)}, \ldots, \mathbf{u}^{(j)}]$, $V_j = [\mathbf{v}^{(j)}, \ldots, \mathbf{v}^{(k)}]$, E_k.

$\beta_1 \leftarrow \|\mathbf{b}\|_{A^{-1}}$, $\mathbf{u}^{(1)} \leftarrow \mathbf{b}/\beta_1$;
$\mathbf{v}^{(1)} \leftarrow B^T A^{-1}\mathbf{u}^{(1)}$, $\alpha_1 \leftarrow \|\mathbf{v}^{(1)}\|$, $\mathbf{v}^{(1)} \leftarrow \mathbf{v}^{(1)}/\alpha_1$;
for $j = 1, 2, \ldots, k$ **do**
    $\mathbf{u}^{(j+1)} \leftarrow B\mathbf{v}^{(j)} - \alpha_j \mathbf{u}^{(j)}$, $\beta_{j+1} \leftarrow \|\mathbf{u}^{(j+1)}\|_{A^{-1}}$;
    if $\beta_{j+1} \neq 0$ **then**
        $\mathbf{u}^{(j+1)} \leftarrow \mathbf{u}^{(j+1)}/\beta_{j+1}$
    else
        break;
    $\mathbf{v}^{(j+1)} \leftarrow B^T A^{-1}\mathbf{u}^{(j+1)} - \beta_{j+1}\mathbf{v}^{(j)}$, $\alpha_{j+1} \leftarrow \|\mathbf{v}^{(j+1)}\|$;
    if $\alpha_{j+1} \neq 0$ **then**
        $\mathbf{v}^{(j+1)} \leftarrow \mathbf{v}^{(j+1)}/\alpha_{j+1}$
    else
        break;

---

again express the recurrence relationship in matrix form as

$$\begin{aligned} U_{k+1}\,(\beta \mathbf{e}_1) =&\mathbf{b}, \\ BV_k =&U_{k+1}E_k, \\ B^T A^{-1} U_{k+1} =&V_k E_k^T + \alpha_{k+1}\mathbf{v}_{k+1}\mathbf{e}_{k+1}^T. \end{aligned}$$

Finally, we can write the Algorithm LSQR(A^{-1}) needed for solving the generalized least square problem of equation (4.62).

Algorithm 4.7: Generalized LSQR algorithm, LSQR(A^{-1})

Input: Matrices A,B, vector $\mathbf{b}$.
Output: Solution $\mathbf{y} = \arg\min_{\mathbf{u}} \|\mathbf{b} - B^T\mathbf{u}\|_{A^{-1}}$.

$\mathbf{y}^{(0)} = 0$;
$\beta_1 \leftarrow \|\mathbf{b}\|_{A^{-1}}$, $\mathbf{w}_1 \leftarrow \mathbf{b}/\beta_1$;
$\mathbf{v}^{(1)} \leftarrow B^T A^{-1}\mathbf{w}^{(1)}$, $\alpha_1 \leftarrow \|\mathbf{v}^{(1)}\|$, $\mathbf{v}^{(1)} = \mathbf{v}^{(1)}/\alpha_1$;
$\mathbf{d}^{(1)} \leftarrow \mathbf{v}^{(1)}$, $\bar{\varphi}_1 \leftarrow \beta_1$, $\bar{\rho}_1 \leftarrow \alpha_1$;
for $j = 1, 2, \ldots$ **do**
  `/* Bidiagonalization step: GKLB(`A^{-1}`) */`
  $\mathbf{w}^{(j+1)} \leftarrow B\mathbf{v}^{(j)} - \alpha_j \mathbf{w}^{(j)}$, $\beta_{j+1} \leftarrow \|\mathbf{w}^{(j+1)}\|_{A^{-1}}$;
  if $\beta_{j+1} \neq 0$ **then**
    $\mathbf{w}^{(j+1)} \leftarrow \mathbf{w}^{(j+1)}/\beta_{j+1}$
  else
    break;
  $\mathbf{v}^{(j+1)} \leftarrow B^T A^{-1}\mathbf{w}^{(j+1)} - \beta_{j+1}\mathbf{v}^{(j)}$, $\alpha_{j+1} \leftarrow \|\mathbf{v}^{(j+1)}\|$;
  if $\alpha_{j+1} \neq 0$ **then**
    $\mathbf{v}^{(j+1)} \leftarrow \mathbf{v}^{(j+1)}/\alpha_{j+1}$
  else
    break;
  `/* Orthogonal transformation step: Givens rotations */`
  $\rho_j \leftarrow \left(\bar{\rho}_j^2 + \beta_{j+1}^2\right)^{1/2}$;
  $c_j \leftarrow \bar{\rho}_j/\rho_j$, $s_j \leftarrow \beta_{j+1}/\rho_j$;
  $\theta_{j+1} \leftarrow s_j\alpha_{j+1}$, $\bar{\rho}_{j+1} \leftarrow -c_j\alpha_{j+1}$;
  $\varphi_j \leftarrow c_j\bar{\varphi}_j$, $\bar{\varphi}_{j+1} \leftarrow s_j\bar{\varphi}_j$;
  `/* Update of the solution */`
  $\mathbf{y}^{(j)} = \mathbf{y}^{(j-1)} + (\varphi_j/\rho_j)\mathbf{d}^{(j)}$;
  $\mathbf{d}^{(j+1)} = \mathbf{v}^{(j+1)} - (\theta_{j+1}/\rho_j)\mathbf{d}^{(j)}$;
  $\|B^T A^{-1}\mathbf{r}^{(j)}\|_2 \leftarrow \bar{\varphi}_{j+1}\alpha_{j+1}|c_j|$;`// Norm of the residual`

Remark 4.10. Observe that we could have chosen for M the matrix L^{-T}. This cancels all the L^{-T} operations and, again, the ML^{-1} products become products with A^{-1}. We can repeat the construction of the Algorithm 4.7 with this choice. In this case, at each step we then need to solve linear systems of the form $A\mathbf{f} = \mathbf{g}$. Now if $\mathbf{f}^{(k)} = U_k\mathbf{h}^{(k)}$ is an approximation of the vector $\mathbf{f}$ from the space spanned by the column of V_k we can suppose reasonably that $AU_k\mathbf{h}^{(k)}$ approximates $\mathbf{g}$ in some sense, so that we may not need to solve the auxiliary linear systems with matrix A. Although this observation seems promising, numerical experiments suggest not.

We can further modify Algorithm 4.7 to build alongside the $\mathbf{y}$ vector also an approximation of $\mathbf{x} = A^{-1}(\mathbf{b} - B\mathbf{y})$. Thus, let us define $A\mathbf{x}^{(j)} = \mathbf{b} - B\mathbf{y}^{(j)}$. then, the update is obtained as

$$
\begin{aligned}
A^{-1}B\mathbf{d}^{(j)} =& A^{-1}B\mathbf{v}^{(j)} - \frac{\theta_j}{\rho_{j-1}} A^{-1}B\mathbf{d}^{(j-1)},\\
\mathbf{x}^{(j)} =& \mathbf{x}^{(j-1)} - \frac{\varphi_j}{\rho_j} A^{-1}B\mathbf{d}^{(j)}.
\end{aligned}
$$

Observe that $A^{-1}B\mathbf{v}^{(j)}$ is already available. therefore, we can update the quantity $A^{-1}B\mathbf{d}^{(j)}$ without performing any extra operations with A^{-1}. We augment the number of matrix vectors products and increase the storage of three vectors of the same size of A. We stress that computing the solution as $\mathbf{x} = A^{-1}(\mathbf{b} - B\mathbf{y})$ at the end of the standard LSQR(A^{-1}) algorithm can be numerically more stable.

The stopping criteria for the generalized LSQR(A^{-1}) method can be built upon the evaluation of the norm of the generalized residual $\|B^TA^{-1}\mathbf{r}^{(j)}\|_2$, that can be computed directly and without further effort as $\|B^TA^{-1}\mathbf{r}^{(j)}\|_2 = \bar{\varphi}_{j+1}\alpha_{j+1}|c_j|$; see [417].

For problems in this form, i.e., for which **S1**, **S3** and **S5** hold, together with $\mathbf{g} = 0$, there are also connections with the MinRes method that we have discussed in § 2.1.3. This reduces to solve

$$
B^TA^{-1}B\mathbf{y} = B^TA^{-1}\mathbf{f},
$$

and then assemble the $\mathbf{x}$ part of the solution again as $\mathbf{x} = A^{-1}(f - B^T\mathbf{y})$.

4.4.2 Spectrum and conditioning of saddle point matrices

To uncover the features of generalized LSQR algorithms and to have hints on the convergence of iterative Krylov methods for solving saddle point linear systems (recall that the eigenvalues analysis only is not enough for this in general; see, e.g., [280] and § 2.2.8), we need some preliminary results characterizing the eigenvalues of saddle point matrices.

Let us start from the symmetric case, i.e., **S1**, **S3** and **S4** hold, with C that can be the zero matrix. From the block triangular factorization of the saddle point matrix $\mathcal{M}$ we get

$$
\begin{bmatrix} I & 0 \\ -BA^{-1} & I \end{bmatrix}
\begin{bmatrix} A & B^T \\ B & -C \end{bmatrix}
\begin{bmatrix} I & -A^{-1}B^T \\ 0 & I \end{bmatrix}
=
\begin{bmatrix} A & 0 \\ 0 & S \end{bmatrix},
$$

where S is again the Schur complement $S = -(C+BA^{-1}B^T)$ and is symmetric and negative definite. Thus $\mathcal{M}$ is *similar* to the block diagonal matrix

$$\begin{bmatrix} A & 0 \\ 0 & S \end{bmatrix}, \quad A \in \mathbb{R}^{n\times n}.$$

By the Sylvester Law of Inertia (cfr. Theorem A.10), $\mathcal{M}$ is indefinite, with n positive and m negative eigenvalues. More specifically, the following results regarding the inertia (denotedf by In()) can be established.

Proposition 4.8 (Gould [275]). *Let $\mathcal{M}$ be the matrix satisfying properties* ***S1****,* ***S3*** *and* ***S5****. If $r = \text{rank}(B)$ and N is a matrix whose column generates* $\ker(B)$*, then*

$$\text{In}(\mathcal{M}) = \text{In}(N^T AN) + (r, r, m - r).$$

For the full symmetric case we can compute the inertia of $\mathcal{M}$ with the following result.

Theorem 4.29 (Forsgren [241]). *Let $\mathcal{M}$ be the saddle point matrix satisfying properties* ***S1****,* ***S3*** *and* ***S4****, and let $r = \text{rank}([B, -C])$. Let U_0 be a matrix whose columns generate* $\ker(C)$ *and N a matrix whose columns generate* $\ker(U_0^T B)$*. Then,*

$$\text{In}(\mathcal{M}) = \text{In}(N^T(A + B^T C^\dagger B)N) + (m_0 - m + r, r, m - r),$$

where $m_0 = \dim\ker(C)$ and $\text{rank}(U_0^T B) = m_0 - m + r$ and $C^\dagger$ is the pseudoinverse of C.

Other characterizations relative to the inertia in the KKT case of Example 4.10 can be found in [242, 264]. What we want to investigate now is the relationship between the spectrum of $\mathcal{M}$ and the spectrum of matrices A and B.

Theorem 4.30 (Rusten and Winther [449]). *Let $\mu_1 \geq \mu_2 \geq \ldots \geq \mu_n > 0$ be the eigenvalues of A, $\sigma_1 \geq \sigma_2 \geq \ldots \geq \sigma_m > 0$ the singular values of B. If we denote as $\sigma(\mathcal{M})$ the spectrum of $\mathcal{M}$, then*

$$\sigma(\mathcal{M}) \subset I = I^- \cup I^+,$$

where

$$I^- = \left[\frac{1}{2}\left(\mu_n - \sqrt{\mu_n^2 + 4\sigma_1^2}\right), \frac{1}{2}\left(\mu_1 - \sqrt{\mu_1^2 + 4\sigma_m^2}\right)\right],$$
$$I^+ = \left[\mu_n, \frac{1}{2}\left(\mu_1 + \sqrt{\mu_1^2 + 4\sigma_1^2}\right)\right].$$

Remark 4.11. Observe that the two intervals I^- and I^+ are located on each side of the origin. Therefore, if the condition numbers $\kappa_2(A) = \mu_1/\mu_n$ and $\kappa_2(B) = \sigma_1/\sigma_m$ are such that $\kappa_2(A), \kappa_2(B) \to 1$, then the interval I^- degenerates to a single point while the measure of I^+ goes to a value that is less than σ_1.This can be a good framework for the convergence of Krylov subspace methods; see also Example 4.13.

Example 4.13. Let us consider a mixed finite element discretization (see [10, 110] for details) for the bidimensional elliptic boundary value problems, called the Darcy problem:

$$\begin{cases} \nabla \cdot (-\mathbf{K}\nabla p) \equiv \nabla \cdot \mathbf{u} = f, & \mathbf{x} \in \Omega, \quad \Omega \subset \mathbb{R}^2, \\ p = p_D, & \mathbf{x} \in \Gamma^D, \quad \partial\Omega = \Gamma^D \cup \Gamma^N, \\ \mathbf{u} \cdot \mathbf{n} = u_N, \ \mathbf{x} \in \Gamma^N. \end{cases}$$

$\Omega \subset \mathbb{R}^2$ is a a rectangular domain with 4 corners, 4 sides with unit outward normals $\mathbf{n}$, $\mathbf{K}$ is the hydraulic permeability tensor that is uniformly symmetric positive definite, f is a source/sink term, p_D and u_N are Dirichlet and Neumann boundary data on the nonoverlapping decomposition of the domain $\Omega = \Gamma^D \cup \Gamma^N$, respectively. We compute the mixed finite element discretization of this problem with the DarcyLite package by Liu et al. [363]. As in the standard procedure for mixed finite element discretization, we define a Sobolev subspace and a manifold for vector–valued functions

$$H_{N,0}(\text{div}, \Omega) = \{\mathbf{v} \in \mathbb{L}^2(\Omega)^2 \,:\, \text{div}(\mathbf{v}) \in \mathbb{L}^2(\Omega), \ \mathbf{v}|_{\Gamma^N} = 0\},$$
$$H_{N,u_n}(\text{div}, \Omega) = \{\mathbf{v} \in \mathbb{L}^2(\Omega)^2 \,:\, \text{div}(\mathbf{v}) \in \mathbb{L}^2(\Omega), \ \mathbf{v}|_{\Gamma^N} \cdot \mathbf{n} = u_N\}.$$

By considering a rectangular mesh $\mathcal{E}_h$ over Ω, we denote a pair of finite element spaces as (V_h, W_h) such that $V_h \subset H(\text{div}, \Omega)$, $W_h \subset \mathbb{L}^2(\Omega)$ and they satisfy the inf–sup condition; see [110]. Then, we denote the spaces $U_h = V_h \cap H_{N,u_n}$ and $V_h^0 = V_h \cap H_{N,0}$. With these notations, the mixed finite element scheme can be stated as: find $\mathbf{u}_h \in U_h$ and $p_h \in W_h$ such that,

$$\begin{cases} \displaystyle\sum_{E \in \mathcal{E}_h} \int_E \mathbf{K}^{-1}\mathbf{u}_h \cdot \mathbf{v} - \sum_{E \in \mathcal{E}_h} \int_E p_h(\nabla \cdot \mathbf{v}) = -\sum_{\gamma \in \Gamma_h^D} \int_\gamma p_D(\mathbf{v} \cdot \mathbf{n}), & \forall \mathbf{v} \in V_h^0, \\ \displaystyle -\sum_{E \in \mathcal{E}_h} \int_E (\nabla \cdot \mathbf{u}_h) q = -\sum_{E \in \mathcal{E}_h} \int_E f q, & \forall q \in W_h. \end{cases}$$

Mixed finite element schemes guarantee local mass conservation (by taking $q = 1$ in the second equation) and normal flux continuity. By assembling the matrices and imposing the Neumann boundary conditions, the problem is reduced to the solution of the following saddle point linear system

$$\begin{bmatrix} A & B^T \\ B & 0 \end{bmatrix} \begin{bmatrix} \mathbf{x} \\ \mathbf{y} \end{bmatrix} = \begin{bmatrix} \mathbf{g} \\ 0 \end{bmatrix}.$$

In Figure 4.14 we reported both the pattern of the discretization matrix and the bound on the set I^- and I^+ obtained with Theorem 4.30. We can observe that the obtained bounds are very accurate and give a detailed information on the localization of the spectra of the matrix $\mathcal{M}$. We choose an example of this kind of discretization because eigenvalue bounds are needed for the assessment of the (inf–sup) stability of mixed finite element discretizations, see, e.g., [88, 372]. ✓

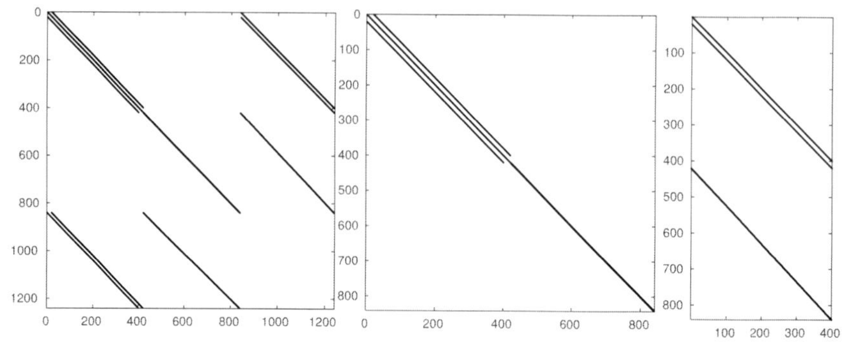

(a) Spy of the matrices $\mathcal{M}$, A and B^T.

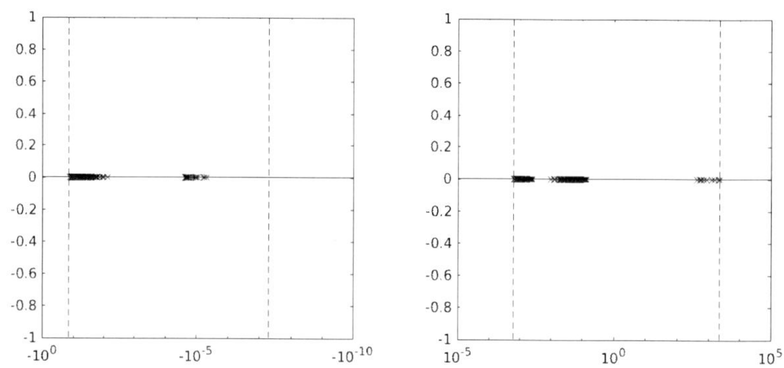

(b) Bound (dashed line) given by Theorem 4.30 for I^- and I^+.

Figure 4.14: Example 4.13. Spy of the saddle matrices from the MFEM discretization of Darcy's equation together with the bound on the spectrum obtained with Theorem 4.30.

There is a generalization of the bound in Theorem 4.30 for the case in which we exchange hypothesis **S5** with **S4**, i.e., $C \neq 0$ symmetric and positive semidefinite.

Proposition 4.9 (Silvester and Wathen [479]). *Let C be such that*

$$\max_{\mathbf{p}\in\mathbb{R}^m\setminus\{0\}} \frac{\mathbf{p}^T C \mathbf{p}}{\mathbf{p}^T \mathbf{p}} \leq \Delta,$$

and let us denote as $\sigma(\tilde{\mathcal{M}})$ the spectrum of $\tilde{\mathcal{M}}$, where $\tilde{\mathcal{M}}$ is given by

$$\begin{bmatrix} P_A^{-1/2} & \\ & P_C^{-1/2} \end{bmatrix} \begin{bmatrix} A & B^T \\ B & -C \end{bmatrix} \begin{bmatrix} P_A^{-1/2} & \\ & P_C^{-1/2} \end{bmatrix} = \begin{bmatrix} \tilde{A} & \tilde{B}^T \\ \tilde{B} & -\tilde{C} \end{bmatrix} = \tilde{\mathcal{M}},$$

where P_A and P_C are both symmetric positive definite matrices. Then, given

$\mu_1 \geq \mu_2 \geq \ldots \geq \mu_n > 0$, *the eigenvalues of* $\tilde{A}$, $\sigma_1 \geq \sigma_2 \geq \ldots \geq \sigma_m > 0$ *the singular values of* $\tilde{B}$, *the following bound holds*

$$\frac{1}{2}\left(\mu_1 - \Delta - \sqrt{(\mu_1 + \Delta)^2 + 4\sigma_m^2}\right) \leq \min_{\lambda \in I^-} \lambda.$$

In the general non symmetric case, we do not have a characterization for the eigenvalues like the one in Theorem 4.30. Moreover, in many cases of interest, the convex hull of the spectrum of $\mathcal{M}$ contains the origin. From the convergence analysis in Chapter 2, this is a case in which we do not expect a fast convergence of iterations for a unpreconditioned Krylov subspace method.

Example 4.14. Consider the mixed finite element discretization of one of the classical test problems in fluid dynamics: the *driven–cavity flow* (see also Example 4.13) :

$$\begin{cases} -\nu\nabla^2\mathbf{u} + (\mathbf{v}\cdot\nabla)\mathbf{u} + \nabla p = \mathbf{f}, & \mathbf{x} \in \Omega, \\ \nabla\cdot\mathbf{u} = 0, & \mathbf{x} \in \Omega, \\ \mathcal{B}\mathbf{u} = \mathbf{g}, & \mathbf{x} \in \Gamma = \partial\Omega. \end{cases},$$

This a the model of a flow in a square cavity Ω with the lid moving from left to right. In our case, Dirichlet no–flow conditions are applied on the side and bottom boundaries while we consider the so–called *leaky cavity*, that is

$$\{y = 1;\ -1 \leq x \leq 1 \,|\, u_x = 1\},$$

that defines the boundary operator $\mathcal{B}$, $\nu > 0$ is the kinematic viscosity coefficient. The mixed finite element discretization with Q1–P0 elements of these equations, produced with the IFISS package [478], give a saddle point matrix with $A \neq A^T$, $B_1 = B_2 = B$ and $0 \neq C \geq 0$, as in Figure 4.15. Moreover, as

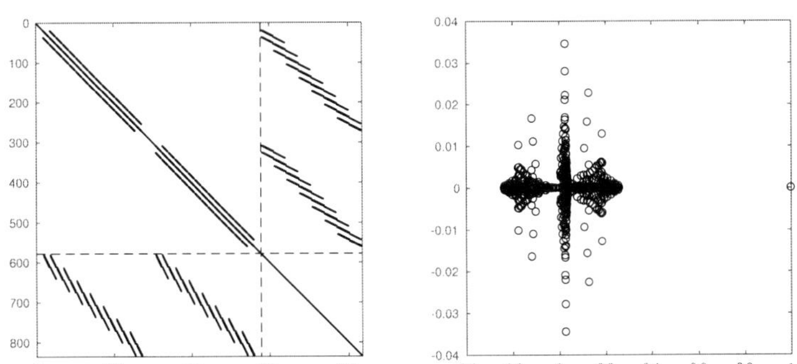

Figure 4.15: Example 4.14. Pattern of the discretization saddle matrix and eigenvalues for the Lid Cavity Flow problem with leaky conditions.

we can observe from the right hand side of Figure 4.15, the eigenvalues of the whole saddle point matrix $\mathcal{M}$ appear on both sides of the complex plane and thus the convex hull of its spectrum encloses the origin. ✓

Nevertheless, there exists a simple transformation able to get an equivalent linear system with a coefficient matrix whose spectrum is entirely contained in the right half plane, i.e., such that the real part of the eigenvalues is always non negative or, equivalently, such that the hermitian part of the transformed matrix is positive semidefinite. To obtain this results we need to modify the system

$$\begin{bmatrix} A & B^T \\ B & -C \end{bmatrix} \begin{bmatrix} \mathbf{x} \\ \mathbf{y} \end{bmatrix} = \begin{bmatrix} \mathbf{f} \\ \mathbf{g} \end{bmatrix}, \ \mathcal{M}\mathbf{u} = \mathbf{b},$$

into

$$\begin{bmatrix} A & B^T \\ -B & C \end{bmatrix} \begin{bmatrix} \mathbf{x} \\ \mathbf{y} \end{bmatrix} = \begin{bmatrix} \mathbf{f} \\ -\mathbf{g} \end{bmatrix}, \ \hat{\mathcal{M}}\mathbf{u} = \hat{\mathbf{b}},$$

where $\hat{\mathcal{M}} = J\mathcal{M}$ and $\hat{\mathbf{b}} = J\mathbf{b}$,

$$J = \begin{bmatrix} I_n & 0 \\ 0 & -I_m \end{bmatrix}.$$

$\hat{\mathcal{M}}$ is non–singular if $\mathcal{M}$ is. Moreover,

Theorem 4.31. *Let $\hat{\mathcal{M}}$ be the modified coefficient matrix obtained as $\hat{\mathcal{M}} = J\mathcal{M}$ and $\sigma(\hat{\mathcal{M}})$ be its spectrum. If we assume that the symmetric part of A, $H = (A + A^T)/2$ is positive semidefinite, $B_1 = B_2 = B$ has full rank, $C = C^T$ is positive semidefinite and $\ker(H) \cap \ker(B) = \{0\}$, then*

1. *$\hat{\mathcal{M}}$ is positive semidefinite, i.e., $\mathbf{v}\hat{\mathcal{M}}\mathbf{v} \geq 0 \ \forall \mathbf{v} \in \mathbb{R}^{n+m}$;*
2. *$\hat{\mathcal{M}}$ is* positive semistable, *i.e., the eigenvalues of $\hat{\mathcal{M}}$ have non negative real part;*
3. *If, in addition, H is positive definite, then $\hat{\mathcal{M}}$ is* positive stable, *i.e., the eigenvalues of $\hat{\mathcal{M}}$ have a positive real part.*

Proof. $\hat{\mathcal{M}}$ is positive semidefinite because H and C are:

$$\mathbf{v}^T\hat{\mathcal{M}}\mathbf{v} = \frac{1}{2}\mathbf{v}^T(\hat{\mathcal{M}} + \hat{\mathcal{M}}^T)\mathbf{v} = \mathbf{v}^T \begin{bmatrix} H & 0 \\ 0 & C \end{bmatrix} \mathbf{v} \geq 0, \ forall \mathbf{v} \in \mathbb{R}^{n+m}.$$

Now let $\mathbf{v}$ be a norm one eigenvalue, i.e., $\mathcal{M}\mathbf{v} = \lambda\mathbf{v}$, then $\mathbf{v}^H\hat{\mathcal{M}}\mathbf{v} = \lambda$ and $\left(\mathbf{v}^H\hat{\mathcal{M}}\mathbf{v}\right)^H = \mathbf{v}^H\hat{\mathcal{M}}^T\mathbf{v} = \overline{\lambda}$. Therefore the real part of λ is obtained as $1/2\mathbf{v}^H(\hat{\mathcal{M}}+\hat{\mathcal{M}})\mathbf{v} = (\lambda+\overline{\lambda})/2$ and $\mathbf{v}^H(\hat{\mathcal{M}}+\hat{\mathcal{M}})\mathbf{v}$ is a real nonnegative quantity. To prove the last one, let us consider again an eigenpair $(\mathbf{v}, \lambda)$ of $\hat{\mathcal{M}}$ with $\mathbf{v} = [\mathbf{x}^T, \mathbf{y}^T]^T$ and proceed by contradiction.

$$\Re(\Lambda) = \mathbf{x}^H H\mathbf{x} + \mathbf{y}^H C\mathbf{y},$$

is a nonnegative quantity, and is zero only if $\mathbf{x} = 0$ and $C\mathbf{y} = 0$, since H is positive definite by hypothesis. But if $\mathbf{x} = 0$ then, from the first block equation of $\hat{\mathcal{M}}\mathbf{v} = \lambda\mathbf{v}$, we find $B^T\mathbf{y} = 0$, thus $\mathbf{y} = 0$ since B has full column rank. Hence $\mathbf{v} = 0$ that is a contradiction since $\mathbf{v}$ is an eigenvector. □

Example 4.15. We take again the discretization matrix from Example 4.14 and apply to it the transformation J. In this way we obtain the spectrum in Figure 4.16. The results of Theorem 4.31 apply and we have a spectrum

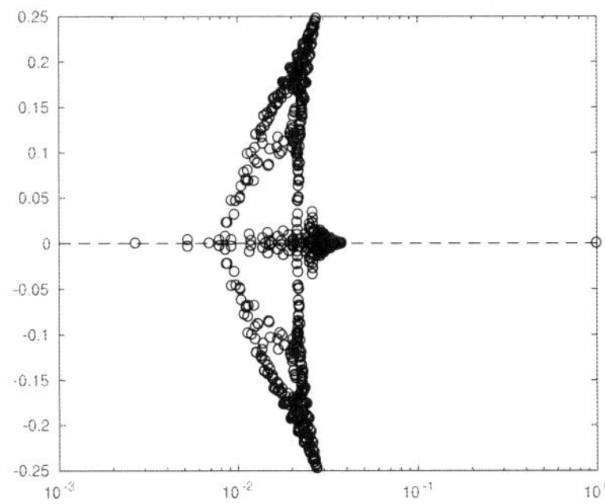

Figure 4.16: Example 4.15. Spectrum of the transformed matrix $\hat{\mathcal{M}}$ for the Lid Cavity Flow problem with leaky conditions.

contained in the right part of the complex plane. Nevertheless, we stress that also this case is not the best for the convergence of Krylov subspace methods since the eigenvalues are far from being clustered. ✓

A stronger result can be obtained for saddle point matrices $\mathcal{M}$ in the following form:

$$\mathcal{A}_\eta^{\pm} = \begin{bmatrix} \eta I & B^T \\ \pm B & 0 \end{bmatrix}, \ \mathbb{R} \ni \eta > 0, \tag{4.65}$$

Theorem 4.32 (Fischer et al. [234]). *Suppose that the matrix B has rank $m-r$ and that we order the $m-r$ nonzero singular values of B as $\sigma_{m-r} \geq \ldots \geq \sigma_1$. We have:*

- *The $n+m$ eigenvalues of the matrix $\mathcal{A}_\eta^+$ are*
  1. *$\lambda = 0$ with multiplicity r,*
  2. *$\lambda = \eta$ with multiplicity $n-m+r$,*
  3. *$\lambda_k = 1/2\left(\eta \pm \sqrt{4\sigma_k^2 + \eta^2}\right)$ for $k = 1, \ldots, m-r$.*
- *If $\sigma_1 \leq \ldots \leq \sigma_t \leq \eta/2 < \sigma_{t+1} \leq \ldots \leq \sigma_{m-r}$, then the $n+m$ eigenvalues of $\mathcal{A}_\eta^-$ are given by*
  1. *$\lambda = 0$ with multiplicity r,*
  2. *$\lambda = \eta$ with multiplicity $n-m+r$,*
  3. *$\lambda_k = \left(\eta \pm \sqrt{\eta^2 - 4\sigma_k^2}\right)/2$ for $k = 1, \ldots, t$,*
  4. *$\lambda_k = \left(\eta \pm i\sqrt{4\sigma_k^2 - \eta^2}\right)/2$ for $k = t+1, \ldots, m-r$.*

Example 4.16. Let us consider again the matrix from the mixed finite element discretization of the Darcy problem in Example 4.13. We use the same matrix B and substitute the matrix A with ηI. We report the resulting spectrum in Figure 4.17. The eigenvalues of $\mathcal{A}_\eta^+$ all lie in two intervals that are symmetric

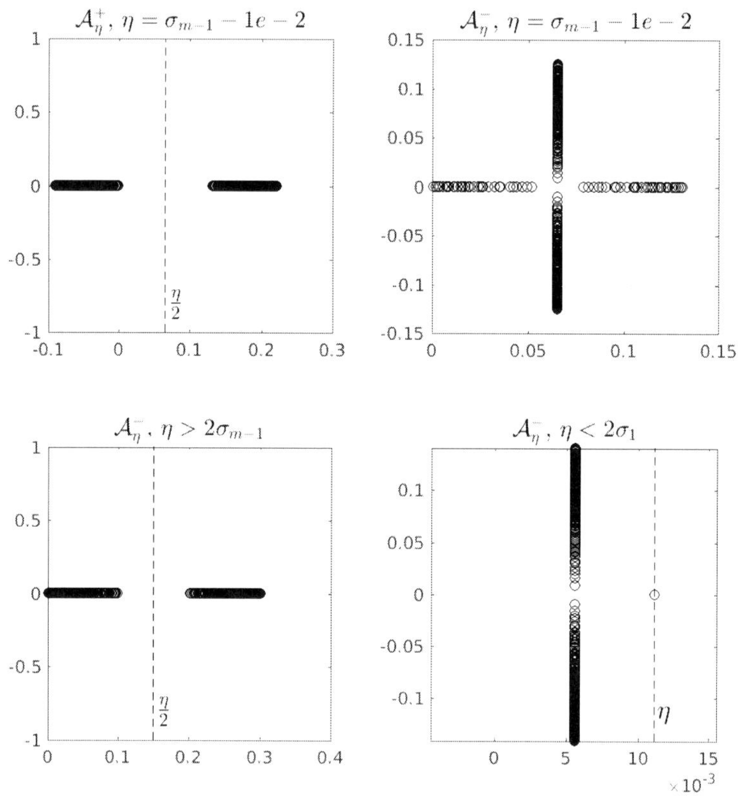

Figure 4.17: Example 4.16. Eigenvalue distribution of $\mathcal{A}_\eta^\pm$ for various η and B from the MFEM discretization of the Darcy's problem.

about the point $\eta/2$ and changing η simply changes the scale. On the other hand, for $\mathcal{A}_\eta^+$, if $\eta < 2\sigma_1$, all of the eigenvalues lie on a vertical segment in the complex plane (excepting the eigenvalues at 0 and η), while if $\eta > 2\sigma_{m-1}$ all the eigenvalues of $\mathcal{A}_\eta^-$ are real and we again have two distinct intervals that are symmetric about $\eta/2$. For all the other intermediate values of η, the eigenvalues of $\mathcal{A}_\eta^-$ lie on a cross in the complex plane. ✓

We return to these results when discussing the preconditioners for Krylov methods.

Let us conclude trying to identify the ones characterizing the condition number of these matrices.

Since we are focused on the solution of the saddle point linear system (4.55), if one between **f** and **g** is the zero vector, we are not interested in the condition number of all the blocks of the saddle point matrix.

Let us start from the case in which **S1**, **S3** and **S5** hold together with the requirement of A being positive definite. Recall that in this case the saddle point matrix $\mathcal{M}$ is symmetric and its spectral condition number is

$$\kappa_2(\mathcal{M}) = \frac{\max_j |\lambda_j(\mathcal{M})|}{\min_j |\lambda_j(\mathcal{M})|}.$$

Therefore, $\kappa_2(\mathcal{M})$ grows unboundedly whenever $\mu_1 = \lambda_{\min}(A)$ or $\sigma_1 = \sigma_{\min}(B)$ goes to zero while $\lambda_{\max}(A)$ and $\sigma_{\max}(B)$ are held fixed. This is the case, e.g., of mixed finite elements. As the discretization parameter h goes to zero, the condition number of $\mathcal{M}$ grows as $O(h^{-p})$ for some positive value of p depending on the finite elements.

A different type of ill–conditioning we need to face is the one coming from interior point methods; see [240, 401, 513]. In this case, the matrix A is a diagonal matrix with positive entries that tend to zero as the iterates of the interior points method reach convergence. Therefore, the overall matrix $\mathcal{M}$ becomes more and more ill–conditioned as the optimization method progresses toward a solution, since A is becoming singular. The following proposition characterizes the norm of the inverse of the Schur complement, the *weighted pseudoinverse* of B^T and the *oblique projector* on the column space of B^T.

Proposition 4.10 (Stewart [491]). *Let B^T be of full column rank and let $A \in \mathcal{D}^+$ where $\mathcal{D}^+$ is the set of diagonal matrices with positive entries. Then if $(B_A^T)^\dagger$ is the* weighted pseudoinverse *of B^T, that is*

$$(B_A^T)^\dagger = (BAB^T)^{-1}BA,$$

and P_A is the oblique projector *on the column space of B^T, that is*

$$P_A = B^T(B_A^T)^\dagger,$$

then there exists a number $\rho > 0$ such that

$$\sup_{A\in\mathcal{D}^+} \|P_A\| \leq \rho^{-1} \text{ and } \sup_{A\in\mathcal{D}^+} \|(B_A^T)^\dagger\| \leq \rho^{-1}\|(B^T)^\dagger\|.$$

A similar result holds for the more general case when the A block is singular. The latter can be treated with the use of the *augmented Lagrangian method.* If **S1**, **S3** and **S5** hold with B with full rank we can replace the saddle point system (4.55) with the equivalent system

$$\begin{bmatrix} A + B^TWB & B^T \\ B & 0 \end{bmatrix} \begin{bmatrix} \mathbf{x} \\ \mathbf{y} \end{bmatrix} = \begin{bmatrix} \mathbf{f} + B^TW\mathbf{g} \\ \mathbf{g} \end{bmatrix}, \; W \in \mathbb{R}^{m\times m},\; W = W^T,\; W \geq 0.$$

W can be suitably determined. Consider the following three possible choices for W from [270].

Scaling based on norms. We can choose $W = \gamma I$ with $\gamma = \|A\|_2/\|B\|_2^2$ to force the norm of A to be of the same magnitude of the norm of BWB^T. This is expected to cause a significant change in the spectrum and in the condition number of the matrix $A + BWB^T$ with respect to the condition of A.

Sparsity consideration. The sparsity pattern of B can be considerably different from the one of A. Thus, it can be desirable to make the first block non singular by reducing the possible changes to its sparsity pattern. To achieve this, we can choose a matrix W with only 1 and 0 entries. In this way, we add a correction term $BWB^T = \tilde{B}\tilde{B}^T$, where $\tilde{B}$ is a matrix composed with some columns of B.

Desired numerical properties. Choose W to obtain a positive definite W block. This is an approach tuned on KKT matrices that uses Lagrange multipliers and penalty terms, e.g., [302].

To estimate the condition number of the modified matrix, we start delineating a connection between the matrix of the original problem and the modified matrix

$$\mathcal{K}(W) = \begin{bmatrix} A + B^TWB & B^T \\ B & 0 \end{bmatrix}, \; W \in \mathbb{R}^{m\times m}, \; W = W^T, \; W \geq 0. \qquad (4.66)$$

To simplify the notation, denote $\mathcal{K}(0)$ simply as $\mathcal{K}$.

Proposition 4.11 (Fletcher [237]). *Let $A \in \mathbb{R}^{n\times n}$, $B \in \mathbb{R}^{m\times n}$ be of full column rank and $W \in \mathbb{R}^{m\times m}$. Then, for any $W \neq 0$ such that $\mathcal{K}(W)$ is nonsingular,*

$$\mathcal{K}^{-1}(W) = \mathcal{K}^{-1} - \begin{bmatrix} 0 & 0 \\ 0 & W \end{bmatrix}.$$

We get a bound for the condition number for the matrix $\mathcal{K}(W)$ by the following corollary.

Corollary 4.2 (Golub and Greif [270]). *The condition number of $\mathcal{K}(W)$ satisfies*

$$\kappa_2(\mathcal{K}(W)) \leq \kappa_2(\mathcal{K}) + \|W\|_2\|B\|_2^2\left(\|\mathcal{K}^{-1}\|_2 + \|W\|_2\right) + \|\mathcal{K}\|_2\|W\|_2.$$

Proof. We have $\|\mathcal{K}(W)\|_2 \leq \|\mathcal{K}\|_2 + \|W\|_2\|B\|_2^2$. From Proposition 4.11, $\|\mathcal{K}^{-1}(W)\|_2 \leq \|\mathcal{K}^{-1}\|_2 + \|W\|_2$. By putting the two inequalities together we obtain the claim. □

By means of these results we can readily observe that if $W = \gamma I$ then,

$$\frac{\kappa_2(\mathcal{K}(\gamma))}{\gamma^2} \overset{\gamma\to+\infty}{\longrightarrow} \|B\|_2^2.$$

If **S1** property holds together with A being positive definite, we can write

$\mathcal{K}(\gamma)$ in a useful way. Let $A = FF^T$. The singular value decomposition of $F^{-1}B^T$ can be expressed as

$$F^{-1}B^T = U\Sigma V^T,$$

where $U \in \mathbb{R}^{n\times n}$, $\Sigma \in \mathbb{R}^{n\times m}$ with its last $n-m$ rows identically zero and $V \in \mathbb{R}^{m\times m}$. If we let $G = FU$, we have $B^T = G\Sigma V^T$ and thus $B^TB = G\Sigma\Sigma^TG^T$. Since U is orthogonal,

$$A = FF^T = FUU^TF^T = GG^T,$$

and

$$(F^{-1}B^T)(F^{-1}B^T)^T = F^{-1}B^TBF^{-T} = U\Sigma\Sigma^TU^T.$$

Then, since the eigenvalues of $F^{-1}B^TBF^{-T}$ are equal to those of $F^{-1}F^{-T}B^TB = A^{-1}B^TB$, we have the generalized eigenvalues σ^2 of

$$\sigma^2 A\mathbf{x} = B^TB\mathbf{x}. \tag{4.67}$$

Since Σ is an $m\times n$ matrix, the matrix $\Sigma\Sigma^T$ cannot have more than m nonzeros ($\ker(B)$ forms a linear space for the zero generalized eigenvalues). With this we can now write $\mathcal{K}(\gamma)$ as

$$\begin{bmatrix} A+\gamma B^TB & B^T \\ B & 0 \end{bmatrix} = \begin{bmatrix} GG^T+\gamma G\Sigma\Sigma^TG^T & G\Sigma V^T \\ V\Sigma^TG^T & 0 \end{bmatrix}$$
$$= \begin{bmatrix} G & \\ & V \end{bmatrix} \begin{bmatrix} I+\gamma\Sigma\Sigma^T & \Sigma^T \\ \Sigma & 0 \end{bmatrix} \begin{bmatrix} G^T & \\ & V^T \end{bmatrix}$$

since

$$\left\| \begin{bmatrix} G & \\ & V \end{bmatrix} \right\| = \left\| \begin{bmatrix} FU & \\ & V \end{bmatrix} \right\| = \max\{\|FU\|, \|V\|\} = \max\{\|F\|, 1\},$$

and similarly for the inverse, we find that

$$\kappa_2(\mathcal{K}(\gamma)) \le \alpha\kappa_2\left(\begin{bmatrix} I+\gamma\Sigma\Sigma^T & \Sigma^T \\ \Sigma & 0 \end{bmatrix} \right),$$

where α depends on $\|F\|$ and $\|F^{-1}\|$ but not on γ. The structure of the matrix

$$T(\gamma) = \begin{bmatrix} I+\gamma\Sigma\Sigma^T & \Sigma^T \\ \Sigma & 0 \end{bmatrix},$$

is quite simple: on the main diagonal the first m entries are $\{1+\gamma\sigma_i^2\}$ for $i = 1,\ldots,m$, while the next $n-m$ are all equal to 1 and the last m are zero. On the nth super and subdiagonal we have the values σ_i, for $i = 1,\ldots,m$. From this decomposition its easy to determine a bound for its condition number.

Lemma 4.2 (Golub and Greif [270]). *Let σ_{min}^2 and σ_{max}^2 be the minimal and maximal generalized eigenvalues of (4.67). For γ sufficiently large, the condition number of $T(\gamma)$ satisfies*

$$\kappa_2(T(\gamma)) \approx \frac{(1+\gamma\sigma_{min}^2)((1+\gamma\sigma_{max}^2)^2+\sigma_{max}^2)}{\sigma_{min}^2(1+\gamma\sigma_{max}^2)}.$$

Moreover, the m negative eigenvalues of $T(\gamma)$ cluster around $-1/\gamma$ as γ increases.

Remark 4.12. To study the inertia of saddle point matrices, or to provide an estimate of their eigenvalues, we could compute a specialized factorization of the indefinite matrix that puts into evidence these properties. Several approaches of this kind have been proposed during the years, like the specialized LDL^T factorization with L lower triangular and D a diagonal block matrix with 1×1 and 2×2 blocks, from which the inertia can be readily computed, see, e.g., [116, 117]. The latter requires a careful handling of the pivoting strategy to be backward stable. Other approaches use the Sturm sequences to give exactly the inertia of the matrix starting from the matrix T being either the tridiagonal matrix of the UTU^T decomposition, for U orthogonal, or the lower triangular matrix of the LTL^T factorization, for L unit lower triangular. Nevertheless, to get an approximation of the eigenvalues, a viable factorization among these is the UTU^T, since it is a similarity transform. Another approach, introduced by Mastronardi and Van Dooren [379], develops a backward stable algorithm for achieving a symmetric factorization QMQ^T for any symmetric matrix, where Q is orthogonal and M block lower antitriangular, i.e., blocks above the main antidiagonal are zeros. This is particularly appealing since it solves and updates are achieved in $O(n^2)$ in the dense case while, in the sparse case, $O(n)$ complexity can be enough, see [346, 380].

Before focusing on iterative solvers for saddle point problems, let us recall that their solution algorithms can be subdivided into two subcategories.

Segregated methods we compute the two unknown vectors, $\mathbf{x}$ and $\mathbf{y}$, separately (first $\mathbf{x}$ and then $\mathbf{y}$ or conversely). Usually this approach involves the solution of two smaller linear systems of size less than $n+m$, one for each of $\mathbf{x}$ and $\mathbf{y}$, while in some other cases a reduced system for an intermediate quantity is required. We have already seen a first example of this class with the Generalized LSQR method (Algorithm 4.7);

Coupled methods we deal with the saddle point system as a whole without recurring to reduced systems. Then an approximation of $\mathbf{x}$ and $\mathbf{y}$ is obtained simultaneously.

We observe that, in general, the distinction is not always sharp. As we have already remarked, often direct solvers can be used to construct effective preconditioners, e.g., see §§ 3.4 and 3.5, and also preconditioners for coupled iterative schemes are frequently based on segregated solvers.

Exercise 4.4. Complete the proof for the eigenvalues of $\mathcal{A}_\eta^-$ of Theorem 4.32.

Exercise 4.5. Prove Proposition 4.11 by using either the Sherman-Morrison-Woodbury formula or the explicit expression of the inverse of the block matrix.

4.4.3 Stationary iterations for saddle point systems

Let us begin our discussion on iterative solvers for saddle point linear systems with stationary iterative methods. Recall that today usually the latter are used mostly as preconditioners for other iterative methods, see, e.g., § 3.3, and as a smoothers for some multigrid algorithms as well, see, e.g., § 4.2, even though they have been popular for many years as standalone solvers. Among the solution methods we have discussed in § 2.1.1 both the SOR and block–SOR methods have been proposed for the solution of saddle point problems, see, e.g., [34, 46, 494]. Nevertheless, since we discussed them in the general context of § 2.1.1 we start from a couple of stationary methods specialized for the saddle point problem case.

The first solver of this class is the *Arrow–Hurwicz–Uzawa* iteration [15], that was originally developed for solving a concave programming problem in economics. In its basic form this algorithm can be expressed as a fixed point iteration coming from a matrix splitting, see § 2.1.1. Let us start from the case in which **S1**, **S3** and **S5** hold with B with full rank and A nonsingular. Consider the following splitting for the saddle point matrix $\mathcal{M}$, where $\omega > 0$ is a relaxation parameter:

$$\mathcal{M} = \mathcal{P} - \mathcal{Q} = \begin{bmatrix} A & 0 \\ B & -\frac{1}{\omega}I \end{bmatrix} - \begin{bmatrix} 0 & -B^T \\ 0 & -\frac{1}{\omega}I \end{bmatrix}.$$

We get the fixed point iteration

$$\mathbf{u}^{(k+1)} = \mathcal{P}^{-1}\mathcal{Q}\mathbf{u}^{(k)} + \mathcal{P}^{-1}\mathbf{b}, \quad \mathbf{u}^{(k)} = \begin{bmatrix} \mathbf{x}^{(k)} \\ \mathbf{y}^{(k)} \end{bmatrix}.$$

Therefore, the iteration matrix for this splitting method is

$$\mathcal{G} = \mathcal{P}^{-1}\mathcal{Q} = \begin{bmatrix} 0 & -A^{-1}B^T \\ 0 & I - \omega BA^{-1}B^T \end{bmatrix},$$

and thus its eigenvalues are all real and (at least) n of them are exactly zero. If we rewrite it into its components, we derive the following coupled iteration for $\mathbf{x}^{(k)}$ and $\mathbf{y}^{(k)}$

$$\begin{cases} \mathbf{x}^{(k+1)} = A^{-1}(\mathbf{f} - B^T\mathbf{y}^{(k)}), \\ \mathbf{y}^{(k+1)} = \mathbf{y}^{(k)} + \omega(B\mathbf{x}^{(k+1)} - \mathbf{g}). \end{cases} \tag{4.68}$$

If we eliminate $\mathbf{x}^{(k+1)}$ from the second recurrence, we get

$$\mathbf{y}^{(k+1)} = \mathbf{y}^{(k)} + \omega(BA^{-1}\mathbf{f} - \mathbf{g} - BA^{-1}B^T\mathbf{y}^{(k)}),$$

that is the Richardson iteration (see § 2.1.3.4) applied to the Schur complement system

$$BA^{-1}B^T\mathbf{y} = BA^{-1}\mathbf{f} - \mathbf{g}.$$

We know how to choose the optimal ω parameter for this method, see (2.16), in terms of the eigenvalues of the matrix $BA^{-1}B^T$, and how to implement a strategy with a variable choice of the parameter (2.19). Obviously, these choices bring with them the limitation we have discussed in the application of the Richardson method. However, there is a noticeable example in which this method is fairly effective, the Stokes problem, see, e.g., [345], in which the Arrow–Hurwicz–Uzawa method converges with a rate that is independent from the discretization step h. For details see Example 4.17.

Example 4.17 (The Stokes Problem). Let us consider the following differential problem

$$\begin{cases} -\nu\nabla^2\mathbf{u} + \nabla p = \mathbf{f}, & \text{in } \Omega \subset \mathbb{R}^2, \\ \nabla \cdot \mathbf{u} = 0, & \text{in } \Omega, \\ \mathbf{u} = 0, & \text{on } \Gamma_1 = \partial\Omega. \end{cases},$$

where $\mathbf{u} = (u_1, u_2)$ and p denote, respectively, the velocity field and the pressure of a flow, $\mathbf{f} = (f_1, f_2)$ is the density of the body forces. The viscosity ν is here assumed constant. This problem admits the weak formulation

$$\text{find } u \in V, p \in M \; : \; \begin{cases} a(\mathbf{u},\mathbf{v}) + b(\mathbf{v},p) =< \mathbf{f},\mathbf{v} >, & \forall v \in V, \\ b(\mathbf{u},q) =< g, q >_M, & \forall q \in M. \end{cases},$$

where $V = \{\mathbf{v} = (v_1, v_2) \; : \; v_i \in \mathbb{H}^1(\Omega),\, v_i = 0 \text{ on } \partial\Omega\}$ and $M = \mathbb{L}^2(\Omega)/\mathbb{R}$. Thus, $< \cdot,\cdot >$ represents the duality pairing between V and its dual V^*, and $< \cdot,\cdot >_M$ coincides with the usual $\mathbb{L}^2$ inner product since M can be identified with its dual M^*. For our Stokes problem

$$a(\mathbf{u},\mathbf{v}) = \sum_{i=1}^{2} \nu \int_\Omega \nabla u_i \cdot \nabla v_i dx = \sum_{i,j=1}^{2} \int_\Omega \frac{\partial u_i}{\partial x_j}\frac{\partial v_i}{\partial x_j} dx,$$

$$b(\mathbf{v},q) = -\int_\Omega q\nabla\cdot\mathbf{v}dx, \quad < f,\mathbf{v} >= \sum_{i=1}^{2}\int_\Omega f_i v_i\, dx, \quad g = 0.$$

We get the saddle point linear system by the Galerkin approach for finite element discretization. Fix two Euclidean vector spaces of approximation V_h and M_h for the vectors $\mathbf{v}_h$ and $\mathbf{q}_h$ to get

$$\begin{bmatrix} A_h & B_h^T \\ B_h & 0 \end{bmatrix} \begin{bmatrix} \mathbf{u}_h \\ \mathbf{p}_h \end{bmatrix} = \begin{bmatrix} \mathbf{f}_h \\ 0 \end{bmatrix},$$

where, given the usual global Lagrangian finite element basis $\{\phi_h^i\}_{i=1}^n$ and $\{\psi_h^k\}_{k=1}^m$ such that $V_h = \text{span}\{\phi_h^i\}_{i=1}^n$ and $W_h = \text{span}\{\psi_h^k\}_{k=1}^m$,

$$(A_h)_{i,j} = a(\phi_h^i, \phi_h^j), \;\; i,j = 1,\ldots,n, \;\; (B_h^T)_{i,k} = b(\phi_h^i, \psi_h^k), \;\; \begin{matrix} i = 1,\ldots,n, \\ k = 1,\ldots,m \end{matrix}$$

$$(\mathbf{f}_h)_i =< \mathbf{f}, \phi_h^i >, \;\; i = 1,\ldots,n.$$

In this case we can give an estimate for the smallest and the largest eigenvalues, $\lambda_{\min}$ and $\lambda_{\max}$, of $BA^{-1}B^T$ with two constant $\Delta_1, \Delta_2 > 0$ as $\lambda_{\min} \geq \Delta_1 h^2$ and $\lambda_{\max} \leq \Delta_2 h^2$ which are independent from h. Let us start estimating the largest eigenvalue. Using the definition of the operator A and B^T and the continuity of the bilinear form $b(\cdot,\cdot)$ we have

$$\begin{aligned} \exists\, c_2 > 0 : < BA^{-1}B^T\mathbf{q}, \mathbf{q} >^2 = & < B^T\mathbf{q}, A^{-1}B^T\mathbf{q} >^2 = b^2\left(A^{-1}B^T q_h, q_h\right) \\ \leq & c_2^2 \|q_h\|_{0,\Omega}^2 \|A^{-1}B^T q_h\|_{1,\Omega}^2. \end{aligned}$$

Then, by using the fact that $a(\cdot,\cdot)$ is coercive,

$$< A\mathbf{u}, \mathbf{u} > \geq c_3 \|u_h\|_{1,\Omega}^2,$$

and that the discrete Babuška–Brezzi condition (see [109]), that is

$$\exists\, c_4 > 0 : \sup_{\substack{v_h \in V_h \\ v_h \neq 0}} \frac{b(v_h, q_h)}{\|v_h\|_{1,\Omega}} \geq c_4 \|q_h\|_{0,\Omega}, \quad \forall q_h \in M_h,$$

$$\begin{aligned} \|A^{-1}B^T\mathbf{q}_h\|_{1,\Omega}^2 \leq & c_3^{-1} < AA^{-1}B^T\mathbf{q}, A^{-1}B^T\mathbf{q} > \\ = & c_3^{-1} < BA^{-1}B^T\mathbf{q}, \mathbf{q} > . \end{aligned}$$

Thus, combining all together, we find

$$< BA^{-1}B^T\mathbf{q}, \mathbf{q} > \leq c_2^2 c_3^{-1} c_5 h^2 < \mathbf{q}, \mathbf{q} > \ \forall \mathbf{q} \in M^E.$$

Using the usual variational definition of the largest eigenvalues we get

$$\lambda_{\max} = \sup_{\substack{\mathbf{q} \in M \\ \mathbf{q} \neq 0}} \frac{< BA^{-1}B^T\mathbf{q}, \mathbf{q} >}{< \mathbf{q}, \mathbf{q} >} \leq c_2^2 c_3^{-1} c_5 h^2 = \Delta_2 h^2.$$

We can proceed similarly also for the smallest eigenvalue. Start observing that for any $\mathbf{q} \in M$ there exists $c_7 > 0$ such that

$$\begin{aligned} c_7 h^2 < \mathbf{q}, \mathbf{q} > \leq \|q_h\|_{0,\Omega}^2 \leq & \left[c_4^{-1} \sup_{\substack{v_h \in V_h \\ v_h \neq 0}} \frac{b(v_h, q_h)}{\|v_h\|_{1,\Omega}} \right]^2 \leq c_4^{-2} \sup_{\substack{v_h \in V_h \\ v_h \neq 0}} \frac{b(v_h, q_h)^2}{c_1^{-1} a(v_h, v_h)} \\ \leq & c_1 c_4^{-2} \sup_{\substack{\mathbf{v} \in V \\ \mathbf{v} \neq 0}} \frac{< B\mathbf{v}, \mathbf{q} >^2}{< A\mathbf{v}, \mathbf{v} >} = c_1 c_4^{-2} \sup_{\substack{\mathbf{w} \in V \\ \mathbf{w} \neq 0}} \frac{< BA^{-1}\mathbf{w}, \mathbf{q} >^2}{< A^{-1}\mathbf{w}, \mathbf{w} >} \end{aligned}$$

where the last equality is obtained by substituting $\mathbf{v} = A^{-1}\mathbf{w}$. Now, decompose the space V into two subspaces, $\text{Span}(B^T)$ and its orthogonal with respect to A^{-1}. Observe that this also induces the inner energy product with respect to the matrix A^{-1}. Thus, any vector admits a unique (Riesz) representation $\mathbf{w} \in V$ as the sum of a vector in span(B^T) and one in its orthogonal, that is $\mathbf{w} = \mathbf{w}_1 + \mathbf{w}_2$ with $< \mathbf{w}_1, \mathbf{w}_2 >_{A^{-1}} = 0$. Therefore,

$$\frac{<BA^{-1}\mathbf{w}, \mathbf{q}>^2}{<A^{-1}\mathbf{w}, \mathbf{w}>} = \frac{<BA^{-1}(\mathbf{w}_1 + \mathbf{w}_2), \mathbf{q}>^2}{<A^{-1}\mathbf{w}_1, \mathbf{w}_1> + <A^{-1}\mathbf{w}_2, \mathbf{w}_2>} \leq \frac{<A^{-1}\mathbf{w}_1, B^T\mathbf{q}>^2}{<A^{-1}\mathbf{w}_1, \mathbf{w}_1>}.$$

Substituting again $\mathbf{w}_1 = B^T\mathbf{r}$ and using Cauchy's inequality we obtain

$$\begin{aligned}\frac{< A^{-1}\mathbf{w}_1, B\mathbf{q} >^2}{< A^{-1}\mathbf{w}_1, \mathbf{w}_1 >} &= \frac{< A^{-1}B^T\mathbf{r}, B\mathbf{q} >^2}{< A^{-1}B^T\mathbf{r}, B^T\mathbf{r} >} = \frac{< BA^{-1}B^T\mathbf{r}, \mathbf{q} >^2}{< BA^{-1}B^T\mathbf{r}, \mathbf{r} >} \\ &= \frac{< \left(BA^{-1}B^T\right)^{1/2}\mathbf{r}, \left(BA^{-1}B^T\right)^{1/2}\mathbf{q} >^2}{< \left(BA^{-1}B^T\right)^{1/2}\mathbf{r}, \left(BA^{-1}B^T\right)^{1/2}\mathbf{r} >} \\ &\leq \frac{\| \left(BA^{-1}B^T\right)^{1/2}\mathbf{r}\|^2 \| \left(BA^{-1}B^T\right)^{1/2}\mathbf{q}\|^2}{\| \left(BA^{-1}B^T\right)^{1/2}\mathbf{r}\|^2}\end{aligned}$$

Therefore we proved that

$$\frac{< BA^{-1}\mathbf{w}, \mathbf{q} >^2}{< A^{-1}\mathbf{w}, \mathbf{w} >} \leq \| \left(BA^{-1}B^T\right)^{1/2}\mathbf{q}\|^2 =< BA^{-1}B^T\mathbf{q}, \mathbf{q} > .$$

Using again the variational definition of the smallest eigenvalue, we find

$$\lambda_{\min} = \inf_{\substack{\mathbf{q}\in M \\ \mathbf{q}\neq 0}} \frac{< BA^{-1}B^T\mathbf{q}, \mathbf{q} >}{< \mathbf{q}, \mathbf{q} >} \geq c_1^{-1}c_7c_4^2h^2 = \Delta_1 h^2.$$

Therefore, the condition number of the Schur complement associated to the underlying saddle point problem is a $O(1)$. We can select the optimal parameter for the Arrow–Hurwicz–Uzawa method, i.e., for the underlying stationary Richardson method, and obtain that it converges at a rate that is independent of h. ✓

When the rate of convergence is not optimal (the number of iterations depends on the discretization), differently to the Stokes problem in the example above, we can use the *augmented Lagrangian method* from the previous section to get rid of the ill–conditioning of the matrix A. That is, we can use the results on the matrix $\mathcal{K}(\gamma)$ from Corollary 4.2 and Lemma 4.2.

Let us start by observing that an implementation of the Uzawa iteration (4.68), requires, at each step, the solution of a linear system with coefficient matrix A. We can decide to tackle this problem by using one of the iterative Krylov methods we have seen so far and indeed this has been done often, see, e.g., [32, 101, 219]. We describe our Krylov iterative algorithm as a nonlinear map Ψ that gives an approximation to the solution of the linear system $A\boldsymbol{\xi} = \boldsymbol{\phi}$ and consider a generic symmetric preconditioner Q_B on the outer iteration of the method, e.g., $Q_B = (1/\omega)I$. In this way we derive the following *incomplete*, *nonlinear* and *preconditioned* version of the Uzawa error iteration related to (4.68)

$$\begin{cases} \mathbf{x}^{(k+1)} = \mathbf{x}^{(k)} + \psi(\mathbf{f} - (A\mathbf{x}^{(k)} + B^T\mathbf{y}^{(k)})), \\ \mathbf{y}^{(k+1)} = \mathbf{y}^{(k)} + Q_B^{-1}(B\mathbf{x}^{(k+1)} - \mathbf{g}). \end{cases}, \tag{4.69}$$

For which we have the following results.

Proposition 4.12 (Bramble et al. [101]). *Consider the iteration* (4.69) *and assume*

$$< BA^{-1}B^T\mathbf{w}, \mathbf{w} > \leq < Q_B, \mathbf{w}, \mathbf{w} > \quad \forall\, \mathbf{w} \in \mathbb{R}^m,$$

and

$$\exists\, \delta < 1 : \; \|\Psi(\boldsymbol{\phi}) - A^{-1}\boldsymbol{\phi}\|_A \leq \delta \|\boldsymbol{\phi}\|_{A^{-1}} \;\; \forall\, \boldsymbol{\phi} \in \mathbb{R}^n.$$

Let γ be the convergence rate of the complete Uzawa iteration

$$\begin{cases} \tilde{\mathbf{x}}^{(k+1)} = \tilde{\mathbf{x}}^{(k)} + A^{-1}(\mathbf{f} - (A\tilde{\mathbf{x}}^{(k)} + B^T\tilde{\mathbf{y}}^{(k)})), \\ \tilde{\mathbf{y}}^{(k+1)} = \tilde{\mathbf{y}}^{(k)} + Q_B^{-1}(B\tilde{\mathbf{x}}^{(k+1)} - \mathbf{g}). \end{cases}$$

Then, $\{\mathbf{x}^{(k)}, \mathbf{y}^{(k)}\}$ *converges to the solution* $\{\mathbf{x}, \mathbf{y}\}$ *if*

$$\delta < \frac{1-\gamma}{3-\gamma}. \tag{4.70}$$

By using the convergence estimate for the Conjugate Gradient method, see Section 2.2.4, we can conclude that the *incomplete* Uzawa iteration is convergent if in the internal cycle the CG method is used. If we have an indefinite matrix A, by applying the technique with $K(W)$ from (4.66) with $W = \gamma I$, we obtain an iteration of the form (4.68)

$$\begin{cases} \mathbf{x}^{(k+1)} = (A + \gamma B^T B)^{-1}(\mathbf{f} - B^T\mathbf{y}^{(k)}), \\ \mathbf{y}^{(k+1)} = \mathbf{y}^{(k)} + \omega(B\mathbf{x}^{(k+1)} - \mathbf{g}). \end{cases}$$

Now we need to investigate the spectrum of the Schur complement.

Proposition 4.13 (Golub and Greif [270]). *Let* $S(\gamma) = B(A + \gamma B^T B)^{-1}B^T$. *If there exists a scalar* $\gamma \geq \nu \geq 0$ *such that* $A + \nu B^T B$ *is positive definite and the eigenvalues of* $S(\nu)$ *are* $\{\mu_i(\nu)\}_i$, *then the eigenvalues of* $S(\gamma)$ *are*

$$\mu_i(\gamma) = \frac{\mu_i(\nu)}{1 + (\gamma - \nu)\mu_i(\nu)}.$$

This result can be proved again with the Sherman–Morrison–Woodbury formula, see Exercise 4.5. For the convergence we need that the spectral radius $\rho(I - \omega S(\gamma)) < 1$ and from Proposition 4.13 the eigenvalues of $I - \omega S(\gamma)$ are given by

$$\frac{1 - (\beta - \omega)\mu_i(\nu)}{1 + \beta\mu_i(\nu)}, \quad \beta = \gamma - \nu.$$

Thus, if $\mu_i(\nu) > 0$ for all i, and hence $1 + \beta\mu_i(\nu) > 0$ (assuming B has full rank), we find that

$$0 < \omega < 2\beta + \frac{2}{\mu_{\max}(\nu)},$$

that is satisfied for any arbitrary set of eigenvalues $\{\mu_i(\nu)\}_i$ if $0 < \omega \leq 2\beta$. Obviously, one wants to find the minimal value of ν that yields a positive definite matrix $A + \nu B^T B$ to get a range for ω that is as large as possible. Observe that if A is positive semidefinite, since the matrix $\mathcal{K}(0)$ is nonsingular, then $A + \nu B^T B$ is positive definite for any $\nu > 0$. On the other hand, if A is singular, we just need to solve a smaller problem that excludes its null spaces. Therefore, we look for an orthogonal transformation $Q = [Q_1, Q_2]$ such that $Q_1^T A Q_1$ is nonsingular, then compute the solution of the reduced system

$$\begin{bmatrix} \hat{\mathcal{K}} & U^T \\ U & \hat{A}_2 \end{bmatrix} \begin{bmatrix} \begin{bmatrix} Q_1^T \mathbf{x} \\ \mathbf{y} \end{bmatrix} \\ Q_2^T \mathbf{y} \end{bmatrix} = \begin{bmatrix} \begin{bmatrix} Q_1^T \mathbf{f} \\ \mathbf{g} \end{bmatrix} \\ Q_2^T \mathbf{f} \end{bmatrix}, \text{with } \hat{\mathcal{K}} = \begin{bmatrix} Q_1^T A Q_1 & Q_1^T B^T \\ B Q_1 & 0 \end{bmatrix},$$

where

$$U^T = \begin{bmatrix} Q_1^T A Q_2 \\ B Q_2 \end{bmatrix}.$$

By eliminating $\mathbf{x}_2 = Q_2^T \mathbf{y}$ through the solution of the Schur complement system, we find

$$(Q_2^T A Q_2 - U^T \hat{\mathcal{K}}^{-1} U)\mathbf{x}_2 = Q_2^T \mathbf{y} - U^T \hat{\mathcal{K}}^{-1} [Q_1^T \mathbf{f}, \mathbf{g}]^T$$

obtaining $\mathbf{z} = [Q_1^T \mathbf{x}, \mathbf{y}]$ as the solution of

$$\hat{\mathcal{K}}\mathbf{z} = [Q_1^T \mathbf{f}, \mathbf{g}]^T - U\mathbf{x}_2,$$

and then $\mathbf{x}$ and $\mathbf{y}$. In this way we eliminate the singularity of the A block by reducing its size by exact its nullity, a deflation that leaves the solution space unchanged. Thus, the smaller the nullity, the lower is the computational cost of this procedure. Moreover, we can solve the reduced problem by using the *incomplete* Uzawa iteration we have discussed before, since $U^T \hat{\mathcal{K}}^{-1} U$ is a quadratic form for which the Lanczos procedure is feasible, see § 2.2.2. The choice of a preconditioner for the inner iteration depends greatly on the problem at hand, i.e., on the spectral properties of the A matrix and of the overall Schur complement system. Some possible choices are the preconditioner by Cahouet and Chabard [121] and using Fourier analysis to investigate the properties of the Green function associated with the Stokes operator to produce a preconditioner, see also [265, Chapter 4.2.1] for further details, or the preconditioner by Kobelkov and Olshanskii [331] for the generalized Stokes problem which requires only the solution of a Poisson equation with Neumann boundary conditions for each iteration.

We conclude the discussion on the Arrow, Hurwicz and Uzawa method by working again on Example 4.17 with the incomplete version and the Conjugate Gradient algorithm.

Example 4.18. Let us consider again the steady Stokes problem from Example 4.17. We consider its solution by choosing as the P2–P1 finite elements for the approximation of the velocity and pressure functions, respectively and the domain Ω is the square $\Omega = [0, 1] \times [0, 1]$. All the computations are performed with the FreeFem++ software [296]. We use the Conjugate Gradient algorithm for the solution of the linear system within the internal cycle and give in Table 4.10 the overall number of iteration for mesh with increasing number of degree of freedom (dof), i.e., for linear systems of increasing size. As we can observe, the theoretical results in Example 4.17 hold and the number of iterations needed to reach convergence is a constant independent from the size of the discretization. The pattern of the saddle point matrix is given in Figure 4.18. ✓

Table 4.10: Example 4.18. Incomplete Uzawa iteration with unpreconditioned CG algorithm for steady state Stokes problem

dof	iteration	dof	iteration
121	41	3721	42
441	44	5041	41
961	44	6561	41
1681	44	8821	41
2601	43	10201	36

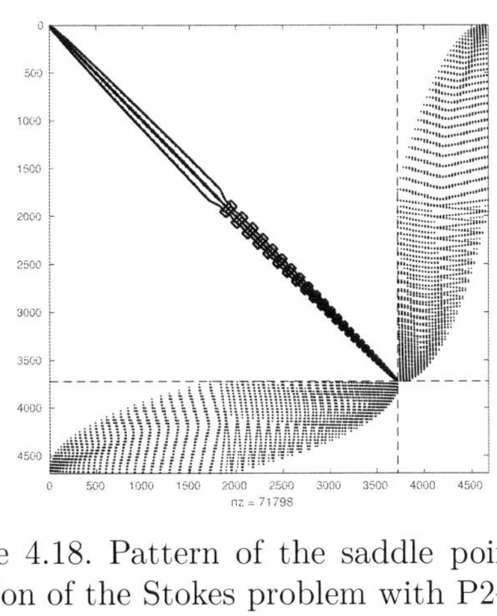

Figure 4.18: Example 4.18. Pattern of the saddle point matrix for the discretization of the Stokes problem with P2–P1 finite elements.

Applications and results obtained with the use of preconditioned Krylov methods for the internal iterations of the incomplete Uzawa iteration (4.69) can be found in [80] for the solution of the 3D steady–state and unsteady–state Stokes equations. The use of the GMRES method for a non symmetric matrix A is discussed in [99], where a generalization of Proposition 4.12 to the non symmetric case is discussed. An application similar to the one we discussed in Example 4.17 but with spectral element discretization is faced in [371] and again in [402] for stable finite elements. For other variants of this algorithm see also [27, 128, 278, 426, 539, 558].

When the solution of linear systems with coefficient matrix A is too expensive and we want to remain in the setting of stationary iteration methods, we can use the *Arrow–Hurwicz method* [13, 14], that is obtained by the splitting

$$\mathcal{M} = \mathcal{P} - \mathcal{Q} = \begin{bmatrix} {}^{1}/_{\alpha} I & 0 \\ B & -{}^{1}/_{\omega} I \end{bmatrix} - \begin{bmatrix} {}^{1}/_{\alpha} I - A & -B^T \\ 0 & -{}^{1}/_{\omega} I \end{bmatrix}, \; \alpha, \omega > 0.$$

The resulting method can be expressed by the iteration

$$\begin{cases} \mathbf{x}^{(k+1)} = \mathbf{x}^{(k)} + \alpha(\mathbf{f} - A\mathbf{x}^{(k)} - B^T \mathbf{y}^{(k)}), \\ \mathbf{y}^{(k+1)} = \mathbf{y}^{(k)} + \omega(B\mathbf{x}^{(x+1)} - \mathbf{g}). \end{cases} \tag{4.71}$$

Comparing (4.71) with the Uzawa iteration (4.68), we observe that the changes amount to the minimizer of the objective function

$$\Phi(\mathbf{x}) = \frac{1}{2} \mathbf{x}^T A \mathbf{x} - \mathbf{x}^T (\mathbf{f} - B^T \mathbf{y}^{(k)}),$$

with a less computationally expensive method consisting of taking one step in the direction of $-\nabla\Phi(\mathbf{x})$ with a fixed step length α. In practice we are using a linear approximation instead of the quadratic one of the Arrow, Hurwicz and Uzawa method. By this observation, we could expect a slower rate of convergence of the Arrow–Hurwicz method. To perform the analysis for this algorithm we introduce the following notation from [440].

- Define on $V = \mathbb{R}^{nm} = \mathbb{R}^n \times \mathbb{R}^m$ a scalar product obtained by combining the inner products on $\mathbb{R}^n$ and $\mathbb{R}^m$ as the sum $< \mathbf{u}_1, \mathbf{u}_2 >_V = < \mathbf{x}_1, \mathbf{y}_1 > + < \mathbf{x}_2, \mathbf{y}_2 >$.. Define a norm on V as the energy norm for a positive definite block matrix $\mathcal{D}$ in the scalar product $< \cdot, \cdot >_V$.
- If $\mathbf{u} = [\mathbf{x}, \mathbf{y}]^T$ is the solution of the saddle point problem, define the error at the kth step of iteration (4.71) as $\mathbf{z}^{(k)} = \mathbf{x}^{(k)} - \mathbf{x}$ and $\mathbf{q}^{(k)} = \mathbf{y}^{(k)} - \mathbf{y}$. Define the vectors $\mathbf{e}^{(k)} = [\mathbf{z}^{(k)}, \mathbf{q}^{(k)}]^T$ and $\mathbf{r}^{(k)} = D^{1/2} \mathbf{e}^{(k)}$.

Lemma 4.3 (Queck [440]). *The following relations for the error* $\mathbf{e}^{(k)}$ *and* $\mathbf{r}^{(k)}$ *of the Arrrow–Hurwicz iteration* (4.71) *hold:*

$$\mathbf{e}^{(k)} = S\mathbf{e}^{(k-1)}, \quad \mathbf{r}^{(k)} = S_1 \mathbf{r}^{(k-1)}, \quad k = 1, 2, \ldots$$

where

$$S_1 = \mathcal{D}^{1/2} S \mathcal{D}^{-1/2} = I - \omega C_1, \quad S = I - \omega C, \quad C_1 = \mathcal{D}^{1/2} C \mathcal{D}^{-1/2},$$

and the operator C *is defined as the saddle point matrix*

$$C = \begin{bmatrix} {}^{\alpha}/_{\omega} A & {}^{\alpha}/_{\omega} B^T \\ -B(I - \alpha A) & \alpha B B^T \end{bmatrix}.$$

The convergence of iterations in (4.71) is equivalent to having either $\|S\| < 1$ or $\|S_1\| < 1$. Let us call $\lambda_{\min}$, $\lambda_{\max}$, $\mu_{\min}$ and $\mu_{\max}$ the minimum and maximum eigenvalues of the symmetric matrix A and $B^T B$, respectively. Then, we define the symmetric and positive definite operator $\mathcal{D}$ as

$$\mathcal{D} = \begin{bmatrix} D_n & \\ & I \end{bmatrix}, \; D_n = \begin{cases} \frac{\omega}{\alpha}(I - \alpha A), & 0 < \alpha < \lambda_{\max}^{-1}, \\ -\frac{\omega}{\alpha}(I - \alpha A), & \lambda_{\min}^{-1} < \alpha < 2\lambda_{\max}^{-1}, \end{cases} \quad \omega > 0.$$

The estimates for the norm of the amplification matrix S and S_1 are in the following Proposition; see Queck [440] for a proof.

Proposition 4.14 (Queck [440]). *Let us assume that $0 < \alpha < \lambda_{max}^{-1}$ with $\lambda_{min} < \lambda_{max}$. If there exist constants $\gamma_1, \gamma_2 > 0$ such that*

$$\begin{aligned} &< C_1\mathbf{u}, \boldsymbol{u} >_V \geq \gamma_1 < \mathbf{u}, \mathbf{u} >_V, && \forall \mathbf{u} \in V, \\ &< C_1\mathbf{u}, C_1\boldsymbol{u} >_V \leq \gamma_2 < C_1\mathbf{u}, \mathbf{u} >_V, && \forall \mathbf{u} \in V, \\ &2 - \omega\gamma_2 > 0, && \omega > 0, \end{aligned}$$

then the following estimate holds

$$\|S_1\| \leq \left[1 - \omega\gamma_1\left(2 - \omega\gamma_2\right)\right]^{1/2}.$$

Moreover, the constants γ_1 and γ_2 are such that

$$\begin{aligned} \gamma_1 \geq &\alpha \min\{\mu_{min}, \lambda_{min}\omega^{-1}\}, \\ \gamma_2 \leq &2 \max\left\{\frac{\alpha}{\omega}\lambda_{max} + \frac{\mu_{max}}{\lambda_{min}}(1 - \alpha\lambda_{min}), \alpha\mu_{max} + \frac{1}{\omega}(1 - \alpha\lambda_{min})\right\}. \end{aligned}$$

By means of this proposition, it is therefore possible to obtain the optimal choice for the parameters ω and α. Let us observe that for $0 < \alpha < \lambda_{\max}^{-1}$, $(1 - \alpha\lambda_{\max})/(1 - \omega\lambda_{\min}) < 1$. Therefore, for both the possible values of γ_2 from Proposition 4.14, $\omega < \lambda_{\min}/\mu_{\max}$ and, consequently, $\mu_{\min} \leq \mu_{\max} < \lambda_{\min}/\omega$. Thus, the condition on ω in Proposition 4.14 rewrites as $\gamma_1 \geq \omega\mu_{\min}$. By assembling all together, we can compute the values of α and ω such that the amplification function $\phi(\alpha, \omega)$, i.e., $\|S_1\| \leq \phi(\alpha, \omega)$, is strictly smaller than 1 and all is collected in Table 4.11. For the first choice of γ_2 from Table 4.11

Table 4.11: Amplification factors and parameter choices for convergent Arrow–Hurwicz iterations

$\gamma_2/2$	α	ω	ϕ
$\alpha\mu_{\max} + \frac{1}{\omega}(1 - \alpha\lambda_{\min})$	$< \frac{1}{\lambda_{\min}+\lambda_{\max}}$	$\leq \frac{\lambda_{\min}}{\mu_{\max}} \frac{1-\alpha(\lambda_{\min}+\lambda_{\max})}{1-2\alpha\lambda_{\min}} < \frac{\lambda_{\min}}{\mu_{\max}}$	ϕ_1
$\frac{\alpha}{\omega}\lambda_{\max} + \frac{\mu_{\max}}{\lambda_{\min}}(1 - \alpha\lambda_{\min})$	$< \frac{1}{\lambda_{\max}}$	< 1 $\geq \max\left\{0, \frac{\lambda_{\min}}{\mu_{\max}} \frac{1-\alpha(\lambda_{\min}+\lambda_{\max})}{1-2\alpha\lambda_{\min}}\right\}$	ϕ_2

the value of $\phi(\alpha, \omega)$ is given by

$$\phi_1(\alpha, \omega) = \left[1 - 2\omega\alpha^2\mu_{\min}(\lambda_{\min} - \omega\mu_{\max})\right]^{1/2},$$

For the second choice of γ_2 from Table 4.11 we find

$$\phi_2(\alpha,\omega) = \left[1 - 2\omega\alpha\mu_{\min}\left((1-\alpha\lambda_{\max}) - \frac{\mu_{\max}}{\lambda_{\min}}(1-\alpha\lambda_{\min})\right)\right]^{1/2}.$$

The optimal choice of the parameters, for both the choices of γ_2, is given by

$$\alpha_{\text{opt}} = \frac{3\lambda_{\min} + 2\lambda_{\max} - (4\lambda_{\max}^2 - 3\lambda_{\min}^2)^{1/2}}{4\lambda_{\min}^2 + 4\lambda_{\min}\lambda_{\max}},$$
$$\omega_{\text{opt}} = \frac{\lambda_{\min}}{\mu_{\max}}\frac{1-\alpha_{\text{opt}}(\lambda_{\min}+\lambda_{\max})}{1-2\alpha_{\text{opt}}\lambda_{\min}}.$$

the optimal parameter is therefore obtained as

$$\phi = \left[1 - 2\left(1 - \frac{\lambda_{\min}}{\lambda_{\max}}\right)\lambda_{\max}^3\alpha_{\text{opt}}^3\frac{1-\alpha_{\text{opt}}(\lambda_{\min}+\lambda_{\max})}{(1-2\alpha_{\text{opt}}\lambda_{\min})^2}\left(\frac{\lambda_{\min}}{\lambda_{\max}}\right)^2\frac{\mu_{\min}}{\mu_{\max}}\right]^{\frac{1}{2}}.$$

Therefore given $\varepsilon > 0$, the number of iterations for reducing the relative error by a factor of ε with the Arrow–Hurwicz iteration (4.71) in the optimal case is $O((\lambda_{\max}/\lambda_{\min})(\mu_{\max}/\mu_{\min})\ln(\varepsilon^{-1}))$. This implies that convergence for this method, see also the beginning of our discussion, is indeed quite slow. To overcome this issue, a preconditioned version can be adopted. The latter is based on modifying the iteration (4.71) as

$$\begin{cases} Q_A\mathbf{x}^{(k+1)} = \mathbf{x}^{(k)} + \alpha(\mathbf{f} - A\mathbf{x}^{(k)} - B^T\mathbf{y}^{(k)}), \\ Q_B\mathbf{y}^{(k+1)} = \mathbf{y}^{(k)} + \omega(B\mathbf{x}^{(x+1)} - \mathbf{g}). \end{cases},$$

with Q_A and Q_B symmetric and positive definite preconditioning matrices. Analysis obtained through Proposition 4.14 can be easily adapted to this case by substituting the original A and B^T with their preconditioned version given by $\overline{A} = Q_A^{-1/2}AQ_A^{-1/2}$ and $\overline{B}^T = Q_A^{-1/2}B^TQ_B^{-1/2}$. The choice of the preconditioning matrix, i.e, of a possible nested stationary iterative method, depends on the problem. In [163], a preconditioning strategy based on the use of the discretized Laplacian operator for the solution of the biharmonic equation with mixed finite elements is proposed. Other approaches for the Navier–Stokes problem can be found in [1, 440, 446, 492].

4.4.4 Preconditioners and Krylov subspace methods

We discussed preconditioners based on factorization § 3.4 and on multilevel methods, § 4.2. Thus, one would be tempted to extend these approaches also to the treatment of *saddle point* linear systems. Nevertheless, as we have seen in Proposition 4.9 and Theorem 4.32, matrices of saddle point problems can be indefinite and this often does not help; see also § 3.4.4. In general, there is also an absence of decay in the inverse of the underlying matrices. Therefore, *approximate inverse preconditioners* also tend to be unsatisfactory. In the following, some block preconditioners specialized for these cases are considered.

4.4.4.1 Block diagonal and block triangular preconditioners

Let us start by considering a block diagonal preconditioner for the case of an invertible (possibly nonsymmetric) saddle point matrix $\mathcal{M}$ with property **S5**, i.e., $C = 0$.

Proposition 4.15 (Murphy et al. [393]). *If $\mathcal{M}$ is preconditioned by the block diagonal preconditioner*

$$\mathcal{P} = \begin{bmatrix} A & 0 \\ 0 & B_2 A^{-1} B_1^T \end{bmatrix}, \tag{4.72}$$

then the preconditioned matrix $\mathcal{T} = \mathcal{P}^{-1}\mathcal{M} (= \mathcal{M}\mathcal{P}^{-1} = \mathcal{P}_1^{-1}\mathcal{M}\mathcal{P}_2^{-1})$, with $\mathcal{P}_1\mathcal{P}_2 = \mathcal{P}$, satisfies

$$\mathcal{T}(\mathcal{T} - I)(\mathcal{T}^2 - \mathcal{T} - I) = 0$$

and $\mathcal{T}$ is diagonalizable and has at most four distinct eigenvalues $\{0, 1, 1/2 \pm \sqrt{5}/2\}$.

By using Theorem 2.11, for any vector $\mathbf{v}$ the Krylov subspace $\mathcal{K}_m(\mathcal{T}, \mathbf{v})$ is at most of dimension three and thus any Krylov subspace iterative method satisfying an optimality or a Galerkin condition converges in at most three steps.

Similarly, one can consider a tridiagonal version of the above preconditioner.

$$\mathcal{P} = \begin{bmatrix} A & B_1^T \\ 0 & B_2 A^{-1} B_1^T \end{bmatrix}, \tag{4.73}$$

yielding only two different eigenvalues ± 1, see also [263]. Both choices yield a preconditioner that cannot be used, being more expensive than the solution by block Gaussian–elimination of the underlying saddle point system.

To reduce the cost for the block diagonal preconditioner (4.72), several approaches exist. As a first choice, exploiting what was discussed in § 3.3, we can split matrix A as in [182, 234]; see Definition 2.2. So, let us consider the splitting of the $(1,1)$–block of $\mathcal{M}$ given by

$$A = D - E, \ \det(D) \neq 0.$$

If both D and $B_2 D^{-1} B_1^T$ are invertible we can consider the preconditioner

$$\mathcal{P}_D^{-1} = \begin{bmatrix} D^{-1} & 0 \\ 0 & (B_2 D^{-1} B_1^T)^{-1} \end{bmatrix}$$

where

$$\mathcal{P}_D^{-1}\mathcal{M} = \begin{bmatrix} I_n - D^{-1}E & D^{-1}B_1^T \\ (B_2 D^{-1} B_1^T)^{-1} B_2 & 0 \end{bmatrix},$$

$$\mathcal{M}\mathcal{P}_D^{-1} = \begin{bmatrix} I_n - ED^{-1} & B_1^T (B_2 D^{-1} B_1^T)^{-1} \\ B_2 D^{-1} & 0 \end{bmatrix}.$$

Therefore we are interested in looking at the solution and the properties of the block system

$$\mathcal{B}\begin{bmatrix}\mathbf{x}\\ \mathbf{y}\end{bmatrix} \equiv \begin{bmatrix} I_n - S & N \\ M & 0\end{bmatrix}\begin{bmatrix}\mathbf{x}\\ \mathbf{y}\end{bmatrix} = \begin{bmatrix}\mathbf{f}\\ \mathbf{g}\end{bmatrix}, \quad \begin{matrix} MN = I, & (NM)^2 = NM, \\ S \in \mathcal{R}^{n\times n}, & M, N^T \in \mathcal{R}^{m\times n}, \end{matrix} \quad n \geq m,$$

having modified the original vectors to account for the application of the preconditioners. Since we obtain the "best preconditioner" (4.72) for the choice of $S = 0$, i.e., $A = D$ and $E = 0$, we can hope that the eigenvalues of $\mathcal{B}$ form a cluster. Observe that $\mathcal{B}$ is singular if and only if 1 is an eigenvalue of the matrix $(I_n - NM)S$. Indeed, if $(I_n - NM)S$ is singular, then there exists a vector $\mathbf{v} = [\mathbf{x}^T, \mathbf{y}^T]^T$ such that $(I_n - NM)S\mathbf{v} = 0$, or, equivalently, $(I_n - S)\mathbf{x} + N\mathbf{y} = 0$ and $M\mathbf{x} = 0$. Since $S\mathbf{x} - N\mathbf{y} = \mathbf{x}$ implies $\mathbf{y} = MS\mathbf{x}$, hence $\mathbf{x} = 0$ implies $\mathbf{y} = 0$. Conversely, $\mathbf{y} = MS\mathbf{x}$ implies that $(I_n - NM)S\mathbf{x} = \mathbf{x}$ and the singularity appears if and only if $(I_n - NM)S$ has an eigenvalue 1. Having clarified the existency conditions, the underlying preconditioning strategy reduces to the choice of a splitting that leads to inverting efficiently matrices D and $B_2 D^{-1} B_1^T$ preserving the properties of the preconditioner $\mathcal{P}$ from Proposition 4.15.

Proposition 4.16 (de Sturler and Liesen [182]). *Let $\mathcal{B}$ be the preconditioned matrix obtained with preconditioner $\mathcal{P}_D$ and $\mathcal{T}$ the matrix obtained with the preconditioner $\mathcal{P}$ discussed in Proposition 4.15. Let $[U_1, U_2]$ be the eigenvector of the NM matrix in $\mathcal{B}$. Then, for each matrix S and each eigenvalue λ_S of $\mathcal{B}$, there exists an eigenvalue λ of $\mathcal{T}$ such that*

$$|\lambda_S - \lambda| \leq \left[2 + \frac{2}{5}\left(\frac{1-\sqrt{5}}{2}\right)^2\right]^{1/2} \|[U_1, U_2]^{-1} S [U_1, U_2]\|.$$

Moreover, with the same notation, if σ_1 is the largest singular values of $U_1^T U_2$,

$$|\lambda_S - \lambda| \leq \left[2 + \frac{2}{5}\left(\frac{1-\sqrt{5}}{2}\right)^2\right]^{1/2} \left(\frac{1+\sigma_1}{1-\sigma_1}\right)^{1/2} \|S\|.$$

However, the preconditioned matrix $\mathcal{B}$ now has some distinct eigenvalues, potentially $n + m$, that can be on both sides of the complex plane. Therefore, some care is needed when using this preconditioner with Krylov methods. To overcome this issue, in [182] the use of an iteration of Uzawa type is proposed (see § 4.4.3) that is based again on a splitting of the preconditioned matrix $\mathcal{B} = \mathcal{B}_1 - \mathcal{B}_2$, where the inverse of $\mathcal{B}_1$ is explicitly known and, with a suitable choices of D, efficiently computed. The problem is therefore reduced to a fixed point iteration of the form:

$$\begin{bmatrix}\mathbf{x}^{(k+1)}\\ \mathbf{y}^{(k+1)}\end{bmatrix} = \mathcal{B}_1^{-1}\mathcal{B}_2 \begin{bmatrix}\mathbf{x}^{(k)}\\ \mathbf{y}^{(k)}\end{bmatrix} + \mathcal{B}_1^{-1}\begin{bmatrix}\mathbf{f}\\ \mathbf{g}\end{bmatrix}, \quad \mathcal{B}_1^{-1}\mathcal{B}_2 = \begin{bmatrix}(I + D^{-1}B_1^T \hat{S}^{-1} B_2)S & 0\\ -\hat{S}^{-1}B_2 S & 0\end{bmatrix}$$

where $\hat{S} = -B_2 D^{-1} B_1^T$ to shorten the notation. Moreover, one can further simplify the computations by considering only the component $\mathbf{x}^{(k+1)}$:

$$\begin{aligned}\mathbf{x}^{(k+1)} &= (I_n + D^{-1} B_1^T \hat{S}^{-1} B_2) S \mathbf{x}^{(k)} + \left(\mathcal{P}_1^{-1} \begin{bmatrix}\mathbf{f}\\ \mathbf{g}\end{bmatrix}\right)_{1:n}\\ &= (I_n - NM) S \mathbf{x}^{(k)} + \hat{f}.\end{aligned}$$

This is a *segregated* approach, in which, first we compute an approximate solution for $\mathbf{x}$ as the solution of the reduced problem

$$\begin{aligned}(I - (I + D^{-1} B_1^T \hat{S}^{-1} B_2) S) \to \mathbf{x} &= \left(\mathcal{P}_1^{-1} \begin{bmatrix}\mathbf{f}\\ \mathbf{g}\end{bmatrix}\right)_{1:n}\\ R\mathbf{x} = (I_n - F)\mathbf{x} = (I_n - (I_n - NM)S)\mathbf{x} &= \hat{f},\end{aligned} \tag{4.74}$$

and then we use it to compute an approximation of $\mathbf{y}$. Indeed, the reduced problem (4.74) has gone from size $n + m$ to size n but, more importantly, its spectrum tends to be clustered around the point one, instead of the three points from the full system, since the spectral radius of F is such that

$$\rho(F) \leq \frac{\|S\|}{(1 - \sigma_1^2)^{1/2}},$$

where σ_1 is the largest singular value of $U_1^T U_2$ in Proposition 4.16. Most of the efforts are concentrated on the search of an efficient splitting for the $(1,1)$–block of $\mathcal{M}$.

Another possible choice for a block diagonal preconditioner can be done when properties **S3**, **S5** hold together and the $(1,1)$–block of $\mathcal{M}$, i.e., A, is positive definite. It is based on the result for the eigenvalues of the saddle point matrix (4.65) from Theorem 4.32. Since the goal is obtaininig a matrix of the form (4.65), in [234] the following block preconditioner was proposed:

$$\mathcal{P}_{\pm}(\tilde{S}) = \begin{bmatrix}\eta^{-1} A & 0\\ 0 & \pm\tilde{S}\end{bmatrix}, \quad \eta > 0, \quad \tilde{S} \approx B A^{-1} B^T.$$

As mentioned at the beginning of Chapter 3, in general a reasonable preconditioner for a symmetric indefinite matrix should be indefinite, but, in order to apply a symmetric iterative method such as MinRes, the preconditioner should be symmetric and positive definite. In this particular case, the best choice seems to be $\mathcal{P}_+(\tilde{S})$ since, as we have seen from Theorem 4.32 and Figure 4.17, the symmetric system matrix $\mathcal{A}_\eta^+$ and, moreover, the following theorem guarantees that, for a suitable choice of the starting vector, MinRes and GMRES produce the same Krylov subspace. Thus, MinRes is preferable because it has a lower computational cost.

Theorem 4.33 (Fischer et al. [234]). *Let $A = LL^T$ and the starting vector*

$$\mathbf{x}^{(0)} = [\eta^{-1}(L^{-1}\mathbf{f})^T, 0^T]^T,$$

we obtain that the starting residual for $\mathcal{P}_{\pm}(\tilde{S})$ *is given by*

$$\mathbf{r}_{\pm}^{(0)} = \begin{bmatrix} 0 \\ \mp\eta^{-1}BL^{-1}\mathbf{f} \end{bmatrix}$$

and $\forall k = 0, 1, \ldots, \forall \eta > 0$*:*

$$\|\mathbf{r}_{+,MinRes}^{(2k+1)}\|_2 = \|\mathbf{r}_{+,MinRes}^{(2k)}\|_2 = \|\mathbf{r}_{-,GMRES}^{(2k+1)}\|_2 = \|\mathbf{r}_{-,GMRES}^{(2k)}\|_2.$$

As well as proving the equivalence of two methods in terms of residual norms, Theorem 4.33 shows that both methods stagnate on every even step. Even if this exact equivalence is lost when a random initial vector is used, numerical experiments in Fischer et al. [234] suggest that the above mentioned behavior is still comparable.

Block diagonal preconditioners are also used when property **S5** does not hold, i.e., $C \neq 0$. Some examples, tailored for the *stabilized* discrete Stokes problem, are in [479, 540]. Other approaches for specific problems are in [38, 164, 368], in particular for optimal control and optimization problems, Stokes and related problem [156, 157, 337, 377, 422, 423], FEM for second order differential problems [340, 341, 436, 515, 535], and data interpolation [369].

As for the block diagonal preconditioner (4.72), we need to use an approximate version also for the block triangular "preconditioner"in (4.73). Even if the preconditioned matrix has a minimum polynomial of degree two (GMRES converges in two step in exact arithmetic; see § 2.2.8), the application cost is the same as solving the original system with block Gaussian elimination. The other potential flaw of this kind of preconditioner is that it can be nonnormal. Nevertheless, we remark that each application of the inverse of the block triangular preconditioners is only marginally more expensive than the block diagonal preconditioner. Indeed, both preconditioners require (an approximation for) A^{-1} and S^{-1}. In the block triangular case, also only one application of either B_2, or B_1^T.

Similarly to what was done for the diagonal case, let us investigate the effects of the approximation of A, and thus of the relative Schur complement $S = -B_2A^{-1}B_1^T$, with some matrices $\tilde{A}$ and $\tilde{S}$. In [100], assuming properties from **S1** to **S4**, the following approach for the symmetric case is proposed. Consider a preconditioner $\tilde{A}$ for the matrix A such that applying $\tilde{A}^{-1}$ has a reasonable computational cost and

$$\exists \alpha_0, \alpha_1 > 0 : \alpha_0 < A\mathbf{u}, \mathbf{u} > \leq < \tilde{A}\mathbf{u}, \mathbf{u} > \leq \alpha_1 < A\mathbf{u}, \mathbf{u} > \ \forall \mathbf{u} \in \mathbb{R}^n,$$

with α_1/α_0 *small* and, possibly, bounded independently from n. If we make the additional assumption that $\alpha_1 < 1$, then for $\alpha = 1 - \alpha_0$,

$$< (A - \tilde{A})\mathbf{u}, \mathbf{u} > \leq \alpha < A\mathbf{u}, \mathbf{u} > \ \forall \mathbf{u} \in \mathbb{R}^n \setminus \{0\}.$$

Straightforward manipulations on the block system gives the following preconditioned system

$$M \begin{bmatrix} \mathbf{x} \\ \mathbf{y} \end{bmatrix} \equiv \begin{bmatrix} \tilde{A}^{-1}A & \tilde{A}^{-1}B^T \\ B\tilde{A}^{-1}(A - \tilde{A}) & C + B\tilde{A}^{-1}B^T \end{bmatrix} \begin{bmatrix} \mathbf{x} \\ \mathbf{y} \end{bmatrix} = \begin{bmatrix} \tilde{A}^{-1}\mathbf{f} \\ B\tilde{A}^{-1}\mathbf{f} - \mathbf{g} \end{bmatrix}.$$

Thus, in the following inner product on $\mathbb{R}^n \times \mathbb{R}^m$,

$$\langle \begin{bmatrix} \mathbf{u} \\ \mathbf{v} \end{bmatrix}, \begin{bmatrix} \mathbf{w} \\ \mathbf{x} \end{bmatrix} \rangle =< A\mathbf{u}, \mathbf{w} > - < \tilde{A}\mathbf{u}, \mathbf{w} > + < \mathbf{v}, \mathbf{x} >,$$

the operator M is symmetric and, moreover, if we define the matrix preconditioned with the full preconditioner (4.73) as

$$\tilde{M} = \begin{bmatrix} I & 0 \\ 0 & C + BA^{-1}B^T \end{bmatrix}$$

the following inequalities hold

$$\lambda_0 \langle \tilde{M}\mathbf{u}, \mathbf{u} \rangle \leq \langle M\mathbf{u}, \mathbf{u} \rangle \leq \lambda_1 \langle \tilde{M}\mathbf{u}, \mathbf{u} \rangle,$$

where

$$\lambda_0 = \left(1 + \frac{\alpha}{2} + \sqrt{\alpha + \frac{\alpha^2}{4}}\right)^{-1}, \; \lambda_1 = \frac{1 + \sqrt{\alpha}}{1 - \alpha}.$$

Thus, the problem of constructing a preconditioner for M is reduced to constructing it for the reduced system $(C + BA^{-1}B^T)\mathbf{y} = BA^{-1}\mathbf{f} - \mathbf{g}$. Observe that as $\alpha \to 0$ then $\lambda_0, \lambda_1 \to 1$ and thus $\sigma(\tilde{M}^{-1}M)$ tends uniformly to one.

Note that symmetric, or symmetrizing, approaches, like the one we have just discussed, are seldom strictly necessary in practice. When good approximations of A, and thus of the Schur complement S, are available, methods for nonnormal systems like GMRES or BiCGstab (or for their variants) can converge in a sufficiently small number of iterations. On the other hand, the analysis of the spectral properties of the preconditioners in these cases is not easy to obtain. Moreover, convergence estimates can be not sharp, see the discussions in §§ 2.2.8 and 2.3.4. In this perspective, we briefly mention preconditioners based on incomplete block triangular factorizations. If we can produce a cheap and accurate approximation of $\tilde{A} \approx A$ and of the Schur complement $\tilde{S} \approx S$ for the matrix of the underlying saddle point problem, we can consider a preconditioner in the following block form

$$\mathcal{P} = \begin{bmatrix} \tilde{A} & 0 \\ B & \tilde{S} \end{bmatrix} \begin{bmatrix} I & \tilde{A}^{-1}B^T \\ 0 & I \end{bmatrix} = \begin{bmatrix} \tilde{A} & B^T \\ B & -\tilde{C} \end{bmatrix},$$

where $\tilde{C} = -B\tilde{A}^{-1}B^T - \tilde{S}$. Observe that $\mathcal{P}$ is symmetric if $\tilde{A}$ and $\tilde{S}$ are and, moreover, it is, in general, indefinite.

One interesting example of a block triangular preconditioner is the so–called semi–implicit method for pressure–linked equations, proposed in [420, 421]. The latter is devised for the solution of systems of equations arising from the steady–state Navier–Stokes equations with stable finite difference discretizations. Other approaches based on block factorization of the saddle point matrix can be found in [158, 244, 258, 424, 439, 462, 480, 536].

4.4.4.2 Constrained preconditioners

In this section we investigate preconditioners that have the same constraints of the original problem, i.e., we impose on the preconditioning matrix the same block structure of the problem while looking for a system that should be computationally cheaper to solve.

Let us start by observing a difference between block–triangular and constrained preconditioners. For the first, the structure imposes that they should satisfy the first one of the two equations in the saddle point system (4.55). The second requires satisfying the equation for the constraint, i.e., the second one in (4.55). If, e.g., we consider the settings of fluid dynamic problems seen sometimes in our examples, block–triangular preconditioners respect the conservation of momentum, whereas constrained ones impose a mass balance condition on the preconditioned system.

Constrained preconditioners have been intensely studied and several possibilities and choices are available. We briefly examine some of them while keeping an eye to the fields of application they originate from.

Let us start from the cases in which hypotheses **S1**–**S3** and **S5** hold. As an example, in the case of an equality constrained nonlinear programming problem solved by the Newton method, see again Example 4.10. The original system has $n+m$ linear equations whose matrix is always indefinite. This fact can lead to some difficulties when standard positive definite preconditioners are used. It can be advantageous to transform (after the usual freeze of the matrix function of the Newton step) the original KKT matrix

$$M = \begin{bmatrix} G(x,u) & B(x)^T \\ B(x) & 0 \end{bmatrix}$$

to have a positive definite matrix with in the left-upper corner the matrix $A = G + \rho B^T B$ with $\rho > 0$, instead of G from the Hessian, and then work with the following saddle point matrix:

$$\mathcal{M} = \begin{bmatrix} A & B^T \\ B & 0 \end{bmatrix}, \quad \begin{array}{lll} A \in \mathbb{R}^{n\times n}, & A = A^T,\ A > 0, & \\ A = G + \rho B^T B, & \rho > 0, & m \leq n\ , \\ B \in \mathbb{R}^{m\times n}, & \operatorname{rank}(B) = m & \end{array} \tag{4.75}$$

Following [368], we consider the indefinite preconditioner for $\mathcal{M}$, based on a diagonal approximation D of the Schur complement:

$$\mathcal{P} = \begin{bmatrix} LL^T & B^T \\ B & B(LL^T)^{-1}B - D \end{bmatrix}, \tag{4.76}$$

where LL^T is either the complete (for the sake of the analysis) or the incomplete Cholesky factorization of A, see § 3.4.1, and D is a positive definite diagonal matrix. Then, by using the expression for the inverse of a saddle point matrix from (4.60),

$$\mathcal{P}^{-1} = \begin{bmatrix} (LL^T)^{-1} - (LL^T)^{-1}BD^{-1}B^T(LL^T)^{-1} & (LL^T)^{-1}BD^{-1} \\ D^{-1}B^T(LL^T)^{-1} & -D^{-1} \end{bmatrix}.$$

Since we are working with symmetric matrices, for obtaining general spectral results we can work, equivalently, in terms of left or right preconditionining. For the sake of presentation, we consider here right preconditioning. By using the (complete) Cholesky factorization, we find

$$\mathcal{M}\mathcal{P}^{-1} = \begin{bmatrix} I & 0 \\ (I - M)B^T A^{-1} & M \end{bmatrix}, \ M = B^T A^{-1} B D^{-1},$$

Theorem 4.34 (Lukšan and Vlček [368]). *Consider the preconditioner $\mathcal{P}$ in (4.76) with $LL^T = A$ and D positive definite. Then, the right preconditioned matrix $\mathcal{M}\mathcal{P}^{-1}$ has at least n unit eigenvalues with a full system of linearly independent eigenvectors. The other eigenvalues are positive and, if for some $\mathbf{v} \in \mathbb{R}^m$ it holds that $\mathbf{v}^T D\mathbf{v} = \mathbf{v}^T B^T A^{-1} B\mathbf{v}$, then all the eigenvalues of the matrix $\mathcal{M}\mathcal{P}^{-1}$ are contained in the interval determined by the extremal eigenvalues of the matrix $B^T A^{-1} BD$.*

This preconditioner is viable whenever an approximation D of the Schur complement $B^T A^{-1} B$ can be easily obtained. We can always compute D as $D = \text{diag}(B^T A^{-1} B)$. Unfortunately, this procedure costs m back substitutions and thus is very time consuming and unusable for large scale problems and parallel machines. One can takes $D = \pm\varepsilon I$ for $\varepsilon > 0$, that gives back an indefinite and a positive definite preconditioner, respectively. Some sharper estimates on the extremal eigenvalues of the matrix for the positive definite preconditioner are in [535].

Instead of approximating the Schur complement, another possible choice is using a positive definite diagonal perturbation of the Schur complement. In particular, consider the perturbed Schur complement $B^T A^{-1} B + D$. This preconditioner was first introduced with $D = \varepsilon I$ and $\varepsilon > 0$ in [19]. The following analysis is for a general D starting from the Cholesky decomposition LL^T of $ABD^{-1}B^T$ giving the following indefinite preconditioner:

$$\mathcal{P}^{-1} = \begin{bmatrix} (LL^T)^{-1} & (LL^T)^{-1} BD^{-1} \\ D^{-1}B^T(LL^T)^{-1} & D^{-1}B^T(LL^T)^{-1}BD^{-1} - D^{-1} \end{bmatrix}. \tag{4.77}$$

Similar to the previous case, if we are using the complete Cholesky factorization, the right preconditioned matrix can be expressed as

$$\mathcal{M}\mathcal{P}^{-1} = \begin{bmatrix} I & 0 \\ (I - M)B^T A^{-1} & M \end{bmatrix}, \ M = B^T A^{-1} B(B^T A^{-1} B + D)^{-1}.$$

Theorem 4.35 (Lukšan and Vlček [368]). *Consider the preconditioner $\mathcal{P}$ in (4.77) with $LL^T = A + BD^{-1}B^T$ and D positive definite. Then, the right preconditioned matrix $\mathcal{M}\mathcal{P}^{-1}$ has at least n unit eigenvalues with a full system of linearly independent eigenvectors. If the matrix $B^T A^{-1} B$ is positive definite, then the eigenvalues $\lambda \in \sigma(\mathcal{M}\mathcal{P}^{-1})$ are such that*

$$0 < \frac{\lambda_{min}(B^T A^{-1} B)}{\lambda_{min}(B^T A^{-1} B) + \|D\|} \leq \lambda \leq 1.$$

Within a similar approach, we consider also the following preconditioner with indefinite spectrum introduced in [224], in which a diagonal approximation D of the matrix B is used

$$\mathcal{P}^{-1} = \begin{bmatrix} D^{-1} - D^{-1}B(R^TR)^{-1}B^TD^{-1} & D^{-1}B(R^TR)^{-1} \\ (R^TR)^{-1}B^TD^{-1} & -(R^TR)^{-1} \end{bmatrix}, \tag{4.78}$$

where R^TR is the Cholesky decomposition of the matrix $B^TD^{-1}B$, either complete or incomplete. If we are using the complete Cholesky factorization, the right preconditioned matrix can be expressed as

$$\mathcal{K}\mathcal{P}^{-1} = \begin{bmatrix} I + MP & MB(B^TD^{-1}B)^{-1} \\ 0 & I \end{bmatrix}, \quad \begin{array}{l} M = AD^{-1} - I \\ P = I - B(B^TD^{-1}B)^{-1}B^TD^{-1}, \end{array}$$

Let Z be a matrix whose columns form an orthonormal basis in the null space of B^T so that $B^TZ = 0$ and $Z^TZ = I$ (if Z^TGZ is positive definite, this is a sufficient condition for the existence of a solution; see [368] for details).

Theorem 4.36 (Lukšan and Vlček [368]). *Consider the preconditioner $\mathcal{P}$ in (4.78) with $R^TR = B^TD^{-1}B$ and D positive definite. Then, the right preconditioned matrix $\mathcal{M}\mathcal{P}^{-1}$ has at least $2m$ unit eigenvalues. If $M = AD^{-1}$ is nonsingular, there exist m linearly independent eigenvectors for these eigenvalues. The other eigenvalues of $\mathcal{M}\mathcal{P}^{-1}$ are also eigenvalues of the matrix $Z^TAZ(Z^TDZ)^{-1}$. Moreover, if Z^TAZ is positive definite, then all the eigenvalues of the matrix $\mathcal{M}\mathcal{P}^{-1}$ are contained in the interval determined by the extremal eigenvalues of the matrix $Z^TAZ(Z^TDZ)^{-1}$. If CG is applied and no breakdowns occur, the solution of the saddle point linear system is obtained in at most $n - m + 2$ iterations in exact arithmetic.*

Observe that since $Z^TAZ = Z^T(G+\rho BB^T)Z = Z^TGZ$, the eigenvalues of $\mathcal{K}\mathcal{P}^{-1}$ are unaffected by the parameter ρ. Then, as confirmed by experiments in [368], the effectiveness of the preconditioner is almost independent from ρ. In the case in which on the right–hand side the values relative to the second equation are 0, as it happens in the Stokes problem, see, e.g, Example 4.17, an estimate of the error reduction in the style of § 2.2.4 can be obtained.

Corollary 4.3 (Lukšan and Vlček [368]). *Consider the preconditioner $\mathcal{P}$ in (4.78) with $R^TR = B^TD^{-1}B$ and D positive definite, applied to the linear system*

$$\mathcal{M}\begin{bmatrix} \boldsymbol{d} \\ \boldsymbol{v} \end{bmatrix} = -\begin{bmatrix} \mathbf{b} \\ \mathbf{0} \end{bmatrix}.$$

If the matrix Z^TAZ is positive definite, then CG computes $\mathbf{d} \in \mathbb{R}^n$ after at most $n - m$ iterations and the following estimate holds:

$$\|\mathbf{d}^{(j)} - \mathbf{d}\| \leq 2\sqrt{\kappa}\left(\frac{1-\sqrt{\kappa}}{1+\sqrt{\kappa}}\right)^j \|\mathbf{d}^{(0)} - \mathbf{d}\|,$$

where $\kappa = \kappa_2(Z^TAZ(Z^TDZ)^{-1})$, the 2-norm condition number of $(Z^TAZ(Z^TDZ)^{-1})$.

Preconditioner (4.78) can be generalized by considering a symmetric matrix $\tilde{A}$ approximating the symmetric and positive definite matrix A, i.e.,

$$\mathcal{P} = \begin{bmatrix} \tilde{A} & B^T \\ B & 0 \end{bmatrix} \approx \mathcal{M}, \tag{4.79}$$

and giving the following generalization of Theorem 4.36.

Theorem 4.37 (Keller et al. [327]). *Let $\mathcal{M}$ be a symmetric and indefinite matrix of the form*

$$\mathcal{M} = \begin{bmatrix} A & B^T \\ B & 0 \end{bmatrix}, \ \mathbb{R}^{n\times n} \ni A = A^T, \ B \in \mathbb{R}^{m\times n}, \ \text{rank}(B) = m,$$

and $Z \in \mathbb{R}^{n\times(n-m)}$ a basis for the null space of B. If we consider the preconditioner (4.79) with $\tilde{A} \neq A$, $\mathcal{P}^{-1}\mathcal{M}$ has $2m$ eigenvalues equal to one and $n-m$ eigenvalues λ which are defined by the generalized eigenvalue problem $Z^T AZ\mathbf{x} = \lambda Z^T \tilde{A} Z\mathbf{x}$. If $Z^T \tilde{A} Z$ is positive definite, there are $m+i+j$ linearly independent eigenvectors such that:

1. *m eigenvectors related to the eigenvalue $\lambda = 1$ are of the form $[0^T, 0^T, \mathbf{y}^T]^T$;*

2. *i linearly independents eigenvectors related to the eigenvalue $\lambda = 1$ are of the form $[\mathbf{x}_z^T, \mathbf{x}_y^T, \mathbf{y}^T]^T$ arising from $A\mathbf{w} = \sigma\tilde{A}\mathbf{w}$ with $\mathbf{w} = [\mathbf{x}_z^T, \mathbf{x}_y^T]$, $\sigma = 1$ for $i = 0, \ldots, n$,*

3. *j eigenvectors of the form relative to the case $\lambda \neq 1$ are of the form $[\mathbf{x}_z^T, \mathbf{0}^T, \mathbf{y}^T]^T$ for $j = 0, \ldots, n-m$.*

Moreover, the dimension of the Krylov subspace $\mathcal{K}(\mathcal{P}^{-1}\mathcal{M}, \mathbf{b})$ for the problem $\mathcal{M}\mathbf{x} = \mathbf{b}$ is at most $n-m+2$.

We emphasize that, as usual, neither the matrix $\mathcal{P}^{-1}$ nor the matrix $\mathcal{M}\mathcal{P}^{-1}$ are formed explicitly in practical implementations. The block LDL^T factorization can be used coupled with the standard preconditioned GMRES, i.e.,

$$\begin{aligned} \mathcal{P}^{-1} =& \begin{bmatrix} I & -\tilde{A}^{-1}B^T \\ 0 & I \end{bmatrix} \begin{bmatrix} \tilde{A}^{-1} & 0 \\ 0 & -(B\tilde{A}^{-1}B^T)^{-1} \end{bmatrix} \begin{bmatrix} I & 0 \\ -B\tilde{A}^{-1} & I \end{bmatrix} \\ =& L^{-1}D^{-1}L^{-T}. \end{aligned} \tag{4.80}$$

Thus, any application of the preconditioner $\mathcal{P}$ is carried out as $\mathcal{P}^{-1}\mathbf{v} = L^{-1}(D^{-1}(L^{-T}\mathbf{v}))$. The latter factorization of the preconditioner can be much less computationally expensive than the factorization of the initial coefficient matrix because, instead of the matrix A, we are using a cheaper approximation. Otherwise, the segregated approach is possible by exploiting the structure to apply the CG to the reduced linear system; see [193].

Under the hypothesis of either $Z^T AZ$ or $Z^T \tilde{A} Z$ being positive definite, we have a preconditioned matrix with a real spectrum. A particular choice for the matrix $\tilde{A}$ is $\tilde{A} = I$, as suggested in [250, 327, 426, 427] for magnetostatic and general optimization problems. With this choice the preconditioner (4.79) projects the problem onto the kernel of the constrained operator. We can write the orthogonal projector onto $\text{span}(B^T)$ as $\Pi = B^T(BB^T)^{-1}B$, so that the right preconditioned matrix can be expressed in term of the projector as

$$\mathcal{M}\mathcal{P}^{-1} = \begin{bmatrix} A(I-\Pi)+\Pi & (A-I)B^T(BB^T)^{-1} \\ 0 & I \end{bmatrix},$$

from which we can discuss the properties of the preconditioner again by means of Theorem 4.37. Given a particular choice for $\tilde{B}$, additional information can be given on the generated Krylov subspaces and the convergence estimate. Left and right preconditioning, as observed in § 3.2.2, generate different spaces used to compute the approximate solution. Nevertheless, the first block of $\mathbf{d}^{(j)} \approx \mathbf{d}$, using the same notation of Corollary 4.3, always belongs to the same space.

Proposition 4.17 (Rozložník and Simoncini [447]). *Let* $\mathbf{x}^{(j)} = [\mathbf{d}^{(j)}, \mathbf{v}^{(j)}]^T$ *be the approximate solution of* $\mathcal{M}\mathbf{x} = \mathbf{b}$ *in either* $\mathcal{K}_j(\mathcal{M}\mathcal{P}^{-1}, \mathbf{r}^{(0)})$ *or* $\mathcal{K}_j(\mathcal{P}^{-1}\mathcal{M}, \mathcal{P}^{-1}\mathbf{r}^{(0)})$. *Then,* $\mathbf{d}^{(j)} = \phi((A(I-\Pi)+\Pi)^T)\tilde{\mathbf{r}}_1^{(0)}$, $\tilde{\mathbf{r}}^{(0)} = \mathcal{P}^{-1}\mathbf{r}^{(0)} = [\tilde{\mathbf{r}}_1^{(0)}, \tilde{\mathbf{r}}_2^{(0)}]^T$, *for some polynomial* $\phi(z)$ *of degree* $j-1$, *possibly different for the two spaces.*

Within this framework, an estimate for the convergence with GMRES can be obtained. By using the same notation and polynomials of Proposition 4.17, the right preconditioned GMRES residual satisfies

$$\|\mathbf{r}_{\text{GMRES}}^{(k)}\|^2 = \min_{\phi\in\mathbb{P}_k,\, \phi(0)=1} \left(\|\phi(G)\mathbf{r}_1^{(0)} + \psi(G)S\mathbf{r}_2^{(0)}\|^2 + |\phi(1)|^2\|\mathbf{r}_2^{(0)}\|^2 \right),$$

and the left preconditioned residual

$$\|\mathbf{r}_{\text{GMRES}}^{(k)}\|^2 = \min_{\phi\in\mathbb{P}_{\leq k},\, \phi(0)=1} \left(\|\phi(G^T)\tilde{\mathbf{r}}_1^{(0)}\|^2 + \|S^T\psi(G^T)\tilde{\mathbf{r}}_1^{(0)} + \phi(1)\tilde{\mathbf{r}}_2^{(0)}\|^2 \right).$$

As we did in Corollary 4.3, we can specialize the analysis for the case in which the second block of the right–hand side is $\mathbf{0}$. Nevertheless, this setting is not as restrictive as with a general matrix $\tilde{A}$. As observed in [447], the starting guess for GMRES can be chosen to belong to this case even if the second block is not the zero vector. For the right preconditioning framework with a diagonalizable G, with a starting guess such that the initial residual is $\mathbf{r}^{(0)} = [\mathbf{s}^{(0)}, \mathbf{0}]^T$, we have the upper bound

$$\|\mathbf{r}_{\text{GMRES}}^{(k)}\| \leq 2\zeta \left(\frac{\sqrt{\kappa}-1}{\sqrt{\kappa}+1}\right)^k \|\mathbf{r}^{(0)}\|,$$

where κ is the ratio $\lambda_{\max}(G)/\lambda_{\min}(G)$ and ζ is a bound on the 2-norm condition number of the matrix Z of the eigenvectors of G. With a similar notation, for the left preconditioned residual we obtain the following bound

$$\|\mathbf{r}^{(k)}_{\text{GMRES}}\| \leq \|\mathcal{P}^{-1}\| \|\mathbf{r}^{(0)}\| \kappa_2(Z) \min_{\phi \in \mathbb{P}_{\leq k},\, \phi(0)=1} \max_{i=1,\ldots,n} |\phi(\lambda_i(G))|.$$

For other non symmetric solvers and details, see [447].

We stress that application of the GMRES method might not be the best approach to solve the underlying saddle point preconditioned linear systems. Sometimes, the segregated approach, based on the reduced linear system and CG seems to give better results; see [193, 276, 327]. A specialized version of the Conjugate Gradient method in Algorithm 4.8 is based on the idea of computing an implicit basis Z of the null space of B used to remove the constraints from the linear system. In this way, one has to solve a reduced positive definite linear system by the standard Conjugate Gradient algorithm.

Algorithm 4.8: PCG for a Reduced System

Input: Saddle matrix $\mathcal{M}$ as in 4.75, right–hand side $\mathbf{b} = [\mathbf{b}_1^T, \mathbf{b}_2^T]^T$, preconditioner $\mathcal{P}$ from (4.79), Z basis of the null space of B, tolerance $\varepsilon > 0$.

Output: Candidate solution $\mathbf{x} = [\mathbf{x}_1^T, \mathbf{x}_2^T]^T$ of $\mathcal{M}\mathbf{x} = \mathbf{b}$.

1 Choose an initial point $\mathbf{x}_1$ satisfying $A\mathbf{x}_1 = \mathbf{b}_2$;
2 Choose an arbitrary vector $\mathbf{y}$;
3 $\mathbf{r} \leftarrow A\mathbf{x}_1 - \mathbf{b}_1$;
4 $\mathbf{r} \leftarrow \mathbf{r} - B^T\mathbf{y}$;
5 $\mathbf{g} \leftarrow Z(Z^T\tilde{A}Z)^{-1}Z^T\mathbf{r}$;
6 $\mathbf{p} \leftarrow -\mathbf{g}$;
7 **while** $| < \mathbf{r}^+, \mathbf{g}^+ > | > \varepsilon$ **do**
8 $\quad \alpha \leftarrow \dfrac{< \mathbf{r}, \mathbf{g} >}{< \mathbf{p}^T A\mathbf{p} >}$;
9 $\quad \mathbf{x}_1 \leftarrow \mathbf{x}_1 + \alpha\mathbf{p}$;
10 $\quad \mathbf{r}^+ \leftarrow \mathbf{r} + \alpha A\mathbf{p}$;
11 $\quad \mathbf{r}^+ \leftarrow \mathbf{r}^+ - B^T\mathbf{y}$;
12 $\quad \mathbf{g}^+ \leftarrow Z(Z^T\tilde{A}Z)^{-1}Z^T\mathbf{r}^+$;
13 $\quad \beta \leftarrow \dfrac{< \mathbf{r}^+, \mathbf{g}^+ >}{< \mathbf{r}^T, \mathbf{g} >}$;
14 $\quad \mathbf{p} \leftarrow -\mathbf{g}^+ + \beta\mathbf{p}$;
15 $\quad \mathbf{g} \leftarrow \mathbf{g}^+$;
16 $\quad \mathbf{r} \leftarrow \mathbf{r}^+$;

The vectors $\mathbf{y}$ in line 4 and 11 of Algorithm 4.8 should be selected to have $\mathbf{r}$ and $\mathbf{r}^+$ close to $A^T\mathbf{y}$. Note that $Z(Z^T\tilde{A}Z)^{-1}Z^T$ is independent from the particular choice of the basis. Moreover, the preconditioned residual $\mathbf{g}^+$ is in the null space of B by construction. Therefore, all the search directions $\mathbf{p}$ lie in the null space of B, ensuring that the iterates $\mathbf{x}_1$, at least in exact arithmetic, satisfy $B\mathbf{x}_1 = \mathbf{b}_2$. Two issues of this approach are that numerical instabilities on the computation of the projection can generate iterates that fail to satisfy the constraint, leading to difficulties in the convergence. The matrix Z, when it is not given by the definition of the problem, has to be computed in a setup phase, see, e.g., [261]. When the preconditioner is the one with $\tilde{A} = I$ from Proposition 4.17, the projection on the null space of B is $\Pi_B = Z(Z^T Z)^{-1} Z^T$ and can be expressed as $\Pi_B = I - B^T(BB^T)^{-1}B$. Thus, we can replace the computation of the preconditioned residual $\mathbf{g}^+$ in line 12 by

$$\mathbf{g}^+ \leftarrow \Pi_B \mathbf{r}^+ = (I - B^T(BB^T)^{-1}B)\mathbf{r}^+ = \mathbf{r}^+ - B^T\mathbf{v}^+,$$

where $\mathbf{v}^+$ is obtained as the solution of $BB^T\mathbf{v}^+ = B\mathbf{r}^+$, that is equivalent to looking for the residual $\mathbf{g}^+$ of the least squares problem

$$\min_{\mathbf{v}} \|\mathbf{r}^+ - B^T\mathbf{v}\|.$$

Otherwise, for a generic matrix $\tilde{A}$, i.e., for the preconditioner $\mathcal{P}$ (4.79), the preconditioned residual can be computed as the solution of

$$\begin{bmatrix} \tilde{A} & B^T \\ B & 0 \end{bmatrix} \begin{bmatrix} \mathbf{g}^+ \\ \mathbf{v}^+ \end{bmatrix} = \begin{bmatrix} \mathbf{r}^+ \\ \mathbf{0} \end{bmatrix}$$

obtained with the $L^{-1}D^{-1}L^{-T}$ factorization (4.80). Both approaches for the computation of $\mathbf{g}^+$ can give rise, in finite precision arithmetic, to significant rounding errors that prevent the iterates from remaining into $\ker(B)$. This is due to the fact that $\mathbf{g}^+$ becomes increasingly small as the Conjugate Gradient approaches the solution, while $\mathbf{r}^+$ does not. Therefore, even if $\mathbf{r}^+$ is bounded away from zero, it becomes closer to the range of B^T, i.e., $\mathbf{r}^+$ tends to become orthogonal to Z. Therefore, $\mathbf{g}^+$ converge to zero as long as $\lambda_{\min}(Z^T\tilde{A}Z) > 0$. More refined estimates for this phenomenon are given in [276], together with an approach base on fixed precision iterative refinement to refine the residual $\mathbf{r}^+$ in line 11 of Algorithm 4.8 so that its norm is closer to that of $\mathbf{g}^+$. This often results in a sensible reduction of the rounding errors; see again [276] for details.

As we have seen for both GMRES and Conjugate Gradient approaches, the most expensive operation per step is clearly the application of the preconditioner, and in particular the solution of the auxiliary linear systems it requires. Thus, to further reduce the computational cost, one could also attempt for an inexact application of the various constrained preconditioners $\mathcal{P}$. We briefly mention that for applying Jacobi, SOR and SSOR splitting methods (see § 2.1.1) to generate $\tilde{A}$ for the preconditioner $\mathcal{P}$ from (4.79), one can also consider the case of using a nonsymmetric matrix $\tilde{A}$; see [214]. This might be interesting if we want to focus on possible smoothers for multigrid algorithms; see § 4.2.

Let us consider the case in which the saddle point matrix $\mathcal{M}$ satisfies hypothesis **S4** instead of **S5** or, more in general, in which the $(2,2)$ block is different from the zero matrix. Also this case has been studied since sometimes it can be easier to solve for iterative methods because the auxiliary linear systems with the modified Schur complement $S+BA^{-1}B^T$ can be more well–conditioned than the case $C=0$. Then, let us modify the preconditioner (4.79) as

$$\mathcal{P}_C = \begin{bmatrix} \tilde{A} & B^T \\ B & -C \end{bmatrix} \approx \mathcal{M} = \begin{bmatrix} A & B^T \\ B & -C \end{bmatrix}. \tag{4.81}$$

For the case $\tilde{A}=I$, a spectral analysis similar to that in Theorem 4.37 can be performed.

Theorem 4.38 (Perugia and Simoncini [426]). *Let $\mathcal{M}$ be an indefinite matrix of the form*

$$\mathcal{M} = \begin{bmatrix} A & B^T \\ B & -C \end{bmatrix}, \ \mathbb{R}^{n\times n} \ni A = A^T,\ B \in \mathbb{R}^{m\times n},\ \text{rank}(B) = m, C = C^T,\ C > 0$$

Consider the preconditioner (4.81) with $\tilde{A}=I$. The right preconditioned matrix reads as

$$\mathcal{M}\mathcal{P}^{-1} = \begin{bmatrix} A(I-\Pi)+\Pi & (A-I)B^T(BB^T+C)^{-1} \\ 0 & I \end{bmatrix}$$

where $\Pi = B^T(BB^T+C)^{-1}B$. If $\lambda \in \sigma(\mathcal{M}\mathcal{P}^{-1})$, then λ is either $\lambda=1$ or a nonzero eigenvalue of $(I-\Pi)A(I-\Pi)$.

With the same approach, we can look at the general case in which the preconditioner (4.81) is used with an approximation $\tilde{A}$ of A. In this case, the eigenvalues of the preconditioned matrix $\mathcal{P}_C^{-1}\mathcal{M}$ are exactly those of the generalized eigenproblem $\mathcal{M}\mathbf{x} = \lambda\mathcal{P}_C\mathbf{x}$, that are of the form $\lambda = \mu+1$, μ is the generalized eigenvalue of

$$\mu \begin{bmatrix} \tilde{A} & B^T \\ B & -C \end{bmatrix} \begin{bmatrix} \mathbf{x} \\ \mathbf{y} \end{bmatrix} = \begin{bmatrix} A-\tilde{A} & 0 \\ 0 & 0 \end{bmatrix} \begin{bmatrix} \mathbf{x} \\ \mathbf{y} \end{bmatrix} \Leftrightarrow \mu \begin{bmatrix} I & \hat{B}^T \\ \hat{B} & -C \end{bmatrix} \begin{bmatrix} \hat{\mathbf{x}} \\ \mathbf{y} \end{bmatrix} = \begin{bmatrix} E & 0 \\ 0 & 0 \end{bmatrix} \begin{bmatrix} \hat{\mathbf{x}} \\ \mathbf{y} \end{bmatrix},$$

where $\hat{B} = B\tilde{A}^{-1/2}$, $E = \tilde{A}^{-1/2}A\tilde{A}^{-1/2} - I$, $\hat{\mathbf{x}} = \tilde{A}^{-1/2}\mathbf{x}$. The generalized eigenvalues of this problem can be characterized by the following proposition.

Proposition 4.18 (Axelsson and Neytcheva [22]). *Let $B \in \mathbb{R}^{n\times m}, C \in \mathbb{R}^{m\times m}, E \in \mathbb{R}^{n\times n}$ with $\text{rank}(B) = m$, $C = C^T \geq 0$ and $E = E^T$. Then, the eigenvalues of the generalized eigenproblem*

$$\mu \begin{bmatrix} I & B^T \\ B & -C \end{bmatrix} \begin{bmatrix} \mathbf{x} \\ \boldsymbol{y} \end{bmatrix} = \begin{bmatrix} E & 0 \\ 0 & 0 \end{bmatrix} \begin{bmatrix} \mathbf{x} \\ \boldsymbol{y} \end{bmatrix},\ \|\mathbf{x}\| + \|\mathbf{y}\| \neq 0,$$

are such that

1. $\mu = \dfrac{\mathbf{x}^H E\mathbf{x}}{\mathbf{x}^H(I+B^TC^{-1}B)\mathbf{x}} \neq 0$ *if $E\mathbf{x} \neq 0$ and C is positive definite,*

2. $\mu = 0$ *if and only if* $E\mathbf{x} = 0$, $\mathbf{y} \neq 0$; *the dimension of the corresponding eigenspace is* $m + \dim \ker(E)$,

3. *If* $\mu \neq 0$, *then* $\mu \in [\min\{0, \lambda_{min}(E)\}, \lambda_{max}(E)]$.

Therefore, also in this case, the eigenvalues of the preconditioned matrix are more clustered around 1 as $\tilde{A}$ better approximates the matrix A. These features of the *stabilized* systems, i.e., saddle point linear systems with $C \neq 0$, suggests the possibility of using a preconditioner $\mathcal{P}_C$ with a suitable choice of C also for saddle point matrices with property **S5**, i.e., (4.75); consider, e.g., [19, 61, 211, 212].

Notes on other preconditioning approaches

From all the preconditioners we mentioned, Multigrid and, more generally, multilevel methods, have been left out. Indeed, they can be used in the segregated approaches, based on the stationary iteration solver in § 4.4.3 and also in the block preconditioners in § 4.4.4.1 to solve the reduced systems. These strategies are indeed viable when the reduced problems are analogous to the discretization of a scalar elliptic PDE, e.g., in the case of an incompressible flow problem. The choice of the smoothers and coarsening strategies for these cases has been widely studied and can accommodate several discretizations, coefficients and problem geometries, see, e.g., [222, 520].

Appendix A

A Review of Numerical Linear Algebra

Linear Algebra goes back to the second century BC, although some hints can be seen two centuries before[1]. The study of linear algebra emerged again in the solution of systems of linear simultaneous equations in the work by Leibniz in 1693 and Seki in 1683. The discovery of *Cramer's Rule*, by Gabriel Cramer, is dated to 1750. The latter generalizes the first step made by Cardan in the *Ars Magna* (1545) and MacLaurin in *Treatise of algebra* (1748). Later, Gauss developed further this theory introducing the *Gaussian elimination*. In a very similar form was already present in the "counting board" described in China between 200 BC and 100 BC. Jacobi, Kronecker and Weierstrass in the years from 1830 to 1860 also looked at matrix results in the special context of linear transformations. Further steps were made by Eisenstein, that considered *linear substitution*. Sylvester was probably the first to use the term *matrix* in 1850. Cayley gave the first abstract definition of a matrix, while Hamilton and Frobenius worked on matrix polynomials. In the first thirty years of the 20th Century, matrices take their place in Mathematics with the books by Bôcher, Turnbull and Aitken. The classical book by Mirsky (1955) gave an account of matrix theory that is not far from the one that can be found in today's university teaching.

In the following we focus on some definitions and basic facts that are used in the books' chapters.

Definition A.1 (Metric space). Let M be a nonempty set and let $d(x,y)$: $M \times M \to \mathbb{R}$ be a nonnegative function such that

1. $d(x,y) \geq 0$, $\forall x,y \in M$ and $d(x,y) = 0$ if and only if $x = y$,
2. $d(x,y) = d(y,x)$ $\forall x,y \in M$,

[1] See the book by Goldstine [268] for an history of Numerical Analysis and the article on "Matrices and determinants" by O'Connor and Robertson [410].

3. $d(x,z) \leq d(x,y) + d(y,z)$, $\forall x,y,z \in M$ (triangular inequality).

The function $d(\cdot,\cdot)$ is called a *metric*, or a *distance*, over the set M. The couple (M,d) is called *metric space.*

Very often the set M is more than just a set. Let us consider the notion of *vector space.*

Definition A.2 (Vector Space). A *vector space* over the field $\mathbb{K}$ is a set V, whose elements are called *vectors*, together with an operation $V \times V \to V$, called *addition*, and denoted as $(\mathbf{x},\mathbf{y}) \mapsto \mathbf{x}+\mathbf{y}$, and another operation $\mathbb{K} \times V \to V$ called *scalar multiplication*, $(c,\mathbf{x}) \to c\mathbf{x}$, which satisfy the following axioms:

Commutativity $\forall\, \mathbf{x},\mathbf{y} \in V$, $\mathbf{x}+\mathbf{y} = \mathbf{y}+\mathbf{x}$,

Associativity (addition) $\forall\, \mathbf{x},\mathbf{y},\mathbf{z} \in V$, $(\mathbf{x}+\mathbf{y})+\mathbf{z} = \mathbf{x}+(\mathbf{y}+\mathbf{z})$,

Identity (addition) there exists a *zero vector* in V such that $\mathbf{0}+\mathbf{x} = \mathbf{x}$ $\forall\, \mathbf{x} \in V$,

Inverse (addition) $\forall \mathbf{x} \in V\ \exists\, (-\mathbf{x}) \in V$ such that $-\mathbf{x}+\mathbf{x} = \mathbf{0}$,

Distributivity I $\forall\, a \in \mathbb{K}$ and $\forall\, \mathbf{x},\mathbf{y} \in V$, $a(\mathbf{x}+\mathbf{y}) = a\mathbf{x}+a\mathbf{y}$,

Distributivity II $\forall\, a,b \in \mathbb{K}$ and $\forall\, \mathbf{x} \in V$, $(a+b)\mathbf{x} = a\mathbf{x}+b\mathbf{x}$,

Associativity ((multiplication)) $\forall\, a,b \in \mathbb{K}$ and $\forall\, \mathbf{x} \in V$, $(ab)\mathbf{x} = a(b\mathbf{x})$,

Identity (multiplication) there exists scalar $1 \in \mathbb{K}$ such that $\forall \mathbf{x} \in V$, $1\mathbf{x} = \mathbf{x}$.

A vector space is called *real* or *complex* if $\mathbb{K} = \mathbb{R}$ or $\mathbb{K} = \mathbb{C}$, respectively. A *subspace* of a vector space V over the field $\mathbb{K}$ is a subset of V, which is itself a vector space over $\mathbb{K}$.

Definition A.3 (Matrix). An $m \times p$ *matrix* over $\mathbb{K}$ is an $m \times p$ rectangular array $A = (a_{i,j})_{\substack{i=1,\ldots,m\\ j=1,\ldots,p}}$ with entries in $\mathbb{K}$. The *size* of A is $m \times p$ and A is said to be square if $m = p$.

Two matrices A and B are said to be equal if they have the same size and $a_{i,j} = b_{i,j}\ \forall\, i,j$. We can build $\mathbb{K}^{m\times p}$ as the set of all the matrices of size $m \times p$ over the field $\mathbb{K}$ and define an addition and a scalar multiplication over $\mathbb{K}$ by $A+B \triangleq (a_{i,j}+b_{i,j})_{i,j}$ and the product with a scalar $c \in \mathbb{K}$ as $cA = (ca_{i,j})_{i,j}$, for all $c \in \mathbb{K}$.

Exercise A.1. Prove that $\mathbb{K}^{n\times n}$ is a vector space over $\mathbb{K}$ in the sense of Definition A.2.

Definition A.4 (Matrix–vector product and matrix product). Let $A = (a_{i,j}) \in \mathbb{K}^{m\times p}$ and let $\mathbf{b} \in \mathbb{K}^{p\times 1} \equiv \mathbb{K}^p$. The *matrix–vector product* of A and $\mathbf{b}$ is the vector $\mathbf{c} = A\mathbf{b}$ given elementwise by $c_i = \sum_{j=1}^{p} \mathbf{a}_{i,j} b_j$.

Let $A \in \mathbb{K}^{m\times p}$, $B \in \mathbb{K}^{p\times n}$ we define the *matrix product* of A and $B = [\mathbf{b}_{:,1}, \ldots, \mathbf{b}_{:,n}]$ as $AB = [A\mathbf{b}_{:,1}, \ldots, A\mathbf{b}_{:,n}]$ and, elementwise as

$$C = AB, \quad C = (c_{i,j}), \quad c_{i,j} = \sum_{k=1}^{p} a_{i,k} b_{k,j}, \quad i = 1, \ldots, m, \ j = 1, \ldots, n.$$

Now let $(V, \mathbb{K})$ be a vector space. To characterize its elements and properties, consider the following definitions.

Definition A.5 (Linear combination, span). A *linear combination* of the vectors $\mathbf{v}_1, \ldots, \mathbf{v}_k \in V$ is a sum of scalar multiples of these vectors, i.e., $c_1\mathbf{v}_1 + \ldots + c_k\mathbf{v}_k$, for some scalar coefficients $\{c_j \in \mathbb{K}\}_{=1}^{k}$. If W is a set of vectors in V, a linear combination of vectors in W is a vector of the form $\sum_{j=1}^{k} c_j\mathbf{v}_j$ with $c_j \in \mathbb{K}$, $\mathbf{v}_j \in W$ and $k \in \mathbb{N}$.

The *span* of the vectors $\mathbf{v}_1, \ldots, \mathbf{v}_k \in V$ is the set of all linear combinations of these vectors and is denoted by $span\{\mathbf{v}_1, \ldots, \mathbf{v}_k\}$. If $W \subset V$ is a set (either finite or infinite) of vectors in V, the *span* of W, denoted by $\text{span}(W)$, is the set of all linear combinations of vectors in W. If $V = \text{span}(W)$ then W spans the vector space V.

Definition A.6 (Linear independence, basis and dimension). A set of vectors $W \subset V$ (either finite or infinite) is *linearly independent* if and only if the only linear combination of distinct vectors in W that produces the zero vector is a *trivial linear combination*, i.e., for distinct $\mathbf{v}_i$ in W, $\sum_{i=1}^{k} c_i\mathbf{v}_i = \mathbf{0}$ implies $c_1 = c_2 = \ldots = c_k = 0$. Vectors that are *not* linearly independent are *linearly dependent.*

A set of vectors $B \subset V$ in a vector space V is a *basis* for V if B is a linearly independent set and $\text{span}(B) = V$. Moreover, the number of (distinct) vectors in a basis for a vector space V is the *dimension* of V and is denoted by $\dim(V)$. If a basis for V contains a finite number of vectors, then V is *finite dimensional.* Otherwise we write $\dim(V) = \infty$ and we say that it is *infinite dimensional.*

If we consider two nonzero, finite dimensional vector spaces $(V, \mathbb{K})$ and $(W, \mathbb{K})$ we can consider the space of all the *linear transformation* T from V to W and call it $\text{Hom}_{\mathbb{K}}(V, W)$. A *linear transformation* is a mapping $T : V \to W$ such that $\forall \mathbf{u}, \mathbf{v} \in V$ and $\forall\, c \in \mathbb{K}$, $T(\mathbf{u}+\mathbf{v}) = T(\mathbf{u})+T(\mathbf{v})$ and $T(c\mathbf{u}) = cT(\mathbf{u})$. If $W = V$ we call the mapping a *linear operator* and denote the space as $\text{Hom}_{\mathbb{K}}(V)$.

Exercise A.2. Given two nonzero, finite dimensional vectors spaces $(V, \mathbb{K})$ and $(W, \mathbb{K})$ prove that $\text{Hom}_{\mathbb{K}}(V, W)$ is a vector space in the sense of Definition A.2.

Definition A.7 (Range, ker, rank). For any $T \in \text{Hom}_{\mathbb{K}}(V, W)$, i.e., for any $A \in \mathbb{K}^{m\times n}$ the *range* of A, i.e., the *image* of T, is the set of all linear combinations of the columns of A. The *kernel*, i.e., the *null space* or the *nullity*, of A, denoted by $\ker(A)$, is the set of all solutions of the homogeneous linear system $A\mathbf{x} = \mathbf{0}$. The *rank* of A, denoted by $\text{rank}(A)$, is the dimension of the range of A.

Exercise A.3. Given $T \in \text{Hom}_{\mathbb{K}}(V, W)$ prove that $\ker(T)$ is a subspace of V while the range is a subspace of W and that $\dim\ker(T) + \text{rank}(T) = \dim(V)$.

The operator $T \in \mathrm{Hom}_{\mathbb{K}}(V, W)$ is *invertible*, i.e., an *isomorphism*, if there exists a function $S \in \mathrm{Hom}_{\mathbb{K}}(W, V)$ such that $ST = I_V$ and $TS = I_W$, where I_V and I_W are the identity operator on V and W, respectively. If such S exists, it is called the *inverse* of T and is denoted by T^{-1}.

Exercise A.4. Prove that $T \in \mathrm{Hom}_{\mathbb{K}}(V, W)$ is nonsingular if and only if $\ker(T) = \{\mathbf{0}\}$ and moreover that T^{-1} is a linear transformation, invertible and such that $(T^{-1})^{-1} = T$.

Definition A.8 (Determinant). The *determinant* of a matrix $A \in \mathbb{K}^{n\times n}$, i.e., the determinant of a linear operator T, is an element in $\mathbb{K}$ defined inductively as follows

1. $\forall\, a \in \mathbb{K}$, $\det(a) = a$,
2. $\forall\, i, j \in \{1, 2, \ldots, n\}$, the ijth minor of A corresponding to $a_{i,j}$ is defined by $m_{i,j} = \det(A(\{i\}, \{j\}))$.
3. The i, jth cofactor of $a_{i,j}$ is $c_{i,j} = (-1)^{i+j} m_{i,j}$.
4. $\det(A) = \sum_{j=1}^{n} (-1)^{i+j} a_{i,j} m_{i,j} = \sum_{j=1}^{n} a_{i,j} c_{i,j}$ for $i \in \{1, 2, \ldots, n\}$.

We have taken into account a very algorithmic way to define the determinant that is usually obtained as the result of the Laplace Theorem for the computation of the determinant by exploiting the minors along the ith row, starting from a more abstract definition. Observe that the determinant is independent of the row i used to evaluate it.

Definition A.9 (Transpose, Hermitian adjoint and trace). Let $A \in \mathbb{R}^{n\times m}$. The *transpose* of the matrix $A = (a_{i,j})$ is the matrix $A^T \triangleq (a_{j,i})_{j,i} \in \mathbb{R}^{m\times n}$. Similarly, the Hermitian of the matrix $A \in \mathbb{C}^{n\times m}$ is the *Hermitian adjoint* of A, i.e., $A^H \triangleq (\overline{a}_{j,i})_{j,i}$. Given $A \in \mathbb{K}^{n\times n}$, the *trace* of A is the function $\mathrm{tr} : \mathbb{K}^{n\times n} \to \mathbb{K}$, the sum of all the diagonal entries of A, i.e., $\mathrm{tr}(A) = \sum_i a_{i,i}$.

Observe that by means of Definition A.8 we can also prove that if $A \in \mathbb{K}^{n\times n}$ then:

- $\det(A) = \det(A^T)$ if $\mathbb{K} = \mathbb{R}$. Moreover, if $\mathbb{K} = \mathbb{C}$, $\det(A^H) = \overline{\det(A)}$,
- If B is obtained from A by multiplying ore row or column by a scalar $\alpha \in \mathbb{F}$, then $\det(B) = c \det(A)$,
- For any $\alpha \in \mathbb{K}$, $\det(\alpha A) = \alpha^n \det(A)$,
- $\det(AB) = \det(A)\det(B)$. If A is invertible, $\det(A^{-1}) = \det(A)^{-1}$,
- If A is invertible and $\mathbf{x}, \mathbf{y} \in \mathbb{K}^n$ then $\det(A + \mathbf{x}\mathbf{y}^T) = \det(A)(1 + \mathbf{y}^T A^{-1}\mathbf{x})$.

Definition A.10 (Symmetric, Hermitian and skew–Hermitian matrices). A matrix $A \in \mathbb{R}^{n\times n}$ such that $A = A^T$ is called *symmetric*. If $A = -A^T$, it is called *skew–symmetric*. Similarly, a matrix $A \in \mathbb{C}^{n\times n}$ such that $A = A^H$ is called *Hermitian*. If $A = -A^H$, it is called *skew–Hermitian*.

A.1 Vector and matrix norms

Definition A.11 (Norm, normed space). Let $(V, \mathbb{K})$ be a vector space with $\mathbb{K} = \mathbb{R}$ or $\mathbb{C}$ and $f : V \to \mathbb{R}$ a function such that

1. $f(\mathbf{v}) \geq 0 \ \forall \, \mathbf{v} \in V$ and $f(\mathbf{v}) = 0$ if and only if $\mathbf{v} = \mathbf{0}$;
2. $f(\alpha \mathbf{v}) = |\alpha| f(\mathbf{v}), \ \forall \, \alpha \in \mathbb{K}, \ \forall \, \mathbf{v} \in V$;
3. $f(\mathbf{v} + \mathbf{w}) \leq f(\mathbf{v}) + f(\mathbf{w}) \ \forall \, \mathbf{v}, \mathbf{w} \in V$.

The function f is a *norm* for the vector space V. The triple $(V, \mathbb{K}, f)$ is called a *normed space.*

For any normed space $(V, \mathbb{K}, f)$, it is customary to use the notation $\|\mathbf{v}\| = f(\mathbf{v})$ to denote the norm of a vector $\mathbf{v}$. A metric can be introduced by the distance $d(\mathbf{u}, \mathbf{v}) \triangleq \|\mathbf{u} - \mathbf{v}\|$. In this way we have a vector space that is both a normed and a metric space where the crucial notion of *convergence* can be set.

Definition A.12 (Convergence, Cauchy sequences). Given a normed space $(V, \mathbb{K}, f)$ we say that a sequence of vectors $\{\mathbf{x}^{(n)}\}_n \subset V$ is *convergent* (to $\mathbf{x}^*$) if

$$\exists \, \mathbf{x}^* \in V \ \text{ such that } \ \lim_{n \to +\infty} d(\mathbf{x}^{(n)}, \mathbf{x}^*) = 0.$$

Moreover, a sequence $\{\mathbf{x}^{(n)}\}_n \subset V$ is a *Cauchy sequence* if

$$\forall \, \varepsilon > 0 \ \exists N_\varepsilon \ \text{ such that } \ \forall n, m \geq N_\varepsilon \ d(\mathbf{x}^{(n)}, \mathbf{x}^{(m)}) \leq \varepsilon.$$

A metric space is called *complete* if any Cauchy sequence in it is convergent. If the metric is induced from a norm (i.e., the space is normed and complete), then the space is a *Banach space.*

Theorem A.1 (*p*–norm). *Let $V = \mathbb{C}^n$ or $\mathbb{R}^n$. If $p \geq 1$ we have*

$$\|\mathbf{x}\|_p \triangleq \left(\sum_{i=1}^{n} |x_i|^p \right)^{1/p}, \quad \forall \mathbf{x} \in V,$$

is a norm and is called p-norm of the vector $\mathbf{x}$. *Moreover,*

$$\|\mathbf{x}\|_\infty \triangleq \max_i |x_i| = \lim_{p \to \infty} \|\mathbf{x}\|_p,$$

is a norm and is called ∞-norm of the vector $\mathbf{x}$.

Exercise A.5. Prove that if $p, q \geq 1$ are such that $1/p + 1/q = 1$ then, for any vectors $\mathbf{x}, \mathbf{y} \in V$,

$$\left| \sum_{i=1}^{n} x_i y_i \right| \leq \|\mathbf{x}\|_p \|\mathbf{y}\|_q.$$

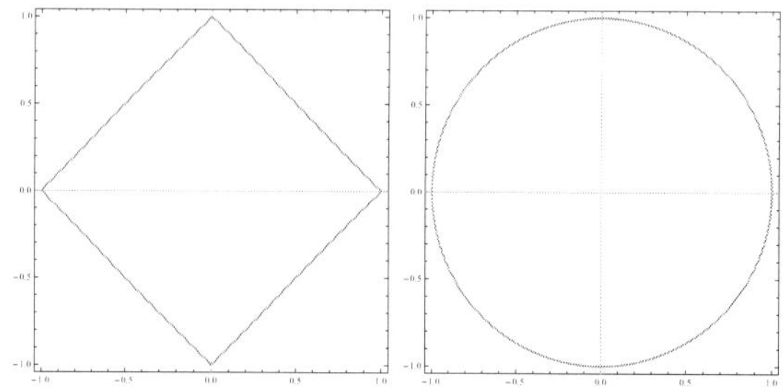

(a) $\{\mathbf{x} \in \mathbb{R}^2 \ : \ \|\mathbf{x}\|_1 \leq 1\}$ (b) $\{\mathbf{x} \in \mathbb{R}^2 \ : \ \|\mathbf{x}\|_2 \leq 1\}$

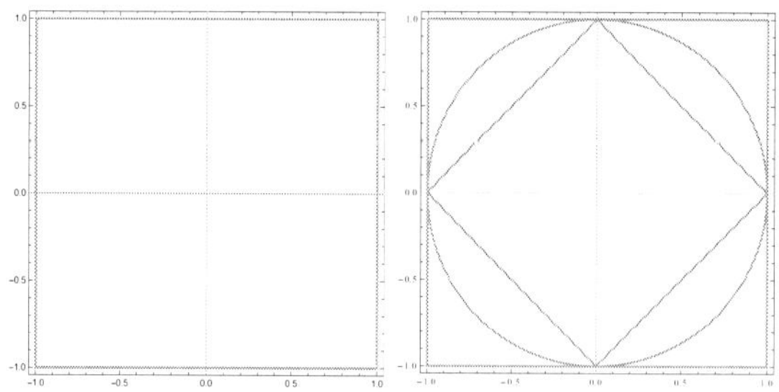

(c) $\{\mathbf{x} \in \mathbb{R}^2 \ : \ \|\mathbf{x}\|_\infty \leq 1\}$ (d) Comparison of the unit balls

Figure A.1: Unit balls for some vector norms.

In Figure A.1 the *unit balls* for the 1,2,∞–norm are depicted.

Since many norms can be defined on the same vectorial space, one needs a way to compare them.

Definition A.13 (Equivalent norms). Norms $\|\cdot\|_{n_1}$ and $\|\cdot\|_{n_2}$ on the same vector space V are called *equivalent* if there exist constants $c_1, c_2 > 0$ such that

$$c_1\|\mathbf{v}\|_{n_1} \leq \|\mathbf{v}\|_{n_2} \leq c_2\|\mathbf{v}\|_{n_1} \quad \forall\, \mathbf{v} \in V.$$

Moreover, for finite dimensional vectorial spaces a very strong characterization of equivalent norms is available.

Theorem A.2. *Any two norms on a finite dimensional vectorial space are equivalent.*

We can consider also norms for matrices. Observe that by taking all matrices of the same size we can treat them as a finite dimensional vector space (see Exercise A.1) and therefore we can define a matrix norm through the use of any vector norm.

Definition A.14 (Matrix norm). Let us consider a function $f \,:\, \mathbb{K}^{m\times n} \to \mathbb{R}$ such that

- $f(A) \geq 0\ \forall\, A \in \mathbb{K}^{m\times n}$ and $f(A) = 0$ if and only if A is the null matrix;

- $f(\alpha A) = |\alpha| f(A)$, $\forall\, \alpha \in \mathbb{K}$, $\forall\, A \in \mathbb{K}^{m\times n}$;
- $f(A+B) \leq f(A) + f(B)$ $\forall\, A,\, B \in \mathbb{K}^{m\times n}$;
- for any A, B such that AB exists, the *submultiplicative property* holds, i.e., $f(AB) \leq f(A)f(B)$.

Then f is called a matrix norm and we write $f(A)$ as $\|A\|$.

One of the most used matrix norms is the *Frobenius norm*.

Definition A.15 (Frobenius norm). Given a matrix $A \in \mathbb{K}^{m\times n}$, with $\mathbb{K} = \mathbb{R}$ or $\mathbb{C}$, the *Frobenius norm* of A is defined as

$$\|A\|_F \triangleq \left(\sum_{i=1}^{m}\sum_{j=1}^{n} |a_{i,j}|^2\right)^{1/2}.$$

There are very important norms, called *operator norms*, that are able always to give $||I|| = 1$. They can be interpreted as an operator $T_A \in \mathrm{Hom}_{\mathbb{K}}(V, W)$.

Definition A.16 (Operator norm). Let us consider a matrix $A \in \mathbb{K}^{m\times n}$ with $\mathbb{K} = \mathbb{R}$ or $\mathbb{C}$, and the normed spaces $(\mathbb{K}^m, \|\cdot\|_{n_1})$ and $(\mathbb{K}^n, \|\cdot\|_{n_2})$. The norm of the operator T_A, i.e., the norm of A, is defined as

$$\|A\| \triangleq \max_{\mathbf{x}\neq \mathbf{0}} \frac{\|A\mathbf{x}\|_{n_1}}{\|\mathbf{x}\|_{n_2}}.$$

Exercise A.6. Prove that the $\|\cdot\|$ in Definition A.16 is a norm and moreover that $\|A\mathbf{v}\|_{n_1} \leq \|A\|\|\mathbf{v}\|_{n_2}$.

By means of Definition A.16, the matrix norms associated to the p–vector norms defined in Theorem A.1, usually called *induced matrix norms*, can be defined as

$$\|A\|_p \triangleq \max_{\mathbf{v}\neq \mathbf{0}} \frac{\|A\mathbf{v}\|_p}{\|\mathbf{v}\|_p}. \tag{A.1}$$

Note that

$$\|A\|_1 \triangleq \max_{1\leq j\leq n} \sum_{i=1}^{m} |a_{i,j}|, \quad \|A\|_\infty \triangleq \max_{1\leq i\leq m} \sum_{j=1}^{n} |a_{i,j}|.$$

We postpone a characterization and more details on the matrix 2–norm to appendix A.2.1.

By definition A.13 one can easily get the following equivalence relations for all $A \in \mathbb{K}^{n\times n}$:

$$\frac{1}{\sqrt{n}}\|A\|_\infty \leq \|A\|_2 \leq \sqrt{n}\|A\|_\infty, \quad \frac{1}{\sqrt{n}}\|A\|_1 \leq \|A\|_2 \leq \sqrt{n}\|A\|_1$$
$$\|A\|_2 \leq \|A\|_F \leq \sqrt{n}\|A\|_2.$$

A.1.1 Inner products, induced norms, angles

Two other very important concepts that can be built upon a vector space are the concept of *length*, and *angle* between vectors. This feat is obtained by adding a *scalar product* to the underlying vector space.

Definition A.17 (Scalar product). Assume that $(V, \mathbb{K})$ is a vector space, with $\mathbb{K} = \mathbb{R}$ or $\mathbb{C}$, A function $< \cdot, \cdot >: V \times V \to \mathbb{K}$ such that

1. $< \mathbf{v}, \mathbf{v} >$ is real and ≥ 0, $\forall \mathbf{v} \in V$, $< \mathbf{v}, \mathbf{v} >= 0$ if and only if $\mathbf{v} = \mathbf{0}$,
2. $< \mathbf{u}, \mathbf{v} >= \overline{< \mathbf{v}, \mathbf{u} >}$, $\forall \mathbf{u}, \mathbf{v} \in V$,
3. $\forall \alpha \in \mathbb{K} < \alpha\mathbf{u}, \mathbf{v} >= \alpha < \mathbf{u}, \mathbf{v} > \forall \mathbf{u}, \mathbf{v} \in V$,
4. $< \mathbf{u} + \mathbf{v}, \mathbf{w} >=< \mathbf{u}, \mathbf{w} > + < \mathbf{v}, \mathbf{w} >$,

is called a *scalar product* over the vector space V. The triple $(V, \mathbb{R}, < \cdot, \cdot >)$ is an *Euclidean vector space*, while the triple $(V, \mathbb{C}, < \cdot, \cdot >)$ is a *unitary vector space*. Moreover, if $< \mathbf{u}, \mathbf{v} >= 0$, the two vectors $\mathbf{u}$ and $\mathbf{v}$ are *orthogonal*, written also as $\mathbf{u} \perp \mathbf{v}$.

For any matrix A, there exists a unique *Hermitian/skew-Hermitian decomposition*

$$A = Re(A) + iIm(A), \quad i^2 = -1, \quad Re(A) = \frac{A + A^H}{2}, \quad Im(A) = \frac{A - A^H}{2i}.$$

The Hermitian part of A, $Re(A)$, is also called the *real part* of A, while the skew-Hermitian part of A, $Im(A)$, the *imaginary part*. If the matrix A has only real entries, the real (imaginary) part of A is also called the *symmetric part* (the skew-symmetric) of A.

Definition A.18 (Positive definite matrix). A matrix A is called *positive definite* if

$$\mathbf{x}^H Re(A)\mathbf{x} > 0, \quad \forall \mathbf{x} \neq \mathbf{0}; \quad \mathbf{x}^H Re(A)\mathbf{x} = 0 \Leftrightarrow \mathbf{x} = 0.$$

Similarly, A matrix A is called *positive semidefinite* if the above relation holds with $\geq$ instead of $>$.

In some books a matrix A is called *positive definite* if it is Hermitian and positive definite.

A Hermitian (symmetric) positive definite matrix induces naturally a scalar product as follows.

$$< \mathbf{u}, \mathbf{v} >_A =< A\mathbf{u}, \mathbf{v} > \text{ for } \mathbf{u}, \mathbf{v} \in V. \tag{A.2}$$

On the other hand, if T is any nonsingular matrix, then we can also define a scalar product based on T by

$$< \mathbf{u}, \mathbf{v} >_T =< T^{-1}\mathbf{u}, T^{-1}\mathbf{v} > \text{ for } \mathbf{u}, \mathbf{v} \in V. \tag{A.3}$$

Exercise A.7. Prove that the expression in Equation (A.2) is a scalar product.

A Euclidean or a unitary vector space defines a norm induced naturally by its scalar product:

$$\|\mathbf{v}\| = <\mathbf{v}, \mathbf{v}>^{1/2}.$$

This norm represents the *length* of the vector $\mathbf{v} \in V$ and is usually called *induced vector norm* or simply *induced norm.*

Theorem A.3. *A norm on a vector space $(V, \|\cdot\|)$ is induce by a scalar product $<\cdot,\cdot>$ if and only if the parallelogram identity holds, i.e.,*

$$\|\mathbf{u}+\mathbf{v}\|^2 + \|\mathbf{u}-\mathbf{v}\|^2 = 2\|\mathbf{u}\|^2 + 2\|\mathbf{v}\|^2, \quad \mathbf{u}, \mathbf{v} \in V$$

holds.

Definition A.19 (Angle between vectors). Given a Euclidean vector space V, the *angle* between any two nonzero vectors $\mathbf{u}, \mathbf{v} \in V$ is the real number $\theta \in (0, \pi)$ such that $<\mathbf{u}, \mathbf{v}> = \|\mathbf{u}\|\|\mathbf{v}\| \cos\theta$.

The vector spaces $\mathbb{R}^n$ and $\mathbb{C}^n$ can be endowed by an *inner product*, or *dot product*, defined as

$$<\mathbf{u}, \mathbf{v}> = \mathbf{u}^T\mathbf{v}, \ \mathbf{u}, \mathbf{v} \in \mathbb{R}^n, \quad \text{and} \quad <\mathbf{u}, \mathbf{v}> = \mathbf{u}^H\mathbf{v}, \ \mathbf{u}, \mathbf{v} \in \mathbb{C}^n, \qquad \text{(A.4)}$$

respectively.

Exercise A.8. What relation can be established between the 2–norm and the standard inner product?

Definition A.20 (Isometric matrices). Let us consider a linear operator T represented by the matrix $Q \in \mathbb{C}^{n\times n}$. T is called *isometric* and Q is called a *norm–preserving* matrix if

$$\|Q\mathbf{v}\| = \|\mathbf{v}\|, \ \forall \mathbf{v} \in \mathbb{C}^n.$$

If we consider the set of isometric matrices for the p–norm, i.e., for $p = 2$, then we find that the preservation of the 2–norm implies the preservation of the scalar products (see Exercise A.8). Therefore, the columns of the matrix Q in Definition A.20 are orthonormal, i.e., $\mathbf{q}_{:,i}^H\mathbf{q}_{:,j} = \delta_{i,j}$ or $Q^HQ = I$. A matrix $Q \in \mathbb{C}^{n\times n}$ that is such that $Q^H = Q^{-1}$ is called *unitary* (or *orthogonal* if it has real entries).

Orthogonality can be extended to subsets $S \subset V$ for V an Euclidean, respectively unitary, vector space. S is an *orthogonal set* if $\mathbf{u} \perp \mathbf{v} \ \forall \mathbf{u}, \mathbf{v} \in S$ such that $\mathbf{u} \neq \mathbf{v}$. Moreover, we call it an *orthonormal set* if is *orthogonal* and each $\mathbf{v} \in S$ has Euclidean norm equal to one. This definition can be extended to account also for orthogonality between subsets $U, W \subset V$ for which we write that $U \perp W$ if and only if $\mathbf{u} \perp \mathbf{w}$ for all $\mathbf{u} \in U$ and for all $\mathbf{w} \in W$ and therefore we can consider the *orthogonal complement* of a subset $S \subset V$ called $S^\perp$ and such that $S^\perp = \{\mathbf{w} \in V \,|\, \mathbf{w} \perp \mathbf{v}, \ \forall \mathbf{v} \in S\}$. An *orthogonal basis* of a vector space $(V, <\cdot,\cdot>)$ is a set S of orthogonal vectors such that $V = \text{span}(S)$.

Exercise A.9. Prove that if W is a subspace of a finite dimensional vector space $(V, <\cdot,\cdot>)$ then $(W^\perp)^\perp = W$ and $\dim(W^\perp) = \dim(V) - \dim(W)$.

A.2 Eigenvalues and singular values

Definition A.21 (Eigenvalues and eigenvectors of a matrix). A scalar $\lambda \in \mathbb{K}$ is an *eigenvalue* for a matrix $A \in \mathbb{K}^{n\times n}$ if there exists a nonzero vector $\mathbf{x} \in \mathbb{K}^n$ such that $A\mathbf{x} = \lambda\mathbf{x}$. Such vector $\mathbf{x}$ is an *eigenvector* of A corresponding to the eigenvalue λ.

Nonzero row vectors $\mathbf{y}$ such that $\mathbf{y}^H A = \lambda \mathbf{y}^H$ are called lefthand eigenvectors for A.

Any matrix $A \in \mathbb{K}^{n\times n}$, with $\mathbb{K} = \mathbb{R}$ or $\mathbb{C}$, has n eigenvalues $\lambda_1, \lambda_2, \ldots, \lambda_n$ that are the n roots of the *characteristic polynomial*

$$p_A(z) = \det(A - zI) = \prod_{j=1}^{n}(z - \lambda_j). \tag{A.5}$$

If all eigenvalues (roots of the characteristic polynomial) are distinct, the roots are *distinct* and they have *algebraic multiplicity* $m_a(\lambda) = 1$. On the other hand, if $p_A(z)$ has roots that are not distinct but multiple, say λ, the *algebraic multiplicity* of λ is the power s in the factor $(z - \lambda)^s$ of $p_A(z)$ (A.5), written also for short as $m_a(\lambda) = s$. Observe that if λ is an eigenvalue of A, then the matrix $A - \lambda I$ is a singular matrix and the null space of this matrix, i.e., $\ker(A - \lambda I)$, is called the *eigenspace* of A corresponding to λ. The dimension of this eigenspace, i.e., the nullity of $A - \lambda I$, is called the *geometric multiplicity* $m_g(\lambda)$ of the eigenvalue λ. Note that

$$1 \leq m_g(\lambda) \leq m_a(\lambda). \tag{A.6}$$

and we say that A has a *complete set* of eigenvectors for the eigenvalue λ if $m_g(A) = m_a(A)$, otherwise this eigenvalues is said to be *defective*. The set of all the eigenvalues of a matrix A is called its *spectrum* and is denoted by $\sigma(A)$ or also $\lambda(A)$. The *spectral radius* $\rho(A)$ is

$$\rho(A) = \max_{\lambda \in \sigma(A)} |\lambda|.$$

Definition A.22 (Minimal polynomial). The *minimal polynomial* of a matrix $A \in \mathbb{K}^{n\times n}$ is the monic polynomial $\mu_A(z)$ of minimal degree over $\mathbb{K}$ such that $\mu_A(A) = \mathbf{0}$.

If λ is a root of μ_A, then λ is a root of the characteristic polynomial and thus an eigenvalue of A. Moreover, the multiplicity m of the root λ is the largest power such that $\ker((A - \lambda I)^m) \supset \ker((A - \lambda I)^{m-1})$ *strictly*. The following relationship between *minimal* and *characteristic* polynomial holds.

Theorem A.4 (Hamilton–Cayley). *The minimal polynomial $\mu_A(z)$ always divides the characteristic polynomial $p_A(z)$.*

A transformation that preserves the spectrum of a matrix is called a *similarity transformation.* Given any nonsingular matrix S, $B = S^{-1}AS$ is a similarity transformation because B has the the same eigenvalues of A (but not necessarily the same eigenvectors).

Exercise A.10. Prove that a *similarity transformation* preserves the spectrum of a matrix A.

Theorem A.5 (Schur Canonical Form). *For any matrix $A \in \mathbb{C}^{n\times n}$ with the spectrum $\sigma(A) = \{\lambda_1, \lambda_2, \ldots, \lambda_n\}$, there exists a unitary matrix Q such that $Q^H AQ$ is upper triangular with the eigenvalues on the main diagonal.*

If every eigenvalue of A has a complete set of eigenvectors, then there exists a similarity transformation X such that $D = X^{-1}AX$ is the diagonal matrix $D = \text{diag}(\lambda_1, \ldots, \lambda_n)$, $X = [\mathbf{x}_1, \ldots, \mathbf{x}_n]$, where $A\mathbf{x}_j = \lambda_j \mathbf{x}_j$, i.e., X is the matrix whose columns are the eigenvectors of A and D is a diagonal matrix whose nonzero entries are given by the eigenvalues of A.

When a matrix is not diagonalizable, i.e., if the matrix A has not a complete set of eigenvectors, another useful canonical form is the *Jordan canonical form.*

Theorem A.6 (Jordan Canonical Form). *Any matrix $A \in \mathbb{C}^{n\times n}$ with s distinct eigenvalues $\lambda_1, \ldots, \lambda_s$ can be decomposed as follows:*

$$A = RJR^{-1},$$

where J is a block diagonal matrix of the form

$$J = \begin{bmatrix} J(\lambda_1, k_1) & & & \\ & J(\lambda_1, k_1) & & \\ & & \ddots & \\ & & & J(\lambda_s, k_s) \end{bmatrix},$$

and $J(\lambda_i, k_i)$ is called a Jordan block of order k_i such that

$$J(\lambda_i, k_i) = \lambda_i I_{k_i} + \begin{bmatrix} 0 & 1 & & & \\ & 0 & 1 & & \\ & & \ddots & \ddots & \\ & & & 0 & 1 \\ & & & & \end{bmatrix}_{k_i \times k_i}, \quad \sum_{i=1}^{s} k_i = n.$$

If λ is an eigenvalue of A with algebraic multiplicity m_a and geometric multiplicity m_g, then λ appears in m_g blocks and the sum of the orders of these blocks is m_a.

Observe that the columns of R are the eigenvectors of A and, in the defective case, also other vectors called *principal vectors* that are needed to complete the set of the underlying eigenvectors to form a base for $\mathbb{C}^n$.

A class of matrices that can be always be diagonalized is the class of *normal matrices.*

Definition A.23 (Normal matrix). A matrix $A \in \mathbb{K}^{n\times n}$, with $\mathbb{K} = \mathbb{R}$ or $\mathbb{C}$, is called *normal* if $A^T A = AA^T$ or $A^H A = AA^H$, respectively.

Hermitian/symmetric and unitary/orthogonal matrices are normal.

Theorem A.7 (characterization of normal matrices). *The eigenvalues of real symmetric and Hermitian matrices are real and those of real skew-symmetric and skew-Hermitian matrices are purely imaginary. A normal matrix is unitary if and only if its eigenvalues have modulus equal to 1. Moreover, eigenvectors corresponding to distinct eigenvalues are mutually orthogonal.*

Exercise A.11. Give a characterization of the eigenvalues of skew–Hermitian/skew–symmetric matrices.

For Hermitian matrices some other useful characterization of the eigenvalues are available, namely they can be expressed through a variational formulation.

Theorem A.8 (Courant–Fischer). *Given $A \in \mathbb{C}^{n\times n}$ Hermitian, i.e., $A = A^H$, and $L \subseteq \mathbb{C}^n$ a subset of $\mathbb{C}^n$. Then,*

$$\lambda_k = \max_{\dim L = k} \min_{\substack{\mathbf{x}\in L,\\ \mathbf{x}\neq 0}} \frac{\mathbf{x}^H A\mathbf{x}}{\mathbf{x}^H\mathbf{x}}, \qquad \text{and} \qquad \lambda_k = \min_{\dim L = n-k+1} \max_{\substack{\mathbf{x}\in L,\\ \mathbf{x}\neq 0}} \frac{\mathbf{x}^H A\mathbf{x}}{\mathbf{x}^H\mathbf{x}}.$$

In particular,

$$\lambda_1 \leq \frac{\mathbf{x}^H A\mathbf{x}}{\mathbf{x}^H\mathbf{x}} \leq \lambda_n.$$

With this result, an estimate of the eigenvalues of a generic Hermitian matrix can be obtained by the eigenvalues of its leading principal submatrices.

Theorem A.9 (Cauchy Interlacing Property). *Given $A \in \mathbb{C}^{n\times n}$ Hermitian, i.e., $A = A^H$, and B any of its leading principal submatrices of order $n-1$. By denoting $\lambda_1,\ldots, \lambda_n$ the eigenvalues of A, $\lambda_1 \geq \ldots \geq \lambda_n \in \sigma(A)$ and by $\mu_1,\ldots, \mu_{n-1}$ the eigenvalues of B, $\mu_1 \geq \ldots \geq \mu_{n-1} \in \sigma(B)$, the following* interlacing property *holds:*

$$\lambda_k \geq \mu_k \geq \lambda_{k+1}, \; k = 1, \ldots, n-1.$$

Another useful information that can be extracted via a canonical form is the so–called *inertia*, or *signature*, of a matrix.

Theorem A.10 (Sylvester Law of Inertia, [500]). *For any symmetric square matrix $A \in \mathbb{R}^{n\times n}$ there exists a matrix $S \in \mathbb{R}^{n\times n}$ such that $D = SAS^T$ with D a diagonal matrix whose entries are* 0, +1 *and* −1, *the number of diagonal entries of each kind being an invariant for A.*

Definition A.24 (Index of inertia). The number of n_+, n_- and n_0 +1, −1 and 0 given by Theorem A.10 are the *positive index of inertia*, the *negative index of inertia* and the dimension of $\ker(A)$. Moreover, the triple $\text{In}(A) = (n_0, n_+, n_-)$ is the *signature* or *inertia* of A.

Observe also that Theorem A.10 can be expressed in term of the *eigenvalues* of A. Indeed, by the Spectral Theorem, any symmetric real matrix A has an eigendecomposition $Q\Lambda Q^T$ where Λ is a diagonal matrix whose entries are the eigenvalues of A, and Q is the orthonormal matrix of the eigenvectors of A. The matrix S giving the transformation for the *signature* is given by

$$\Lambda = WDW^T, \quad w_{i,i} = \sqrt{\lambda_i}, \quad S = QW, \; D = SAS^T.$$

Therefore, the positive and negative indices of the symmetric matrix A are also the number of positive and negative eigenvalues of A.

Theorem A.11 (SVD). *Let $A \in \mathbb{K}^{m\times n}$, with $\mathbb{K} = \mathbb{R}$ or $\mathbb{C}$, $r = \mathrm{rank}(A)$. There exist uniquely determined numbers, called* singular values, $\sigma_1 \geq \ldots \geq \sigma_r > 0$ *and unitary matrices $U \in \mathbb{C}^{m\times m}$, $V \in \mathbb{C}^{m\times m}$ called, respectively, the matrix of right and left singular vectors for which there holds*

$$A = U\Sigma V^H = \sum_{i=1}^{r} \sigma_i \mathbf{u}_i \mathbf{v}_i^H, \quad \Sigma = \mathrm{diag}(\sigma_1, \ldots, \sigma_r, 0, \ldots, 0) \in \mathbb{C}^{m\times n},$$

where both left and right singular vectors are determined uniquely up to a factor equal to 1 in modulus. Moreover,

$$\begin{array}{ll} \ker(A) = \mathrm{span}\{\mathbf{v}_{r+1}, \ldots, \mathbf{v}_n\}, & \mathrm{range}(A) = \mathrm{span}\{\mathbf{u}_1, \ldots, \mathbf{u}_r\}, \\ \ker(A^H) = \mathrm{span}\{\mathbf{u}_{r+1}, \ldots, \mathbf{u}_m\}, & \mathrm{range}(A^H) = \mathrm{span}\{\mathbf{v}_1, \ldots, \mathbf{v}_r\}. \end{array}$$

Let $A \in \mathbb{K}^{n\times n}$. A is singular if and only if $\det(A) = 0$ and if and only if $0 \in \sigma(A)$, i.e., zero is a singular value for A. Moreover, $|\det(A)| = \prod_{j=1}^{n} \sigma_j$ and $\det(A) = \prod_{j=1}^{n} \lambda_j$.

A.2.1 Eigenvalues and induced norms

Eigenvalues and norms induced by scalar or inner products are related, since, for example, the following results hold.

Theorem A.12. *For any square matrix A we have:*

$$0 \leq \rho(A) \leq \|A\|$$

for a matrix norm $\|\cdot\|$ induced by a scalar product.

Let $A \in \mathbb{K}^{n\times n}$, with $\mathbb{K} = \mathbb{R}$ or $\mathbb{C}$, and $\mathbf{u} \in \mathbb{K}^n$. The 2–norm of $\mathbf{u}$, see Theorem A.1, is defined as

$$\|\mathbf{u}\|_2 = \left(\mathbf{u}^H\mathbf{u}\right)^{1/2} = \left(\sum_{i=1}^{m} |u_i|^2\right)^{1/2} = < \mathbf{u}, \mathbf{u} >^{1/2},$$

where $< \cdot, \cdot >$ is the standard inner product (A.4), see Exercise A.8.

By Definition A.16, the 2–norm of the matrix A is defined as

$$\|A\|_2 = \max_{\mathbf{u}\neq\mathbf{0}} \frac{\|A\mathbf{u}\|_2}{\|\mathbf{u}\|_2} = \max_{\mathbf{u}\,:\,\|\mathbf{u}\|_2=1} \|A\mathbf{u}\|_2 = \max_{\mathbf{u}\,:\,\|\mathbf{u}\|_2=1} \left(\mathbf{u}^H A^H A\mathbf{u}\right)^{1/2}.$$

Note that if $\mathbf{u}$ is an eigenvector of A, $A\mathbf{u} = \lambda\mathbf{u}$, then $\left(\mathbf{u}^H A^H A\mathbf{u}\right)^{1/2} = |\lambda|$, and $\|A\|_2 \geq \rho(A)$. Now, observe that the matrix $B = A^H A$ is always Hermitian and so it is diagonalizable with a unitary eigenvector matrix, Theorem A.7, so $B = XMX^H$, with $M = \text{diag}(\mu_1, \ldots, \mu_n)$, $\mu_j \geq 0$ for all j and $X^H X = I$. Therefore,

$$\begin{aligned}
\|A\|_2 &= \max_{\mathbf{u}\,:\,\|\mathbf{u}\|_2=1} \left(\mathbf{u}^H A^H A\mathbf{u}\right)^{1/2} = \max_{\mathbf{u}\,:\,\|\mathbf{u}\|_2=1} \left(\mathbf{u}^H B\mathbf{u}\right)^{1/2} \\
&= \max_{\mathbf{w}\,:\,\|\mathbf{w}\|_2=1} \left(\mathbf{w}^H X^H BX\mathbf{w}\right)^{1/2} = \max_{\mathbf{w}\,:\,\|\mathbf{w}\|_2=1} \left(\mathbf{w}^H M\mathbf{w}\right)^{1/2} \\
&= \max_{j=1,2,\ldots,n} |\mu_j|^{1/2} = \sqrt{\rho(A^H A)}.
\end{aligned}$$

If A is normal then $\rho(A^H A) = \rho(A^2) = \rho(A)^2$ and thus $\|A\|_2 = \rho(A)$. On the the other hand, if A is not normal, then typically $\|A\|_2 > \rho(A)$. Observe that $\|A\|_2 = \sigma_1$ where σ_1 is the largest singular value of the matrix A. Moreover,

$$\forall\, A, B \in \mathbb{K}^{n\times n} \quad \|A\|_2 \leq \sqrt{\|A\|_1 \|A\|_\infty}, \; \|AB\|_F \leq \{\|A\|_F \|B\|_2, \|A\|_2 \|B\|_F\}.$$

A.2.2 Diagonal dominance and eigenvalues localization

If a matrix $A \in \mathbb{K}^{n\times n}$ is diagonal, then its eigenvalues coincide with the diagonal elements. If the off–diagonal elements are of small modulus with respect those on the diagonal, we can expect that diag(A) is a good approximations to $\sigma(A)$.

Definition A.25 (Diagonal dominance). A matrix $A \in \mathbb{K}^{n\times n}$, with $\mathbb{K} = \mathbb{R}$ or $\mathbb{C}$, is *rowwise diagonally dominant* if

$$|a_{i,i}| > r_i \triangleq \sum_{j=1,\, j\neq i}^{n} |a_{i,j}|, \quad i = 1, \ldots, n,$$

and *columnwise diagonally dominant* if

$$|a_{j,j}| > c_j \triangleq \sum_{i=1,\, i\neq j}^{n} |a_{i,j}|, \quad j = 1, \ldots, n,$$

If the previous inequalities hold with the "$\geq$" the matrix A is said to be, respectively, *weakly row–wise diagonally dominant* or *weakly column–wise diagonally dominant.*

Theorem A.13 (Levy–Desplanques). *If a matrix is row–wise or column–wise diagonally dominant then it is nonsingular.*

The Gerschgörin Theorem(s) quantifies how much the diagonal elements are good approximations for the eigenvalues providing some general bounds, i.e., bounds that are also valid in the non diagonally dominant case.

Theorem A.14 (Gerschgörin). *For a matrix $A \in \mathbb{K}^{n\times n}$, with $\mathbb{K} = \mathbb{R}$ or $\mathbb{C}$, assuming the same notation of Definition A.25, consider the Gerschgörin discs*

$$D_i^r \triangleq \{z \in \mathbb{C} : |a_{i,i} - z| \leq r_i\}, \quad D_j^c \triangleq \{z \in \mathbb{C} : |a_{j,j} - z| \leq c_j\}, \quad i, j = 1, \ldots, n.$$

The following results hold

1. *If $\lambda \in \sigma(A)$ is an eigenvalue of A then $\lambda \in (\cup_{i=1}^n D_i^r) \cap \left(\cup_{j=1}^n D_j^c\right)$.*
2. *If G is the union of m Gerschgörin discs and is isolated from the other $n - m$ discs, then G contains exactly m eigenvalues.*
3. *If the Gerschgörin discs are pairwise disjoint, then each one captures exactly one eigenvalue.*

Example A.1 (Hogben [306]). To visualize an example of application of Theorem A.14 we consider the matrix A given by

$$A = \begin{bmatrix} 3i & 1 & 0.5 & -1 & 0 \\ -1 & 2i & 1.5 & 0 & 0 \\ 1 & 2 & -7 & 0 & 1 \\ 0 & -1 & 0 & 10 & i \\ 1 & 0 & 1 & -1 & 1 \end{bmatrix},$$

for which the row Gerschgörin discs are depicted in Figure A.2. Observe three

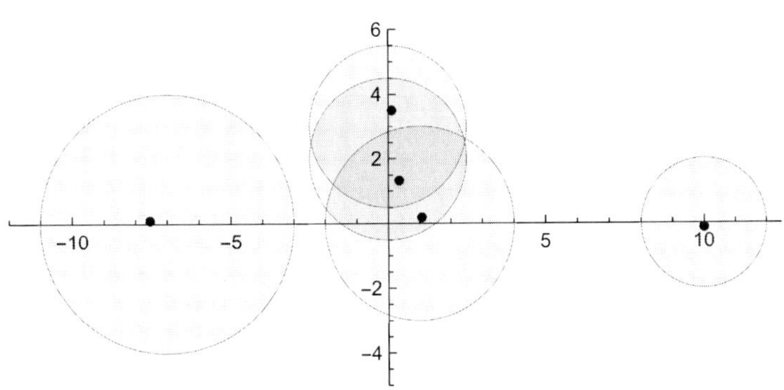

Figure A.2: Example A.1. The Row Gerschgörin circle for a matrix A and its eigenvalues (black dots).

groups of isolated Gerschgörin discs, two composed of one disc and that contain one eigenvalues each and one made of three discs that contain three eigenvalues. ✓

We recall briefly the *numerical range* for a matrix $A \in \mathbb{C}^{n\times n}$.

Definition A.26 (Numerical range). Given a matrix $A \in \mathbb{C}^{n\times n}$ we define the *numerical range* of A as the following subset of $\mathbb{C}$:

$$W(A) = \{\mathbf{x}^H A\mathbf{x} \,:\, \mathbf{x} \in \mathbb{C}^n,\ \mathbf{x}^H\mathbf{x} = 1\}.$$

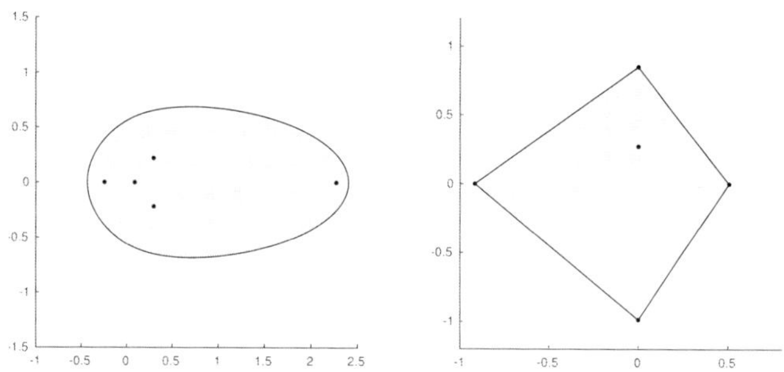

Figure A.3: Numerical range of a random matrix (on the left) and of a random normal matrix (on the right). Eigenvalues are represented as black dots.

The numerical range is such that $\sigma(A) \subset W(A)$. If A is normal, then the numerical range gives the convex hull of the eigenvalues, see, e.g., Figure A.3.

The results from § 1.1 can be extended to the spectrum of a matrix. The Bauer–Fike Theorem characterizes the perturbations on a matrix and its effects on the eigenvalues.

Theorem A.15 (Bauer–Fike). *Let A be diagonalizable, $X^{-1}AX = \Lambda$. Given $E \in \mathbb{K}^{n\times n}$, if $\mu \in \sigma(A+E)$ but $\mu \notin \sigma(A)$ then*

$$\min_{1\le i\le n} |\mu - \lambda_i| \le \kappa_2(X)\|E\|_2.$$

Note that the sensitivity of the spectrum to small perturbations can be influenced by the condition number of the matrix of the eigenvector of A.

When we know more information on the matrix and/or the underlying perturbations, more refined results and bound can be obtained.

Theorem A.16. *Let $A \in \mathbb{C}^{n\times n}$ be Hermitian, i.e., $A^H = A$ with $\lambda_1 \ge \ldots \ge \lambda_n$ its eigenvalues. Given the perturbation of A*

$$B = A + \varepsilon\mathbf{e}\mathbf{e}^H, \quad \varepsilon \in \mathbb{R},\ \mathbf{e} \in \mathbb{C}^n,\ \|\mathbf{e}\|_2 = 1,$$

where the eigenvalues of B are $\mu_1 \geq \ldots \geq \mu_n$. If $\varepsilon > 0$, then $\mu_1 \geq \lambda_1 \geq \mu_2 \geq \ldots \geq \mu_n \geq \lambda_n$, while if $\varepsilon \leq 0$ $\lambda_1 \geq \mu_1 \geq \lambda_2 \geq \ldots \geq \lambda_n \geq \mu_n$. Moreover, independently from the sign of ε,

$$\mu_k = \lambda_k + t_k \varepsilon, \quad t_k \geq 0, \quad \sum_{i=1}^{n} t_i = 1.$$

If B is a generic perturbation of A we have the following result.

Theorem A.17 (MinMax property [271]). *Given $A, B \in \mathbb{R}^{n \times n}$,*

$$\lambda_k(A) + \lambda_n(B) \leq \lambda_k(A+B) \leq \lambda_k(A) + \lambda_1(B), \quad k = 1, 2, \ldots, n.$$

Appendix B

Data sets and software codes

Numerical tests and the effective performance of a strategy are of fundamental importance for comparing algorithms. This book provides the source codes for most of the performed examples. In this appendix we give some information on them and on some popular test sets of matrices and other useful source codes.

B.1 Test matrices and data sets

In this section we collect information on test matrices used in the book.

Among the most popular collections for sparse matrices that are available on the web we can find:

- The Matrix Market [91]:
  `http://math.nist.gov/MatrixMarket/`,
- The SuiteSparse Matrix Collection [177]:
  `https://sparse.tamu.edu/`.

Both contain matrix from several problems from science and engineering that can be used as test data in comparative studies of algorithms for numerical linear algebra. The *Matrix Market* provides also generators for building some classes of matrices from classical tests like, e.g., the Wilkinson matrix, for which Gaussian elimination with partial pivoting gives huge growth factors, or, e.g., the generator for random sparse matrices with prescribed size and density.

B.1.1 Generating functions for Toeplitz matrices

A Toeplitz matrix is a matrix of the form

$$T_n = \begin{bmatrix} t_0 & t_{-1} & \dots & t_{2-n} & t_{1-n} \\ t_1 & t_0 & t_{-1} & \dots & t_{2-n} \\ \vdots & t_1 & t_0 & \ddots & \vdots \\ t_{n-2} & \dots & \ddots & \ddots & t_{-1} \\ t_{n-1} & t_{n-2} & \dots & t_1 & t_0 \end{bmatrix},$$

i.e., its entries are constant along the diagonal and can be interpreted as the *Fourier Coefficients* of a function $f(\theta) = \sum_{n=-\infty}^{+\infty} t_n e^{i \cdot n\theta}$. In Table B.1 we report the Fourier coefficients for some test functions we used in Chapter 4.

Another important class of examples of Toeplitz matrices is given by the banded Toeplitz. For the latter, the functions $f(\theta)$ are trigonometric polynomials. To get the coefficients $\{t_n\}_n$, several algorithms are available, see, e.g., [2, 169, 239, 351]. If we work with an uniform grid of step–size h, to approximate the derivative of order m of a function f, we can use the expression

$$f^{(m)}(x) \approx \sum_{i=-s}^{n-s} c_s f(x+sh), \quad n \geq m,$$

by choosing the coefficients $\{c_s\}$ such that the above approximation is accurate as desired. To obtain this, we can substitute $f(x) = e^{i\omega x}$ and expand the expression locally around $h = 0$. Then, by substituting $e^{i\omega h} = \xi$, it remains to compute the coefficients $\{c_s\}_s$ as the ones that best approximate the fraction $(\log \xi/h)^m$ in the polynomial expansion. As an example, compute the expansion for $m = n = 2$, $s = 1$:

$$\left(\frac{\log \xi}{h}\right)^2 \approx \frac{(x-1)^2}{h^2} + O\left((x-1)^3\right) = c_{-1} f(x-h) + c_0 f(x) + c_1 f(x+h)$$

therefore $c_{-1} = c_1 = 1$, $c_0 = -2$. The above procedure can be done with just a line of *Mathematica* (Mathematica is the mathematical frontend by Wolfram Research inc., see `http://www.wolfram.com`).

Listing B.1: Fornberg [239] Algorithm

```
CoefficientList [Normal[Series[x^s*Log[x]^m, {x, 1, n}]/h^m], x]
```

A collection of the coefficients for the underlying trigonometric polynomials $p(\theta)$, and of the related finite difference scheme is in Table B.2.

B.1.2 Saddle point test problems

In Chapter 4 we considered the solution of saddle point problems § 4.4. In particular, the *Quadratic Programming* techniques (QP) requires solving the

Table B.1: Generating functions for some examples of Toeplitz matrices

$f(\theta)$	Fourier coefficients	zeros/order	$\min_{x\in[-\pi,\pi]} f(\theta)$	$\max_{x\in[-\pi,\pi]} f(\theta)$
θ^2	$t_n = \begin{cases} \frac{\pi^2}{3}, & n=0, \\ \frac{2(-1)^n}{n^2}, & n\neq 0. \end{cases}$	$0/2$	0	π^2
θ^2+1	$t_n = \begin{cases} \frac{1}{3}\left(3+\pi^2\right), & n=0, \\ \frac{2(-1)^n}{n^2}, & n\neq 0. \end{cases}$	–	1	π^2+1
θ^4	$t_n = \begin{cases} \frac{\pi^4}{5}, & n=0, \\ \frac{4(-1)^n\left(\pi^2n^2-6\right)}{n^4}, & n\neq 0. \end{cases}$	$0/4$	0	π^4
θ^4+1	$t_n = \begin{cases} \frac{1}{5}\left(5+\pi^4\right), & n=0, \\ \frac{4(-1)^n\left(\pi^2n^2-6\right)}{n^4}, & n\neq 0. \end{cases}$	–	1	π^4+1
$\lvert\theta\rvert^3$	$t_n = \begin{cases} \frac{\pi^3}{4}, & n=0, \\ \frac{3(-1)^n\left(\pi^2n^2+2(-1)^n-2\right)}{\pi n^4}, & n\neq 0. \end{cases}$	$0/3$	0	π^3
$\lvert\theta\rvert^3+0,01$	$t_n = \begin{cases} \frac{1}{100}+\frac{\pi^3}{4}, & n=0, \\ \frac{3(-1)^n\left(\pi^2n^2+2(-1)^n-2\right)}{\pi n^4}, & n\neq 0. \end{cases}$	–	$0,01$	$\pi^3+0,01$
$\theta^2\left(\pi^4-\theta^4\right)$	$t_n = \begin{cases} \frac{4\pi^6}{21}, & n=0, \\ -\frac{4(-1)^n\left(\pi^4n^4-30\pi^2n^2+180\right)}{n^6}, & n\neq 0. \end{cases}$	$0/2, \pm\pi/1, \pm i\pi/1$	0	$\frac{2\pi^6}{3\sqrt{3}}$
$\theta^2\left(\pi^2-\theta^2\right)$	$t_n = \begin{cases} \frac{2\pi^4}{15}, & n=0, \\ -\frac{2(-1)^n\left(\pi^2n^2-12\right)}{n^4}, & n\neq 0. \end{cases}$	$0/2, \pm\pi/1$	0	$\frac{\pi^4}{4}$

Table B.2: Coefficients of trigonometric polynomial of few finite differences schemes

		$p(\theta) = \sum t_n e^{in\theta}$									
Order	Accuracy	t_{-4}	t_{-3}	t_{-2}	t_{-1}	t_0	t_1	t_2	t_3	t_4	t_5
	2				$-1/2$	0	$1/2$				
1	4			$1/12$	$-2/3$	0	$+2/3$	$-1/12$			
	6		$-1/60$	$3/20$	$-3/4$	0	$3/4$	$-3/20$	$1/60$		
	2				1	-2	1				
2	4			$-1/12$	$3/3$	$-5/2$	$+4/3$	$-1/12$			
	6		$1/90$	$-3/20$	$3/2$	$-49/18$	$3/2$	$-3/20$	$1/90$		
	2			$-1/2$	1	0	-1	$1/2$			
3	4		$1/8$	-1	$13/8$	0	$-13/8$	1	$-1/8$		
	6	$-7/240$	$3/10$	$-169/120$	$61/30$	0	$-61/30$	$169/120$	$-3/10$	$7/240$	
	1					-1	1				
1	2					$-3/2$	2	$-1/2$			
	3					$-11/6$	3	$-3/2$		$1/3$	
	1					1	-2	1			
2	2					2	-5	4	-1		
	3					$35/12$	$-26/3$	$19/2$	$-14/3$	$11/12$	
	1					-1	3	-3	1		
3	2					$-5/2$	9	-12	7	$-3/2$	
	3					$-17/4$	$71/4$	$-59/2$	$49/2$	$-41/4$	$7/4$

constrained optimization problem

$$\begin{cases} \min f(x) = & c_0 + \mathbf{c}^T\mathbf{x} + \frac{1}{2}\mathbf{x}^T Q\mathbf{x}, \\ \text{subject to} & A\mathbf{x} = \mathbf{b}, \\ & \mathbf{l} \leq \mathbf{x} \leq \mathbf{u}. \end{cases} \quad , Q = Q^T, Q \geq 0.$$

Refer to the collection[1] from Maros and Mészáros [378]; we include the related MATLAB (MATLAB is the mathematical frontend by Mathworks inc. `http://www.mathworks.com`) compatible files with this book. The MATLAB files have the same name of the problem and, when loaded, they include the definition of the saddle point matrix, the coefficient vectors and the right–hand sides.

To generate other saddle point matrices, see the following *MATLAB* packages: IFISS [478] by Elman et al. [220, 221] or to the DarcyLite package by Liu et al. [363]. Both packages provide the discretization of several differential problems with mixed finite elements; see, e.g., Examples 4.14, 4.15, 4.17 and 4.18.

Another source of interesting saddle point point test matrices is the `OPTPDE` collection; see [301, 413]. It is a library designed to provide a reference framework of prototypical optimization problems with PDE constraints

$$\begin{cases} \min J(y,u) = \frac{1}{2}\|y - z_d\|_2^2 + \frac{\lambda}{2}\|u\|_2^2, \\ \text{subject to } e(y,u) = 0. \end{cases} \tag{B.1}$$

where J and e are two continuously Fréchet differentiable functionals such that

$$J : Y \times U \to \mathbb{R}, \quad e : Y \times U \to W,$$

with Y, U and W reflective Banach spaces; see [181, 350] for an introduction.

B.2 Numerical Linear Algebra software

`http://www.netlib.org/`: is a collection of mathematical software, papers, and links to other relevant databases for scientific computing. It includes more than just numerical linear algebra, since it covers bibliographical indexes, codes for statistics and more computer science oriented sources.

Another very useful reference is the SourceForge repository of *free* and *open–source* software projects; see `https://sourceforge.net/`. The latter serves as a centralized hub for hosting many software projects also for scientific purposes. It is usually a good starting point to search for the existence of libraries one would like to use.

[1] The collection in the original `QPS` format can be downloaded from the three subdirectories in the anonymous ftp `ftp://ftp.sztaki.hu/pub/oplab/QPDATA`.

Other software packages used in this book have been quoted in previous chapters throughout the book, see, e.g., §§ 1.3, 3.7 and 4.2 and § 4.2.3.

Today most programming languages and frameworks (like the most popular and powerful *Matlab* and *Mathematica*) support libraries suitable for dealing with linear algebra tasks.

Just to give an hint on how to orient among the various possibilities, we could divide between two extrema:

- codes for productions and large scale benchmarks;
- codes and tests for academic and prototyping uses.

In the first case, one often can use a (relatively) low–level programming language, such as `C/C++` or one of the various version of `FORTRAN`, joined with the high–efficiency libraries like, e.g., *PETSc* [28–30], *Trilinos* [342], `CUDA` [463] and so on. The pros for the user/programmer are mainly a high efficiency, portability and scalability. The cons are mainly a less intuitive way of programming and debugging. One should take this route when having a clear objective in mind, in terms of both methods and algorithms, and for producing fast codes. Of course, the situation change in the prototyping phase. Indeed, in the second case, there exist several candidates. In this regard the `MATLAB` suite represents nowadays a standard, either for academic or industrial uses. It implements all the BLAS operations and permits some degree of interfacing with high–end libraries. Another somewhat related choice is the `GNU-OCTAVE` project [215] that is a *free*[2] mathematical frontend with some drop-in compatibility with `MATLAB` scripts. Another new and promising project of an a high-level, high-performance dynamic programming language for numerical (and then numerical linear algebra) computing is the package `Julia` [82]. Among the various features, it can be flexible enough to support some degree of parallelism and permits to integrate some `C` based functions. Finally, we mention the `SciPy` project [322], pronounced as "Sigh Pie". This is a `Python`–based *open–source* softwares and tools for various mathematics tasks that includes more than numerical linear algebra codes and can interface with various other low–level libraries, achieving both flexibility and ease of use.

B.2.1 Codes included with the book

This book is shipped with most of the codes needed for generating the included examples and how to use the underlying algorithms in some academic contexts. Most of the codes is not optimized to make it more understandable and reusable. In the spirit of what have been discussed at the beginning of this section, our codes are in various languages and uses various of the mentioned

[2] "Roughly, it means that **the users have the freedom to run, copy, distribute, study, change and improve the software**. Thus, "free software" is a matter of liberty, not price. To understand the concept, you should think of "free" as in "free speech," not as in "free beer"." – `https://www.gnu.org/philosophy/free-sw.en.html`

tools. Code for MATLAB/OCTAVE are reported with the label "M" and "O" if they can run only with one of them.

Chapter 1

matrixspy.m use of both the `spy` and `mesh` command to show the pattern and the cityplot of a sparse matrix;

sparse_mv.py Python benchmark of sparse matrix–vector product for several sparse storage formats;

Chapter 2 MATLAB, OCTAVE and SciPy contain implementations of Krylov iterative methods, so most of the following codes can be adapted to fit in these frameworks. For implementations from scratch of standard Krylov methods see the *Matlab Suite* by Y. Saad at `http://www-users.cs.umn.edu/~saad/software/`. For the original code of BiCGstab(ℓ) method, needed for running the examples on oblique projections, see the code by G. Sleijpen at `http://www.staff.science.uu.nl/~sleij101/`.

jacobi.m Jacobi fixed point iteration, see (2.5).

ex_jacobi.m Code to generate Example 2.1. Works on M, matrix for the example not included in O.

steepestdescent.m Steepest Descent from Algorithm 2.1.

cgvsonedimensional.m code for Example 2.2 comparing CG and one dimensional projection methods. Works on M. Matrix for the example not included in O.

fgmres.m FGMRES method for Algorithm 3.5. Inner variable preconditioner is GMRES(m) from Algorithm 2.8. Can be changed easily to other variable preconditioner.

ex_gmres.m Example for the GMRES and FGMRES methods, see Example 3.1.

ex_obliqueprojection.m code for the Example 2.3 comparing methods based on *oblique projections.*

Chapter 3

UFget The UFget package is a gateway for the *SuiteSparse Matrix Collection* [177]. It includes both a Java program (UFgui) and a MATLAB interface for browsing and downloading the matrices in any format on any platform.

ex_fixedpointprec.m code for Example 3.2. Usage of fixed point preconditioners.

ichol0.m Incomplete Cholesky factorization with 0 level of fill–in; see Example 3.3.

ex_ilut.m Example of application of ILUT preconditioners as preconditioners for the GMRES(m) method.

ainv.m Test code for building an AINV preconditioner, Algorithm 3.18.

ex_symprecond.m Code for Example 3.10. Test of SAINV and IC–type preconditioners for symmetric/Hermitian matrices

iluk.m The ILUT(P) Algorithm 3.10 with P generated by the level–of-fill Algorithm 3.9. The code is by K. Miller and the original can be found at https://goo.gl/AkogmK.

ex_nonsymprecond.m Code for Example 3.6. Test for the linear system obtained from the finite difference discretization of the advection–diffusion equation.

Chapter 4 Some of the codes in this section depend on the FreeFem++ library [296] to build the FEM matrices for the examples and on the Python PyAMG library [40] for the AMG preconditioners.

circmv.m matrix vector product for generic circulant matrices; see Algorithm 4.1.

ex_product.m Example of naïve and FFT–based circulant matrix vector product.

gencircprec.m Circulant preconditioner from Table 4.2.

cprec.m Application of a generic circulant preconditioner.

ex_testcirculantpreconditioner.m Example of the application of a circulant preconditioner from Table 4.2 to some Toeplitz matrix; see Example 4.2.

toepmv.m Matrix–vector product for a generic Toeplitz matrix.

ex_toeplitzdistribution.m Code for Example 4.1, distribution of the eigenvalues of a Toeplitz matrix.

ocircmv.m Matrix–vector product for generic ω–circulant matrix.

ocircsol.m Application of a generic ω–circulant preconditioner.

jacksonprec.m Construction of the Jackson circulant preconditioner.

ex_jacksonkernel.m Example of application of the Jackson circulant preconditioner; see Example 4.4.

hankenagy.m Application of the Hanke and Nagy approximate inverse Toeplitz preconditioner for banded matrices.

ex_toeplitzpreconditioner.m Application of Toeplitz and band Toeplitz preconditioners for Toeplitz linear systems; see Example 4.3.

ex_python_amg_1.py Python example for C/F–splitting for AMG algorithms; see Example 4.6 and Figure 4.10.

`ex_python_amg_2.py` Python example for aggregation AMG algorithms; see Example 4.6 Figure 4.11.

`ex_python_amg_3.py` Python example of AMG used as preconditioner; see Example 4.6.

`/toeplitzmultigrid/` This folder contains a simple implementation for the Multigrid code for Toeplitz matrices:

- `jacobi.m` Jacobi smoother for the Multigrid algorithm;
- `projector.m` projector for the Multigrid algorithm;
- `mgm.m` recursive implementation of both a V–cycle and a W–cycle of a multigrid algorithm;
- `mgm_main.m` main function for applying the Toeplitz multigrid as either an iterative solver or a preconditioner;
- `ex_toepmultigrid.m` example of a Toeplitz multigrid for some one–level Toeplitz matrices;

`ex_complexrealspectra.m` Spectral properties of the real equivalent formulations of a complex system; see Figure 4.13.

`ex_complexrealsolve.m` Solution with incomplete factorizations of the real equivalent formulations of a complex system; see Example 4.8.

`ex_helmoltz.m` Code for Example 4.9: preconditioners update for the Helmholtz equation with complex wave number.

`/freefemtomatlab/FFmatrix_fread.m` Interfaces for importing matrices and vectors built with `FreeFem++` [296] into the MATLAB environment. Original code can be recovered from the website `http://www.um.es/freefem/ff++/pmwiki.php` that contains also some example of applications of the `FreeFem++` library.

`ex_stokesuzawa.m` Code for Examples 4.17 and 4.18. Application of the incomplete Uzawa iteration for the steady state Stokes problem. **Remark**: the code depends on the `StokesUzawa.edp` FreeFem++ script to assemble the discretization matrices.

Bibliography

[1] Aboulaich, R. and M. Fortin (1989). Iterative methods for the solution of Stokes equations. *Comput. Methods Appl. Mech. Engrg. 75*(1-3), 317–324. (One citation at page 260.)

[2] Abramowitz, M. and I. A. Stegun (1964). *Handbook of mathematical functions: with formulas, graphs, and mathematical tables*, Volume 55 of *Applied mathematics series*. Courier Corporation. (2 citations in pages 7 and 294.)

[3] Achdou, Y. and F. Nataf (2007). Low frequency tangential filtering decomposition. *Numer. Linear Algebra Appl. 14*(2), 129–147. (One citation at page 210.)

[4] Adams, M. F. and R. Taylor (2000). Parallel multigrid solvers for 3D unstructured finite element problems in large deformation elasticity and plasticity. *Int. J. Numer. Meth. Eng. 48*(8), 1241–1262. (One citation at page 209.)

[5] Alber, D. M. and L. N. Olson (2007). Parallel coarse-grid selection. *Numer. Linear Algebra Appl. 14*(8), 611–643. (One citation at page 216.)

[6] Alefeld, G. and R. S. Varga (1975). Zur konvergenz des symmetrischen Relaxationsverfahrens. *Numer. Math. 25*(3), 291–295. (One citation at page 29.)

[7] Ament, M., G. Knittel, D. Weiskopf, and W. Strasser (2010). A parallel preconditioned conjugate gradient solver for the poisson problem on a multi–GPU platform. In *2010 18th Euromicro Conference on Parallel, Distributed and Network-based Processing*, pp. 583–592. IEEE. (One citation at page 124.)

[8] Ammar, G. S. and W. B. Gragg (1988). Superfast solution of real positive definite Toeplitz systems. *SIAM J. Matrix Anal. Appl. 9*(1), 61–76. (One citation at page 170.)

[9] Anzt, H., E. Chow, J. Saak, and J. Dongarra (2016). Updating incomplete factorization preconditioners for model order reduction. *Numerical Algorithms 73*(3), 611–630. (One citation at page 151.)

[10] Arnold, D. N. (1990). Mixed finite element methods for elliptic problems. *Comput.Methods in Appl.Mech.Eng. 82*(1-3), 281–300. (One citation at page 241.)

[11] Arnoldi, W. E. (1951). The principle of minimized iterations in the solution of the matrix eigenvalue problem. *Quart. Appl. Math. 9*, 17–29. (2 citations in pages 38 and 47.)

[12] Arridge, S., H. Egger, and M. Schlottbom (2013). Preconditioning of complex symmetric linear systems with applications in optical tomography. *Appl. Numer. Math. 74*, 35–48. (One citation at page 221.)

[13] Arrow, K. and L. Hurwicz (1958a). Gradient method for concave programming, III: Further global results and application to resource allocation. In *Studies in linear and non-linear programming*, Volume 8. Stanford University Press Stanford, CA. (One citation at page 258.)

[14] Arrow, K. J. and L. Hurwicz (1958b). Gradient method for concave programming, I: Local results. *Studies in Linear and Nonlinear Programming. Stanford University Press, Stanford, CA 31*, 322–338. (One citation at page 258.)

[15] Arrow, K. J., L. Hurwicz, and H. Uzawa (1958). Iterative methods for concave programming. *Studies in linear and nonlinear programming 6*, 154–165. (One citation at page 251.)

[16] Astrakhantsev, G. (1971). An iterative method of solving elliptic net problems. *Z. Vycisl. Mat. i. Mat. Fiz. 11*(2), 171–182. (One citation at page 207.)

[17] Aubert, G. and P. Kornprobst (2006). *Mathematical problems in image processing: partial differential equations and the calculus of variations*, Volume 147. Springer Science & Business Media. (One citation at page 143.)

[18] Avram, F. (1988). On bilinear forms in Gaussian random variables and Toeplitz matrices. *Probab. Theory Related Fields 79*(1), 37–45. (2 citations in pages 166 and 167.)

[19] Axelsson, O. (1979). Preconditioning of indefinite problems by regularization. *SIAM J. Numer. Anal. 16*(1), 58–69. (2 citations in pages 267 and 274.)

[20] Axelsson, O. (1985). A survey of preconditioned iterative methods for linear systems of algebraic equations. *BIT 25*(1), 165–187. (One citation at page 161.)

[21] Axelsson, O. and V. A. Barker (2001). *Finite element solution of boundary value problems: theory and computation* (illustrated edition ed.).

Classics in Applied Mathematics. Society for Industrial Mathematics. (One citation at page 104.)

[22] Axelsson, O. and M. Neytcheva (2003). Preconditioning methods for linear systems arising in constrained optimization problems. *Numer. Linear Algebra Appl. 10*(1-2), 3–31. (One citation at page 273.)

[23] Axelsson, O. and P. S. Vassilevski (1991). A black box generalized conjugate gradient solver with inner iterations and variable–step preconditioning. *SIAM J. Matrix Anal. Appl. 12*(4), 625–644. (One citation at page 85.)

[24] Bai, Z.-Z., M. Benzi, F. Chen, and Z.-Q. Wang (2012). Preconditioned MHSS iteration methods for a class of block two–by–two linear systems with applications to distributed control problems. *IMA J. Numer. Anal. 33*(1), 343–369. (One citation at page 222.)

[25] Bai, Z.-Z., G. H. Golub, and M. K. Ng (2003). Hermitian and skew-Hermitian splitting methods for non-Hermitian positive definite linear systems. *SIAM J. Matrix Anal. Appl. 24*(3), 603–626. (One citation at page 185.)

[26] Bai, Z.-Z., G. H. Golub, and M. K. Ng (2007). On successive-overrelaxation acceleration of the Hermitian and skew–Hermitian splitting iterations. *Numer. Linear Algebra Appl. 14*(4), 319–335. (One citation at page 186.)

[27] Bai, Z.-Z. and Z.-Q. Wang (2008). On parameterized inexact Uzawa methods for generalized saddle point problems. *Linear Algebra Appl. 428*(11), 2900–2932. (One citation at page 258.)

[28] Balay, S., S. Abhyankar, M. F. Adams, J. Brown, P. Brune, K. Buschelman, L. Dalcin, V. Eijkhout, W. D. Gropp, D. Kaushik, M. G. Knepley, L. C. McInnes, K. Rupp, B. F. Smith, S. Zampini, H. Zhang, and H. Zhang (2016a). PETSc users manual. Technical Report ANL-95/11 - Revision 3.7, Argonne National Laboratory. (One citation at page 298.)

[29] Balay, S., S. Abhyankar, M. F. Adams, J. Brown, P. Brune, K. Buschelman, L. Dalcin, V. Eijkhout, W. D. Gropp, D. Kaushik, M. G. Knepley, L. C. McInnes, K. Rupp, B. F. Smith, S. Zampini, H. Zhang, and H. Zhang (2016b). PETSc Web page. `http://www.mcs.anl.gov/petsc`. (No citations.)

[30] Balay, S., W. D. Gropp, L. C. McInnes, and B. F. Smith (1997). Efficient management of parallelism in object oriented numerical software libraries. In E. Arge, A. M. Bruaset, and H. P. Langtangen (Eds.), *Modern Software Tools in Scientific Computing*, pp. 163–202. Birkhäuser Press. (One citation at page 298.)

[31] Bank, R. E. and T. Dupont (1981). An optimal order process for solving finite element equations. *Math. Comp. 36*(153), 35–51. (One citation at page 207.)

[32] Bank, R. E., B. D. Welfert, and H. Yserentant (1989). A class of iterative methods for solving saddle point problems. *Numer. Math. 56*(7), 645–666. (One citation at page 254.)

[33] Barbieri, D., V. Cardellini, and S. Filippone (2009). Generalized GEMM applications on GPGPUs: Experiments and applications. In *Proc. of 2009 Int'l Conf. on Parallel Computing (ParCo 2009)*. IOS Press. (One citation at page 162.)

[34] Barlow, J. L., N. Nichols, and R. J. Plemmons (1988). Iterative methods for equality–constrained least squares problems. *SIAM J. Sci. Statist. Comput. 9*(5), 892–906. (One citation at page 251.)

[35] Barnard, S. T., L. M. Bernardo, and H. D. Simon (1999). An MPI implementation of the SPAI preconditioner on the T3E. *Int. J. High Perform. C. 13*(2), 107–123. (One citation at page 124.)

[36] Barrett, R., M. Berry, T. F. Chan, J. Demmel, J. Donato, J. Dongarra, V. Eijkhout, R. Pozo, C. Romine, and H. Van der Vorst (1994). *Templates for the solution of linear systems: building blocks for iterative methods*. SIAM. (One citation at page 12.)

[37] Barrowes, B. E., F. L. Teixeira, and J. A. Kong (2001). Fast algorithm for matrix–vector multiply of asymmetric multilevel block–Toeplitz matrices in 3–D scattering. *Microw. Opt. Techn. Let. 31*(1), 28–32. (One citation at page 172.)

[38] Battermann, A. and M. Heinkenschloss (1998). Preconditioners for Karush-Kuhn-Tucker matrices arising in the optimal control of distributed systems. In *Control and Estimation of Distributed Parameter Systems*, pp. 15–32. Springer. (One citation at page 264.)

[39] Bell, N. and M. Garland (2009). Implementing sparse matrix-vector multiplication on throughput-oriented processors. In *Proceedings of the Conference on High Performance Computing Networking, Storage and Analysis*, pp. 18. ACM. (One citation at page 19.)

[40] Bell, W. N., L. N. Olson, and J. B. Schroder (2015). PyAMG: Algebraic multigrid solvers in Python v3.0. Release 3.2. (4 citations in pages xiv, 211, 219, and 300.)

[41] Bellavia, S., D. Bertaccini, and B. Morini (2011). Nonsymmetric Preconditioner Updates in Newton–Krylov Methods for Nonlinear Systems. *SIAM J. Sci. Comput. 33*(5), 2595–2619. (4 citations in pages 144, 149, 150, and 151.)

[42] Bellavia, S., V. De Simone, D. Di Serafino, and B. Morini (2012). A preconditioning framework for sequences of diagonally modified linear systems arising in optimization. *SIAM J. Num. Anal. 50*(6), 3280–3302. (One citation at page 143.)

[43] Bellavia, S., V. De Simone, D. Di Serafino, and B. Morini (2015). Updating constraint preconditioners for KKT systems in quadratic programming via low-rank corrections. *SIAM J. on Optimization 25*(6), 1787–1808. (One citation at page 151.)

[44] Bellavia, S., V. D. Simone, D. Di Serafino, and B. Morini (2011). Efficient preconditioner updates for shifted linear systems. *SIAM J. Sci. Comput. 33*(4), 1785–1809. (2 citations in pages 143 and 144.)

[45] Benbow, S. J. (1999). Solving generalized least–squares problems with LSQR. *SIAM J. Matrix Anal. Appl. 21*(1), 166–177. (2 citations in pages 234 and 235.)

[46] Benzi, M. (1993). Solution of equality-constrained quadratic programming problems by a projection iterative method. *Rend. Mat. Appl.(7) 13*, 275–296. (One citation at page 251.)

[47] Benzi, M. (2002). Preconditioning Techniques for Large Linear Systems: A Survey. *J. Comp. Phys. 182*, 418–477. (2 citations in pages 114 and 162.)

[48] Benzi, M. (2016). *Localization in Matrix Computations: Theory and Applications*, pp. 211–317. Number 2173 in Lecture Notes in Mathematics. Springer. (2 citations in pages 10 and 163.)

[49] Benzi, M. and D. Bertaccini (2003). Approximate Inverse Preconditioning for Shifted Linear Systems. *BIT 43*, 231–244. (4 citations in pages 144, 150, 151, and 153.)

[50] Benzi, M. and D. Bertaccini (2008). Block preconditioning of real–valued iterative algorithms for Complex Linear Systems. *IMA J. of Num. Anal. 28*, 598–618. (2 citations in pages 222 and 223.)

[51] Benzi, M., J. K. Cullum, and M. Tüma (2000). Robust approximate inverse preconditioning for the conjugate gradient method. *SIAM J. Sci. Comput. 22*(4), 1318–1332. (4 citations in pages 131, 132, 134, and 139.)

[52] Benzi, M. and G. H. Golub (2004). A preconditioner for generalized saddle point problems. *SIAM J. Matrix Anal. Appl. 26*(1), 20–41. (One citation at page 233.)

[53] Benzi, M., G. H. Golub, and J. Liesen (2005). Numerical solution of saddle point problems. *Acta Numer. 14*, 1–137. (2 citations in pages 231 and 232.)

[54] Benzi, M., J. Haws, and M. Tüma (2000a). Preconditioning Highly Indefinite and Nonsymmetric Matrices. *SIAM J. Sci. Comput. 22*(4), 1333–1353. (One citation at page 142.)

[55] Benzi, M., J. C. Haws, and M. Tüma (2000b). Preconditioning highly indefinite and nonsymmetric matrices. *SIAM J. Sci. Comput. 22*(4), 1333–1353. (2 citations in pages 131 and 161.)

[56] Benzi, M., C. Meyer, and M. Tüma (1996). A Sparse Approximate Inverse Preconditioner for the Conjugate Gradient Method. *SIAM J. Sci. Comput. 17*(5), 1135–1149. (4 citations in pages 132, 134, 136, and 139.)

[57] Benzi, M., D. B. Szyld, and A. Van Duin (1999). Orderings for incomplete factorization preconditioning of nonsymmetric problems. *SIAM J. Sci. Comput. 20*(5), 1652–1670. (One citation at page 124.)

[58] Benzi, M. and M. Tüma (1998). A Sparse Approximate Inverse Preconditioner for Nonsymmetric Linear Systems. *SIAM J. Sci. Comput. 19*(3), 968–994. (2 citations in pages 132 and 134.)

[59] Benzi, M. and M. Tüma (2000). Orderings for factorized sparse approximate inverse preconditioners. *SIAM J. Sci. Comput. 21*(5), 1851–1868. (One citation at page 146.)

[60] Benzi, M. and M. Tüma (2001). A robust incomplete factorization preconditioner for positive definite matrices. *Numer. Linear Alg. Appl. 99*, 1–20. (One citation at page 161.)

[61] Bergamaschi, L., J. Gondzio, and G. Zilli (2004). Preconditioning indefinite systems in interior point methods for optimization. *Comput. Optim. Appl. 28*(2), 149–171. (One citation at page 274.)

[62] Berman, A. and R. J. Plemmons (1979). CHAPTER 6 - M–matrices. In A. Berman and R. J. Plemmons (Eds.), *Nonnegative Matrices in the Mathematical Sciences*, pp. 132–164. Academic Press. (One citation at page 94.)

[63] Bernstein, S. (1926). *Leçons sur les propriétés extrémales et la meilleure approximation des fonctions analytiques d'une variable réelle.* Paris. (One citation at page 117.)

[64] Bertaccini, D. (2000). A circulant preconditioner for the systems of LMF–based ODE codes. *SIAM J. Sci. Comp. 22*(3), 767–786. (2 citations in pages 143 and 171.)

[65] Bertaccini, D. (2001). Reliable preconditioned iterative linear solvers for some numerical integrators. *Numer. Linear Algebra Appl. 8*(2), 111–125. (One citation at page 181.)

[66] Bertaccini, D. (2002). The spectrum of circulant–like preconditioners for some general linear multistep formulas for linear boundary value problems. *SIAM J. Numer. Anal. 40*(5), 1798–1822. (One citation at page 181.)

[67] Bertaccini, D. (2004). Efficient preconditioning for sequences of parametric complex symmetric linear systems. *Electron. Trans. Numer. Anal. 18*, 49–64. (9 citations in pages 144, 151, 221, 222, 224, 225, 226, 227, and 228.)

[68] Bertaccini, D. (2007). Fast simulation of solid tumors thermal ablation treatments with a 3D reaction diffusion model. *Comput. Biol. Med. 37*, 1173–1182. (One citation at page 151.)

[69] Bertaccini, D., C. Di Fiore, and P. Zellini (2013). *Complessità e iterazione numerica. Percorsi, matrici e algoritmi veloci nel calcolo numerico*. Programma di mat. fisica elettronica. Bollati Boringhieri. (One citation at page 23.)

[70] Bertaccini, D., M. Donatelli, F. Durastante, and S. Serra-Capizzano (2017). Optimizing a multigrid Runge–Kutta smoother for variable–coefficient convection–diffusion equations. *Linear Algebra Appl. 533*, 507–535. (One citation at page 219.)

[71] Bertaccini, D. and F. Durastante (2016). Interpolating preconditioners for the solution of sequence of linear systems. *Comput. Math. Appl. 72*(4), 1118 – 1130. (4 citations in pages 144, 153, 155, and 156.)

[72] Bertaccini, D. and F. Durastante (2017). Solving mixed classical and fractional partial differential equations using short–memory principle and approximate inverses. *Numer. Algorithms 74*(4), 1061–1082. (One citation at page 151.)

[73] Bertaccini, D. and S. Filippone (2016). Sparse approximate inverse preconditioners on high performance GPU platforms. *Comput. Math. Appl. 71*(3), 693–711. (9 citations in pages 114, 127, 131, 135, 136, 137, 139, 141, and 159.)

[74] Bertaccini, D., G. H. Golub, S. Serra-Capizzano, and C. Tablino-Possio (2005). Preconditioned HSS methods for the solution of non-Hermitian positive definite linear systems and applications to the discrete convection-diffusion equation. *Numer. Math. 99*(1), 441–484. (One citation at page 186.)

[75] Bertaccini, D. and M. K. Ng (2000). Skew–circulant preconditioners for systems of LMF–based ODE codes. In *International Conference on Numerical Analysis and Its Applications*, pp. 93–101. Springer. (One citation at page 181.)

[76] Bertaccini, D. and M. K. Ng (2003). Block $\{\omega\}$-circulant preconditioners for the systems of differential equations. *Calcolo 40*(2), 71–90. (One citation at page 171.)

[77] Bertaccini, D., M. Popolizio, and F. Durastante (2017). Adaptive updating techniques for the approximation of functions of large matrices. *arXiv preprint arXiv:1709.06351*. Preprint. (One citation at page 149.)

[78] Bertaccini, D. and F. Sgallari (2010). Updating preconditioners for nonlinear deblurring and denoising image restoration. *Appl. Numer. Math. 60*, 994–1006. (2 citations in pages 143 and 151.)

[79] Bertaccini, D. and R. Sisto (2011). Fast numerical solution of nonlinear nonlocal cochlear models. *J. Comput. Phys. 230*, 2575–2587. (One citation at page 151.)

[80] Bertrand, F. and P. A. Tanguy (2002). Krylov–based Uzawa algorithms for the solution of the Stokes equations using discontinuous–pressure tetrahedral finite elements. *J. Comp. Phys. 181*(2), 617–638. (One citation at page 257.)

[81] Bevilacqua, R., D. Bini, M. Capovani, and O. Menchi (1992). *Metodi numerici.* Zanichelli. (One citation at page 4.)

[82] Bezanson, J., A. Edelman, S. Karpinski, and V. B. Shah (2017). Julia: A fresh approach to numerical computing. *SIAM Rev. 59*(1), 65–98. (One citation at page 298.)

[83] Bianchi, D., A. Buccini, M. Donatelli, and S. Serra-Capizzano (2015). Iterated fractional Tikhonov regularization. *Inverse Problems 31*(5), 055005. (One citation at page 171.)

[84] Bini, D., M. Capovani, and O. Menchi (1988). *Metodi Numerici per l'Algebra Lineare.* Zanichelli. (One citation at page 4.)

[85] Bini, D. and P. Favati (1993). On a matrix algebra related to the discrete Hartley transform. *SIAM J. Matrix Anal. Appl. 14*(2), 500–507. (One citation at page 187.)

[86] Birken, P., J. Duintjer Tebbens, A. Meister, and M. Tüma (2008). Preconditioner updates applied to CFD model problems. *Appl. Numer. Math. 58*(11), 1628–1641. (2 citations in pages 143 and 151.)

[87] Bisseling, R. H. and J. G. van de Vorst (1991). Parallel Triangular System Solving on a mesh network of Transputers. *SIAM J. Sci. Comput. 12*(4), 787–799. (One citation at page 158.)

[88] Bitar, L. and C. Vincent (2000). Eigenvalue upper bounds for the discretized Stokes operator. *Commun. Numer. Meth. Engng. 16*(7), 449–457. (One citation at page 241.)

[89] Bitmead, R. R. and B. D. Anderson (1980). Asymptotically fast solution of Toeplitz and related systems of linear equations. *Linear Algebra Appl. 34*, 103–116. (One citation at page 170.)

[90] Björck, A. (1996). *Numerical methods for least squares problems.* SIAM. (One citation at page 184.)

[91] Boisvert, R. F., R. Pozo, K. Remington, R. F. Barrett, and J. J. Dongarra (1997). Matrix Market: a web resource for test matrix collections. In *Quality of Numerical Software*, pp. 125–137. Springer. (One citation at page 293.)

[92] Boman, E. and I. Koltracht (1995). Fast transform based preconditioners for Toeplitz equations. *SIAM J. Matrix Anal. Appl. 16*(2), 628–645. (One citation at page 187.)

[93] Böttcher, A. and S. M. Grudsky (2005). *Spectral properties of banded Toeplitz matrices.* SIAM. (One citation at page 164.)

[94] Böttcher, A. and S. M. Grudsky (2012). *Toeplitz matrices, asymptotic linear algebra, and functional analysis.* Birkhäuser. (One citation at page 168.)

[95] Böttcher, A. and B. Silbermann (2012). *Introduction to large truncated Toeplitz matrices.* Springer Science & Business Media. (2 citations in pages 164 and 168.)

[96] Bozzo, E. and C. Di Fiore (1995). On the use of certain matrix algebras associated with discrete trigonometric transforms in matrix displacement decomposition. *SIAM J. Matrix Anal. Appl. 16*(1), 312–326. (One citation at page 187.)

[97] Braess, D. (1995). Towards algebraic multigrid for elliptic problems of second order. *Computing 55*(4), 379–393. (One citation at page 216.)

[98] Braess, D. and W. Hackbusch (1983). A new convergence proof for the multigrid method including the v-cycle. *SIAM J. Numer. Anal. 20*(5), 967–975. (One citation at page 207.)

[99] Bramble, J., J. Pasciak, and A. Vassilev (2000). Uzawa type algorithms for nonsymmetric saddle point problems. *Math. Comp. 69*(230), 667–689. (One citation at page 257.)

[100] Bramble, J. H. and J. E. Pasciak (1988). A preconditioning technique for indefinite systems resulting from mixed approximations of elliptic problems. *Math. Comp. 50*(181), 1–17. (One citation at page 264.)

[101] Bramble, J. H., J. E. Pasciak, and A. T. Vassilev (1997). Analysis of the inexact Uzawa algorithm for saddle point problems. *SIAM J. Numer. Anal. 34*(3), 1072–1092. (2 citations in pages 254 and 255.)

[102] Brandt, A. (1986). Algebraic multigrid theory: The symmetric case. *Appl. Math. Comput. 19*(1-4), 23–56. (4 citations in pages 206, 208, 215, and 216.)

[103] Brent, R. P., F. G. Gustavson, and D. Y. Yun (1980). Fast solution of Toeplitz systems of equations and computation of Padé approximants. *J. Algorithms 1*(3), 259–295. (2 citations in pages 164 and 170.)

[104] Brezinski, C. and M. Redivo-Zaglia (1994). Treatment of near–breakdown in the CGS algorithm. *Numer. Algorithms 7*(1), 33–73. (One citation at page 64.)

[105] Brezinski, C. and M. Redivo-Zaglia (1995). Look–ahead in Bi-CGSTAB and other product methods for linear systems. *BIT 35*(2), 169–201. (One citation at page 60.)

[106] Brezinski, C., M. Redivo Zaglia, and H. Sadok (1999). New look–ahead Lanczos–type algorithms for linear systems. *Numer. Mathematik 83*(1), 53–85. (One citation at page 60.)

[107] Brezinski, C., M. R. Zaglia, and H. Sadok (1991). Avoiding breakdown and near-breakdown in Lanczos type algorithms. *Numer. Algorithms 1*(2), 261–284. (One citation at page 59.)

[108] Brezinski, C., M. R. Zaglia, and H. Sadok (1992). A breakdown–free Lanczos type algorithm for solving linear systems. *Numer. Math. 63*(1), 29–38. (One citation at page 59.)

[109] Brezzi, F. (1974). On the existence, uniqueness and approximation of saddle–point problems arising from Lagrangian multipliers. *RAIRO 8*(2), 129–151. (One citation at page 253.)

[110] Brezzi, F. and M. Fortin (2012). *Mixed and hybrid finite element methods*, Volume 15. Springer Science & Business Media. (One citation at page 241.)

[111] Briggs, W. L., V. E. Henson, and S. F. McCormick (2000). *A multigrid tutorial*. SIAM. (One citation at page 198.)

[112] Bröker, O. and M. J. Grote (2002). Sparse approximate inverse smoothers for geometric and algebraic multigrid. *Appl. Numer. Math. 41*(1), 61–80. (One citation at page 210.)

[113] Bröker, O., M. J. Grote, C. Mayer, and A. Reusken (2001). Robust parallel smoothing for multigrid via sparse approximate inverses. *SIAM J. Sci. Comput. 23*(4), 1396–1417. (One citation at page 210.)

[114] Bruun, G. (1978). z–transform DFT filters and FFT's. *IEEE T. Acoust. Speech 26*(1), 56–63. (One citation at page 169.)

[115] Bunch, J. R. (1985). Stability of methods for solving Toeplitz systems of equations. *SIAM J. Sci. Comput. 6*(2), 349–364. (One citation at page 170.)

[116] Bunch, J. R. and L. Kaufman (1977). Some stable methods for calculating inertia and solving symmetric linear systems. *Math. Comp. 31*, 163–179. (One citation at page 250.)

[117] Bunch, J. R. and B. N. Parlett (1971). Direct methods for solving symmetric indefinite systems of linear equations. *SIAM J. Numer. Anal. 8*(4), 639–655. (One citation at page 250.)

[118] Buttari, A., P. D'Ambra, D. di Serafino, and S. Filippone (2004). Extending PSBLAS to build parallel Schwarz preconditioners. In *International Workshop on Applied Parallel Computing*, pp. 593–602. Springer. (2 citations in pages 210 and 219.)

[119] Buttari, A., P. D'Ambra, D. Di Serafino, and S. Filippone (2007). 2LEV-D2P4: a package of high-performance preconditioners for scientific and engineering applications. *Appl. Algebra Engrg. Comm. Comput. 18*(3), 223–239. (2 citations in pages 210 and 219.)

[120] Byun, J.-H., R. Lin, K. A. Yelick, and J. Demmel (2012). Autotuning sparse matrix-vector multiplication for multicore. Technical report, EECS, UC Berkeley. (One citation at page 19.)

[121] Cahouet, J. and J.-P. Chabard (1988). Some fast 3D finite element solvers for the generalized Stokes problem. *Internat. J. Numer. Methods Fluids 8*(8), 869–895. (One citation at page 256.)

[122] Cai, M.-C. and X.-Q. Jin (2005). BCCB preconditioners for solving linear systems from delay differential equations. *Comput. Math. Appl. 50*(1), 281–288. (One citation at page 171.)

[123] Cai, X.-C. and Y. Saad (1996). Overlapping domain decomposition algorithms for general sparse matrices. *Numer. Linear Algebra Appl. 3*(3), 221–237. (One citation at page 210.)

[124] Cai, X.-C. and O. B. Widlund (1992). Domain decomposition algorithms for indefinite elliptic problems. *SIAM J. Sci. Statist. Comput. 13*(1), 243–258. (One citation at page 210.)

[125] Calgaro, C., J.-P. Chehab, and Y. Saad (2010). Incremental incomplete LU factorizations with applications. *Numer. Linear Algebra Appl. 17*(5), 811–837. (One citation at page 143.)

[126] Calvetti, D. and L. Reichel (2003). Tikhonov regularization of large linear problems. *BIT 43*(2), 263–283. (One citation at page 143.)

[127] Canuto, C., V. Simoncini, and M. Verani (2014). On the decay of the inverse of matrices that are sum of Kronecker products. *Linear Algebra Appl. 452*, 21–39. (One citation at page 118.)

[128] Cao, Z.-H. (2004). Fast Uzawa algorithms for solving non-symmetric stabilized saddle point problems. *Numer. Linear Algebra Appl. 11*(1), 1–24. (One citation at page 258.)

[129] Carayannis, G., N. Kalouptsidis, and D. Manolakis (1982). Fast recursive algorithms for a class of linear equations. *IEEE Trans. Acoust., Speech, Signal Process. 30*(2), 227–239. (One citation at page 173.)

[130] Cardellini, V., S. Filippone, and D. Rouson (2014). Design patterns for sparse-matrix computations on hybrid CPU/GPU platforms. *Scientific Programming 22*(1), 1–19. (One citation at page 162.)

[131] Carpentieri, B., I. S. Duff, and L. Giraud (2000). Sparse pattern selection strategies for robust Frobenius-norm minimization preconditioners in electromagnetism. *Numer. Linear Algebra Appl. 7*(7-8), 667–685. (One citation at page 120.)

[132] Chan, R. H. (1989). Circulant preconditioners for Hermitian Toeplitz systems. *SIAM J. Matrix Anal. Appl. 10*(4), 542–550. (One citation at page 176.)

[133] Chan, R. H. (1991). Toeplitz preconditioners for Toeplitz systems with nonnegative generating functions. *IMA J. Numer. Anal. 11*(3), 333–345. (3 citations in pages 188, 192, and 193.)

[134] Chan, R. H. and T. F. Chan (1992). Circulant preconditioners for elliptic problems. *J. Numer. Lin. Alg. and Appl. 1*, 77–101. (2 citations in pages 171 and 175.)

[135] Chan, R. H., Q.-S. Chang, and H.-W. Sun (1998). Multigrid method for ill–conditioned symmetric Toeplitz systems. *SIAM J. Sci. Comput. 19*(2), 516–529. (One citation at page 217.)

[136] Chan, R. H. and X.-Q. Jin (1991). Circulant and skew-circulant preconditioners for skew–Hermitian type Toeplitz systems. *BIT 31*(4), 632–646. (One citation at page 184.)

[137] Chan, R. H., X.-Q. Jin, and M.-C. Yeung (1991a). The circulant operator in the Banach algebra of matrices. *Linear Algebra Appl. 149*, 41–53. (One citation at page 180.)

[138] Chan, R. H., X.-Q. Jin, and M.-C. Yeung (1991b). The spectra of super–optimal circulant preconditioned Toeplitz systems. *SIAM J. Numer. Anal. 28*(3), 871–879. (One citation at page 180.)

[139] Chan, R. H. and K.-P. Ng (1993a). Fast iterative solvers for Toeplitz–plus–band systems. *SIAM J. Sci. Comput. 14*(5), 1013–1019. (One citation at page 173.)

[140] Chan, R. H. and K.-P. Ng (1993b). Toeplitz preconditioners for Hermitian Toeplitz systems. *Linear Algebra Appl. 190*, 181–208. (One citation at page 188.)

[141] Chan, R. H., M. K. Ng, and R. J. Plemmons (1996). Generalization of Strang's preconditioner with applications to Toeplitz least squares problems. *Numer. Linear Algebra Appl. 3*(1), 45–64. (One citation at page 181.)

[142] Chan, R. H., M. K. Ng, and C. Wong (1996). Sine transform based preconditioners for symmetric Toeplitz systems. *Linear Algebra Appl. 232*, 237–259. (One citation at page 187.)

[143] Chan, R. H., M. K. Ng, and A. M. Yip (2002). The best circulant preconditioners for hermitian Toeplitz systems II: The multiple-zero case. *Numer. Math. 92*(1), 17–40. (2 citations in pages 182 and 183.)

[144] Chan, R. H., D. Potts, and G. Steidl (2001). Preconditioners for non–Hermitian Toeplitz systems. *Numer. Linear Algebra Appl. 8*(2), 83–98. (One citation at page 185.)

[145] Chan, R. H. and P. T. P. Tang (1994). Fast band–Toeplitz preconditioners for Hermitian Toeplitz systems. *SIAM J. Sci. Comput. 15*(1), 164–171. (2 citations in pages 188 and 196.)

[146] Chan, R. H. and M.-C. Yeung (1992a). Circulant preconditioners constructed from kernels. *SIAM J. Numer. Anal. 29*(4), 1093–1103. (3 citations in pages 175, 176, and 179.)

[147] Chan, R. H. and M.-C. Yeung (1992b). Circulant preconditioners for Toeplitz matrices with positive continuous generating functions. *Math. Comput. 58*(197), 233–240. (One citation at page 177.)

[148] Chan, R. H. and M.-C. Yeung (1993). Circulant preconditioners for complex Toeplitz matrices. *SIAM J. Numer. Anal. 30*(4), 1193–1207. (One citation at page 184.)

[149] Chan, R. H., A. M. Yip, and M. K. Ng (2001). The best circulant preconditioners for Hermitian Toeplitz systems. *SIAM J. Numer. Anal. 38*(3), 876–896. (One citation at page 182.)

[150] Chan, T. F. (1988). An optimal circulant preconditioner for Toeplitz systems. *SIAM J. Sci. Stat. Comp. 9*(4), 766–771. (2 citations in pages 175 and 176.)

[151] Chan, T. F. and P. C. Hansen (1992). A look–ahead Levinson algorithm for general Toeplitz systems. *IEEE Transactions on signal processing 40*(5), 1079–1090. (One citation at page 170.)

[152] Chan, T. F. and T. P. Mathew (1994). Domain decomposition algorithms. *Acta Numer. 3*, 61–143. (One citation at page 210.)

[153] Chan, T. F. and H. A. Van der Vorst (1997). Approximate and incomplete factorizations. In *Parallel numerical algorithms*, pp. 167–202. Springer. (One citation at page 111.)

[154] Chen, Z., R. E. Ewing, R. D. Lazarov, S. Maliassov, and Y. A. Kuznetsov (1996). Multilevel preconditioners for mixed methods for second order elliptic problems. *Numer. Linear Algebra Appl. 3*(5), 427–453. (One citation at page 219.)

[155] Ching, W. K. (2013). *Iterative methods for queuing and manufacturing systems*. Springer Science & Business Media. (One citation at page 164.)

[156] Chizhonkov, E. (2001). On solving an algebraic system of Stokes type under block diagonal preconditioning. *Comput. Math. Math. Phys. 41*(4), 514–521. (One citation at page 264.)

[157] Chizhonkov, E. V. (2002). Improving the convergence of the lanczos method in solving algebraic saddle point problems. *Comput. Math. Math. Phys. 42*(4), 483–491. (One citation at page 264.)

[158] Chorin, A. J. (1968). Numerical solution of the Navier-Stokes equations. *Math. Comp. 22*(104), 745–762. (One citation at page 265.)

[159] Chow, E. (2000). A priori sparsity patterns for parallel sparse approximate inverse preconditioners. *SIAM J. Sci. Comput. 21*(5), 1804–1822. (One citation at page 120.)

[160] Chow, E. (2001). Parallel implementation and practical use of sparse approximate inverse preconditioners with a priori sparsity patterns. *Int. J. High Perform C. 15*(1), 56–74. (One citation at page 124.)

[161] Chow, E. and Y. Saad (1997). Experimental study of ILU preconditioners for indefinite matrices. *J. Comput. Appl. Math. 86*(2), 387–414. (One citation at page 161.)

[162] Chow, E. and Y. Saad (1998). Approximate inverse preconditioners via sparse-sparse iterations. *SIAM J. Sci. Comput. 19*(3), 995–1023. (3 citations in pages 124, 125, and 126.)

[163] Ciarlet, P. G. and P.-A. Raviart (1974). A mixed finite element method for the biharmonic equation. In *Proceedings of Symposium on Mathematical Aspects of Finite Elements in PDE*, pp. 125–145. (One citation at page 260.)

[164] Cipolla, S. and F. Durastante (2018). Fractional PDE constrained optimization: An optimize-then-discretize approach with L–BFGS and approximate inverse preconditioning. *Appl. Numer. Math. 123*, 43–57. (2 citations in pages 151 and 264.)

[165] Cleary, A. J., R. D. Falgout, V. E. Henson, and J. E. Jones (1998). *Coarse-grid selection for parallel algebraic multigrid*, pp. 104–115. Berlin, Heidelberg: Springer Berlin Heidelberg. (One citation at page 216.)

[166] Cleary, A. J., R. D. Falgout, V. E. Henson, J. E. Jones, T. A. Manteuffel, S. F. McCormick, G. N. Miranda, and J. W. Ruge (2000). Robustness and scalability of algebraic multigrid. *SIAM J. Sci. Comput. 21*(5), 1886–1908. (One citation at page 216.)

[167] Clemens, M. and T. Weiland (1996). Iterative methods for the solution of very large complex symmetric linear systems of equations in electrodynamics. In E. Manteuffel and S. McCormick (Eds.), *Eleventh Copper Mountain Conference on Iterative Methods.* (One citation at page 221.)

[168] Clemens, M., T. Weiland, and U. Van Rienen (1998). Comparison of Krylov–type methods for complex linear systems applied to high–voltage problems. *IEEE Trans. Magn. 34*(5), 3335–3338. (One citation at page 221.)

[169] Cohen, J. K. and D. R. DeBaun (1992). Discrete approximation of linear functions. *Mathematica Journal 2*(2), 62–65. (One citation at page 294.)

[170] Cooley, J. W. and J. W. Tukey (1965). An algorithm for the machine calculation of complex fourier series. *Math. Comput. 19*(90), 297–301. (2 citations in pages 169 and 170.)

[171] Cosgrove, J., J. Díaz, and A. Griewank (1992). Approximate inverse preconditionings for sparse linear systems. *Int. J. Comput. Math. 44*(1-4), 91–110. (One citation at page 126.)

[172] Dalton, S., N. Bell, L. Olson, and M. Garland (2014). Cusp: Generic Parallel Algorithms for Sparse Matrix and Graph Computations. Version 0.5.0. (2 citations in pages 19 and 161.)

[173] D'Ambra, P., D. D. Serafino, and S. Filippone (2010). MLD2P4: a package of parallel algebraic multilevel domain decomposition preconditioners in Fortran 95. *ACM Trans. Math. Softw. 37*(3), 30. (One citation at page 219.)

[174] D'Ambra, P., D. D. Serafino, and S. Filippone (2015). Mld2p4 multilevel domain decomposition parallel preconditioners package based on psblas. Version 2.0. (One citation at page 219.)

[175] Davis, J. D. and E. S. Chung (2012). SpMV: A memory-bound application on the GPU stuck between a rock and a hard place. Technical report, Microsoft Research Silicon Valley. (One citation at page 19.)

[176] Davis, T. A. (2006). *Direct methods for sparse linear systems*, Volume 2. SIAM. (One citation at page 2.)

[177] Davis, T. A. and Y. Hu (2011). The University of Florida sparse matrix collection. *ACM Trans. Math. Software 38*(1), 1. (8 citations in pages 87, 90, 102, 107, 135, 223, 293, and 299.)

[178] Day, D. and M. A. Heroux (2001). Solving complex-valued linear systems via equivalent real formulations. *SIAM J. Sci. Comput. 23*(2), 480–498. (2 citations in pages 221 and 222.)

[179] De Cecchis, D., H. López, and B. Molina (2009). FGMRES preconditioning by symmetric/skew–symmetric decomposition of generalized Stokes problems. *Math. Comput. in Simulat. 79*(6), 1862–1877. (One citation at page 85.)

[180] de Hoog, F. (1987). A new algorithm for solving Toeplitz systems of equations. *Linear Algebra Appl. 88*, 123–138. (One citation at page 170.)

[181] De los Reyes, J. C. (2015). *Numerical PDE–constrained optimization.* Springer. (One citation at page 297.)

[182] de Sturler, E. and J. Liesen (2005). Block–diagonal and constraint preconditioners for nonsymmetric indefinite linear systems. part I: Theory. *SIAM J. Sci. Comput. 26*(5), 1598–1619. (2 citations in pages 261 and 262.)

[183] Dehnavi, M. M., F. D. M., J. L. Gaudiot, and D. D. Giannacopoulos (2013, Sept). Parallel Sparse Approximate Inverse Preconditioning on Graphic Processing Units. *IEEE Trans. Parallel Distrib. Syst. 24*(9), 1852–1862. (One citation at page 124.)

[184] Delves, L. M. and J. Mohamed (2008). *Computational methods for integral equations.* CUP. (One citation at page 173.)

[185] Demko, S., W. F. Moss, and P. W. Smith (1984). Decay rates for inverses of band matrices. *Math. Comp. 43*(168), 491–499. (4 citations in pages 116, 117, 118, and 145.)

[186] Demmel, J., J. Dongarra, and J. Langou (2016). LAPACK. (One citation at page 159.)

[187] Dennis, Jr., J. and J. Moré (1977). Quasi–Newton Methods, Motivation and Theory. *SIAM Rev. 19*(1), 46–89. (One citation at page 142.)

[188] Deshpande, V., M. J. Grote, P. Messmer, and W. Sawyer (1996). Parallel implementation of a sparse approximate inverse preconditioner. In *International Workshop on Parallel Algorithms for Irregularly Structured Problems*, pp. 63–74. Springer. (One citation at page 124.)

[189] Di Benedetto, F. (1998). Solution of Toeplitz normal equations by sine transform based preconditioning. *Linear Algebra Appl. 285*(1), 229–255. (One citation at page 187.)

[190] Di Benedetto, F. and S. Serra-Capizzano (1999). A unifying approach to abstract matrix algebra preconditioning. *Numer. Math. 82*(1), 57–90. (One citation at page 187.)

[191] Di Fiore, C. and P. Zellini (2001). Matrix algebras in optimal preconditioning. *Linear Algebra Appl. 335*(1), 1–54. (One citation at page 187.)

[192] Dobrev, V. A., R. D. Lazarov, P. S. Vassilevski, and L. T. Zikatanov (2006). Two–level preconditioning of discontinuous Galerkin approximations of second–order elliptic equations. *Numer. Linear Algebra Appl. 13*(9), 753. (One citation at page 219.)

[193] Dollar, H. S. and A. J. Wathen (2006). Approximate factorization constraint preconditioners for saddle-point matrices. *SIAM J. Sci. Comput. 27*(5), 1555–1572. (2 citations in pages 269 and 271.)

[194] Donatelli, M. (2010). An algebraic generalization of local Fourier analysis for grid transfer operators in multigrid based on Toeplitz matrices. *Numer. Linear Algebra Appl. 17*(2-3), 179–197. (One citation at page 217.)

[195] Donatelli, M., A. Dorostkar, M. Mazza, M. Neytcheva, and S. Serra-Capizzano (2017). Function-based block multigrid strategy for a two-dimensional linear elasticity-type problem. *Comput. Math. Appl. 74*(5), 1015–1028. SI: SDS2016 – Methods for PDEs. (One citation at page 219.)

[196] Donatelli, M., C. Garoni, M. Mazza, S. Serra-Capizzano, and D. Sesana (2014). Spectral behavior of preconditioned non-Hermitian multilevel block Toeplitz matrices with matrix–valued symbol. *Appl. Math. Comput. 245*, 158–173. (One citation at page 186.)

[197] Donatelli, M., C. Garoni, M. Mazza, S. Serra-Capizzano, and D. Sesana (2016). Preconditioned HSS method for large multilevel block Toeplitz linear systems via the notion of matrix–valued symbol. *Numer. Linear Algebra Appl. 23*(1), 83–119. (One citation at page 186.)

[198] Donatelli, M., M. Mazza, and S. Serra-Capizzano (2016). Spectral analysis and structure preserving preconditioners for fractional diffusion equations. *J. Comp. Phys. 307*, 262–279. (One citation at page 188.)

[199] Dongarra, J., J. Bunch, C. Moler, and G. Stewart (1970). LINPACK. (One citation at page 159.)

[200] Dongarra, J., M. Gates, A. Haidar, J. Kurzak, P. Luszczek, S. Tomov, and I. Yamazaki (2014). *Accelerating Numerical Dense Linear Algebra Calculations with GPUs*, Chapter 1, pp. 3–28. Springer International Publishing. (2 citations in pages 19 and 159.)

[201] Douglas, C. C. (1984). Multi–grid algorithms with applications to elliptic boundary value problems. *SIAM J. Numer. Anal. 21*(2), 236–254. (One citation at page 207.)

[202] Du, I. S., R. G. Grimes, and J. G. Lewis (1992). Users' guide for the Harwell–Boeing sparse matrix collection (Release I). Technical report, Report RAL-92-086, Atlas Centre, Rutherford Appleton Laboratory, Didcot, Oxon, UK. (One citation at page 134.)

[203] Duff, I. S., A. M. Erisman, and J. K. Reid (1986). *Direct methods for sparse matrices.* Clarendon Press Oxford. (2 citations in pages 2 and 12.)

[204] Duff, I. S., R. G. Grimes, and J. G. Lewis (1989). Sparse matrix test problems. *ACM Trans. Math. Soft. 15*(1), 1–14. (One citation at page 13.)

[205] Duff, I. S. and J. Koster (1999). The Design and Use of Algorithms for Permuting Large Entries to the Diagonal of Sparse Matrices. *SIAM J. Matrix Anal. Appl. 20*, 889–901. (One citation at page 142.)

[206] Duff, I. S. and J. Koster (2001). On Algorithms For Permuting Large Entries to the Diagonal of a Sparse Matrix. *SIAM J. Matrix Anal. Appl. 22*, 973–996. (One citation at page 142.)

[207] Duff, I. S. and G. A. Meurant (1989). The effect of ordering on preconditioned conjugate gradients. *BIT 29*(4), 635–657. (One citation at page 124.)

[208] Dupont, T., R. P. Kendall, and H. Rachford, Jr (1968). An approximate factorization procedure for solving self-adjoint elliptic difference equations. *SIAM J. Numer. Anal. 5*(3), 559–573. (One citation at page 111.)

[209] Durastante, F. Preconditioned fast solvers for large linear systems with specific sparse and/or Toeplitz–like structures and applications. (4 citations in pages 143, 151, 153, and 219.)

[210] Durastante, F. (2015). Interpolant Update of Preconditioners for Sequences of Large Linear Systems. In *Mathematical Methods, Computational Techniques and Intelligent Systems (MAMECTIS '15)*, Volume 41, pp. 40–47. WSEAS Press. (One citation at page 153.)

[211] Durazzi, C. and V. Ruggiero (2003a). Indefinitely preconditioned conjugate gradient method for large sparse equality and inequality constrained quadratic problems. *Numer. Linear Algebra Appl. 10*(8), 673–688. (One citation at page 274.)

[212] Durazzi, C. and V. Ruggiero (2003b). Numerical solution of special linear and quadratic programs via a parallel interior–point method. *Parallel Comput. 29*(4), 485–503. (One citation at page 274.)

[213] Dym, H. and H. McKean (1972). *Fourier Series and Integrals*. Probability and mathematical statistics. Academic Press. (One citation at page 164.)

[214] Dyn, N. and W. E. Ferguson (1983). The numerical solution of equality constrained quadratic programming problems. *Math. Comp. 41*(163), 165–170. (One citation at page 272.)

[215] Eaton, J. W. and et Al. (2017). GNU octave. `https://www.gnu.org/software/octave/`. (One citation at page 298.)

[216] Efstathiou, E. and M. J. Gander (2003). Why restricted additive Schwarz converges faster than additive Schwarz. *BIT 43*(5), 945–959. (One citation at page 210.)

[217] Elman, H. C. (1986). A Stability Analysis of Incomplete LU Factorizations. *Math. Comp. 47*(175), 191–217. (One citation at page 111.)

[218] Elman, H. C., O. G. Ernst, and D. P. O'leary (2001). A multigrid method enhanced by Krylov subspace iteration for discrete Helmholtz equations. *SIAM J. Sci. Comput. 23*(4), 1291–1315. (One citation at page 209.)

[219] Elman, H. C. and G. H. Golub (1994). Inexact and preconditioned Uzawa algorithms for saddle point problems. *SIAM J. Numer. Anal. 31*(6), 1645–1661. (One citation at page 254.)

[220] Elman, H. C., A. Ramage, and D. Silvester (2007). Algorithm 866: IFISS, a Matlab toolbox for modelling incompressible flow. *ACM Trans. Math. Softw. 33*, 2–14. (One citation at page 297.)

[221] Elman, H. C., A. Ramage, and D. Silvester (2014). IFISS: A computational laboratory for investigating incompressible flow problems. *SIAM Review 56*, 261–273. (One citation at page 297.)

[222] Elman, H. C., D. J. Silvester, and A. J. Wathen (2014). *Finite elements and fast iterative solvers: with applications in incompressible fluid dynamics*. Oxford University Press (UK). (One citation at page 274.)

[223] Elsner, J. B. and A. A. Tsonis (2013). *Singular spectrum analysis: a new tool in time series analysis*. Springer Science & Business Media. (One citation at page 164.)

[224] Ewing, R. E., R. D. Lazarov, P. Lu, and P. S. Vassilevski (1990). *Preconditioning indefinite systems arising from mixed finite element discretization of second-order elliptic problems*, pp. 28–43. Berlin, Heidelberg: Springer Berlin Heidelberg. (One citation at page 268.)

[225] Falgout, R. D. and U. M. Yang (2002). HYPRE: A library of high performance preconditioners. In *International Conference on Computational Science*, pp. 632–641. Springer. (One citation at page 220.)

[226] Fan, K. (1960). Note on M–matrices. *Q. J. Math. 11*(1), 43–49. (2 citations in pages 98 and 99.)

[227] Fernandes, P. and P. Girdinio (1989). A new storage scheme for an efficient implementation of the sparse matrix-vector product. *Parallel Comput. 12*(3), 327–333. (One citation at page 12.)

[228] Field, M. (1997). An efficient parallel preconditioner for the conjugate gradient algorithm. Technical report, Hitachi Dublin Laboratory Technical Report HDL-TR-97-175, Dublin, Ireland. (One citation at page 124.)

[229] Filippone, S. (2012). Approximate Inverse Preconditioners for Solving Large and Sparse Linear Systems. Ph.d. thesis, Universitàă di Roma "Tor Vergata". (One citation at page 107.)

[230] Filippone, S. and M. Colajanni (2000). PSBLAS: A library for parallel linear algebra computation on sparse matrices. *ACM Trans. on Math. Software 26*(4), 527–550. (3 citations in pages 19, 159, and 219.)

[231] Fiorentino, G. and S. Serra-Capizzano (1991). Multigrid methods for Toeplitz matrices. *Calcolo 28*(3), 283–305. (2 citations in pages 209 and 217.)

[232] Fiorentino, G. and S. Serra-Capizzano (1996a). Multigrid methods for indefinite Toeplitz matrices. *Calcolo 33*(3-4), 223–236. (One citation at page 217.)

[233] Fiorentino, G. and S. Serra-Capizzano (1996b). Multigrid methods for symmetric positive definite block Toeplitz matrices with nonnegative generating functions. *SIAM J. Sci. Comput. 17*(5), 1068–1081. (2 citations in pages 209 and 217.)

[234] Fischer, B., A. Ramage, D. J. Silvester, and A. J. Wathen (1998). Minimum residual methods for augmented systems. *BIT 38*(3), 527–543. (4 citations in pages 245, 261, 263, and 264.)

[235] Fischer, B. and L. Reichel (1989). A stable Richardson iteration method for complex linear systems. *Numer. Math. 54*(2), 225–242. (One citation at page 222.)

[236] Fischer, R. and T. Huckle (2003). Using ω-circulant matrices for the preconditioning of Toeplitz systems. *Selçuk J. Appl. Math 4*, 71–88. (5 citations in pages 174, 180, 181, 190, and 191.)

[237] Fletcher, R. (1975). An ideal penalty function for constrained optimization. *IMA J. Appl. Math. 15*(3), 319–342. (One citation at page 248.)

[238] Flórez, E., M. Garcıa, L. González, and G. Montero (2002). The effect of orderings on sparse approximate inverse preconditioners for nonsymmetric problems. *Adv. Eng. Softw. 33*(7), 611–619. (One citation at page 124.)

[239] Fornberg, B. (1998). Classroom note: Calculation of weights in finite difference formulas. *SIAM review 40*(3), 685–691. (One citation at page 294.)

[240] Forsgren, A. (1996). On linear least–squares problems with diagonally dominant weight matrices. *SIAM J. Matrix Anal. Appl. 17*(4), 763–788. (One citation at page 247.)

[241] Forsgren, A. (2002). Inertia–controlling factorizations for optimization algorithms. *Appl. Numer. Math. 43*(1), 91–107. (One citation at page 240.)

[242] Forsgren, A. and W. Murray (1993). Newton methods for large-scale linear equality-constrained minimization. *SIAM J. Matrix Anal. Appl. 14*(2), 560–587. (One citation at page 240.)

[243] Fox, L., H. D. Huskey, and J. H. Wilkinson (1948). Notes on the solution of algebraic linear simultaneous equations. *Quart. J. Mech. and Applied Math. 1*, 149–173. (One citation at page 132.)

[244] Frangioni, A. and C. Gentile (2004). New preconditioners for KKT systems of network flow problems. *SIAM J. Optim. 14*(3), 894–913. (One citation at page 265.)

[245] Freund, R. W. (1990). On conjugate gradient type methods and polynomial preconditioners for a class of complex non-hermitian matrices. *Numer. Math. 57*(1), 285–312. (One citation at page 222.)

[246] Freund, R. W. (1992). Conjugate gradient–type methods for linear systems with complex symmetric coefficient matrices. *SIAM J. Sci. Statist. Comput. 13*(1), 425–448. (One citation at page 221.)

[247] Freund, R. W. (1993). A transpose–free quasi minimal residual algorithm for non-Hermitian linear systems. *SIAM J. Sci. Stat. Comput. 14*(1), 470–482. (2 citations in pages 62 and 71.)

[248] Freund, R. W., M. H. Gutknecht, and N. M. Nachtigal (1993). An implementation of the Look-Ahead Lanczos algorithm. *SIAM J. Sci. Stat. Comput. 14*(2), 470–482. (One citation at page 61.)

[249] Freund, R. W. and N. M. Nachtigal (1991). QMR: a quasi-minimal residual method for non-Hermitian linear systems. *Numer. Math. 60*(1), 315–339. (One citation at page 62.)

[250] Freund, R. W. and N. M. Nachtigal (1995). Software for simplified Lanczos and QMR algorithms. *Appl. Numer. Math. 19*(3), 319–341. (One citation at page 270.)

[251] Frigo, M. and S. Johnson (2005). The Design and Implementation of FFTW3. *Proceedings of the IEEE 93*(2), 216–231. Special issue on "Program Generation, Optimization, and Platform Adaptation". (One citation at page 169.)

[252] Frobenius, G. F. (1908). Über matrizen aus positiven elementen, 1. *B. Preuss. Akad. Wiss. Berlin, Germany.*, 471–476. (3 citations in pages 28, 95, and 96.)

[253] Frobenius, G. F. (1909). Über matrizen aus positiven elementen, 2. *B. Preuss. Akad. Wiss. Berlin, Germany.*, 514–518. (No citations.)

[254] Frobenius, G. F. (1912). *Über Matrizen aus nicht negativen Elementen.* Königliche Akademie der Wissenschaften. (3 citations in pages 28, 95, and 96.)

[255] Gansterer, W., J. Schneid, and C. Ueberhuber (2003). Mathematical properties of equilibrium systems. Technical report, Department of Distributed and Multimedia Systems, University of Vienna. (One citation at page 233.)

[256] Garoni, C. and S. Serra-Capizzano. *Generalized Locally Toeplitz Sequences: Theory and Applications* (1 ed.). Springer International Publishing. (One citation at page 173.)

[257] Gauss, C. F. (1903). Werke, 12 vols. *Göttingen: Köninglichen Gesellschaft der Wissenschaften. 9*, 279. (One citation at page 25.)

[258] Gauthier, A., F. Saleri, and A. Veneziani (2004). A fast preconditioner for the incompressible Navier Stokes Equations. *Comput. Visual. Sci. 6*(2-3), 105–112. (One citation at page 265.)

[259] Gijzen, M. B., G. L. Sleijpen, and J.-P. M. Zemke (2015). Flexible and multi–shift induced dimension reduction algorithms for solving large sparse linear systems. *Numer. Linear Algebra Appl. 22*(1), 1–25. (One citation at page 87.)

[260] Gilbert, J. R. (1994). Predicting structure in sparse matrix computations. *SIAM J. Matrix Anal. Appl. 15*(1), 62–79. (One citation at page 77.)

[261] Gilbert, J. R. and M. T. Heath (1987). Computing a sparse basis for the null space. *SIAM J. Algebraic Discrete Methods 8*(3), 446–459. (One citation at page 272.)

[262] Gilbert, J. R., C. Moler, and R. Schreiber (1992). Sparse matrices in MATLAB: Design and implementation. *SIAM J. Matrix Anal. Appl. 13*(1), 333–356. (One citation at page 12.)

[263] Gill, P. E., W. Murray, D. B. Ponceleón, and M. A. Saunders (1992). Preconditioners for indefinite systems arising in optimization. *SIAM J. Matrix Anal. Appl. 13*(1), 292–311. (One citation at page 261.)

[264] Gill, P. E., W. Murray, M. A. Saunders, and M. H. Wright (1991). Inertia-controlling methods for general quadratic programming. *SIAM Rev. 33*(1), 1–36. (One citation at page 240.)

[265] Glowinski, R. (2003). *Finite element methods for incompressible viscous flow.* Number 9 in Handbook of numerical analysis. Elsevier. (One citation at page 256.)

[266] Gohberg, I. and I. Koltracht (1989). Efficient algorithm for Toeplitz plus Hankel matrices. *Integral Equations Operator Theory 12*(1), 136–142. (One citation at page 173.)

[267] Gohberg, I. and A. Semencul (1972). On the inversion of finite Toeplitz matrices and their continuous analogs. *Mat. issled 2*, 201–233. (2 citations in pages 170 and 172.)

[268] Goldstine, H. (2012). *A History of Numerical Analysis from the 16th through the 19th Century.* Studies in the History of Mathematics and Physical Sciences. Springer New York. (One citation at page 275.)

[269] Golinskii, L. and S. Serra-Capizzano (2007). The asymptotic properties of the spectrum of nonsymmetrically perturbed Jacobi matrix sequences. *J. Approx. Theory 144*(1), 84–102. (One citation at page 78.)

[270] Golub, G. H. and C. Greif (2003). On solving block–structured indefinite linear systems. *SIAM J. Sci. Comput. 24*(6), 2076–2092. (6 citations in pages xi, 79, 247, 248, 250, and 255.)

[271] Golub, G. H. and C. F. Van Loan (1996). *Matrix computations* (3rd ed ed.). Johns Hopkins studies in the mathematical sciences. Johns Hopkins University Press. (14 citations in pages 2, 9, 23, 24, 38, 39, 50, 77, 93, 108, 120, 133, 236, and 291.)

[272] Good, I. J. (1958). The interaction algorithm and practical Fourier analysis. *J. Royal Statist. Soc. 20*(2), 361–372. (One citation at page 169.)

[273] Goodman, J., P. Flatau, and B. Draine (1991). Application of fast–Fourier–transform techniques to the discrete–dipole approximation. *Opt. Let. 16*(15), 1198–1200. (One citation at page 172.)

[274] Goossens, S., K. Tan, and D. Roose (1998). An efficient FGMRES solver for the shallow water equations based on domain decomposition. In *Domain Decomposition Methods in Sciences and Engineering*, pp. 350–358. (One citation at page 85.)

[275] Gould, N. I. (1985). On practical conditions for the existence and uniqueness of solutions to the general equality quadratic programming problem. *Math. Program. 32*(1), 90–99. (One citation at page 240.)

[276] Gould, N. I., M. E. Hribar, and J. Nocedal (2001). On the solution of equality constrained quadratic programming problems arising in optimization. *SIAM J. Sci. Comput. 23*(4), 1376–1395. (2 citations in pages 271 and 272.)

[277] Gould, N. I. and J. A. Scott (1998). Sparse approximate-inverse preconditioners using norm–minimization techniques. *SIAM J. Sci. Comput. 19*(2), 605–625. (One citation at page 120.)

[278] Gräser, C. and R. Kornhuber (2007). On preconditioned Uzawa–type iterations for a saddle point problem with inequality constraints. In *Domain decomposition methods in science and engineering XVI*, pp. 91–102. Springer. (One citation at page 258.)

[279] Greenbaum, A. (1984). Analysis of a multigrid method as an iterative technique for solving linear systems. *SIAM J. Numer. Anal. 21*(3), 473–485. (One citation at page 207.)

[280] Greenbaum, A. (1997). *Iterative methods for solving linear systems*, Volume 17. SIAM. (9 citations in pages 34, 38, 47, 50, 56, 70, 71, 80, and 239.)

[281] Greenbaum, A., V. Pták, and Z. Strakoš (1996). Any Nonincreasing Convergence Curve is Possible for GMRES. *SIAM J. Matrix Anal. Appl. 17*(3), 465–469. (One citation at page 56.)

[282] Greenbaum, A. and Z. Strakoš (1994). *Matrices that generate the same Krylov residual spaces.* Springer. (One citation at page 56.)

[283] Grenander, U. and G. Szegö (2001). *Toeplitz forms and their applications*, Volume 321. University of California Press. (4 citations in pages 164, 166, 167, and 172.)

[284] Grote, M. J. and T. Huckle (1997). Parallel preconditioning with sparse approximate inverses. *SIAM J. Sci. Comput. 18*(3), 838–853. (2 citations in pages 120 and 121.)

[285] Guo, H., G. A. Sitton, and C. S. Burrus (1998). The quick Fourier transform: an FFT based on symmetries. *IEEE Trans. Signal Process. 46*(2), 335–341. (One citation at page 169.)

[286] Gustafsson, I. (1978). A class of first order factorization methods. *BIT 18*(2), 142–156. (One citation at page 111.)

[287] Hackbusch, W. (1980). Convergence of multigrid iterations applied to difference equations. *Math. Comp. 34*(150), 425–440. (One citation at page 207.)

[288] Hagen, R., S. Roch, and B. Silbermann (2000). *C*–algebras and numerical analysis.* CRC Press. (2 citations in pages 119 and 168.)

[289] Hamming, R. W. (1989). *Digital filters.* Courier Corporation. (One citation at page 176.)

[290] Hanke, M. and J. G. Nagy (1994). Toeplitz approximate inverse preconditioner for banded Toeplitz matrices. *Numer. Algorithms 7*(2), 183–199. (6 citations in pages xiv, 188, 189, 190, 191, and 192.)

[291] Hansen, P. C. (1998). *Rank-deficient and discrete ill-posed problems: numerical aspects of linear inversion.* SIAM. (One citation at page 143.)

[292] Hansen, P. C. (2002). Deconvolution and regularization with Toeplitz matrices. *Numer. Algorithms 29*(4), 323–378. (One citation at page 171.)

[293] Hansen, P. C., J. G. Nagy, and D. P. O'leary (2006). *Deblurring images: matrices, spectra, and filtering*, Volume 3. SIAM. (One citation at page 164.)

[294] Harrod, W. J., J. G. Nagy, and R. J. Plemmons (1994). Image restoration using fast Fourier and wavelet transforms. In *Substance Identification Technologies*, pp. 369–380. International Society for Optics and Photonics. (One citation at page 187.)

[295] Heath, M. T. and C. H. Romine (1988). Parallel solution of triangular systems on distributed–memory multiprocessors. *SIAM J. Sci. Comput. 9*(3), 558–588. (One citation at page 158.)

[296] Hecht, F. (2012). New development in FreeFem++. *J. Numer. Math. 20*(3-4), 251–265. (4 citations in pages 157, 257, 300, and 301.)

[297] Heinig, G. et al. (2013). *Algebraic methods for Toeplitz–like matrices and operators*, Volume 13. Birkhäuser. (One citation at page 170.)

[298] Heinig, G. and A. Bojanczyk (1997). Transformation techniques for Toeplitz and Toeplitz-plus-Hankel matrices. I. Transformations. *Linear Algebra Appl. 254*(1), 193–226. (One citation at page 173.)

[299] Heinig, G. and A. Bojanczyk (1998). Transformation techniques for Toeplitz and Toeplitz-plus-Hankel matrices II. Algorithms. *Linear Algebra Appl. 278*(1), 11–36. (One citation at page 173.)

[300] Hemmingsson, L. (1996). Toeplitz Preconditioners with Block Structure for First-order PDEs. *Numer. Linear Algebra Appl. 3*(1), 21–44. (One citation at page 188.)

[301] Herzog, R., A. Rösch, S. Ulbrich, and W. Wollner (2014). OPTPDE: A collection of problems in PDE-constrained optimization. In G. Leugering, P. Benner, S. Engell, A. Griewank, H. Harbrecht, M. Hinze, R. Rannacher, and S. Ulbrich (Eds.), *Trends in PDE Constrained Optimization*, Volume 165 of *International Series of Numerical Mathematics*, pp. 539–543. Springer International Publishing. (One citation at page 297.)

[302] Hestenes, M. R. (2012). *Conjugate direction methods in optimization*, Volume 12. Springer Science & Business Media. (One citation at page 248.)

[303] Hestenes, M. R. and E. Stiefel (1952). Methods of conjugate gradients for solving linear systems. *J. Res. Nat. Bur. Stand. 49*, 409–436. (One citation at page 39.)

[304] Higham, N. (2002). *Accuracy and Stability of Numerical Algorithms* (Second ed.). Society for Industrial and Applied Mathematics. (One citation at page 4.)

[305] Hochstenbach, M. E. and L. Reichel (2010). An iterative method for Tikhonov regularization with a general linear regularization operator. *J. Integral Equations Appl 22*(3). (One citation at page 143.)

[306] Hogben, L. (2006). *Handbook of Linear Algebra*. Discrete Mathematics and Its Applications. CRC Press. (One citation at page 289.)

[307] Hopcroft, J. E. and J. D. Ullman (1983). *Data structures and algorithms*, Volume 175. Addison-Wesley Boston, MA, USA:. (One citation at page 131.)

[308] Horesh, L., M. Schweiger, M. Bollhöfer, A. Douiri, D. S. Holder, and S. R. Arridge (2006). Multilevel preconditioning for 3D large-scale soft field medical applications modelling. *Int. J. Inf. Syst. Sci 2*, 532–56. (One citation at page 222.)

[309] Huckle, T. (1993). Some aspects of circulant preconditioners. *SIAM J. Sci. Comput. 14*(3), 531–541. (One citation at page 176.)

[310] Huckle, T. (1994). Iterative methods for Toeplitz–like matrices. Manuscript sccm {94–05}, Scientific Computing and Computational Mathematics Program, Stanford University, Stanford, CA. (One citation at page 180.)

[311] Huckle, T. (1999). Approximate sparsity patterns for the inverse of a matrix and preconditioning. *Appl. Numer. Math. 30*(2), 291–303. (One citation at page 120.)

[312] Huckle, T. (2000). Factorized sparse approximate inverses for preconditioning and smoothing. *Selcuk J. Appl. Math 1*, 63. (One citation at page 122.)

[313] Huckle, T. (2003). Factorized sparse approximate inverses for preconditioning. *J Supercomput 25*(2), 109–117. (One citation at page 122.)

[314] Huckle, T. and J. Staudacher (2002). Multigrid preconditioning and Toeplitz matrices. *Electron. Trans. Numer. Anal 13*, 81–105. (One citation at page 218.)

[315] IEEE (2008, Aug). IEEE Standard for Floating–Point Arithmetic. Standard, Institute of Electrical and Electronics Engineers (IEEE). (One citation at page 2.)

[316] Jacobi, C. G. J. (1845). Über eine neue auflösungsart der bei der Methode der kleinsten Quadrate vorkommenden lineären Gleichungen. *Astr. Nachr. 22*(523), 297–306. (One citation at page 25.)

[317] Jaffard, S. (1990). Propriétés des matrices «bien localisées» près de leur diagonale et quelques applications. In *Annales de l'IHP Analyse non linéaire*, Volume 7.5, pp. 461–476. (One citation at page 119.)

[318] Jeffers, J. and J. Reinders (2013). *Intel Xeon Phi coprocessor high–performance programming*. Newnes. (One citation at page 19.)

[319] Jin, X.-Q. (1994). Hartley preconditioners for Toeplitz systems generated by positive continuous functions. *BIT 34*(3), 367–371. (One citation at page 187.)

[320] Jin, X.-Q. (1996). Band Toeplitz preconditioners for block Toeplitz systems. *J. Comput. Appl. Math. 70*(2), 225–230. (One citation at page 188.)

[321] Jing, Y.-F., T.-Z. Huang, Y. Duan, and B. Carpentieri (2010). A comparative study of iterative solutions to linear systems arising in quantum mechanics. *J. Comp. Phys. 229*(22), 8511–8520. (One citation at page 221.)

[322] Jones, E., T. Oliphant, P. Peterson, et al. (2001–). SciPy: Open source scientific tools for Python. [Online; accessed January 29, 2018]. (One citation at page 298.)

[323] Kailath, T. and J. Chun (1994). Generalized displacement structure for block-Toeplitz, Toeplitz-block, and Toeplitz-derived matrices. *SIAM J. Matrix Anal. Appl. 15*(1), 114–128. (One citation at page 171.)

[324] Kaporin, I. E. (1994). New convergence results and preconditioning strategies for the conjugate gradient method. *Numer. Linear Algebra Appl. 1*(2), 179–210. (One citation at page 122.)

[325] Kaporin, I. E. (1998). High quality preconditioning of a general symmetric positive definite matrix based on its RTU–decomposition. *Numer. Linear Algebra Appl. 5*(6), 483–509. (One citation at page 97.)

[326] Kazeev, V. A., B. N. Khoromskij, and E. E. Tyrtyshnikov (2013). Multilevel Toeplitz matrices generated by tensor–structured vectors and convolution with logarithmic complexity. *SIAM J. Sci. Comput. 35*(3), A1511–A1536. (One citation at page 172.)

[327] Keller, C., N. I. Gould, and A. J. Wathen (2000). Constraint preconditioning for indefinite linear systems. *SIAM J. Matrix Anal. Appl. 21*(4), 1300–1317. (3 citations in pages 269, 270, and 271.)

[328] Kelley, C. T. (1995). *Iterative Methods for Linear and Nonlinear Equations*. SIAM. (One citation at page 38.)

[329] Kershaw, D. S. (1978). The incomplete Cholesky–conjugate gradient method for the iterative solution of systems of linear equations. *J. Comput. Phys. 26*(1), 43–65. (2 citations in pages 93 and 161.)

[330] Knoll, D. and D. Keyes (2004). Jacobian–free Newton–Krylov methods: a survey of approaches and applications. *J. Comp. Phys. 193*, 357–397. (One citation at page 142.)

[331] Kobelkov, G. M. and M. A. Olshanskii (2000). Effective preconditioning of Uzawa type schemes for a generalized Stokes problem. *Numer. Math. 86*(3), 443–470. (One citation at page 256.)

[332] Kolotilina, L. Y. and A. Y. Yeremin (1993). Factorized sparse approximate inverse preconditionings I. Theory. *SIAM J. Matrix Anal. Appl. 14*(1), 45–58. (2 citations in pages 121 and 122.)

[333] Kolotilina, L. Y. and A. Y. Yeremin (1995). Factorized sparse approximate inverse preconditioning II: Solution of 3D FE systems on massively parallel computers. *Int. J. High Speed Com. 7*(02), 191–215. (One citation at page 121.)

[334] Körner, T. (1989). *Fourier Analysis.* Cambridge University Press. (One citation at page 164.)

[335] Koschinski, C. (1999). *Properties of approximate inverses and adaptive control concepts for preconditioning.* Ph. D. thesis, Karlsruhe Institute of Technology. (One citation at page 122.)

[336] Krechel, A. and K. Stüben (1998). Operator dependent interpolation in algebraic multigrid. In *Multigrid methods V*, pp. 189–211. Springer. (One citation at page 215.)

[337] Krzyzanowski, P. (2001). On block preconditioners for nonsymmetric saddle point problems. *SIAM J. Sci. Comput. 23*(1), 157–169. (One citation at page 264.)

[338] Ku, T.-K. and C.-C. J. Kuo (1993a). Preconditioned iterative methods for solving Toeplitz–plus–Hankel systems. *SIAM J. Numer. Anal. 30*(3), 824–845. (One citation at page 173.)

[339] Ku, T.-K. and C.-C. J. Kuo (1993b). Spectral properties of preconditioned rational Toeplitz matrices: the nonsymmetric case. *SIAM J. Matrix Anal. Appl. 14*(2), 521–544. (One citation at page 185.)

[340] Kuznetsov, Y. A. (1995). Efficient iterative solvers for elliptic finite element problems on nonmatching grids. *Russian J. Numer. Anal. Math. Modelling 10*(3), 187–212. (One citation at page 264.)

[341] Kuznetsov, Y. A. (2004). Efficient preconditioner for mixed finite element methods on nonmatching meshes. *Russian J. Numer. Anal. Math. Modelling 19*(2), 163–172. (One citation at page 264.)

[342] Laboratories, S. N. (2017). Trilinos home page. (2 citations in pages 220 and 298.)

[343] Lanczos, C. (1950). An iteration method for the solution of the eigenvalue problem of linear differential and integral operators. *J. Res. Nat. Bur. Stand. 45*, 255–282. (2 citations in pages 38 and 44.)

[344] Lanczos, C. (1952). Solution of systems of linear equations by minized iterations. *J. Res. Nat. Bur. Stand. 49*, 33–53. (2 citations in pages 57 and 58.)

[345] Langer, U. and W. Queck (1986). On the convergence factor of Uzawa's algorithm. *J. Comput. Appl. Math. 15*(2), 191–202. (One citation at page 252.)

[346] Laudadio, T., N. Mastronardi, and P. V. Dooren (2017). Numerical issues in computing the antitriangular factorization of symmetric indefinite matrices. *Appl. Numer. Math. 116*, 204–214. (One citation at page 250.)

[347] Lawson, C. L., R. J. Hanson, D. R. Kincaid, and F. T. Krogh (1979). Basic linear algebra subprograms for Fortran usage. *ACM Trans. Math. Softw. 5*(3), 308–323. (One citation at page 159.)

[348] Lee, J., J. Zhang, and C.-C. Lu (2003). Incomplete LU preconditioning for large scale dense complex linear systems from electromagnetic wave scattering problems. *J. Comp. Phys. 185*(1), 158–175. (One citation at page 221.)

[349] Lemeire, F. (1975). Bounds for condition numbers of triangular and trapezoid matrices. *BIT 15*(1), 58–64. (One citation at page 155.)

[350] Leugering, G., P. Benner, S. Engell, A. Griewank, H. Harbrecht, M. Hinze, R. Rannacher, and S. Ulbrich (2014). *Trends in PDE constrained optimization*, Volume 165. Springer. (One citation at page 297.)

[351] LeVeque, R. (2007). *Finite Difference Methods for Ordinary and Partial Differential Equations: Steady-State and Time-Dependent Problems.* SIAM. (7 citations in pages 5, 29, 72, 118, 143, 164, and 294.)

[352] Levinson, N. (1946). The Wiener (root mean square) error criterion in filter design and prediction. *J. Math. Phys. 25*(1), 261–278. (One citation at page 170.)

[353] Li, G. and T. F. Coleman (1988). A parallel triangular solver for a distributed–memory multiprocessor. *SIAM J. Sci. Comput. 9*(3), 485–502. (One citation at page 158.)

[354] Li, G. and T. F. Coleman (1989). A new method for solving triangular systems on distributed–memory message-passing multiprocessors. *SIAM J. Sci. Comput. 10*(2), 382–396. (One citation at page 158.)

[355] Li, N. and Y. Saad (2005). Crout versions of ILU factorization with pivoting for sparse symmetric matrices. *Electron. Trans. Numer. Anal. 20*, 75–85. (One citation at page 112.)

[356] Li, X., A.-L. Yang, and Y.-J. Wu (2014). Lopsided PMHSS iteration method for a class of complex symmetric linear systems. *Numer. Algorithms 66*(3), 555–568. (One citation at page 222.)

[357] Lin, C.-J. and J. J. Moré (1999). Incomplete Cholesky factorizations with limited memory. *SIAM J. Sci. Comput. 21*(1), 24–45. (One citation at page 98.)

[358] Lin, C.-J. and R. Saigal (2000). An incomplete Cholesky factorization for dense symmetric positive definite matrices. *BIT 40*(3), 536–558. (One citation at page 98.)

[359] Lin, F.-R., W.-K. Ching, and M. K. Ng (2003). Discrete wavelet transforms for Toeplitz matrices. *Linear Algebra Appl. 370*, 269–285. (One citation at page 187.)

[360] Lin, F.-R., X. Jin, and S. Lei (2003). Strang–type preconditioners for solving linear systems from delay differential equations. *BIT 43*(1), 139–152. (One citation at page 171.)

[361] Lin, F.-R. and C.-X. Wang (2012). BTTB preconditioners for BTTB systems. *Numer. Algorithms 60*(1), 153–167. (One citation at page 197.)

[362] Linzer, E. (1992). Extended circulant conditioning of Toeplitz systems. Technical report, Thomas J. Watson IBM Research Center. Research Division. (One citation at page 190.)

[363] Liu, J., F. Sadre-Marandi, and Z. Wang (2016). DarcyLite: A Matlab Toolbox for Darcy Flow Computation. *Procedia Comput Sci. 80*, 1301–1312. (2 citations in pages 241 and 297.)

[364] Liu, J. W. (1986). A compact row storage scheme for Cholesky factors using elimination trees. *ACM Trans. Math. Software 12*(2), 127–148. (One citation at page 12.)

[365] Liu, Q.-X., T.-Y. Zhao, W.-Z. Zhang, and F.-H. Yu (2008). Image restoration based on generalized minimal residual methods with antireflective boundary conditions in a wavefront coding system. *Optical Engineering 47*(12), 127005–127005. (One citation at page 173.)

[366] Liu, X., M. Smelyanskiy, E. Chow, and P. Dubey (2013). Efficient sparse matrix-vector multiplication on x86-based many-core processors. In *Proceedings of the 27th international ACM conference on International conference on supercomputing*, pp. 273–282. ACM. (One citation at page 19.)

[367] Lu, T.-T. and S.-H. Shiou (2002). Inverses of 2×2 block matrices. *Comput Math Appl 43*(1), 119–129. (One citation at page 234.)

[368] Lukšan, L. and J. Vlček (1998). Indefinitely preconditioned inexact Newton method for large sparse equality constrained nonlinear programming problems. *Numer. Linear Algebra Appl. 5*(3), 219–247. (4 citations in pages 264, 266, 267, and 268.)

[369] Lyche, T., T. K. Nilssen, and R. Winther (2002). Preconditioned iterative methods for scattered data interpolation. *Adv. Comput. Math. 17*(3), 237–256. (One citation at page 264.)

[370] MacLachlan, S. P. and C. W. Oosterlee (2008). Algebraic multigrid solvers for complex–valued matrices. *SIAM J. Sci. Comput. 30*(3), 1548–1571. (2 citations in pages 221 and 222.)

[371] Maday, Y., D. Meiron, A. T. Patera, and E. M. Rønquist (1993). Analysis of iterative methods for the steady and unsteady Stokes problem: Application to spectral element discretizations. *SIAM J. Sci. Comput. 14*(2), 310–337. (One citation at page 258.)

[372] Malkus, D. (1981). Eigenproblems associated with the discrete LBB condition for incompressible finite elements. *Int. J. Engng. Sci. 19*(10), 1299–1310. (One citation at page 241.)

[373] Mandel, J. (1988). Algebraic study of multigrid methods for symmetric, definite problems. *Appl. Math. Comput. 25*(1), 39–56. (2 citations in pages 204 and 207.)

[374] Mandel, J., S. McCormick, and J. Ruge (1988). An algebraic theory for multigrid methods for variational problems. *SIAM J. Numer. Anal. 25*(1), 91–110. (One citation at page 207.)

[375] Manteuffel, T. A. (1979). Shifted incomplete Cholesky factorization. In *Sparse Matrix Proceedings 1978*, pp. 41–61. SIAM, Philadelphia, PA. (One citation at page 111.)

[376] Manteuffel, T. A. (1980). An incomplete factorization technique for positive definite linear systems. *Math. Comp. 34*, 473–473. (2 citations in pages 100 and 111.)

[377] Mardal, K.-A. and R. Winther (2004). Uniform preconditioners for the time dependent Stokes problem. *Numer. Math. 98*(2), 305–327. (One citation at page 264.)

[378] Maros, I. and C. Mészáros (1999). A repository of convex quadratic programming problems. *Optim. Methods Softw. 11*(1-4), 671–681. (One citation at page 297.)

[379] Mastronardi, N. and P. Van Dooren (2013). The antitriangular factorization of symmetric matrices. *SIAM J. Matrix Anal. Appl. 34*(1), 173–196. (One citation at page 250.)

[380] Mastronardi, N. and P. Van Dooren (2014). An algorithm for solving the indefinite least squares problem with equality constraints. *BIT 54*(1), 201–218. (One citation at page 250.)

[381] Mazzia, F. and R. McCoy (1999). *Numerical experiments with a shifted SSOR preconditioner*, Volume V of *IMACS Series in Computational and Applied Mathematics*, pp. 195–209. IMCAS. (One citation at page 222.)

[382] McCormick, S. (1985). Multigrid methods for variational problems: general theory for the V–cycle. *SIAM J. Numer. Anal. 22*(4), 634–643. (One citation at page 204.)

[383] Meerschaert, M. M. and C. Tadjeran (2004). Finite difference approximations for fractional advection–dispersion flow equations. *J. Comput. Appl. Math. 172*(1), 65–77. (One citation at page 153.)

[384] Meijerink, J. and H. A. Van der Vorst (1977). An iterative solution method for linear systems of which the coefficient matrix is a symmetric M-matrix. *Math. comp. 31*(137), 148–162. (3 citations in pages 93, 96, and 98.)

[385] Meijerink, J. and H. A. Van der Vorst (1981). Guidelines for the usage of incomplete decompositions in solving sets of linear equations as they occur in practical problems. *J. Comput. Phys. 44*(1), 134–155. (One citation at page 161.)

[386] Meinardus, G. and L. L. Schumaker (1967). *Approximation of functions: theory and numerical methods* (1 ed.). Springer Tracts in Natural Philosophy. Springer. (2 citations in pages 43 and 117.)

[387] Melhem, R. and D. Gannon (1987). Toward efficient implementation of preconditioned conjugate gradient methods on vector supercomputers. *Internat. J. Supercomput. Appls. 1*(1), 70–98. (One citation at page 15.)

[388] Merchant, G. A. and T. W. Parks (1980). Efficient solution of a Toeplitz–plus–Hankel coefficient matrix system of equations. Technical Report 8003, Rice University ECE Technical Report. (One citation at page 173.)

[389] Meurant, G. (2000). On the incomplete cholesky decomposition of a class of perturbed matrices. *SIAM J. Sci. Comput. 23*(2), 419–429. (One citation at page 143.)

[390] Meyer, C. D. (2000). *Matrix analysis and applied linear algebra.* SIAM Society for Industrial and Applied Mathematics. (3 citations in pages 2, 95, and 96.)

[391] Miranda, M. and P. Tilli (2000). Asymptotic spectra of Hermitian block Toeplitz matrices and preconditioning results. *SIAM J. Matrix Anal. Appl. 21*(3), 867–881. (One citation at page 197.)

[392] Moore, G. E. (2006). Cramming more components onto integrated circuits, reprinted from electronics, volume 38, number 8, april 19, 1965, pp. 114 ff. *IEEE Solid State Circuits Mag. 3*(20), 33–35. (One citation at page 17.)

[393] Murphy, M. F., G. H. Golub, and A. J. Wathen (2000). A note on preconditioning for indefinite linear systems. *SIAM J. Sci. Comput. 21*(6), 1969–1972. (One citation at page 261.)

[394] Naumov, M. (2011). Incomplete–LU and Cholesky preconditioned iterative methods using CUSPARSE and CUBLAS. Technical report, NVIDIA Corporation. (One citation at page 158.)

[395] Neumann, M. and R. S. Varga (1980). On the sharpness of some upper bounds for the spectral radii of SOR iteration matrices. *Numer. Math. 35*(1), 69–79. (One citation at page 28.)

[396] Ng, M. K. (1994). Fast iterative methods for solving Toeplitz-plus-Hankel least squares problems. *Electron. Trans. Numer. Anal. 2*, 154–170. (One citation at page 173.)

[397] Ng, M. K. (1997). Band preconditioners for block-Toeplitz-Toeplitz-block systems. *Linear Algebra Appl. 259*, 307–327. (One citation at page 197.)

[398] Ng, M. K. (2003). Circulant and skew–circulant splitting methods for Toeplitz systems. *J. Comput. Appl. Math. 159*(1), 101–108. (One citation at page 186.)

[399] Ng, M. K. (2004). *Iterative methods for Toeplitz systems.* Oxford University Press, USA. (One citation at page 179.)

[400] Ng, M. K., R. H. Chan, and W.-C. Tang (1999). A fast algorithm for deblurring models with Neumann boundary conditions. *SIAM J. Sci. Comput. 21*(3), 851–866. (One citation at page 173.)

[401] Nocedal, J. and S. Wright (2006). *Numerical optimization.* Springer Science & Business Media. (3 citations in pages 32, 230, and 247.)

[402] Nochetto, R. H. and J.-H. Pyo (2004). Optimal relaxation parameter for the Uzawa method. *Numer. Math. 98*(4), 695–702. (One citation at page 258.)

[403] Notay, Y. (1998). Optimal V–cycle algebraic multilevel preconditioning. *Numer. Linear Algebra Appl. 5*(5), 441–459. (One citation at page 219.)

[404] Notay, Y. (2000). Flexible conjugate gradients. *SIAM J. Sci. Comput. 22*(4), 1444–1460. (One citation at page 87.)

[405] Notay, Y. (2010). An aggregation–based algebraic multigrid method. *Electron. Trans. Numer. Anal. 37*(6), 123–146. (2 citations in pages 210 and 216.)

[406] Notay, Y. and P. S. Vassilevski (2008). Recursive Krylov–based multigrid cycles. *Numer. Linear Algebra Appl. 15*(5), 473–487. (One citation at page 210.)

[407] Noutsos, D., S. Serra-Capizzano, and P. Vassalos (2006). Block band Toeplitz preconditioners derived from generating function approximations: analysis and applications. *Numer. Math. 104*(3), 339–376. (One citation at page 197.)

[408] Noutsos, D. and P. Vassalos (2002). New band Toeplitz preconditioners for ill–conditioned symmetric positive definite Toeplitz systems. *SIAM J. Matrix Anal. Appl. 23*(3), 728–743. (One citation at page 197.)

[409] Noutsos, D. and P. Vassalos (2011). Band plus algebra preconditioners for two-level Toeplitz systems. *BIT 51*(3), 695–719. (One citation at page 197.)

[410] O'Connor, J. J. and E. F. Robertson (1996). Matrices and determinants. https://goo.gl/BQWm5w. (One citation at page 275.)

[411] Olkin, J. A. (1986). *Linear and nonlinear deconvolution problems.* Ph. D. thesis, Rice University. (One citation at page 173.)

[412] Oosterlee, C. W. (1997). A GMRES–based plane smoother in multigrid to solve 3D anisotropic fluid flow problems. *J. Comp. Phys. 130*(1), 41–53. (One citation at page 209.)

[413] OPTPDE (2017). OPTPDE — a collection of problems in PDE-constrained optimization. `http://www.optpde.net`. (One citation at page 297.)

[414] Ostrowski, A. (1954). *On the linear iteration procedures for symmetric matrices*, Volume 14. "Guido Castelnuovo" Department of Mathematics, Sapienza University of Rome and Istituto Nazionale di Alta Matematica Francesco Severi (Italy). (One citation at page 28.)

[415] Paige, C. C. (1979). Fast numerically stable computations for generalized linear least squares problems. *SIAM J. Numer. Anal. 16*(1), 165–171. (One citation at page 235.)

[416] Paige, C. C. and M. A. Saunders (1975). Solution of sparse indefinite systems of linear equations. *SIAM J. Numer. Anal. 12*(4), 617–629. (3 citations in pages 34, 71, and 237.)

[417] Paige, C. C. and M. A. Saunders (1982). LSQR: An algorithm for sparse linear equations and sparse least squares. *ACM Trans. Math. Softw. 8*(1), 43–71. (One citation at page 239.)

[418] Parlett, B. N. (1980). *The symmetric eigenvalue problem.* Prentice-Hall. (2 citations in pages 23 and 38.)

[419] Parter, S. V. (1986). On the distribution of the singular values of Toeplitz matrices. *Linear Algebra Appl. 80*, 115–130. (2 citations in pages 166 and 167.)

[420] Patankar, S. V. (1980). *Numerical heat transfer and fluid flow.* CRC press. (One citation at page 265.)

[421] Patankar, S. V. and D. B. Spalding (1972). A calculation procedure for heat, mass and momentum transfer in three–dimensional parabolic flows. *Int. J. Heat and Mass Trans. 15*(10), 1787–1806. (One citation at page 265.)

[422] Pavarino, L. F. (1997). Preconditioned conjugate residual methods for mixed spectral discretizations of elasticity and Stokes problems. *Comput. Methods Appl. Mech. Engrg. 146*(1-2), 19–30. (One citation at page 264.)

[423] Pavarino, L. F. (1998). Preconditioned mixed spectral element methods for elasticity and Stokes problems. *SIAM J. Sci. Comput. 19*(6), 1941–1957. (One citation at page 264.)

[424] Perot, J. B. (1993). An analysis of the fractional step method. *J. Comp. Phys. 108*(1), 51–58. (One citation at page 265.)

[425] Perron, O. (1907). Zur Theorie der Matrices. *Math. Ann. 64*, 248–263. (2 citations in pages 95 and 96.)

[426] Perugia, I. and V. Simoncini (2000). Block–diagonal and indefinite symmetric preconditioners for mixed finite element formulations. *Numer. Linear Algebra Appl. 7*(7-8), 585–616. (3 citations in pages 258, 270, and 273.)

[427] Perugia, I., V. Simoncini, and M. Arioli (1999). Linear algebra methods in a mixed approximation of magnetostatic problems. *SIAM J. Sci. Comput. 21*(3), 1085–1101. (One citation at page 270.)

[428] Petitet, A., R. C. Whaley, J. Dongarra, and A. Cleary (2016). HPL - A Portable Implementation of the High-Performance Linpack Benchmark for Distributed-Memory Computers. (One citation at page 159.)

[429] Philippe, B. and Y. Saad (1988). Solving large sparse eigenvalue problems on supercomputers. Technical report, NASA Ames Research Center, RIACS TR 88.38. (One citation at page 15.)

[430] Ping, X., R. Chen, K. Tsang, and E. K. Yung (2006). The SSOR-preconditioned inner outer flexible GMRES method for the FEM analysis of EM problems. *Microw. Opt. Techn. Let. 48*(9), 1708–1712. (One citation at page 85.)

[431] Plank, G., M. Liebmann, R. W. dos Santos, E. J. Vigmond, and G. Haase (2007). Algebraic multigrid preconditioner for the cardiac bidomain model. *IEEE Trans. Biomed. Eng. 54*(4), 585–596. (One citation at page 219.)

[432] Plemmons, R. J. (1977). M–matrix characterizations. nonsingular M–matrices. *Linear Algebra Appl. 18*, 175–188. (One citation at page 94.)

[433] Podlubny, I. (1998). *Fractional differential equations: an introduction to fractional derivatives, fractional differential equations, to methods of their solution and some of their applications*, Volume 198. Academic press. (2 citations in pages 151 and 152.)

[434] Potts, D. and G. Steidl (1999). Preconditioners for ill–conditioned Toeplitz matrices. *BIT 39*(3), 513–533. (One citation at page 182.)

[435] Potts, D. and G. Steidl (2001). Preconditioners for ill–conditioned Toeplitz systems constructed from positive kernels. *SIAM J. Sci. Comput. 22*(5), 1741–1761. (One citation at page 182.)

[436] Powell, C. E. and D. Silvester (2003). Optimal preconditioning for Raviart–Thomas mixed formulation of second–order elliptic problems. *SIAM J. Matrix Anal. Appl. 25*(3), 718–738. (One citation at page 264.)

[437] Press, W. H. (2007). *Numerical recipes 3rd edition: The art of scientific computing*. Cambridge university press. (One citation at page 112.)

[438] Pustyl'nikov, L. (1980). On the algebraic structure of the spaces of Toeplitz and Hankel matrices. In *Soviet Math. Dokl*, Volume 21.1, pp. 141–144. (2 citations in pages 174 and 188.)

[439] Quarteroni, A., F. Saleri, and A. Veneziani (2000). Factorization methods for the numerical approximation of Navier–Stokes equations. *Comput. Methods Appl. Mech. Engrg. 188*(1), 505–526. (One citation at page 265.)

[440] Queck, W. (1989). The Convergence Factor of Preconditioned Algorithms of the Arrow–Hurwicz Type. *SIAM J. Numer. Anal. 26*(4), 1016–1030. (3 citations in pages 258, 259, and 260.)

[441] Rafiei, A. and F. Toutounian (2008). New breakdown–free variant of AINV method for nonsymmetric positive definite matrices. *J. Comput. Appl. Math. 219*(1), 72–80. (One citation at page 141.)

[442] Ramage, A. (1999). A multigrid preconditioner for stabilised discretisations of advection–diffusion problems. *J. Comput. Appl. Math. 110*(1), 187–203. (One citation at page 219.)

[443] Reich, E. (1949). On the convergence of the classical iterative procedures for symmetric matrices. *Ann. Math. Statist 20*, 448–451. (One citation at page 28.)

[444] Remez, E. Y. (1934). Sur la détermination des polynômes d'approximation de degré donnée. *Comm. Soc. Math. Kharkov 10*, 41–63. (One citation at page 196.)

[445] Rice, J. R. (1966). A Theory of Condition. *SIAM J. Numer. Anal. 3*(2), 287–310. (One citation at page 4.)

[446] Robichaud, M. P., P. A. Tanguy, and M. Fortin (1990). An iterative implementation of the Uzawa algorithm for 3-D fluid flow problems. *Int. J. Numer. Meth. Fluids 10*(4), 429–442. (One citation at page 260.)

[447] Rozložník, M. and V. Simoncini (2002). Krylov subspace methods for saddle point problems with indefinite preconditioning. *SIAM J. Matrix Anal. Appl. 24*(2), 368–391. (2 citations in pages 270 and 271.)

[448] Ruge, J. and K. Stüben (1987). *Algebraic Multigrid*, Chapter 4, pp. 73–130. Frontiers In Applied Mathematics. SIAM. (4 citations in pages 198, 207, 208, and 216.)

[449] Rusten, T. and R. Winther (1992). A preconditioned iterative method for saddlepoint problems. *SIAM J. Matrix Anal. Appl. 13*(3), 887–904. (One citation at page 240.)

[450] Saad, Y. (1990). SPARSKIT: A basic tool kit for sparse matrix computations. Technical report, Research Inst. for Advanced Computer Science. (One citation at page 12.)

[451] Saad, Y. (1993). A flexible inner–outer preconditioned GMRES algorithm. *SIAM J. Sci. Comput. 14*(2), 461–469. (2 citations in pages 85 and 87.)

[452] Saad, Y. (1994a). Highly parallel preconditioners for general sparse matrices. In *Recent Advances in Iterative Methods*, pp. 165–199. Springer. (One citation at page 90.)

[453] Saad, Y. (1994b). ILUT: A dual threshold incomplete LU factorization. *Numer. Linear Algebra Appl. 1*(4), 387–402. (One citation at page 104.)

[454] Saad, Y. (1995). *Preconditioned Krylov subspace methods for CFD applications*, pp. 139–158. Wiley. (One citation at page 126.)

[455] Saad, Y. (2003). *Iterative Methods for Sparse Linear Systems: Second Edition*. SIAM. (21 citations in pages 23, 30, 33, 34, 38, 39, 42, 43, 45, 50, 55, 57, 58, 64, 71, 76, 80, 84, 104, 108, and 112.)

[456] Saad, Y. (2011). *Numerical Methods for Large Eigenvalue Problems 2nd edition*. SIAM. (One citation at page 47.)

[457] Saad, Y. and M. H. Schultz (1986). GMRES: A Generalized Minimal Residual Algorithm for Solving Nonsymmetric Linear Systems. *SIAM J. Sci. Stat. Comput. 7*(3), 856Ű869. (One citation at page 50.)

[458] Saad, Y. and M. Sosonkina (1999a). Distributed Schur complement techniques for general sparse linear systems. *SIAM J. Sci. Comput. 21*(4), 1337–1356. (One citation at page 90.)

[459] Saad, Y. and M. Sosonkina (1999b). Enhanced parallel multicolor preconditioning techniques for linear systems. In *PPSC*. Citeseer. (One citation at page 90.)

[460] Saad, Y. and H. A. Van der Vorst (2000). Iterative solution of linear systems in the 20th century. *J. Comput. Appl. Math. 123*, 1–33. (One citation at page 23.)

[461] Sakrison, D. (1969). An extension of the theorem of Kac, Murdock and Szegö to n dimensions (corresp.). *IEEE Trans. Inform. Theory 15*(5), 608–610. (One citation at page 172.)

[462] Saleri, F. and A. Veneziani (2005). Pressure correction algebraic splitting methods for the incompressible Navier–Stokes equations. *SIAM J. Numer. Anal. 43*(1), 174–194. (One citation at page 265.)

[463] Sanders, J. and E. Kandrot (2010). *CUDA by example: an introduction to general–purpose GPU programming.* Addison-Wesley Professional. (2 citations in pages 19 and 298.)

[464] Sawyer, W., C. Vanini, G. Fourestey, and R. Popescu (2012). SPAI preconditioners for HPC applications. *PAMM 12*(1), 651–652. (One citation at page 124.)

[465] Schnabel, R. B. and E. Eskow (1990). A new modified Cholesky factorization. *SIAM J. Sci. Stat. Comput. 11*(6), 1136–1158. (One citation at page 111.)

[466] Schröder, J. (1980). *Operator inequalities.* Academic Press. (One citation at page 209.)

[467] Serra-Capizzano, S. (1997). Optimal, quasi-optimal and superlinear band–Toeplitz preconditioners for asymptotically ill-conditioned positive definite toeplitz systems. *Math. Comp. 66*(218), 651–665. (3 citations in pages 188, 196, and 197.)

[468] Serra-Capizzano, S. (1998). Toeplitz preconditioners constructed from linear approximation processes. *SIAM J. Matrix Anal. Appl. 20*(2), 446–465. (One citation at page 188.)

[469] Serra-Capizzano, S. (2002). Convergence analysis of two-grid methods for elliptic Toeplitz and PDEs matrix–sequences. *Numer. Math. 92*(3), 433–465. (2 citations in pages 209 and 218.)

[470] Serra-Capizzano, S. (2003). Practical band Toeplitz preconditioning and boundary layer effects. *Numer. Algorithms 34*(2-4), 427–440. (One citation at page 197.)

[471] Serra-Capizzano, S. and C. T. Possio (2000). Preliminary remarks on multigrid methods for circulant matrices. In *International Conference on Numerical Analysis and Its Applications*, pp. 152–159. Springer. (One citation at page 218.)

[472] Serra-Capizzano, S. and C. T. Possio (2003). Preconditioning strategies for 2D finite difference matrix sequences. *Electron. Trans. Numer. Anal. 16*, 1–29. (One citation at page 197.)

[473] Serra-Capizzano, S. and C. T. Possio (2004). Multigrid methods for multilevel circulant matrices. *SIAM J. Sci. Comput. 26*(1), 55–85. (One citation at page 218.)

[474] Serra-Capizzano, S. and P. Tilli (1999). Extreme singular values and eigenvalues of non–Hermitian block Toeplitz matrices. *J. Comput. Appl. Math. 108*(1), 113–130. (One citation at page 185.)

[475] Serra-Capizzano, S. and E. E. Tyrtyshnikov (2003). How to prove that a preconditioner cannot be superlinear. *Math. Comp. 72*(243), 1305–1316. (One citation at page 186.)

[476] Shahnaz, R., A. Usman, and I. R. Chughtai (2005). Review of storage techniques for sparse matrices. In *9th International Multitopic Conference, IEEE INMIC 2005*, pp. 1–7. IEEE. (One citation at page 12.)

[477] Sherman, J. and W. J. Morrison (1950, 03). Adjustment of an Inverse Matrix Corresponding to a Change in One Element of a Given Matrix. *Ann. Of Math. Stat. 21*, 124–127. (One citation at page 128.)

[478] Silvester, D., H. C. Elman, and A. Ramage (2016, September). Incompressible Flow and Iterative Solver Software (IFISS) version 3.5. `http://www.manchester.ac.uk/ifiss/`. (2 citations in pages 243 and 297.)

[479] Silvester, D. and A. Wathen (1994). Fast iterative solution of stabilised Stokes systems part II: using general block preconditioners. *SIAM J. Numer. Anal. 31*(5), 1352–1367. (2 citations in pages 242 and 264.)

[480] Simoncini, V. (2004). Block triangular preconditioners for symmetric saddle–point problems. *Appl. Numer. Math. 49*(1), 63–80. (One citation at page 265.)

[481] Simoncini, V. and D. B. Szyld (2002). Flexible inner–outer Krylov subspace methods. *SIAM J. Numer. Anal. 40*(6), 2219–2239. (One citation at page 87.)

[482] Simoncini, V. and D. B. Szyld (2003). Theory of inexact Krylov subspace methods and applications to scientific computing. *SIAM J. Sci. Comput. 25*(2), 454–477. (One citation at page 85.)

[483] Simoncini, V. and D. B. Szyld (2007). Recent computational developments in Krylov subspace methods for linear systems. *Numer. Linear Algebra Appl. 14*(1), 1–59. (2 citations in pages 85 and 221.)

[484] Sleijpen, G. L. and D. R. Fokkema (1993). BiCGstab(l) for linear equations involving unsymmetric matrices with complex spectrum. *Electron. Trans. Numer. Anal. 1*(11), 2000. (One citation at page 68.)

[485] Sleijpen, G. L. G. and H. A. Van der Vorst (1995). Maintaing convergence properties of BiCGstab methods in finite precision arithmetic. *Numer. Algorithms 10*, 203–223. (One citation at page 70.)

[486] Sleijpen, G. L. G., H. A. Van der Vorst, and D. R. Fokkema (1994). BiCGstab(l) and other hybrid Bi–Cg methods. *Numer. Algorithms 7*, 75–109. (2 citations in pages 68 and 70.)

[487] Smith, R. B. (1959). Two theorems on inverses of finite segments of the generalized Hilbert matrix. *Math. Comput. 13*(65), 41–43. (One citation at page 5.)

[488] Sonneveld, P. (1989). CGS, a fast Lanczos–type solver for nonsymmetric linear systems. *SIAM J. Sci Stat. Comput. 10*, 36–52. (One citation at page 64.)

[489] Sonneveld, P. and M. B. Van Gijzen (2008). IDR(s): A family of simple and fast algorithms for solving large nonsymmetric systems of linear equations. *SIAM J. Sci. Comput. 31*(2), 1035–1062. (One citation at page 87.)

[490] Stevenson, R. (1994). Modified ILU as a smoother. *Numer. Math. 68*(2), 295–309. (One citation at page 210.)

[491] Stewart, G. W. (1989). On scaled projections and pseudoinverses. *Linear Algebra Appl. 112*, 189–193. (One citation at page 247.)

[492] Stoyan, G. (2001). Iterative Stokes solvers in the harmonic Velte subspace. *Computing 67*(1), 13–33. (One citation at page 260.)

[493] Strang, G. (1986). A proposal for Toeplitz matrix calculations. *Stud. Appl. Math. 74*(2), 171–176. (2 citations in pages 173 and 176.)

[494] Strikwerda, J. C. (1984). An iterative method for solving finite difference approximations to the Stokes equations. *SIAM J. Numer. Anal. 21*(3), 447–458. (One citation at page 251.)

[495] Strohmer, T. (2002). Four short stories about Toeplitz matrix calculations. *Linear Algebra Appl. 343*, 321–344. (2 citations in pages 188 and 190.)

[496] Stüben, K. (2001). A review of algebraic multigrid. *J. Comput. Appl. Math. 128*(1), 281–309. (3 citations in pages 207, 216, and 219.)

[497] Suárez, A., H. Sarmiento, E. Flórez, M. García, and G. Montero (2011). Updating incomplete factorization preconditioners for shifted linear systems arising in a wind model. *Journal of computational and applied mathematics 235*(8), 2640–2646. (One citation at page 151.)

[498] Suarjana, M. and K. H. Law (1995). A robust incomplete factorization based on value and space constraints. *Internat. J. Numer. Methods Engrg. 38*(10), 1703–1719. (One citation at page 97.)

[499] Sun, H.-W., R. H. Chan, and Q.-S. Chang (1997). A note on the convergence of the two–grid method for Toeplitz systems. *Comput. Math. Appl. 34*(1), 11–18. (One citation at page 218.)

[500] Sylvester, J. J. (1852). XIX. a demonstration of the theorem that every homogeneous quadratic polynomial is reducible by real orthogonal substitutions to the form of a sum of positive and negative squares. *Philos. Mag. 4*(23), 138–142. (One citation at page 286.)

[501] Szyld, D. B. and J. A. Vogel (2001). FQMR: A flexible quasi-minimal residual method with inexact preconditioning. *SIAM Journal on Scientific Computing 23*(2), 363–380. (One citation at page 87.)

[502] Tang, P. T. P. (1988). A fast algorithm for linear complex Chebyshev approximations. *Math. Comp. 51*(184), 721–739. (One citation at page 196.)

[503] Tang, W.-P. and W. L. Wan (2000). Sparse approximate inverse smoother for multigrid. *SIAM J. Matrix Anal. Appl. 21*(4), 1236–1252. (One citation at page 210.)

[504] Tarjan, R. E. and A. C.-C. Yao (1979). Storing a sparse table. *ACM Commun. 22*(11), 606–611. (One citation at page 12.)

[505] Tay, R. (2013). *OpenCL Parallel Programming Development Cookbook*. Packt Publishing Ltd. (One citation at page 19.)

[506] Tchebychev, P. L. (1907). Sur les polynômes réprésentant le mieux les valeurs des fonctions fractionnaires élémentaires pour les valeurs de la variable contenues entre deux limites données. In S. Petersburg (Ed.), *Oeuvres*, Volume II, pp. 669–678. Commissionaires de l'Académie impériale des sciences. (One citation at page 117.)

[507] Tebbens, J. D. and M. Tüma (2007). Efficient Preconditioning of Sequences of Nonsymmetric Linear Systems. *SIAM J. Sci. Comp. 29*, 1918–1941. (One citation at page 143.)

[508] Tebbens, J. D. and M. Tüma (2010). Preconditioner updates for solving sequences of linear systems in matrix-free environment. *Numer. Linear Algebra Appl. 17*, 997–1019. (One citation at page 143.)

[509] Thomas, L. (1963). Using a computer to solve problems in physics. In W. Freiberger and W. Prager (Eds.), *Applications of Digital Computers*, pp. 44–45. Boston: Ginn and Company. (One citation at page 169.)

[510] Tilli, P. (1998). Locally Toeplitz sequences: spectral properties and applications. *Linear Algebra Appl. 278*(1), 91–120. (One citation at page 173.)

[511] Tismenetsky, M. (1991). A new preconditioning technique for solving large sparse linear systems. *Linear Algebra Appl. 154*, 331–353. (2 citations in pages 97 and 180.)

[512] Todd, J. (1960). Computational problems concerning the Hilbert matrix. *J. Res. Nat. Bur. Stand 65*, 19–22. (One citation at page 5.)

[513] Todd, M. J. (1990). A Dantzig-Wolfe-like variant of Karmarkar's interior-point linear programming algorithm. *Oper. Res. 38*(6), 1006–1018. (One citation at page 247.)

[514] Toeplitz, O. (1911). Zur Theorie der quadratischen und bilinearen Formen von unendlichvielen Veränderlichen. *Math. Ann. 70*(3), 351–376. (One citation at page 164.)

[515] Toh, K.-C., K.-K. Phoon, and S.-H. Chan (2004). Block preconditioners for symmetric indefinite linear systems. *Int. J. Numer. Methods Engng. 60*(8), 1361–1381. (One citation at page 264.)

[516] Tomov, S., J. Dongarra, and M. Baboulin (2010, June). Towards dense linear algebra for hybrid GPU accelerated manycore systems. *Parallel Computing 36*(5-6), 232–240. (2 citations in pages 19 and 159.)

[517] Tomov, S., R. Nath, H. Ltaief, and J. Dongarra (2010, April 19-23). Dense Linear Algebra Solvers for Multicore with GPU Accelerators. In *Proc. of the IEEE IPDPS'10*, Atlanta, GA, pp. 1–8. IEEE Computer Society. DOI: 10.1109/IPDPSW.2010.5470941. (2 citations in pages 19 and 159.)

[518] Trench, W. F. (1964). An algorithm for the inversion of finite Toeplitz matrices. *SIAM J. Appl. Math. 12*(3), 515–522. (One citation at page 170.)

[519] Trottenberg, U., C. W. Oosterlee, and A. Schuller (2000). *Multigrid.* Academic press. (4 citations in pages 198, 207, 215, and 216.)

[520] Turek, S. (1999). *Efficient Solvers for Incompressible Flow Problems: An Algorithmic and Computational Approache*, Volume 6. Springer Science & Business Media. (One citation at page 274.)

[521] Tyrtyshnikov, E. E. (1992). Optimal and superoptimal circulant preconditioners. *SIAM J. Matrix Anal. Appl. 13*(2), 459–473. (2 citations in pages 177 and 180.)

[522] Tyrtyshnikov, E. E. (1996). A unifying approach to some old and new theorems on distribution and clustering. *Linear Algebra Appl. 232*, 1–43. (3 citations in pages 166, 167, and 172.)

[523] Tyrtyshnikov, E. E. (1997). *A Brief Introduction to Numerical Analysis.* Birkhauser. (One citation at page 78.)

[524] Tyrtyshnikov, E. E. and N. L. Zamarashkin (1998). Spectra of multilevel Toeplitz matrices: advanced theory via simple matrix relationships. *Linear Algebra Appl. 270*(1), 15–27. (One citation at page 172.)

[525] Umetani, N., S. P. MacLachlan, and C. W. Oosterlee (2009). A multigrid–based shifted Laplacian preconditioner for a fourth–order Helmholtz discretization. *Numer. Linear Algebra Appl. 16*(8), 603–626. (One citation at page 221.)

[526] Van der Vorst, H. A. (1981). Iterative solution methods for certain sparse linear systems with a non-symmetric matrix arising from PDE–problems. *J. Comp. Phys. 44*(1), 1–19. (One citation at page 111.)

[527] Van der Vorst, H. A. (1992). Bi–CGSTAB: A fast and smoothly converging variant of Bi–CG for the solution of nonsymmetric linear systems. *SIAM J. Sci. Stat. Comput. 13*(2), 631–644. (3 citations in pages 62, 64, and 68.)

[528] Van der Vorst, H. A. (2003). *Iterative Krylov Methods for Large Linear Systems.* Cambridge University Press. (3 citations in pages 62, 68, and 85.)

[529] van Duin, A. C. N. (1999). Scalable Parallel Preconditioning with the Sparse Approximate Inverse of Triangular Matrices. *SIAM J. Matrix Anal. Appl. 20*, 987–0. (2 citations in pages 127 and 132.)

[530] Vaněk, P., J. Mandel, and M. Brezina (1994). Algebraic multigrid on unstructured meshes. Technical report, University of Colorado at Denver. (One citation at page 216.)

[531] Vaněk, P., J. Mandel, and M. Brezina (1996). Algebraic multigrid by smoothed aggregation for second and fourth order elliptic problems. *Computing 56*(3), 179–196. (One citation at page 216.)

[532] Varga, R. S. (1962). *Matrix iterative analysis* (1st ed.). Prentice Hall INC. (One citation at page 96.)

[533] Varga, R. S. (1991). *Solving Ordinary Differential Equations II. Stiff and Differential-Algebraic Problems*. Springer-Verlag. (One citation at page 143.)

[534] Vassilevski, P. S. (1992). Hybrid V–cycle algebraic multilevel preconditioners. *Math. Comp. 58*(198), 489–512. (One citation at page 219.)

[535] Vassilevski, P. S. and R. D. Lazarov (1996). Preconditioning Mixed Finite Element Saddle-point Elliptic Problems. *Numer. Linear Algebra Appl. 3*(1), 1–20. (2 citations in pages 264 and 267.)

[536] Veneziani, A. (2003). Block factorized preconditioners for high-order accurate in time approximation of the Navier-Stokes equations. *Numer. Methods Partial Differential Eq. 19*(4), 487–510. (One citation at page 265.)

[537] Vogel, J. A. (2007). Flexible BiCG and flexible Bi–CGSTAB for nonsymmetric linear systems. *Appl. Math. Comput. 188*(1), 226–233. (One citation at page 87.)

[538] Wagner, C. (1997). Tangential frequency filtering decompositions for symmetric matrices. *Numer. Math. 78*(1), 119–142. (One citation at page 210.)

[539] Wang, Z.-Q. (2009). Optimization of the parameterized Uzawa preconditioners for saddle point matrices. *J. Comput. Appl. Math. 226*(1), 136–154. (One citation at page 258.)

[540] Wathen, A. and D. Silvester (1993). Fast iterative solution of stabilised Stokes systems. part I: Using simple diagonal preconditioners. *SIAM J. Numer. Anal. 30*(3), 630–649. (One citation at page 264.)

[541] Wesseling, P. and P. Sonneveld (1980). Numerical experiments with a multiple grid and a preconditioned Lanczos type method. In *Approximation methods for Navier-Stokes problems*, pp. 543–562. Springer. (One citation at page 87.)

[542] Wilkinson, J. H. (1964). *Rounding errors in algebraic processes.* Notes on applied science. Prentice-Hall. (One citation at page 4.)

[543] Wilkinson, J. H. (1965). *The algebraic eigenvalue problem.* Clarendon Press. (4 citations in pages 23, 38, 58, and 59.)

[544] Wilkinson, J. H. and C. Reinsch (1971). *Handbook for Automatic Computation Vol. II – Linear Algebra.* Springer–verlag. (One citation at page 10.)

[545] Williams, S., L. Oliker, R. Vuduc, J. Shalf, K. Yelick, and J. Demmel (2009). Optimization of sparse matrix–vector multiplication on emerging multicore platforms. *Parallel Comput. 35*(3), 178–194. (One citation at page 19.)

[546] Windisch, D. D. G. (1989). *M–matrices in Numerical Analysis.* Teubner–Texte zur Mathematik. Vieweg+Teubner Verlag. (One citation at page 94.)

[547] Winograd, S. (1978). On computing the discrete fourier transform. *Math. Comput. 32*(141), 175–199. (One citation at page 169.)

[548] Wittum, G. (1989). On the robustness of ILU smoothing. *SIAM J. Sci. Comput. 10*(4), 699–717. (One citation at page 210.)

[549] Wu, S.-L. (2015). Several variants of the hermitian and skew–Hermitian splitting method for a class of complex symmetric linear systems. *Numer. Linear Algebra Appl. 22*(2), 338–356. (One citation at page 222.)

[550] Yang, U. M. et al. (2002). BoomerAMG: a parallel algebraic multigrid solver and preconditioner. *Appl. Numer. Math. 41*(1), 155–177. (One citation at page 220.)

[551] Yang, X. I. and R. Mittal (2014). Acceleration of the Jacobi iterative method by factors exceeding 100 using scheduled relaxation. *J. Comp. Phys. 274*, 695–708. (One citation at page 27.)

[552] Yeremin, A. Y., L. Y. Kolotilina, and A. Nikishin (1998). Factorized sparse approximate inverse preconditionings III. Iterative construction of preconditioners. *Zap. Nauchn. Sem. S. Peterburg. Otdel. Mat. Inst. Steklov. (POMI) 248*, 17–48. (One citation at page 124.)

[553] Yeremin, A. Y., L. Y. Kolotilina, and A. Nikishin (2000). Factorized sparse approximate inverse preconditionings. III. Iterative construction of preconditioners. *J. Math. Sci. (N.Y.) 101*(4), 3237–3254. (One citation at page 124.)

[554] Young, D. M. (1954). Iterative methods for solving partial difference equations of elliptic type. *Trans. Amer. Math. Soc. 76*(1), 92–111. (One citation at page 28.)

[555] Young, D. M. (1971). *Iterative solution of large linear systems.* Academic Press. (2 citations in pages 26 and 90.)

[556] Zhang, J. (2000). Sparse approximate inverse and multilevel block ILU preconditioning techniques for general sparse matrices. *Appl. Numer. Math. 35*(1), 67–86. (One citation at page 210.)

[557] Zhang, J. and B. Morini (2013). Solving regularized linear least–squares problems by the alternating direction method with applications to image restoration. *Electronic Transactions on Numerical Analysis 40*, 356–372. (One citation at page 151.)

[558] Zheng, B., Z.-Z. Bai, and X. Yang (2009). On semi-convergence of parameterized Uzawa methods for singular saddle point problems. *Linear Algebra Appl. 431*(5), 808–817. (One citation at page 258.)

[559] Zohar, S. (1974). The solution of a Toeplitz set of linear equations. *J. Assoc. Comput. Mach. 21*(2), 272–276. (One citation at page 170.)

[560] Zygmund, A. (2002). *Trigonometric series*, Volume 1. Cambridge university press. (2 citations in pages 164 and 176.)

Index